BRYOPHYTE SYSTEMATICS

Proceedings of an International Symposium
held at the University College of North
Wales, Bangor

THE SYSTEMATICS ASSOCIATION
SPECIAL VOLUME NO. 14

BRYOPHYTE SYSTEMATICS

Edited by

G. C. S. CLARKE

British Museum (Natural History), London

and

J. G. DUCKETT

*Department of Plant Biology and Microbiology,
Queen Mary College, University of London*

1979

Published for the
SYSTEMATICS ASSOCIATION and the BRITISH BRYOLOGICAL
SOCIETY
by
ACADEMIC PRESS   LONDON   NEW YORK   TORONTO
SYDNEY   SAN FRANCISCO

ACADEMIC PRESS INC. (LONDON) LTD.
24–28 Oval Road
London NW1 7DX

U.S. Edition published by
ACADEMIC PRESS INC.
111 Fifth Avenue
New York, New York 10003

British Library Cataloguing in Publication Data

Bryophyte systematics.—(Systematics Association.
  Special volumes; no. 14 ISSN 0309–2593).
  1. Bryophytes—Classification—Congresses
  I. Clarke, Giles Colin Scott   II. Duckett,
  Jeffrey Graham   III. British Bryological Society
  588'.01'2   QK533.7   79–40897

  ISBN 0–12–175050–7

PRINTED IN GREAT BRITAIN BY
LATIMER TREND & COMPANY LTD PLYMOUTH

Preface

This volume brings together the papers presented at the international symposium on bryophyte systematics which was jointly organized by the British Bryological Society and the Systematics Association and held at the University College of North Wales, Bangor, on 16–19 August 1978. In recent years a number of symposia, such as those held at Lille in 1973 or in Bordeaux in 1977, have demonstrated the quantity and range of current research on bryophytes and it was clearly time to take stock of the achievements and shortcomings of this research as it applied to systematics.

Our main aim in planning the programme for the symposium was therefore to review, in the broadest possible way, modern research and ideas on the systematics and phylogeny of the Bryophyta. We invited a range of distinguished speakers to prepare reviews of particular aspects of the subject, incorporating new data from their own research together with a critical appraisal of the published literature. Thanks to the energy and thought which our contributors have put into their task, we believe that the resulting papers represent a unique synthesis of the current state of bryophyte systematics which will, we hope, provide a useful, and much needed, reference work for students, teachers and research workers in bryology. We would like to think that this could act as a stimulus to research on the bryophytes in the tradition of Frans Verdoorn's classic *Manual of Bryology* published in 1932, although, of course, that work was broader in its scope.

One of the most interesting features of the symposium has been to see the way in which new and traditional techniques have complemented each other in systematic studies. The new tools which are available to the modern bryologist have produced a wealth of fascinating new information, but they have in no way superseded many of the traditional approaches which are as valid today as they ever were. Because of this, the links that have always existed between professional and amateur bryologists are no less important today than they have been in the past. It would be natural to expect that with the increasing sophistication of much bryological work, people without access to expensive equipment would have to play a progressively more passive role. Surprising as it may seem, the reverse is true because as more bryologists specialize in detailed aspects of the subject, they have to rely on others with broader experience to see the general significance of their work. There will never be a substitute for the man who knows his plants in the field.

Although, for obvious reasons, the majority of publications on bryophyte systematics have so far tackled problems at the infrageneric level, there is a good deal of evidence that more attention is now being turned to taxonomy at the higher levels. This is shown, for example, by the contributions of Miller, Schuster and Edwards which concentrate largely on traditional techniques, or of Carothers, Duckett, Neidhart and Suire which rely on more modern techniques. For some of these new disciplines, such as the cytology of conducting tissues (Hébant), spermatogenesis (Carothers, Duckett) and sporogenesis (Neidhart), the most pressing need is for observations on a wider range of taxa, whereas for others such as chromosome cytology (Newton, Ramsay) work on the bryophytes is beginning to approach that on angiosperms in its level of sophistication.

The introductory contribution on bryology in North Wales (Richards) together with those on the boreal region (Steere), Japan (Koponen) and the tropics (Argent) forcibly bring home the unevenness of our knowledge in different parts of the world. Studies which are wide ranging from the taxonomic and geographical point of view, such as those of Gradstein and Eddy, should surely come high on the list of priorities of institutions with extensive herbaria and travel funds.

One of the most fundamental taxonomic propositions put forward independently by several authors (e.g. Carothers, Duckett, Schuster, Suire) is that the Anthocerotales are isolated from the rest of the bryophytes. A further inescapable conclusion from the contributions of Miller, Schuster and Eddy is that the centres of bryophyte evolution and distribution lie in the southern hemisphere where primitive and isolated taxa are particularly well represented.

The reviews by Longton, Smith and Ramsay clearly demonstrate that some taxonomic problems can only be solved by careful and often prolonged experiment. The results of such experiments are often critical for the understanding of systematics at or below the species level where so much effort has, in the past, been concentrated. Another way in which recent work can increase our appreciation of the significance of traditional taxonomic characters is demonstrated by Proctor who has analysed the adaptive significance of a number of characters which have long been used in everyday identification. Appraisals of the taxonomic potential of features such as spores (Clarke), rhizoids (Crundwell) and peristomes (Edwards) not only reveal a surprising wealth of potentially valuable information, but also highlight the importance of taking into account as many different characters as possible in systematic work.

One other impression that we are left with at the end of the symposium is that there is a pressing need for more attention to be paid to the theory of systematics rather than its practice. Our contributors have demonstrated an inexhaustible ability to discover new data about the plants they are studying; what we need now is a sounder method for deciding what the data mean and how they can be used in a logical way to deduce relationships. In practical terms the process by which bryological classifications are arrived at has hardly altered since well before Darwin's time. This is a topic that zoologists have given a good deal of thought to in recent years. Flowering plant botanists are beginning to follow their example but bryologists have been left behind here.

We are very grateful to all those who helped make the symposium the success it was, especially Professor J. L. Harper, F.R.S., for allowing us to use the facilities in the School of Plant Biology at the University College of North Wales, Ms M. O. Allen who assisted with the domestic arrangements and Mr M. O. Hill who led the field excursions which took place after the lectures. Dr A. R. Duckett and Mr B. J. O'Shea kindly helped with proof reading and preparation of the indexes. Finally we would like to thank all the participants for making the symposium such a stimulating and fruitful occasion.

August 1979

G. C. S. CLARKE
J. G. DUCKETT

A Note on Nomenclature

Throughout this volume, the nomenclature used for bryophyte names follows that of the standard British works. These are for Musci: A. J. E. Smith *The Moss Flora of Britain and Ireland* (1978), and for Hepaticae and Anthocerotae: J. A. Paton *Census Catalogue of British Hepatics*, 4th Edition (1965). Author citations for specific names have only been included for non-British plants.

Contents

Systematics Association Publications

LONDON. Published by the Association

Systematics Association Special Volumes

* Published by Academic Press for the Systematics Association
† Published by the Palaeontological Association in conjunction with the Systematics Association

Systematics Association Special Volumes

* Published by Academic Press for the Systematics Association

1 | A Note on the Bryological Exploration of North Wales

(Extract from an Introductory Talk to the Symposium)

14 Wootton Way, Cambridge CB3 9LX, Great Britain

Abstract: Many British bryophytes were first recorded from North Wales. The investigation of its rich bryophyte flora was initiated by Lhuyd and others in the 17th century but really began with the visit of Dillenius, Brewer and Brown in 1726. This pioneer exploration was followed by the work of H. Davies (1739–1821) and J. Wynne Griffith (1763–1834). During the 19th and 20th centuries many more species were discovered, mostly by visitors from outside Wales. The addition of species still continues but a few have not been seen in recent years and may be extinct.

Wales is a small country, about one-sixth of the size of England, and was very thinly populated until quite recent times, but it has been of some importance in the history of science. This is perhaps as much because of its natural features as because of the scientific achievements of Welshmen. For example, the names of no less than three of the world's geological formations, the Cambrian, Ordovician and Silurian, refer to Wales or parts of Wales.

In botany it is especially North Wales (corresponding approximately to the modern counties of Gwynedd and Clwyd) which has played an important part, and this is especially true of bryology. One reason is because the bryophyte flora of North Wales is unusually abundant and diverse. North Wales also lies on the route to Ireland and was relatively accessible in times when most botanical activity in Britain was centred on London and south-eastern England. Hill

Systematics Association Special Volume No. 14, "Bryophyte Systematics", edited by G. C. S. Clarke and J. G. Duckett, 1979, pp. 1–9, Academic Press, London and New York.

1978) has estimated that 10% of the species in the British bryophyte flora were first discovered in Gwynedd, a figure which may well be an underestimate. Some of these species, e.g. *Glyphomitrium daviesii, Oedipodium griffithianum* and *Petalophyllum ralfsii*, were new to science, but most had previously been found outside the British Isles.

Until late in the 19th century Wales, like most mainly rural countries, was poor and educationally backward and it is not surprising that many of the early botanical explorers of Wales were visitors from England. The chief attraction for them was Snowdon (1085 m), Wales's highest mountain: it became well known very early for the rich and beautiful alpine flora of its rock ledges. It is said that Edward Lhuyd (1660–1709) (also spelt Lhwyd and Lloyd), the remarkable Welshman who became Keeper of the Ashmolean Museum at Oxford, was the first person to realize that the mountains of Britain had a distinctive alpine flora (Desmond, 1977). The extraordinarily rich flora of the island of Anglesey and of the Carboniferous limestone of the Creuddyn peninsula (Great Orme's Head) also became botanically famous at an early date. Most of the early botanical explorers such as Thomas Johnson (1604–1644) and John Ray (1627–1705), as well as Lhuyd, though they collected ferns and lycopods, paid comparatively little attention to mosses and hepatics. The species they mention are not easy to identify (except some of Lhuyd's of which herbarium specimens may possibly still exist), though it is easy to recognize *Muscus terrestri vulgari similis lanuginosus* as *Racomitrium lanuginosum*: Lhuyd told Ray (1690) that it was a pest of grasslands in the Welsh mountains (*"vitium graminis in montosis Cambriae"*).

The first botanist to visit North Wales who had a good knowledge of bryophytes was John Jacob Dillenius (1684–1747), the author of the *Historia Muscorum* (1741). When he came to England from Germany in 1721 he was already an expert bryologist and had recorded some 200 species of "Musci" in the neighbourhood of Giessen in Hesse (Druce and Vines, 1907). In August 1726 Dillenius undertook a tour of Wales and the west of England. He was accompanied by Samuel Brewer (1670–1742), an enthusiastic botanist who after making a large amount of money as a woollen manufacturer in Trowbridge, Wiltshire, devoted the latter part of his life to natural history. At Bishop's Castle in Shropshire the two were joined by Littleton Brown, whom Dillenius described as "a young ingenious clergyman"; he also was a skilled botanist. The party first visited Cader Idris and then went to Caernarfon, returning home via Bala, Oswestry and Shrewsbury. They spent in all about three weeks in Gwynedd during which they climbed Snowdon and the Glyders, explored Anglesey and visited Priestholm (Puffin Island). The greater part of Dillenius's diary of this tour is reproduced by Druce and Vines (1907) and a further portion by Hyde (1931) but a small part is apparently lost. Brewer also kept a diary (Hyde, 1931).

The bryophytes collected by Dillenius and Brewer, many of which were probably figured and described in the *Historia Muscorum*, are listed by Druce and Vines (1907) who give modern identifications of the specimens in the Oxford University Herbarium, as well as the names used by Dillenius. Specimens, most of which are localized, are to be found in the herbarium Dillenius made when preparing the revised version (third edition) of Ray's *Synopsis Methodica Stirpium Britannicarum* (1724); these were identified by H. Boswell. Dillenius made another herbarium of plants described in the *Historia Muscorum*. The bryophytes in this collection were identified by S. O. Lindberg (Lindberg, 1874, 1883) and others: many of the specimens may have come from Wales but few of them have localities.

Before he joined Dillenius on this tour Brewer had little knowledge of plants in the field, according to his own account (Hyde, 1931), and probably none of mosses, but he proved an invaluable member of the party. Dillenius said of him, "I am sure I shall never meet a better searcher especially for mosses. When we travelled together in Wales in all the badeness and violency of the weather and rain he would stop and pick up mosses" (Hyde, 1931, p. 6). Brewer became so interested in North Wales and its plants that he returned in 1727 and spent more than a year at Bangor botanizing in the neighbourhood, sometimes with two companions from Anglesey, the Rev. W. Green of Holyhead and William Jones of Llanfaethlu (Hyde, 1931). Brewer and Green found various mosses not seen during the tour in 1726; these were sent to Dillenius and included in his herbaria.

Many of the bryophytes found by Dillenius and Brewer in North Wales were new records for Britain or for Wales and some were new species not previously described. Except for *Scapania ornithopodioides* on Snowdon (Dillenius, 1741, p. 493), they do not seem to have found any of the greatest rarities, but they collected many of the western species that are so characteristic of the area such as *Andreaea alpina*, *Campylopus atrovirens*, *Breutelia chrysocoma*, *Dicranum scottianum* and *Herberta adunca*. Some of the specimens are not closely localized, e.g. *Ptychomitrium polyphyllum* "upon ye walls about Bangor", but others have quite precise localities, e.g. *Eucladium verticillatum* "from the rocks facing the sea near Bangor" (where it still grows), and *Splachnum ampullaceum* "amongst ye heath in the Turbery at Pentir, always upon cow-dung". One record seems surprising, *Weissia controversa* "from within the cathedral of Bangor, Saml. Brewer 1728", because the cathedral was not ruinous at that time (M. L. Clarke, personal communication). One or two of Boswell's identifications of specimens in Dillenius's *Synopsis* herbarium seem improbable if the localities are correct, and should be checked, e.g. *Bryum pendulum surculis teretibus viridibus* "from ye walls about Bangor", identified by Boswell as *Bryum filiforme* Dicks. (*Anomobryum filiforme*) (Druce and Vines, 1907, p. 139), a species which in North Wales is usually found on moist mountain rocks.

After Dillenius and Brewer the next important contributors to the bryology of Gwynedd were two Welshmen, Hugh Davies (1739–1821) and J. Wynne Griffith (1763–1834), both of whom made interesting discoveries. Davies, a clergyman born at Llandegfan, Anglesey, and educated at Jesus College, Oxford (Owen, 1961) (not Peterhouse, Cambridge, as stated in the *Dictionary of National Biography* (Stephen, 1888) and the *Dictionary of Welsh Biography* (Jenkins, 1959)), held church livings at Llandegfan and at Aber, Caernarfonshire, but spent the last years of his life at Beaumaris, Anglesey. He spent most of his life in botanical and other scholarly studies and is said to have owed his initiation into natural history to the celebrated zoologist, Thomas Pennant (1726–1798) of Downing near Holywell (Clwyd): Davies assisted Pennant in writing his *Indian Zoology* (1769). In the preface to his *Welsh Botanology* (1813) Davies says, "A constitutional nervous sensibility, which increases with years to an oppressive degree, having rendered me unequal to the duties of my profession, it occurred to me to amuse myself in laying together the contents of the following pages." Davies was an excellent botanist and became a friend of Hudson, author of the *Flora Anglica* (1762). He corresponded with many botanists in Britain and abroad and contributed to Smith's *English Botany* and other works as well as writing a few papers of his own. Davies's most important publication was the curiously titled *Welsh Botanology* (1813). It is in two parts of which the first is sub-titled "A systematic catalogue of the native plants of the Isle of Anglesey in Latin, English and Welsh; with the habitats of the rarer species and a few observations." Part Two is "An alphabetical catalogue of the Welsh names of vegetables rendered into Latin and English with some account of the qualities, economical or medicinal of the most remarkable." The text of Part Two is entirely in Welsh.

It is clear from the preface that Davies's main object in writing this book was to provide a reliable list of the Welsh names of plants: in this he was successful and his list has not yet been superseded. But Davies seems to have been a very acute observer and Part One of the *Welsh Botanology*, which is arranged on the Linnean system and includes algae, fungi, lichens and bryophytes as well as vascular plants, is a remarkably good local flora of Anglesey, or at least of the parts of the island which were accessible to him (mainly the southern half and Holyhead island). Nearly 200 species of bryophytes are included: the localities are precisely stated for all but the commoner species and with some exceptions the identifications seem to be reliable. Davies's herbarium is in the British Museum (Natural History).

Davies's most interesting bryological discovery was *Glyphomitrium daviesii*, a moss which Dickson named after him (as *Bryum daviesii*) in 1790. He found it in two places, quite near together, which he gives as, "At Carreg Onnan, and maritime rocks below Llanfihangel Dinsylwi". Carreg Onnen and Llanfihangel-Din-Sylwi (modern spellings) are close to Bwrdd Arthur, a hill near the eastern extremity of Anglesey; Llanfihangel-Din-Sylwi is an isolated church (grid reference 23/588815), not a village. J. E. Griffith (1895, p. 257) found *Glypho-mitrium* on "rocks at Carreg Onen, very sparingly", but nobody seems to have found the plant in Anglesey after that, though it is known from many localities elsewhere in Britain and may still exist on the mainland of Gwynedd.

J. Wynne Griffith was an active collector of bryophytes about whom very little is known; like Davies he was an early member of the Linnean Society. He corresponded with Withering who gives many of his localities for mosses in the third edition of his *Arrangement of British Plants* (1796). Griffith discovered *Oedipodium griffithianum* which was at first erroneously identified as *Splachnum froehlichianum* (*Tayloria froehlichiana*) "on the eastern side of Snowdon, about 150 yards from the summit" (Withering, 1796, p. 794).

From the time of Davies and Wynne Griffith onwards knowledge of the bryophyte flora of North Wales steadily increased, especially after the opening of the railway from Chester to Holyhead in 1849 which made the area much easier to reach from London, Manchester and other English towns. Most of the botanists who worked in Gwynedd during this period came from outside Wales. On the whole they tended to visit localities which were already well known such as Snowdon and a few rich localities in Merioneth, e.g. Cwm Bychan near Harlech and the waterfall Rhaiadr Du near Dolgellau (usually referred to in the older literature as "Tyn-y-groes", the name of a well-known hotel nearby). Almost all the more eminent British bryologists of the 19th century, including William Wilson, E. M. Holmes, W. Mitten, G. A. Holt, J. Ralfs, H. N. Dixon and many others, collected in North Wales and added species to the list. Only one or two of them need be singled out for special mention.

William Wilson (1799–1871), the author of the *Bryologia Britannica* (1855) was one of the most distinguished. A lawyer by profession, he practised as a solicitor in Warrington, Lancashire, where most of his life was spent, but he was a frequent visitor to Anglesey and Snowdonia. He made many important additions to the flora of Gwynedd, such as *Fissidens polyphyllus* which he found at Pont Aberglaslyn near Tremadoc in 1838 (first record for Great Britain, but previously found in Ireland by Wilson). Other discoveries of Wilson's include *Ditrichum zonatum*, *Sematophyllum demissum* and *Tortula cuneifolia*. The study of the Hepaticae progressed more slowly than that of the mosses, but B. Carrington and W. H. Pearson of Manchester made many interesting discoveries. Pearson paid particular attention to the neighbourhood of Dolgellau in Merioneth.

Two notable bryologists resident in North Wales who were active in the late 19th and early 20th centuries were J. E. Griffith (1843–1933) and Daniel A. Jones (1861–1936). The former is well known as the author of the *Flora of Anglesey and Carnarvonshire* (1895) which includes a list of the bryophytes of the two counties giving localities. Griffith was the first discoverer of *Leptodontium recurvifolium* and possibly other species in Snowdonia, but he was mainly interested in vascular plants. D. A. Jones was for many years head of the primary school at Harlech, Merioneth. He made an intensive study of the district in which he lived which is very rich in bryophytes. He also explored Snowdonia, Anglesey and other adjoining areas, adding many species, especially Hepaticae, e.g. *Gymnocolea acutiloba*, *Riccia crozalsii* and *Scapania nimbosa*, to the North Wales flora. His patient collecting, extending over many years, provided much information about the detailed distribution of the rarer species.

Among the few botanists from abroad who visited North Wales in the 19th century, the most famous is W. P. Schimper (1808–1880) who came in 1865. With his usual energy, Schimper climbed both Snowdon and the Glyders as well as visiting Cwm Bychan near Harlech to see *Bartramidula wilsonii* in the locality where it had been discovered by the Rev. T. Salwey in 1841. Although he does not seem to have made any important new discoveries, Schimper in the second edition of his *Synopsis* (1876) refers to finding various well-known North Wales mosses such as *Breutelia chrysocoma* and *Campylopus atrovirens* ("copiosissime"). The Welsh names for his localities seem to have given Schimper some trouble: thus under *Andreaea alpina* he gives, "*magna copia vidi in Cambrovalliae m. Snowdon, et praecipue in m. Glydr versus faucem Twell-du (culina diaboli) ubi magnis caespitibus turgescentibus saxa humida obtegit*", and under *Andreaea falcata* (*A. rothii* var. *falcata*), "*ad rupem valleculae desertae Crib-y-Dscil versus Top-of-Snowdon (ipse 1865)*".

In the 60 years since the end of World War I, and especially since about 1950, much further work has been done on the bryophytes of North Wales. A considerable amount of this has been by bryologists based on the Botany Department (now part of the School of Plant Biology) at the University College of North Wales, Bangor, and on the Nature Conservancy Council and Institute of Terrestrial Ecology at Bangor, but as in earlier years much has been contributed by visitors from outside Wales. The number of visitors interested in bryophytes has greatly increased, some coming as individuals and many more as members of scientific societies or student parties from universities. This enormous upsurge of interest in the bryophyte flora of North Wales has resulted in a steady increase in the number of species known as well as in knowledge of their detailed distribution and ecology. Some of the more recent discoveries have been remarkable, e.g. *Homalothecium nitens, Micromitrium tenerum, Platygyrium repens, Southbya tophacea*. Even now, exploration is by no means finished: though the well-known localities are continually revisited, many remoter places remain little known. An area of such rugged topography and diversified geology and climate is difficult to explore exhaustively, and further new discoveries are to be expected. A bryophyte flora of Gwynedd by J. G. Duckett, M. O. Hill and A. J. E. Smith is in preparation.

The discovery of new and sometimes strikingly distinct species in a region such as North Wales where bryophytes have been collected for more than 200 years obviously raises the question whether such species could have been overlooked or whether they are in fact recent immigrants. In our area there are at least two bryophyte species which are generally recognized as recent immigrants to the British Isles, *Campylopus introflexus* and *Orthodontium lineare*. The former was first found in Gwynedd in 1952 (Richards, 1963) and since then has spread rapidly in Anglesey and mainland Gwynedd, colonizing a variety of habitats (including the top of a stone wall) (Richards and Smith, 1975). *O. lineare*, on the other hand, though discovered on Holyhead Mountain, Anglesey, in 1957, was not found on the mainland until 1966 and is still known in Gwynedd only from five isolated localities. Local conditions must in some way be less favourable for it than in many districts of England where it is now one of the commonest mosses. One other species which may be suspected of being an immigrant is *Atrichum crispum*, a moss which is widespread in North America but in Europe is not found outside the British Isles (where only male plants are found). In Gwynedd it is not uncommon on gravel by streams, on lake shores and other habitats: though quite a large and conspicuous moss it does not seem to have been recorded from Gwynedd before about 1900 and it seems rather unlikely that it could have passed unnoticed in the previous century.

An unfortunate result of the greatly increased attention given to North Wales bryophytes in the last 50 years is that some of the rarer species are in danger from over-collecting. At the same time the growth of tourism, mountain climbing and other leisure activities as well as the agricultural and industrial development of the area are damaging or destroying the habitats of many species. North Wales because of its lack of heavy industry and its situation on the west coast of Great Britain is relatively free from atmospheric pollution, but public pressure both on the mountains and the coast is damaging the vegetation and producing erosion while bogs and fens are being lost by drainage. Though the Nature Conservancy Council and the North Wales Naturalists' Trust are endeavouring to conserve and protect the most important sites much more needs to be done. A number of the rarer bryophyte species have not been seen for many years, e.g. *Bartramidula wilsonii, Conostomum tetragonum* and *Scapania nimbosa*. It is impossible to be sure that these species are extinct, but they and many others will certainly be lost unless there is some restraint on collecting and more effort is given to the conservation of bryophytes and their habitats.

REFERENCES

Davies, H. (1813). "Welsh Botanology". Printed for the author by W. Marchant, London.

Desmond, R. (1977). "Dictionary of British and Irish Botanists and Horticulturalists". Taylor and Francis, London.

Dillenius, J. J. (1741). "Historia Muscorum". E Theatro Sheldoniano, Oxford.

Druce, G. C. and Vines, S. H. (1907). "The Dillenian Herbaria". Clarendon Press, Oxford.

Griffith, J. E. (1895). "The Flora of Anglesey and Carnarvonshire". Nixon and Jarvis, Bangor.

Hill, M. O. (1978). A new flora of Gwynedd. *Welsh Bull. bot. Soc. Br. Isl.* **28**, 5–6.

Hudson, W. (1762). "Flora Anglica". Nourse, London.

Hyde, H. A. (1931). Samuel Brewer's diary. *Rep. bot. Soc. Exch. Club Br. Isl.*, Suppl., **1931**, 1–30.

Jenkins, R. T. (ed.) (1959). "Dictionary of Welsh Biography down to 1940". Blackwell, Oxford.

Lindberg, S. O. (1874). "Historiae muscorum" hepaticae; secundum specimina in herbario Dillenii nunc asservata determinatae. Notis. Sällsk. Faun. Fl. fenn. Förh. **13**, 353–356.

Lindberg, S. O. (1883). "Kritisk Granskning af Mossorna uti Dillenii Historia Muscorum". Frenckell, Helsingfors (Quoted in Druce and Vines, 1907).

Owen, T. J. (1961). Hugh Davies: the Anglesey botanist. *Trans. Anglesey Antiq. Soc. Fld Club* **1961**, 39–52.

Pennant, T. (1769). "Indian Zoology". First edition, London.

Ray, J. (1690). "Synopsis Methodica Stirpium Britannicarum". S. Smith, London.

Ray, J. (1724). "Synopsis Methodica Stirpium Britannicarum". Third edition (Revised by J. J. Dillenius and others). G. and J. Innys, London.

Richards, P. W. (1963). *Campylopus introflexus* (Hedw.) Brid. and *C. polytrichoides* De Not. in the British Isles, a preliminary account. *Trans. Br. bryol. Soc.* **4**, 404–417.

RICHARDS, P. W. and SMITH, A. J. E. (1975). A progress report on *Campylopus intro-flexus* (Hedw.) Brid. and *C. polytrichoides* De Not. in Britain and Ireland. *J. Bryol.* **8,** 293–298.

SCHIMPER, W. P. (1876). "Synopsis Muscorum Europaeorum". Second edition, Vol. 1. E. Schweizerbart, Stuttgart.

STEPHEN, L. (1888). "Dictionary of National Biography". Vol. 14. Smith, Elder, London.

WILSON, W. (1855). "Bryologia Britannica". Longman, Brown, Green and Longmans, London.

WITHERING, W. (1796). "An Arrangement of British Plants According to the Latest Improvements of the Linnean System". Third edition, Vol. 3. M. Swinney, Birmingham and London.

2 | The Phylogeny and Distribution of the Musci

H. A. MILLER

Department of Biological Sciences, University of Central Florida,
Orlando, Florida, 32816 USA

Abstract: Modern classification aims to recognize taxa at several hierarchical levels and to arrange them in such a way that phylogenetic relationships may be inferred. Such a phylogenetic system is useful in the sense that a group with a particular aggregate of primitive characteristics can be considered least evolved. Then one or more advanced groups which show some evidence of derivation may be placed to show putative relationships. The great antiquity of mosses which probably extend back at least to the Silurian, the extinction of many intermediate types, and the fact that gametophyte and sporophyte generations have not evolved in absolute parallel combine to obscure phylogeny among Musci. New approaches including cytology, electron microscopy, phytochemistry, ecophysiology, micropaleontology and computer analyses have provided promising insights to moss origin and evolution. Geobotanical analysis combining data from extant distributions with continental movement and paleoclimatology since Paleozoic time show five distinctive distributional patterns—northern, southern, high latitude–altitude, wet tropical, and widely distributed. Correlations between the ecology of extant plants and paleoclimatic history give further evidence for the existence of these major floristic elements. Inadequacies of the classification system used for over 50 years are pointed out; a new phylogeny must soon be developed which takes into account the abundance of new evidence which cannot be fitted into the current schemes.

CLASSIFICATION

The basic system for classification of mosses today is derived from that produced by Fleischer (1904–1923). This system works reasonably well, is familiar and

Systematics Association Special Volume No. 14, "Bryophyte Systematics", edited by G. C. S. Clarke and J. G. Duckett, 1979, pp. 11–39, Academic Press, London and New York.

fairly well detailed. However, much has happened in bryology during the 20th century, especially during the last 15 years. In fact, this symposium deals with many topics that were almost or totally unknown in the bryological literature of the 1950s. It is this incredible burst of new information, of additional approaches, and the great number of scholars now involved with the study of mosses that has brought about this symposium. As something over half the students of bryophytes, in the broadest sense, since the time of Micheli in the 18th century are alive today, a stock-taking is very much in order if we are to go forward in a reasonably orderly manner. Accordingly, I have sought to outline the state of systematic muscology and perhaps to stimulate new directions for inquiry.

Classification is most useful when related, or at least phenotypically similar, taxa are grouped together to produce a hierarchical system. Implied in all this is a basic concept of genetic relationship stretching back through time to ancestral populations. Rarely, if ever, are populations of the ancestral type extant, assuming we could recognize them as such, and further, direct evidence from fossil remains is quite limited and can never be complete. Thus, the classifier becomes a phylogenist of sorts (Vitt, 1971; Lewinsky, 1976, 1977) in his attempts to construct a natural system which shows possible relationships—i.e., a putative evolutionary history. It is axiomatic that absolute truth cannot be realized and that individual interpretations of the same biological phenomena can produce different systems. Greene (1976) recognized this with reference to genera and species by stating that ". . . most of these arrangements have no real biological basis and are little more than supposition . . . I suggest that it is a waste of time trying to imply relationships and it would be better to settle for an arbitrary e.g. alphabetical order". It is difficult to see how such an arbitrary system, if carried to higher levels of classification, would improve matters or resolve the dilemma of incomplete knowledge. Surely, the posing of questions to the taxa at whatever level is made simpler by having organisms which show some evidence of possible common ancestry placed to reflect that relationship. What neophyte investigator or monographer could move in a potentially meaningful way without some aid from the systematics of the day—however crude it might prove to be in the future? As Cronquist (1975) put it, "development of a taxonomic scheme and a phylogenetic interpretation properly proceed in close association, each influencing the other".

1. Identification

From a practical point of view, systematic bryology is parallel to any other field of science—that is, there is a place for theory and a conceptual scheme to be tested. Revised theories, or even rejected ones, serve to advance the frontiers of knowledge. Paradigms limited to bryology have been few but certainly our discipline has not been immune to the phenomenon so well documented for physics and chemistry by such science philosophers as Kuhn (1970) and Price (1963). We have come a long way since Micheli (1729) established the discreetness of several taxa of bryophytes and Linnaeus incorporated mosses and liverworts into the *Species Plantarum* (1753). The compound microscope provided Hedwig (1801) with the basic tool to observe reproductive structures in mosses as well as something of their areolation. Even so, taxonomic bryology was essentially a magnifying glass science until well into the 19th century. With the production of *Bryologia Europaea* (Bruch *et al.*, 1836–1855), the compound microscope came fully forward as an indispensable tool for bryology. The philosophical base upon which moss taxonomy rested then shifted from the organism–organ level to an organ–tissue base utilizing features of the cells visible only with magnifications in the compound microscope range. As microscope manufacture improved, observations became more sophisticated but the fundamental approach to identification has remained little changed for 150 years. I doubt that basic identification procedure will change greatly for some time to come given today's technology. Scanning Electron Microscopy (SEM) expands the subtleties observable at the cellular level but does not change the basic approach. The SEM plus Transmission Electron Microscopy (TEM) bring our potential for systematic evaluation to the tissue-cell level whereby intracellular structures and/or compounds of diverse chemical makeup can be important aids to determining relationships.

We must rely upon morphological features initially to characterize and identify our basic taxonomic unit, the species, as well as to postulate several levels of higher categories. The process is intuitive in some measure but the fact that it is intuitive in no way compromises its validity. Some taxonomists, of course, seem to have different intuitive reactions but they are usually not so diverse that wide discrepancies exist concerning the identification of a species or genus. It is among the higher categories, where greater generalizations must obtain, that judgement of the value of individual characters and combinations enter, and concensus begins to break down. Part of the problem was addressed by Cronquist (1975) who wrote:

> One of the most fundamental taxonomic principles that most of us are comfortable with is that taxonomy proceeds by the recognition of multiple correlations. A corollary is that individual characters are only as important as they prove to be in marking groups that have been recognised on a larger set of information. It is a natural assumption that once the value of a character in a particular group has been established in this way, it

can be applied fairly uniformly across the board in other groups. This assumption is false and it must be unlearned by each successive generation of taxonomists . . . paraphrasing Orwell, all characters are equal, but some are more equal than others.

2. Systematics

When we are presented with a total package of generalizations which is sensible in its structure—that is, it can be understood as well as followed—good sense dictates that it be used despite individually perceived lacunae. So it was with Fleischer's system and the subsequent near-canonical impact derived from its use by Brotherus in the second edition of *Die natürlichen Pflanzenfamilien* in 1924–1925. Within a few years, Dixon (1932) brought his insights and intuitions, partly as a critic, to bear in the *Manual of Bryology* but proposed few substantive changes, mostly on the basis of peristome structure. The situation then seems to have remained rather static until Reimers (1954) reviewed the bryophytes in the twelfth edition of *Syllabus der Pflanzenfamilien* with minor refinements. Indeed, while individual authors have raised or lowered this group or that and shifted a few things here and there (e.g. Smith, 1978), the bulwark remains the Brotherus treatment based on Fleischer. In a sense, then, the Philibert phylogenetic generalities (Taylor, 1962) based heavily on outer peristome structure still hold considerable sway.

Perhaps the best way to follow phylogenetic thinking and the revisionary pattern since the introduction of Fleischer's system is to compare the arrangements at the family level (given in the appendix to this paper, Tables I–VII) as designated by different authors. There is some hazard in so doing because family concepts in certain groups are quite unequal from author to author but this problem does not seriously compromise our purpose to show the somewhat individual minor modifications employed by a spectrum of bryologists from various parts of the world. The selection is intended to be representative rather than comprehensive and inclusion or omission from the Tables has no special significance.

Comparison among these systems is most interesting on two counts. First, the points of disagreement may indicate an inadequate philosophical or informational base to handle the disparate groups as perceived by the authors, and second, the points of agreement depend upon acceptance of the scheme because it answers the questions put to it in a predictable and consistent way. Do the elements of concensus mean that we are half-way to developing the ultimate phylogeny or natural system, although the latter is not quite the same? Perhaps. But it could also mean that there has been a diminution of interest in phylogeny amongst muscologists who, with an operational system, have tended to direct their energies to naming collections, making floristic lists and producing monographs. This has led to a tendency to "divorce taxonomy from phylogeny, reverting in this respect to pre-Darwinian days . . ." (Cronquist, 1975). This seeming retrenchment has led to a tendency by bryologists to apply newer taxonomic techniques in a most limited way and to overlook potentially useful applications to phylogeny. To complicate matters even further, many promising studies on bryophytes have been by non-taxonomists and without the benefit of insights possible from an experienced bryologist. The absence of voucher specimens for many of these studies compromises the potential value of the work.

Much remains to be learned about every aspect of the biology of bryophytes and for maximum efficiency to be achieved in development of new insights applicable to phylogenetic thinking, taxonomists must be involved. The taxonomist with a strong interest in phylogeny and natural relationships is in a prime position to suggest which organisms might be best suited for a particular type of study. In order to do so, of course, the taxonomist must seek to stay abreast of developments ancillary to purely morphologically based systematic work and this is not an easy task. One often has enough trouble keeping up with taxonomic literature, let alone the experimental and biochemical contributions which are listed with increasing frequency in the current literature lists of our major bryological journals. Practical considerations, too, must enter our deliberations in that it is much easier to obtain research support for studies of a biological or biochemical nature than it is for studies centred upon identification such as conventional monographs or floristics (Richards, 1978). It would seem that personal involvement by a taxonomic bryologist in a biological research team would create the best situation possible for all concerned. The possibilities are many with the wealth of approaches available using high technology for a diversity of analytical procedures. I can neither list nor predict all the opportunities, and surely no one person or team of manageable size can follow up on everything, but many working towards a common goal can accelerate the arrival of the quantum leap in understanding that is just before us.

ORIGIN OF MOSSES

It is in the literature outside the usual purview of bryologists where much that pertains very directly to the origin and evolution of mosses is published. I have previously summarized those contributions which seemed relevant (Miller, 1973–1974, 1974, 1977) and much background information is therefore omitted from the present discussion. None the less, because phylogenetic concepts depend upon outlooks on origin and evolutionary phenomena, I review recent developments and some interpretations of these data appropriate to mosses.

For me, the most significant breakthrough was the discovery of Banks's (1968a, b) papers on classification of early land plants with accompanying insights to possible affinities of bryophytes. A search of literature on paleoclimatology and plate tectonics provided further information. Ultimately, the notion that bryophytes arose with other terrestrial plants and constituted several little-related lines became inescapable. The oldest known macrofossil land plants, the Rhyniophytina and Zosterophyllophytina (Banks, 1973), share many common features with the bryophytes (Crum, 1976) which suggest an origin from a diverse gene pool at a level of evolution sufficient to achieve a partial transmigration to the land. Achievement of successful occupation of the terrestrial habitat depended upon some sort of internal water-conducting system (Raven, 1977). In mosses, the conducting strands did not elaborate quantities of lignins (Hébant, 1977) and so true tracheids were not formed even though the basic cellular organization and function was apparently evolved if we can judge from extant mosses (Scheirer and Goldklang, 1977). Accordingly, the vascular system is best described as "non-generate" as opposed to degenerate, which carries a connotation of reduction.

1. Megafossils of Land Plants

Megafossil evidence for the earliest bryophytes is very limited indeed, and recognizable mosses (moss-like plants) have not yet been reported from Devonian strata (Lacey, 1969). Eventually, they will probably be found there as paleobotanists look for smaller structures and expand the application of maceration and peel techniques to Devonian and Silurian rocks. My confidence in their ultimate discovery is bolstered by Gray and Boucot (1977) who have given us a most penetrating essay "Early vascular land plants: proof and conjecture", duplicating the title of an article by Banks (1975). As students of microfossils, they bring us a point of view derived from a different base. They emphasize that the terms "land plants" and "vascular plants" do not have precisely the same meaning, despite the frequent use of one when the other is meant. A land plant is one that customarily lives on land and whose relations are mainly to other plants living on land; the plant may or may not be vascular. Implied in this is that a time existed when land plants were not vascularized. Hence there is an "evolutionary gap" in the record which must be pre-Devonian. It is to the "evolutionary gap" that Gray and Boucot address themselves in a manner relevant to considerations of moss phylogeny, whether or not all their ideas prove sound.

It is obvious, although not always recognized, that the time of evolution and the time of appearance in the fossil record must be kept conceptually distinct because we cannot be sure how long evolution progressed before the fossils present today were laid down in the rocks. Further, the physical evidence may be microscopic cuticular fragments of a most delicate sort, or spores, or traces of vascular cells. Fossils like these are found in pre-Devonian deposits even though such plant structures may have had limited capability for preservation. The low incidence of bryophytes and of gametophytes of vascular plants reported in the post-Silurian fossil record emphasizes the problem of finding and recognizing remains of small, delicate organisms and structures. Absence of a cuticle on some mosses which were ectohydric surely limited the possibility for preservation but endohydric mosses have a cuticle which can persist as in *Neubergia*, a Permian moss from Antarctica. This suggests, of course, that the fossil bryophytes will be mainly from those groups which are structurally endohydric as, for instance, some Bartramiaceae. Evidence from ectohydric groups may have to be drawn from other sources.

2. *Microfossils of Land Plants*

Those spores that contain sporopollenin, which is resistant to water and decay, have a much greater potential for survival on land and for preservation. Heslop-Harrison (1971) emphasized that the "... development of the protected spore was the key event in the spread of life over the land surface of the earth". Such spores are the type formed by the bryophytes and other extant terrestrial cryptogams.

Durable-walled spores linked permanently in tetrahedral tetrads and displaying a permanent and distinct triradiate mark are the most common plant remains in pre-Devonian samples. Such trilete spores are known from extant bryophytes (though the mark is sometimes weak or lost during maturation in mosses), ferns, lycopods and other vascularized spore plants. In the old fossil record, they may be indicative only of land plants in a prevascular or near-vascular stage. It is important to note that no extant alga produces trilete spores with a durable cell wall (Gray and Boucot, 1977) despite notations in paleobotanical literature that they can be found among some red and brown algae. A much better correlation is found between the trilete spore (land adapted) and the archegonium (also a land-adapted structure) among extant groups. It seems reasonable that such a universal link between every group of archegoniate plants is highly significant and reflects the most successful combination of attributes evolved in the transmigration to the land. That spores and cuticles are found only in shoreline environments of Silurian deposits bolsters this morphological link to transmigration.

Gray and Boucot stressed that the cuticle, trilete spore and tracheid-like tubes are adaptations of terrestrial or emergent aquatic plants. These features show in the fossil record 30 million years before *Cooksonia* and other rhyniophytes, and such fossils are always in shoreline deposits. They wrote:

> We resist the temptation to attribute undue significance to the existing dichotomy between mosses and their relatives and vascular plants in discussing the taxonomic assignment of these remains. If the archegoniates are a natural group stemming from a common land-dwelling ancestor, or if one group of archegoniate land plants has arisen from the other . . . the question of whether the spores and other structures are from vascular or non-vascular land plants would be obviated in the early history of land plant evolution . . . one *cannot rule out* the possibilities of land plants, including vascular plants, in the early Silurian.

The absence of lignin in vascular cells prior to Devonian time could account for the lack of the xylary fossil remains which would be expected whenever the biosynthetic pathways for lignification had evolved and become widespread. Plants at the bryophytic level of evolution which could not develop the capacity for lignin biosynthesis would be limited in their size and would evolve other adaptive strategies for survival on land.

The aggregate of evidence to date suggests an emergence of terrestrial plants in Silurian time with such plants ecologically situated near shorelines perhaps as emergents or as plants of mud-flats subject to periodic flooding. The archegoniate mode of zygote protection which sets the stage for the embryophyte life cycle was surely recognizable. Further, the early separation of the several terrestrial Divisions must already have begun. Although more genetically distinct populations probably existed, an assumption based somewhat upon spore diversity, that the rhyniophyte and allied zosterophyllophyte lines were diverging seems reasonable. In addition, the isolation of mosses with some lingering tendencies toward zosterophyllophytes was begun and the hepatics, with tendencies toward rhyniophytes, were also being differentiated. The hornworts seem to represent a separate response to the transmigratory ecosystems and hold little in common with mosses and hepatics, save the level of evolution achieved. Schuster (1977), among recent authors, has been unable to find any firm link between the hepatics and hornworts.

PHYLOGENETIC CONSIDERATIONS

Acceptance of the viewpoints on the origin of mosses so far presented must affect one's view of phylogeny within the mosses. Because the fossil record is meagre, even though instructive, most phylogenetic arrangements are based necessarily upon the biology of extant plants. The long history of bryophytes has allowed for variation, selection, and extinction of many groups with the result that several systematic disjunctions are sharply defined and the phylogeny proportionately obscured. The nature of the sporophyte and gametophyte of mosses has allowed, within physical and physiological limits, some independence of selection and evolution such that a highly evolved sporophyte may be carried upon a primitive gametophyte or vice versa. This condition also means that wide divergences in concepts of relationship are possible depending upon the weighting given to characters of the generations. Crosby (1974) discussed these problems and ". . . adopted what may be called the Philibert-Dixon Principle: In constructing a classification of the mosses, primary weight must be given to characters of the sporophyte, and particularly the peristome; mosses with similar gametophytes but different peristomes must not be grouped together with mosses which have similar peristomes." He went on to say that the Principle works and that ". . . with study of other sporophytic characters and the gametophyte, it is clear that the overall similarity of groups is much greater than if the gametophyte had been used as the main criterion for classification." The classification proposed by Crosby for the Hookeriaceae (s.l.) demonstrates his application of the Philibert-Dixon Principle. But, successful application in one group does not assure equal success in others, as noted earlier.

1. Principles

Some principles for moss systematics have been presented previously (Miller, 1971) but new approaches should be incorporated. The principles listed can only show major tendencies and in limited groups (e.g. the Pottiaceae, Saito, 1975), some trends may even be reversed.

Principles of Moss Systematics
1. Within any group, larger mosses are generally more primitive than smaller ones.
2. Closely attached forms with all stems leafy are more primitive than stoloniferous forms.
3. Perennial mosses are more primitive than annual or ephemeral species.

4. Both aquatic and xerophytic forms are derived from forms of wet terrestrial habitats.

5. A well-developed central strand in the stem is primitive.

6. Stems with a several-layered cortex of thick-walled cells are more primitive than stems with a unistratose or undifferentiated cortex.

7. Leaf gaps are primitive.

8. Radial leaf arrangement is more primitive than complanate or disticous.

9. A strong costa is more primitive than a weak one with the ecostate conditions most derived.

10. An excurrent costa is an advanced characteristic sometimes associated with blade reduction.

11. Well-developed alar cells may be an advanced condition and correlate with ectohydric conduction.

12. Smooth leaf cells may be primitive with papillate cells the derived condition.

13. Extremely thin-walled or thick-walled cells are derived.

14. Vegetative reproduction by specialized diaspores is more advanced than propagation by simple fragmentation and segmentation.

15. Numerous axillary hairs with hyaline basal cells are primitive; those having few hairs and with brownish basal cells are the most advanced.

16. Numerous gametangia and paraphyses per inflorescence is more primitive than few gametangia per inflorescence.

17. The monoecious condition is normally advanced over dioecism.

18. Sexual dimorphism, expressed in the extreme by the formation of dwarf males and the anisosporous tendency, as in some species of *Macromitrium* and *Homalothecium*, is advanced.

19. Sperm with low numbers of microtubules in the spline may be more advanced than those with high numbers.

20. An elongate seta bearing an exposed capsule is more primitive than a short seta bearing an immersed capsule.

21. A capsule with stomata, especially when they are associated with air chambers, is more primitive than an estomatate capsule.

22. Cleistocarpy is probably a derived condition in the Bryidae.

23. Development of peristome from a fixed number and arrangement of initial cells is usually advanced over variable numbers of initials.

24. A reduced endostome lacking processes on the basal membrane may be advanced over one with processes and a high basal membrane; the presence of cilia may be advanced.

25. A peristome which is reduced or absent is derived from a normal peristome.
26. Retention of the operculum or a portion of it on the columella is an advanced condition.

These morphologically-based principles cannot be applied uniformly across every group but addressing each of them is useful to evaluate some taxa and to organize one's search for diagnostic characters. However, characters other than those mentioned have great value and must not be ignored in classifications. Cytological, chemical and other data have phylogenetic implications and such attributes cannot be ignored. However, chromosomal data suggest contrary explanations in different groups and chemical data may yet be too inconclusive, as are promising results from both TEM and SEM.

Koponen (1978) summarized very well the state of taxonomy and applications of modern methods noting that there is a "... lack of interest in the system of classification above generic level ...". Thus, the primary thrust of his paper is at the generic level although it does deal with phenomena significant at higher taxonomic levels. The recency of Koponen's publication with emphasis on lower taxonomic levels allows us to focus upon certain aspects of modern methods of taxonomy as they may pertain to classification at higher levels.

2. Cytology

Chromosome numbers have been reported for more than 1000 species of mosses, mostly nearctic taxa. Steere (1972) discussed the status of selected families and orders and provided an historical summary of moss cytology. Despite numerous accounts of moss chromosome numbers, cytological studies have sometimes yielded divergent results (Vaarama, 1976).

Although individual moss chromosomes are small, new methods have been developed whereby individual chromosomes can be characterized in such a manner that chromosomal modifications can be ascertained. Among others, Newton (this volume, Chapter 10) has demonstrated the potential value of investigations on chromosome morphology. However, it is still too early to say that precise karyotyping by heterochromatin mapping will contribute to systematics at the higher levels.

On the basis of reported chromosome numbers, Fritsch (1972) noted that seven karyotype groups can be characterized in relation to higher systematic categories. These are:

$n =$	7	Polytrichidae
$n =$	6–8	Mniaceae, Bartramiaceae, Tetraphidales

$n = $ 8–9   Buxbaumiales
$n = $ 10–11   Bryaceae, Hypnobryales, some families in the Isobryales
$n = $ 10–11   Andreaeidae
$n = $ 13–14   Pottiales, Grimmiales
$n = $ 19 + 2 m   Sphagnaceae

However, the Fissidentales, Dicranales, Funariales, Hookeriales and Schisto-stegales cannot be characterized at higher levels by chromosome numbers. One might ask if such cytologically disparate groups are otherwise homogeneous. After all, cytological evidence can be viewed only as confirmatory of aggregate evidence for delimitation of taxa; it cannot affirm a relationship among taxa independently of an aggregate of other characteristics. The rather high base numbers for most mosses suggest the possibility, still incompletely explored, that many moss gametophytes may be diploid or otherwise polyploid (Inoue and Iwatsuki, 1976). Such a situation would go far towards allowing sufficient genetic plasticity as well as monoecism in the gametophyte so that survival might therefore be enhanced. Vaarama (1976) reported occasional nuclear fusions in premeiotic divisions of sporogenous tissue in mosses which produce polyploid spores in the sporangium. A stabilized polyploid, *Bryum corrensii*, was produced experimentally by Wettstein and Straub (1942) from a sporophyte regenerant of *Bryum caespiticium* $(n = 10)$. It was at first a typical polyploid (with enlarged cells) which suddenly ceased multivalent formation, lost ir-regularities in meiosis and cell size was reduced thus becoming a normal sized polyploid plant $(n = 20)$. Such phenomena in nature plus naturally occurring shifts in ploidy may explain why generalizations based upon karyotypes are so difficult to recognize and define. With some higher categories, cytological characterization of a very specific type may prove to be impossible. Much remains to be learned of moss cytology but it will not solve all systematic problems and is not universally applicable to phylogenetic considerations.

3. Electron Microscopy and Ultrastructure

Until recent years, observations upon mosses were limited by the resolving power of the light microscope. With the introduction of the TEM in the early 1960s, new insights were possible. Thin sections of various tissues in the mosses have shown broad agreement with intracellular ultrastructure of other ter-restrial plants. However, Carothers and Duckett (1978) have shown phylo-genetically significant similarities and differences in spermatozoids. Spore structure, many features of which exceed the resolving power of the light microscope, came to the fore as new insights were gained into the complexity of surface ornamentation (e.g. Lewinsky, 1974; Sorsa, 1976) and the spore wall structure. Sorsa and Koponen (1973) found exine thickness to be of some value at the generic level in Mniaceae, and Neidhart (this volume, Chapter 12) shows that spore sections may reveal fundamental differences between higher categories. The potential return seems small for the great effort required to prepare and examine spores using the TEM.

Because of its great depth of field, the SEM has the potential to provide very informative images of surfaces at higher magnification and resolution than optical microscopes can provide. Koponen (1978) has successfully applied SEM to a review of the genera of Splachnaceae from the standpoint of peristome and spore structure demonstrating clearly its value for monographic or revisionary studies. Sorsa (1976) presented a much broader review of representatives from most of the families in Brotherus's Order Eubryales. Some non-conforming features associated with problematic genera were brought closer to understanding by insights derived from the new information.

4. Chemistry

Knowledge of the chemical composition of bryophytes has recently expanded at an almost logarithmic rate as biochemists have discovered that bryophytes were little studied. Huneck (1977) summarized much of this literature while Suire and Asakawa (this volume, Chapter 19) have provided a review based more upon usefulness of the compounds in taxonomy than upon the diversity of compounds themselves which is the primary focus of Huneck's report.

As Suire and Asakawa point out, many compounds thus isolated are substances which are integral to the life processes of all green plants and have little value for evolutionary studies. Some unusual forms of lipids and fatty acids have been reported for the mosses, but these may change in response to environmental conditions (Swanson *et al.*, 1976) so their significance appears to be more biological than phylogenetic.

Of all the compounds so far ascribed to mosses, only the flavonoids seem to have a demonstrated potential for systematic work at higher classification levels. McClure and Miller (1967) provided a provocative survey for putative flavonoids but full identification of compounds such as that by Lindberg *et al.* (1974) has gone very slowly. Several problems of a technical nature hinder a wide ranging survey. These include, among others: (1) difficulty in obtaining sufficiently large samples of clean moss; (2) the nature of binding of some compounds to cell walls which makes them difficult to extract; (3) the necessity for a collaborating biochemist or a sound knowledge of biochemistry; and (4) costs in time and money for supplies, equipment and technical assistance. Most flavonoids so far reported for mosses are of a rather simple type in terms of their biosynthesis (Harborne *et al.*, 1975) although a bi-flavone allied to types expected in such groups as Psilotaceae has been discovered in *Dicranum* (Lindberg *et al.*, 1974). Several investigators are carrying forward flavonoid phytochemistry including samples from groups which occupy phylogenetically critical positions. Much remains to be done, of course, so that a bryologist who can develop a collaboration with a biochemist stands to make significant contributions.

5. Taximetrics

Computers have made possible statistical manipulation of previously indigestible masses of raw data so that population analyses of high sophistication are now possible. Even though such studies have been presented for angiosperms, Koponen (1978) indicated that few have been directed to mosses and only Seki's (1968) treatment of Japanese Sematophyllaceae deals with higher taxa. Evaluation of such studies for mosses is difficult and perhaps premature, none the less, the effort involved in the mechanics of gathering data from moss populations on the scale necessary for comparisons of higher categories is a high price to pay for something of unknown relevance. Bryology already suffers from too few workers (Richards, 1978) to advocate great effort in taximetrics at a time when the field itself, its strengths and weaknesses, has not stabilized among the much larger work force in angiosperm taxonomy. Computers can probably best be used in furtherance of bryology for the next few years in the areas of cataloguing species and distributions and in indexing the diffusely published literature of the field. Taximetrics should and will be developed for mosses but, given the present position of muscology, it is difficult to give high priority to such studies.

6. Experimental Taxonomy

Several approaches have been taken to elucidate the nature of variation in populations of a single or a few related species (Koponen, 1978). Some reports are based upon single spore cultures, upon side-by-side culturing of populations brought in from the field to a controlled environment, upon transplant experiments in nature, and by study of mixed populations in nature. New insights into the limits of species, of population plasticity, and perhaps the process of speciation itself may be gained from these studies, but they are less helpful for understanding of higher categories. An exception may be found in studies of protonematal development such as those by Nehira (1976) who has found promising correlations between such patterns and systematic position. More exploration and refinement are required in this promising field of enquiry.

GEOBOTANY OF MOSSES

Mosses have always been linked to the terrestrial environment and have achieved their greatest diversity in cool, moist, oceanic climates found at higher latitudes or altitudes. Deserts, mediterranean and warm lowland climates are inhospitable to all but a few groups. Diversity among higher categories today is inextricably bound to the history of land masses during and since Paleozoic time (Miller, 1977). Much has been written (Smith, 1972; Miller, 1976; Steere, 1976) on present moss distributions as they have been influenced by events in the recent geological past, but the influences which predate Cenozoic–Tertiary history (Frederikson, 1972) have been more difficult to analyse. It is instructive in development of an understanding of moss geosystematics to review climatic history as well as tectonic movements of major land masses in relation to diversity and distribution of extant orders. Further, we must assume that the suitability of the niches for today's mosses is not greatly different from those of the past. Allowed this element of physiological conservatism, or plasticity depending on the group, we can build a plausible explanation for many bryogeographic phenomena.

On the basis of the Permian fossil record which now includes the famous Angara flora (Neuberg, 1958, 1960), the flora of South Africa (Plumstead, 1966) and of Antarctica (Schopf and Miller, unpublished), we have good evidence that great diversity existed at the time among the mosses and that they had become both widely distributed and well established during the cool and moist glacier-moderated periods of the Carboniferous (Beaty, 1978). The end of Paleozoic time is marked by a shift to much drier mediterranean and desert climates which must have resulted in mass extinctions of mesophilic mosses which failed to adapt or find a microniche in which to persist. There is evidence that only southern South America, South Africa, southern India and a part of Antarctica had a mesophytic or cool oceanic climate during the Permo-Triassic desert episodes. Modern distributions of systematically isolated groups of mosses and, more spectacularly, hepatics reflect the tremendous impact of desiccation of lands outside areas where the *Glossopteris* flora thrived. The few surviving taxa of the northern droughts had to be hardy, indeed, to persist through long dry periods or in deeply shaded sites where moisture was sufficient for their survival.

Schistostega and perhaps *Bryoxiphium* may be the remnants of Permian cave-dwelling bryophytes of Eurasia. *Mittenia* is a parallel case for Australia which was on the opposite side of the *Glossopteris* zone in the Permian.

Mosses were adapted to cool, wet climates which provided a wide spectrum of hospitable edaphic conditions and exposures. They remain today plants of cool, wet climates—the great diversity of species in tropical latitudes occurs because of the orographic rainfall and cool cloud cover over many tropical mountains. It is in the cool cloud forests that mosses thrive in tropical latitudes (Russell *et al.*, 1977). Tropical lowland bryofloras usually lack both the abundance and the diversity of the mountain floras even in regions of high rainfall. At higher latitudes where evaporation stress is less, increased diversity and cover are found at lower elevations. Thus, the northward drift of continents created the barrier of the tropics and dry, warm-temperate climates between the antipodal floras of South America, South Africa, Australia and New Zealand, and the holarctic floras of Eurasia and Northern Africa. The comparatively impoverished bryoflora of west and central Africa reflects its long isolation from the rich antipodal stocks and its intermittent floristic exchange with somewhat less diverse holarctic elements. The strong antipodal ties of the Cape of Good Hope bryoflora are well documented (Schuster, 1969).

We can recognize today some of the stocks which have remained effectively isolated since Paleozoic time. In very crude terms, surely amenable to much refinement, we can recognize northern, southern, high latitude/altitude, wet tropical and widely distributed groups (Miller, 1977).

1. Northern Groups

The Protosphagnales were probably limited in distribution and shared some common ancestry with *Sphagnum*, a few populations of which may have successfully migrated across the tropics into the south although the evidence is inconclusive. The Bryoxiphiales and Schistostegales probably survived in grottos or damp caves. The Fontinalales are aquatic mosses of freshwater rivers and lakes where climatic extremes were ameliorated by the milieu. Both the Tetraphidales and Buxbaumiales are adapted to coniferous forests which may be of long standing and the key to their survival. Whether the Timmiineae belong in this group or in the high latitude/altitude group is difficult to say.

2. Southern Groups

The Rhizogoniineae are represented in the wet tropics but their overall distribution suggests antipodal ties. A few species of Hypopterygiineae have penetrated northern latitudes but the centre of diversity was probably Gondwanan. The long isolation of the Dawsoniales seems clear as does their austral distribution.

3. High Latitude/Altitude Groups

Several bipolar taxa belong here, especially the Andreaeales and Encalyptales. Some families and genera of the Bartramiineae rest comfortably here.

4. Wet Tropical Groups

These are the mosses which are best represented in low to middle altitude tropical forest areas with some tendency for old-world/new-world separations as well. Included here are Syrrhopodontales, Hypnodendrales, Racopilineae, Neckerineae and Hookeriineae.

5. Widely Distributed Groups

Several orders, comprising those which had the potential to adapt to changing climates, belong here. The considerable extant diversity of these groups reflects the genetic plasticity of the ancestral populations and their great success in occupying the many new and edaphically hospitable niches created by the rise of the angiosperms. It seems safe to say that (with the possible exception of the Sphagnales and Archidiales) the Dicranales, Fissidentales, Pottiales, Grimmiales, Orthotrichales, Funariales, Bryineae, Leucodontineae, Hypnobryales and Polytrichales have evolved their numerous taxa during and since Cretaceous time. Whereas many moss groups were lost at the end of the Paleozoic and in the early Mesozoic, as reflected by considerable systematic disjunctions, the angiosperm-linked evolutionary explosion of radiately derived genera and species has given us today perhaps as many or more genera and species than have ever existed at any one time in the past.

ACKNOWLEDGEMENTS

The British Museum (Natural History) has provided both herbarium and library facilities to assist me; thanks are especially extended to R. Ross, J. F. M. Cannon, G. C. S. Clarke and A. Eddy for their concern and support. The manuscript was reviewed by A. J. Harrington who gave constructive suggestions. Other colleagues who have assisted in development and refinement of ideas include H. O. Whittier, D. H. Norris, P. J. Wanstall, C. Hébant, C. Suire, H. A. Crum, J. McClure and Z. B. Carothers.

REFERENCES

BANKS, H. P. (1968a). The stratigraphic occurrence of early land plants and its bearing on their origin. *In* "Proceedings of the International Symposium on the Devonian System" (D. H. Oswald, ed.) pp. 721–730. Calgary, Canada.

BANKS, H. P. (1968b). The early history of land plants. *In* "Evolution and Environment" (E. T. Drake, ed.) pp. 73–107. Yale Univ. Press, New Haven.

BANKS, H. P. (1973). Reclassification of Psilophyta. *Taxon* **24**, 401–413.

BANKS, H. P. (1975). Early vascular plants: proof and conjecture. *BioScience* **25**, 730–737.

BEATY, C. B. (1978). The causes of glaciation. *Am. Scient.* **66**, 452–459.

BROTHERUS, V. F. (1924–1925). Musci (Laubmoose). *In* "Die natürlichen Pflanzenfamilien" (A. Engler, ed.). Second edition Vol. 10, pp. 129–478; Vol. 11, pp. 1–542. W. Engelmann, Leipzig.

BRUCH, P., SCHIMPER, W. P. and GÜMBEL, T. (1836–1855). "Bryologia Europaea". E. Schweizerbart, Stuttgart.

CAROTHERS, Z. B. and DUCKETT, J. G. (1978). A comparative study of the multilayered structure in developing bryophyte spermatozoids. *Bryophyt. Biblthca* **13**, 95–112.

CRONQUIST, A. (1975). Some thoughts on angiosperm phylogeny and taxonomy. *Ann. Mo. bot. Gdn* **62**, 517–520.

CROSBY, M. R. (1974). Toward a revised classification of the Hookeriaceae (Musci). *J. Hattori bot. Lab.* **38**, 129–141.

CROSBY, M. R. and MAGILL, R. E. (1977). "A Dictionary of Mosses". Missouri Botanical Garden, St Louis.

CRUM, H. A. (1976). "Mosses of the Great Lakes Forest". Revised ed. Univ. Michigan Press, Ann Arbor.

DIXON, H. N. (1932). Classification of mosses. *In* "Manual of Bryology" (F. Verdoorn, ed.) pp. 397–412. Martius Nijhoff, The Hague.

FLEISCHER, M. (1904–1923). "Die Musci der Flora von Buitenzorg". E. J. Brill, Leiden.

FREDERIKSEN, N. O. (1972). The rise of the Mesophytic flora. *Geosci. Man* **4**, 17–28.

FREY, W. (1977). Neue Vorstellungen über die Verwandtschaftsgruppen und die Stammesgeschichte der Laubmoose. *In* "Beiträge zur Biologie der niederen Pflanzen" (W. Frey, H. Hurka and F. Oberwinkler, eds) pp. 117–139. Gustav Fischer, Stuttgart.

FRITSCH, R. (1972). Chromosomenzahlen der Bryophyten. Eine Übersicht und Diskussion ihres Aussagewertes für das System. *Wiss. Z. Friedrich-Schiller-Univ., Jena,* Math. Nat. Reihe, **21**, 839–944.

GRAY, J. and BOUCOT, A. J. (1977). Early vascular land plants: proof and conjecture. *Lethaia* **10**, 145–174.

GREENE, S. W. (1976). Are we satisfied with the rate at which bryophyte taxonomy is developing? *J. Hattori bot. Lab.* **41**, 1–6.

HARBORNE, J. B., MABRY, T. J. and MABRY, H. (eds) (1975). "The Flavonoids". Chapman and Hall, London.

HEBANT, C. (1977). The conducting tissues of bryophytes. *Bryophyt. Biblthca* **10**, 1–155.

HEDWIG, J. (1801). "Species Muscorum". Barth, Lipsiae.

HESLOP-HARRISON, J. (1971). Sporopollenin in the biological context. *In* "Sporopollenin" (J. Brooks, P. R. Grant, M. Muir, P. van Gijzel and G. Shaw, eds) pp. 1–30. Academic Press, London and New York.

HUNECK, S. (1977). Neue Ergebnisse zur Chemie der Moose, eine Übersicht. Teil 5. *J. Hattori bot. Lab.* **43**, 1–30.

INOUE, S. and IWATSUKI, Z. (1976). A cytotaxonomic study of the genus *Rhizogonium* Brid. (Musci). *J. Hattori bot. Lab.* **41**, 389–403.

KOPONEN, A. (1978). The peristome and spores in Splachnaceae and their evolutionary and systematic significance. *Bryophyt. Biblthca* **13**, 535–577.

KOPONEN, T. (1978). Modern taxonomical methods and the classification of mosses. *Bryophyt. Biblthca* **13**, 443–481.

KUHN, T. S. (1970). "The Structure of Scientific Revolutions". Second edition. Univ. Chicago Press, Chicago.

LACEY, W. S. (1969). Fossil bryophytes. *Biol. Rev.* **44**, 189–205.

LEWINSKY, J. (1974). The family Plagiotheciaceae in Denmark. *Lindbergia* **2**, 185–217.

LEWINSKY, J. (1976). On the systematic position of *Amphidium* Schimp. *Lindbergia*, **3**, 227–231.

LEWINSKY, J. (1977). The genus *Orthotrichum*. Morphological studies and evolutionary remarks. *J. Hattori bot. Lab.* **43**, 31–61.

LINDBERG, G., ÖSTERDAHL, B. G. and NILSSON, E. (1974). Chemical studies on bryophytes. 16. 5′,8′′-Biluteolin, a new biflavone from *Dicranum scoparium*. *Chemica Scripta* **5**, 140–144.

LINNAEUS, C. (1753). "Species Plantarum" Vol. 2. Laurentii Salvii, Holmii.

McCLURE, J. W. and MILLER, H. A. (1967). Moss chemotaxonomy: a survey for flavonoids and the taxonomic implications. *Nova Hedwigia* **14**, 111–125.

MICHELI, P. A. (1729). "Nova Plantarum Genera". B. Paperini, Florence.

MILLER, H. A. (1971). An overview of the Hookeriales. *Phytologia* **21**, 243–252.

MILLER, H. A. (1973–1974). Hepaticae through the ages. *Revta Fac. Ciênc. Univ. Lisb.*, Ser. C, Ciênc. nat., **17**, 733–745.

MILLER, H. A. (1974). Rhyniophytina, alternation of generations, and the evolution of bryophytes. *J. Hattori Bot. Lab.* **38**, 161–168.

MILLER, H. A. (1977). A geobotanical overview of the Bryophyta. *In* "Geobotany" (R. C. Romans, ed.) pp. 95–107. Plenum, New York.

MILLER, N. G. (1976). Quaternary fossil bryophytes in North America: a synopsis of the record and some phytogeographic implications. *J. Hattori bot. Lab.* **41**, 73–85.

NEHIRA, K. (1976). Protonema development in mosses. *J. Hattori bot. Lab.* **41**, 157–165.

NEUBERG, M. (1958). Permian true mosses of Angaraland. *J. palaeont. Soc. India* **3**, 22–29.

NEUBERG, M. (1960). [Leafy mosses from the Permian deposits of Angarida.] *Trudy-geol. Inst. Leningrad* **19**, 1–104 (in Russian).

PLUMSTEAD, E. P. (1966). Recent palaeobotanical advances and problems in Africa. *In* "Symposium on Floristics and Stratigraphy of Gondwanaland", pp. 1–12. Birbal Sahni Inst. Palaeobotany, Lucknow.

PRICE, D. J. de S. (1963). "Little Science, Big Science". Columbia Univ. Press, New York.

RAVEN, J. A. (1977). The evolution of vascular land plants in relation to supracellular transport processes. *Adv. bot. Res.* **5**, 153–219.

REIMERS, H. (1954). Bryophyta. Moose. *In* "A. Engler's Syllabus der Pflanzen-familien" (H. Melchior and E. Werdermann, eds) Twelfth edition, pp. 218–268. Gebrüder Borntraeger, Berlin.

RICHARDS, P. W. (1978). The taxonomy of bryophytes. *In* "Essays in Plant Taxonomy" (H. E. Street, ed.), pp. 177–209. Academic Press, London and New York.

ROBINSON, H. (1971). A revised classification for the orders and families of mosses. *Phytologia* **21**, 289–293.

RUSSELL, K. W., MILLER, H. A. and WHITTIER, H. O. (1977). The ecology of an elfin forest in Puerto Rico, 17. Mosses. *J. Arnold Arbor.* **58**, 1–25.

SAITO, K. (1975). A monograph of Japanese Pottiaceae (Musci). *J. Hattori bot. Lab.* **39**, 373–537.

SCHEIRER, D. C. and GOLDKLANG, I. J. (1977). Pathway of water movement in hydroids of *Polytrichum commune* Hedw. (Bryopsida). *Am. J. Bot.* **64**, 1046–1047.

SCHUSTER, R. M. (1969). Problems of antipodal distribution in lower land plants. *Taxon* **18**, 46–91.

SCHUSTER, R. M. (1977). The evolution and early diversification of the Hepaticae and Anthocerotae. *In* "Beiträge zur Biologie der niederen Pflanzen" (W. Frey, H. Hurka and F. Oberwinkler, eds), pp. 107–115. Gustav Fischer, Stuttgart.

SEKI, T. (1968). A revision of the family Sematophyllaceae of Japan with special reference to a statistical demarcation of the family. *J. Sci. Hiroshima Univ.*, Ser. B, Div. 2 Bot., **12**, 1–80.

SMITH, A. J. E. (1978). "The Moss Flora of Britain and Ireland". Cambridge Univ. Press, Cambridge.

SMITH, G. L. (1972). Continental drift and the distribution of Polytrichaceae. *J. Hattori bot. Lab.* **35**, 41–49.

SORSA, P. (1976). Spore wall structure in Mniaceae and some adjacent bryophytes. *In* "The Evolutionary Significance of the Exine" (I. K. Ferguson and J. Muller, eds), pp. 211–229. Academic Press, London and New York.

SORSA, P. and KOPONEN, T. (1973). Spore morphology of Mniaceae Mitt. (Bryophyta) and its taxonomic significance. *Annls bot. fenn.* **10**, 187–200.

STEERE, W. C. (1972). Chromosome numbers in bryophytes. *J. Hattori bot. Lab.* **35**, 99–125.

STEERE, W. C. (1976). Ecology, phytogeography and floristics of arctic Alaskan bryophytes. *J. Hattori bot. Lab.* **41**, 47–72.

SWANSON, E. S., ANDERSON, W. H., GELLERMAN, J. L. and SCHLENK, H. (1976). Ultrastructure and lipid composition of mosses. *Bryologist* **79**, 339–349.

TAYLOR, E. C. (1962). The Philibert peristome articles. An abridged translation. *Bryologist* **65**, 175–212.

VAARAMA, A. (1976). The cytotaxonomic approach to the study of bryophytes. *J. Hattori bot. Lab.* **41**, 7–12.

VITT, D. H. (1971). The infrageneric evolution, phylogeny, and taxonomy of the genus *Orthotrichum* (Musci) in North America. *Nova Hedwigia* **21**, 683–711.

WETTSTEIN, F. von and STRAUB, J. (1942). Experimentelle Untersuchungen zum Artbildungsproblem III. Weitere Beobachtungen an polyploiden *Bryum*-Sippen. *Z. indukt. Abstamm. u. VererbLehre* **80**, 271–280.

TABLE I. Classification of the Class Musci given by Fleischer (1904–1923). Fleischer's "Reihengruppe" are here designated "Superorders" because of their position in the system

Subclass Sphagnales
    Sphagnaceae
Subclass Andreaeales
    Andreaeaceae
Subclass Bryales
  Superorder Eubryinales
    Order Fissidentales
      Fissidentaceae
    Order Dicranales
      Suborder Dicranineae
        Archidiaceae
        Ditrichaceae
        Seligeriaceae
        Trematodontaceae
        Rhabdoweisiaceae
        Dicranaceae
        Dicnemonaceae
        Pleurophascaceae
      Suborder Leucobryineae
        Leucobryaceae
        Leucophanaceae
    Order Pottiales
      Suborder Syrrhopodontineae
        Syrrhopodontaceae
        Calymperaceae
      Suborder Encalyptineae
        Encalyptaceae
      Suborder Pottiineae
        Trichostomaceae
        Pottiaceae
    Order Grimmiales
      Grimmiaceae
    Order Funariales
      Suborder Funariineae
        Gigaspermaceae
        Funariaceae
        Disceliaceae
      Suborder Splachnineae
        Oedipodiaceae
        Splachnaceae
    Order Schistostegales
      Schistostegaceae
    Order Tetraphidales

      Georgiaceae
    Order Eubryales
      Suborder Bryineae
        Bryaceae
        Leptostomaceae
        Mniaceae
      Suborder Rhizogoniineae
        Drepanophyllaceae
        Eustichiaceae
        Sorapillaceae
        Mitteniaceae
        Calomniaceae
        Rhizogoniaceae
      Suborder Hypnodendrineae
        Hypnodendraceae
      Suborder Bartramiineae
        Aulacomniaceae
        Meeseaceae
        Catoscopiaceae
        Bartramiaceae
        Spiridentaceae
      Suborder Timmiineae
        Timmiaceae
    Order Isobryales
      Suborder Orthotrichineae
        Erpodiaceae
        Orthotrichaceae
      Suborder Rhacopilineae
        Helicophyllaceae
        Rhacopilaceae
      Suborder Fontinalineae
        Fontinalaceae
        Climaciaceae
      Suborder Leucodontineae
        Hedwigiaceae
        Cryphaeaceae
        Leucodontaceae
        Cyrtopodaceae
        Ptychomniaceae
        Lepyrodontaceae
        Prionodontaceae
        Rutenbergiaceae
        Trachypodaceae

Subclass Bryales—Suborder Leucodontineae—cont.

    Myuriaceae
    Pterobryaceae
    Meteoriaceae
  Suborder Neckerineae
    Phyllogoniaceae
    Neckeraceae
    Lembophyllaceae
    Echinodiaceae
Order Hookeriales
  Suborder Nematacineae
    Nemataceae
  Suborder Hookeriineae
    Pilotrichaceae
    Hookeriaceae
    Symphyodontaceae
    Leucomiaceae
    Hypopterygiaceae
Order Hypnobryales
  Suborder Leskeineae
    Theliaceae

    Fabroniaceae
    Leskeaceae
    Thuidiaceae
    Amblystegiaceae
    Brachytheciaceae
  Suborder Hypnineae
    Entodontaceae
    Plagiotheciaceae
    Sematophyllaceae
    Hypnaceae
    Rhytidiaceae
    Hylocomiaceae
Superorder Buxbaumiinales
Order Buxbaumiales
    Buxbaumiaceae
Order Diphysciales
    Diphysciaceae
Superorder Polytrichinales
Order Dawsoniales
    Dawsoniaceae
Order Polytrichales
    Polytrichaceae

TABLE II. Classification of the Class Musci given by Brotherus (1924–1925)

Subclass Sphagnales
    Sphagnaceae
Subclass Andreaeales
    Andreaeaceae
Subclass Bryales
  Superorder Eubryinales
  Order Fissidentales
    Fissidentaceae
  Order Dicranales
    Suborder Dicranineae
    Archidiaceae
    Ditrichaceae
    Bryoxiphiaceae
    Seligeraceae
    Dicranaceae
    Dicnemonaceae
    Suborder Pleurophascineae
    Pleurophascaceae
    Suborder Leucobryineae
    Leucobryaceae

Order Pottiales
  Suborder Syrrhopodontineae
    Calymperaceae
  Suborder Encalyptineae
    Encalyptaceae
  Suborder Pottiineae
    Pottiaceae
Order Grimmiales
    Grimmiaceae
Order Funariales
  Suborder Funariineae
    Gigaspermaceae
    Disceliaceae
    Ephemeraceae
    Funariaceae
  Suborder Splachnineae
    Oedipodiaceae
    Splachnaceae
Order Schistostegiales
    Schistostegaceae

Subclass Bryales—cont.
  Order Tetraphidales
    Georgiaceae
  Order Eubryales
    Suborder Bryineae
      Bryaceae
      Leptostomaceae
      Mniaceae
    Suborder Rhizogoniineae
      Drepanophyllaceae
      Eustichiaceae
      Sorapillaceae
      Mitteniaceae
      Calomniaceae
      Rhizogoniaceae
    Suborder Hypnodendrineae
      Hypnodendraceae
    Suborder Bartramiineae
      Aulacomniaceae
      Meeseaceae
      Catoscopiaceae
      Bartramiaceae
    Suborder Timmiineae
      Timmiaceae
  Order Isobryales
    Suborder Orthotrichineae
      Erpodiaceae
      Ptychomitriaceae
      Orthotrichaceae
      Helicophyllaceae
    Suborder Rhacopilineae
      Rhacopilaceae
    Suborder Fontinalineae
      Fontinalaceae
      Climaciaceae
    Suborder Leucodontineae
      Hedwigiaceae
      Cryphaeaceae
      Leucodontaceae
      Cyrtopodaceae
      Ptychomniaceae
      Lepyrodontaceae
      Prionodontaceae

      Rutenbergiaceae
      Trachypodaceae
      Myuriaceae
      Pterobryaceae
      Meteoriaceae
    Suborder Neckerineae
      Phyllogoniaceae
      Neckeraceae
      Lembophyllaceae
      Echinodiaceae
  Order Hookeriales
    Suborder Nematacineae
      Nemataceae
    Suborder Hookeriineae
      Pilotrichaceae
      Hookeriaceae
      Symphyodontaceae
      Leucomiaceae
      Hypopterygiaceae
  Order Hypnobryales
    Suborder Leskeineae
      Theliaceae
      Fabroniaceae
      Leskeaceae
      Thuidiaceae
      Amblystegiaceae
      Brachytheciaceae
    Suborder Hypnineae
      Entodontaceae
      Plagiotheciaceae
      Sematophyllaceae
      Hypnaceae
      Rhytidiaceae
      Hylocomiaceae
Superorder Buxbaumiinales
  Order Buxbaumiales
    Buxbaumiaceae
    Diphysciaceae
Superorder Polytrichinales
  Order Polytrichinales (sic!)
    Polytrichaceae
  Order Dawsoniales
    Dawsoniaceae

TABLE III. Classification of the Class Musci given by Dixon (1932)

Subclass Sphagnales
                Sphagnaceae
Subclass Andreaeales
                Andreaeaceae
Subclass Bryales
  Clan Nematodonteae
    Order Tetraphidales
        Georgiaceae
    Order Calomniales
        Calomniaceae
    Order Schistostegales
        Schistostegaceae
    Order Buxbauminales
      Suborder Buxbaumineae
        Buxbaumiaceae
      Suborder Diphyscineae
        Diphysciaceae
    Order Polytrichales
      Suborder Dawsoniineae
        Dawsoniaceae
      Suborder Polytrichineae
        Polytrichaceae
  Clan Arthrodonteae
    Subclan Haplolepideae
      Order Fissidentales
        Archefissidentaceae
        Fissidentaceae
      Order Grimmiales
        Grimmiaceae
      Order Dicranales
        Archidiaceae
        Dicranaceae
        Dicnemonaceae
        Pleurophascaceae
        Leucobryaceae
      Order Syrrhopodontales
        Syrrhopodontaceae
      Order Pottiales
        Pottiaceae
    Subclan Heterolepideae
      Order Encalyptales
        Encalyptaceae
    Subclan Diplolepideae
      Order Orthotrichales

        Erpodiaceae
        Ptychomitriaceae
        Orthotrichaceae
      Order Funariales
        Suborder Funariineae
          Gigaspermaceae
          Funariaceae
          Disceliaceae
          Oedipodiaceae
        Suborder Splachnineae
          Splachnaceae
      Order Eubryales
        Eubryales Acrocarpi
        Suborder Bryineae
          Bryaceae
          Leptostomaceae
          Mniaceae
        Suborder Timmiineae
          Timmiaceae
        Suborder
          Rhizogoniineae
          Meeseaceae
          Aulacomniaceae
          Mitteniaceae
          Drepanophyllaceae
          Sorapillaceae
          Rhizogoniaceae
        Suborder Bartramineae
          Bartramiaceae
        Eubryales Pleurocarpi
        Suborder Hypno-
          dendrineae
          Hypnodendraceae
        Suborder Spiridentineae
          Spiridentaceae
      Order Isobryales
        Suborder Rhacopilineae
          Helicophyllaceae
          Rhacopilaceae
        Suborder Fontinalineae
          Fontinalaceae
          Climaciaceae
        Suborder Leuco-
          dontineae

ubclass Bryales—Order Isobryales—
ont.

Cryphaeaceae
Hedwigiaceae
Leucodontaceae
Ptychomniaceae
Lepyrodontaceae
Cyrtopodaceae
Prionodontaceae
Rutenbergiaceae
Trachypodaceae
Pterobryaceae
Meteoriaceae
Suborder Neckerineae
Phyllogoniaceae
Neckeraceae
Echinodiaceae
Lembophyllaceae
Order Hookeriales
Suborder Nematacineae
Nemataceae

Suborder Hookeriineae
Pilotrichaceae
Hookeriaceae
Hypopterygiaceae
Order Hypnobryales
Suborder Leskeineae
Theliaceae
Thuidiaceae
Leskeaceae
Amblystegiaceae
Brachytheciaceae
Suborder Hypnineae
Fabroniaceae
Symphyso-
dontaceae (sic!)
Entodontaceae
Myuriaceae
Sematophyllaceae
Leucomiaceae
Hypnaceae
Hylocomiaceae

TABLE IV. Classification of the Class Musci given by Reimers (1954)

Subclass Sphagnidae
Order Sphagnales
Sphagnaceae
Subclass Andreaeidae
Order Andreaeales
Andreaeaceae
Subclass Bryidae
Order Archidiales
Archidiaceae
Order Dicranales
Ditrichaceae
Archifissidentaceae
Bryoxiphiaceae
Dicranaceae
Dicnemonaceae
Pleurophascaceae
Leucobryaceae
Order Fissidentales
Fissidentaceae
Order Pottiales
Suborder Syrrhopodontinales

Calymperaceae
Suborder Encalyptinales
Encalyptaceae
Suborder Pottiinales
Pottiaceae
Order Grimmiales
Grimmiaceae
Order Funariales
Gigaspermaceae
Disceliaceae
Ephemeraceae
Funariaceae
Oedipodiaceae
Splachnaceae
Order Schistostegales
Schistostegaceae
Order Tetraphidales
Georgiaceae
Order Eubryales
Suborder Bryinales
Bryaceae

Subclass Bryidae—Order Eubryales—cont.
    Leptostomaceae
    Mniaceae
    Suborder Rhizogoniinales
      Drepanophyllaceae
      Eustichiaceae
      Sorapillaceae
      Mitteniaceae
      Calomniaceae
      Rhizogoniaceae
    Suborder Hypnodendrinales
      Hypnodendraceae
    Suborder Bartramiinales
      Aulacomniaceae
      Meeseaceae
      Catascopiaceae
      Bartramiaceae
    Suborder Spiridentinales
      Spiridentaceae
    Suborder Timmiinales
      Timmiaceae
Order Isobryales
    Suborder Orthotrichinales
      Erpodiaceae
      Ptychomitriaceae
      Orthotrichaceae
    Suborder Rhacopilinales
      Helicophyllaceae
      Rhacopilaceae
    Suborder Leucodontinales
      Hedwigiaceae
      Cryphaeaceae
      Leucodontaceae
      Cyrtopodaceae
      Ptychomniaceae
      Lepyrodontaceae
      Prionodontaceae
      Rutenbergiaceae
      Trachypodaceae
      Myuriaceae

      Pterobryaceae
      Meteoriaceae
    Suborder Neckerinales
      Phyllogoniaceae
      Neckeraceae
      Lembophyllaceae
      Echinodiaceae
    Suborder Fontinalinales
      Fontinalaceae
      Climaciaceae
Order Hookeriales
    Nemataceae
    Pilotrichaceae
    Hookeriaceae
    Symphyodontaceae
    Leucomiaceae
    Hypopterygiaceae
Order Hypnobryales
    Theliaceae
    Fabroniaceae
    Leskeaceae
    Thuidiaceae
    Amblystegiaceae
    Brachytheciaceae
    Entodontaceae
    Plagiotheciaceae
    Sematophyllaceae
    Hypnaceae
    Rhytidiaceae
    Hylocomiaceae
Subclass Buxbaumiidae
Order Buxbaumiales
    Diphysciaceae
    Buxbaumiaceae
Subclass Polytrichidae
Order Polytrichales
    Polytrichaceae
Order Dawsoniales
    Dawsoniaceae

TABLE V. Classification of the Class Bryatae given by Robinson (1971)

ubclass Sphagnidae
  Order Protosphagnales
    Protosphagnaceae (fossil)
    Intiaceae (fossil)
  Order Sphagnales
    Sphagnaceae
ubclass Bryidae
  Order Andreaeales
    Andreaeaceae
  Order Tetraphidales
    Tetraphidaceae (= Georgiaceae)
  Order Polytrichales
    Polytrichaceae
    Dawsoniaceae
  Order Dicranales (= Haplolepidae)
    Archidiaceae
    Ditrichaceae
    Bryoxiphiaceae
    Seligeriaceae
    Grimmiaceae (including Ptycho-
      mitriaceae)
    Fissidentaceae (including Archifissi-
      dentaceae)
    Dicranaceae (including
      Leucobryaceae *p.p.*)
    Dicnemonaceae
    Pleurophascaceae
    Calymperaceae (including
      Leucobryaceae *p.p.*)
    Pottiaceae (including
      Trichostomaceae, Cinclidotaceae,
      Splachnobryum)
    Bryobartramiaceae
    Encalyptaceae
    Buxbaumiaceae
    Diphysciaceae
  Order Bryales (= Diplolepidae)
    Rhacitheciaceae
    Erpodiaceae
    Helicophyllaceae
    Orthotrichaceae
    Gigaspermaceae
    Disceliaceae
    Ephemeraceae

Funariaceae
Splachnaceae
Schistostegaceae
Mitteniaceae
Drepanophyllaceae
Calomniaceae
Eustichiaceae
Sorapillaceae
Timmiaceae
Bryaceae
Leptostomataceae
Mniaceae
Aulacomniaceae
Meeseaceae
Catoscopiaceae
Bartramiaceae
Rhizogoniaceae
Spiridentaceae
Hypnodendraceae
Hypopterygiaceae
Rhacopilaceae
Fontinalaceae
Wardiaceae
Hedwigiaceae
Cryphaeaceae
Leucodontaceae
Cyrtopodaceae
Prionodontaceae
Lepyrodontaceae
Rutenbergiaceae
Trachypodaceae
Myuriaceae
Pterobryaceae
Meteoriaceae
Phyllogoniaceae
Neckeraceae
Lembophyllaceae
Climaciaceae
Pleuroziopsidaceae
Echinodiaceae
Fabroniaceae
Leskeaceae (including Theliaceae,
  Thuidiaceae)
Amblystegiaceae

Subclass Bryidae—Order Bryales—cont.
   Brachytheciaceae (including
     Rigodium)
   Entodontaceae
   Plagiotheciaceae
   Ephemeropsidaceae (= Nemataceae)
   Hookeriaceae (including
     Pilotrichaceae)

   Ptychomniaceae
   Symphyodontaceae
   Leucomiaceae
   Sematophyllaceae
   Hypnaceae (including Rhytidiaceae)
   Hylocomiaceae
   Hydropogonaceae

TABLE VI. Families of the Musci accepted by Crosby and Magill (1977). Crosby and Mag
did not comment on taxa above the rank of family

Sphagnaceae
Andreaeaceae
Andreaeobryaceae
Fissidentaceae
Nanobryaceae
Archidiaceae
Ditrichaceae
Viridivelleraceae
Bryoxiphiaceae
Seligeriaceae
Dicranaceae
Dicnemonaceae
Pleurophascaceae
Calymperaceae
Encalyptaceae
Pottiaceae
Bryobartramiaceae
Grimmiaceae
Gigaspermaceae
Disceliaceae
Ephemeraceae
Funariaceae
Pseudoditrichaceae
Oedipodiaceae
Splachnaceae
Schistostegaceae
Tetraphidaceae
Bryaceae
Mniaceae
Phyllodrepaniaceae (= Drepanophyllaceae)
Eustichiaceae
Sorapillaceae
Mitteniaceae

Calomniacaeae
Rhizogoniaceae
Hypnodendraceae
Aulacomniaceae
Meesiaceae
Catoscopiaceae
Bartramiaceae
Spiridentaceae
Timmiaceae
Erpodiaceae
Rhachitheciaceae
Ptychomitriaceae
Orthotrichaceae
Rhabdoweisiaceae
Helicophyllaceae
Racopilaceae
Fontinalaceae
Wardiaceae
Hydropogonaceae
Climaciaceae
Pleuroziopsaceae
Hedwigiaceae
Cryphaeaceae
Leucodontaceae
Cyrtopodaceae
Ptychomniaceae
Lepyrodontaceae
Prionodontaceae
Rutenbergiaceae
Trachypodaceae
Myuriaceae
Pterobryaceae
Meteoriaceae

ıyllogoniaceae
eckeraceae
altoniaceae
embophyllaceae
chinodiaceae
ookeriaceae
heliaceae
abroniaceae
eskaceae
egmatodontaceae
huidiaceae

Amblystegiaceae
Brachytheciaceae
Entodonaceae
Plagiotheciaceae
Sematophyllaceae
Hypnaceae
Rhytidiaceae
Hylocomiaceae
Buxbaumiaceae
Polytrichaceae

TABLE VII. Classification of the Bryidae given by Frey (1977). Frey did not include all the families in his list

ıperorder Polytrichanae
    Polytrichales
    Dawsoniales
ıperorder Dicrananae
    Dicranales (including Archidiales)
    Fissidentales
    Pottiales
    Grimmiales
ıperorder Bartramianae
        Bartramiaceae
        Timmiaceae
ıperorder Funarianae
    Funariales
ıperorder Eubryanae
    Eubryales *s. str.*

        Bryaceae
        Mniaceae
        Hypnodendraceae
Superorder Hypnobryanae
    Isobryales
    Hookeriales
    Hypnobryales

Groups of uncertain position

Rhizogoniinae
Spiridentineae
Buxbaumiidae
Tetraphidales
Schistostegales

3 | The Phylogeny of the Hepaticae

R. M. SCHUSTER

Department of Botany, University of Massachusetts, Amherst,
Massachusetts, USA

Abstract: The still unresolved problems of the time, place and origin of the Hepaticae
are surveyed. It is concluded that modern evidence almost precludes a direct common
ancestor for Hepaticae and Anthocerotae, the latter being assigned to a Division Anthocerotophyta. Hepaticae, Musci and the various groups of lower Tracheophyta may
well have evolved simultaneously as a number of moves towards colonization of the
land at different places, probably in riverine-estuarine environments. Extrapolating
from the existence of Devonian Hepaticae which appear to belong to Metzgeriales,
it is concluded that major groups (subclasses, orders) of Hepaticae must go back at least
to this time; the separation of the Musci and Hepaticae possibly goes back to Silurian
times, and that the various groups of early land plants are nearly synchronous in their
appearance, representing a simultaneous "explosion" of a land flora probably derived
from a common group of aquatic or amphibious antecedents. The basic features of the
Bryophyta (primitively unisexual gametophytes; permanently epiphytic sporophytes;
lack of vascularization of the sporophyte) appear to have been carried to an ultimate
point early in the evolution of the Hepaticae, in which the following unique features
were established at an early date: "internalization" of the sporophyte to the time when
spores are mature; loss—or lack of development of—stomata of the sporophyte; reliance
to a large extent on asexual reproductive mechanisms by the normally perennial and
often exceedingly long-lived gametophyte. The ancestry of the Hepaticae is emphasized
because ancestors of the group became adapted to the land environment in different ways
and this determined the basic subsequent lines of evolution within the group. Hence
the origin of, and evolution within, the group cannot be effectively separated.

The basic organography of the gametophyte which is derived from merophytes cut
off from a normally tetrahedral apical cell is limited in its complexity, and especially
height, by the aquatic nature of reproduction. Consequently the originally erect gametophyte has evolved by planation into a flattened form on a number of occasions thus

Systematics Association Special Volume No. 14, "Bryophyte Systematics", edited by
G. C. S. Clarke and J. G. Duckett, 1979, pp. 41–82, Academic Press, London and New
York.

producing convergent types showing basic similarities such as leafy prostrate game-tophytes with two rows of leaves, and planate thalli which are often highly simplified in form. This is linked with parallel reduction and simplification of the sporophyte to the point where it retains few features to suggest phylogeny. These two interrelated problems make it difficult to deduce phylogeny and force us to rely upon criteria other than morphology, several of which are discussed.

Using all the available evidence, a modified classification is given which, I hope, reflects phylogeny. Many of the simple types, all in the Jungermanniidae, are concentrated in Gondwanaland and it is possible that this subclass originated there. By contrast, the Marchantiidae show largely ecologically-based rather than historically-based distribution patterns and are concentrated in continental regions; they may represent a series of parallel adaptations at an early date, when Pangaea existed, to inland and continental conditions. Their basic reliance on large spores with considerable longevity and/or on gametophytes able to tolerate desiccation for many months suggests a long history divorced from that of the Jungermanniidae, which show by far the greatest diversity in an anciently oceanic region fringing the ancestral Pacific, or Panthallassa.

INTRODUCTION

No group of plants offers more challenges to students of phylogeny and evolution than the Hepaticae. This assertion must be considered in the context of several facts: (1) Almost two centuries after the group was first formally defined by de Jussieu (1789), the direction in which its evolution has proceeded remains controversial. (2) No other group of land plants has wholly internalized the sporophyte until after spores are fully mature so that it is, in effect, wholly removed from the selection pressures normal in an external existence. (3) There still remains no general agreement as to the perimeters of the group; some workers such as Bold (1973) and Mägdefrau (1978) retain the Anthocerotae in the Hepaticae, others (Schuster, 1977) place them not only outside the Hepaticae, but even exclude them from the Bryophyta. Thus, in effect, there is still controversy as to what should be included in (and what excluded from) the concept of a hepatic, and the direction in which evolution is proceeding. A corollary is that there is no consensus as to what is primitive and what advanced in the Hepaticae. Paradoxes abound; we must acknowledge existence of a genus such as *Plagiochila** with c. 1800 described species of which perhaps no more than 500 are valid, and a family such as the Lejeuneaceae with more than 1500 species in at least 70 genera (Gradstein, this volume, Chapter 4), which both give unambiguous evidence of recent and explosive speciation and genus formation. Yet this phenomenon does not wholly invalidate Kashyap's (1919)

---

* Authorities for generic and suprageneric names are given in the classification outlined on pp. 72–78.

assertion that the Hepaticae is a group sliding into oblivion! More than in any other group of which I am aware, we find here what could be called the "Rashomon syndrome" after the classic Japanese motion picture in which the ambiguity of the perception of reality is examined.

By contrast, the Musci are, in an evolutionary sense, almost dull: one is tempted to paraphrase the aphorism of Gertrude Stein and state "a moss is a moss is a moss". No such cliché can be applied to the Hepaticae.

The following examples amplify this assertion: (1) Austin, many years ago, described a member of the Musci, specifically of the Erpodiaceae, as a species of the hepatic genus *Lejeunea* (Evans, 1902). (2) Montagne in 1838 described as *Anthoceros dissectus* a plant which was long ago shown to be a member of the Podostemaceae (Angiospermae!). (3) In 1870 the German botanist von Martens described the new hepatic genus *Kurzia* (Lepidoziaceae) as a member of the Rhodophyta (Algae). (4) In January 1966 I collected plants in Dominica that I initially believed to be a green alga, but which proved to represent a new genus and family, *Phycolepidozia* (Phycolepidoziaceae, Jungermanniales). I cite this case not because I wish to advertise the fact that at times I find it difficult to separate Hepaticae from Chlorophyta, but because it graphically illustrates one of many problems encountered in the study of the Hepaticae. In this instance, chips of bark on which *Phycolepidozia* grew (associated with "obvious" hepatics such as *Cephalozia* and *Zoopsidella*) were discarded when the *Cephalozia* and *Zoopsidella* were present only as scattered stems. *Phycolepidozia* at first sight resembled interwoven ramified strands of a peculiar, delicate yet poly-seriate, filamentous, green alga. After much time consigning pieces of bark principally covered with the "alga" to the wastebasket, "algal filaments" were seen that gave rise to a perfectly normal hepatic perianth and sporophyte! Hours were spent in carefully screening the contents of the wastebasket in order to recover what was clearly a major new type of liverwort!

One is thus almost forced to look at the Hepaticae and Anthocerotae as unique groups of organisms. In part, the uniqueness of the two groups is a result of several unanswered questions and certain lines of evolution that the Hepaticae, in particular, have followed. A major question still to be answered is the origin of the Anthocerotae; this group has no real evolutionary contact point with the Hepaticae and probably originated from algal ancestors that were cytologically different from ancestors of all other land plants. Both the Hepaticae and Anthocerotae independently developed permanent epiphytism of sporophyte on gametophyte, but the Hepaticae went much further. Even in their most primitive extant members they are very specialized in this respect, to the point where the sporophyte remains surrounded by gametophytic tissue and is hence shielded from the external environment until after it has fulfilled its role. The sporophyte is externalized only after the spores are mature, when external selection pressures no longer have any relevance. If the Devonian *Pallaviciniites* is indeed allied to the Pallaviciniaceae (Metzgeriales), then such internalization appears to go back 350–375 million years at the least.

Since the sporophyte of the Hepaticae was internalized at the very start of their evolution, their phylogeny has been conditioned by this ever since. This, in turn, goes far to explain the relative structural uniformity of hepatic sporophytes and the almost exclusive limitation of evolutionary diversification to the gametophyte state. In this sense, the Hepaticae is a unique group. The Musci and Anthocerotae, with much earlier externalization of the sporophyte, are much more akin to early vascular plants in this regard and evolution within these groups reflects this difference.

THE POSITION OF THE ANTHOCEROTAE

Liverworts are here defined to exclude the Anthocerotae, which were stated in Schuster (1977) to belong to a division, Anthocerotophyta. Carothers and Duckett (this volume, Chapter 18) show that the Anthocerotae possess a bilaterally symmetric spermatozoid with symmetric flagellar insertion. By contrast, both the Hepaticae and Musci possess spermatozoids which are asymmetric with regard to the angle at which the flagella diverge and have flagella which are inserted on the sperm body at different levels. This is another new criterion that seems to exclude the Anthocerotae from the Bryophyta *s. lat.*, while it suggests that the Hepaticae and Musci may form a monophyletic group, and also suggests that the phylogeny of Bold (1973), who proposed a division Hepatophyta which includes two classes, Hepatopsida and Anthocerotopsida and an autonomous division Bryophyta, to include only the mosses, is in error.

Thus cytological criteria, spermatozoid ultrastructure, gametangial ontogeny, sporophyte structure, the nature of the spore-elater division, the presence in some instances of stomata in both *n* and *2n* generations, and the non-synchronous production of spores, collectively suggest that the Anthocerotae represent a parallel and analogous land invasion, independent of that undertaken by true Bryophyta (hepatics, mosses). Of these criteria, the single most salient one may well be the non-synchronous production of spores. No other group of land plants has spores that develop essentially continuously within the sporangium. Excluding the Anthocerotae from further consideration simplifies the problem of devising a phylogeny for the Hepaticae, since after such removal, the latter becomes clearly monophyletic and sharply delimitable. Only when one considers poorly known fossils such as the Silurian–Devonian *Sporogonites*, which has been placed in a separate order, the Sporogonitales (Schuster, 1966a, p. 353), do the limits of the Hepaticae become imprecise. It is exactly at this geological horizon, some 390–420 million years B.P., where the fragmentary evidence suggests that phylogenetic connections between the Hepaticae and the earliest vascular plants, especially the Rhyniophytina, are to be sought. Such common ancestral types have been repeatedly conceptualized, under names like Proto-psilophytales or Anthorhyniaceae.

ANCESTRY, TIME, PLACE, AND MODE OF ORIGIN

"A solution to the problem of the origin of the Hepaticae . . . and the initial steps in their evolution appears to be nearly as remote as it was some fifty years ago" (Schuster, 1966a, p. 257). Equally pessimistic is the evaluation of Watson (1964) who refers to the present "Age of Speculation" and states that "perhaps the only honest conclusion to draw is that we do not know how these organisms are interrelated. Nor do we know from what earlier organisms the remote ancestors . . . came." Watson, indeed, suggests that "the contemporary botanist is wise to withhold judgement". I disagree: it is exactly in situations like this where, as Watson states, there is "insufficient reliable evidence" that informed judgement is our only hope for future progress. In the last 25 years we have also seen the accumulation of new evidence from a wide spectrum of wholly unrelated fields. Thus any judgements as to the origin and evolution of the Hepaticae must take account of recent data derived from the study of a variety of organisms, past and present, which were unknown or unstudied 25 years ago even though such data may seem irrelevant at first. Important organisms in this category include *Takakia*, *Haplomitrium inter-medium*, *Phycolepidozia* and *Pallaviciniites*.

Before the large problems inherent in the above heading are specifically attacked, the following scattered items of recent data and pertinent conclusions, all relevant at several levels, deserve mention. (1) The discovery of the Devonian *Pallaviciniites devonicus* (Hueber) Schust. suggests that Hepaticae of the order Metzgeriales existed by Devonian times. Thus if the interpretation of *Pallaviciniites* is correct, the separation of the subclasses Jungermanniidae and Marchantiidae must predate the middle Devonian, and the origin of the Hepaticae as an autonomous group must be pushed back to the very start of the Devonian, if not considerably earlier. Equally important, since *Pallaviciniites* is thallose and planate, one must almost assume a pre-Devonian origin of the

Hepaticae if, as is done here, it is assumed that radial, erect (but not necessarily leafy) gametophytes preceded planate types! If this is so, one could interpret the Hepaticae as one of many roughly contemporaneous attempts at land invasion. (2) The probability that mature sporophytes of *Sporogonites* arose, in groups, possibly from a flattened gametophyte (Andrews, 1960) suggests that permanent epiphytism of sporophyte on gametophyte evolved, probably repeatedly, as long ago as the earliest Devonian. (3) If Merker (1959, 1961) and Lemoigne (1968) are correct in their interpretation of the prostrate parts of Rhyniophytes as being gametophytic, then the Rhyniophytina are, as regards perhaps the most important criterion we can use to separate Bryophyta from Tracheophyta, bryophytes. (4) The demonstration that *Actinostachys*, a true fern, has a sporophyte permanently epiphytic on the tuberous, persistent gametophyte (Bierhorst, 1968a, b) shows conclusively that epiphytism of sporophyte on gametophyte has evolved repeatedly. Hence, just because *Anthoceros* has a permanently attached sporophyte does not make it any more of a bryophyte than this phenomenon makes *Actinostachys* one. Indeed, one could argue that certain features of *Actinostachys* (e.g. synchronous spore production within any one sporangium) argue for a closer affinity of that genus to the Bryophyta than *Anthoceros* shows! (5) Fine structure of the sperm (Carothers and Duckett, this volume, Chapter 18) shows clearly that in this respect the Anthocerotae are quite divorced from the Hepaticae, but that Hepaticae and Musci show clear affinities. (6) The complex fine structure of *Tetraphis* sperm (Carothers and Duckett, 1978) suggests that if *Tetraphis* is interpreted as a primitive moss, an interpretation common to almost all moss phylogenies of the last 50 years or more, then *Haplomitrium*, which has similarly relatively complex spermatids, is also primitive. (7) The ill-defined organography of *Takakia*, a genus unknown prior to 1957, and of the two most primitive species of *Haplomitrium*, *H. intermedium* Berrie (Berrie, 1962; Schuster, 1967) and *H. ovalifolium* Schust. (Schuster, 1971a), strongly suggests that we do have a logical starting point for some, if not all, hepatic evolution. (8) The extraordinarily generalized—I hesitate to employ the over-used term primitive—sex organ position of the taxa cited under 7 is additional evidence that they are relatively unspecialized (Schuster, 1967). I conclude that the ill-defined organography of *Takakia* and the most primitive species of *Haplomitrium* can hardly be equated with the very specialized phylogenetic position assigned them by Smith (1955) and Grolle (1969) amongst others. Equally, the apparently haphazard production of gametangia in these organisms, and the unique identity in early ontogeny of male and female gametangia in *Haplomitrium*, strongly suggest that they are primitive. (9)

Analysis of the advanced and highly reduced genus *Phycolepidozia* (Schuster, 1966b) suggests that the evolution of non-leafy gametophytes from leafy ones can readily occur—leaves being reduced back to slime papillae. This means that the argument as to whether leafy or thallose gametophytes came first, is very largely an empty debate that diverts attention from true and real phylogenetic problems. In the Hepaticae, leaves probably evolved several times and, in the Jungermanniales at least, probably by the elaboration of slime papillae, to which they may again be reduced. Many underleaves are reduced to slime papillae in *Cephalozia* (Schuster, 1966a, Fig. 44 : 11), while lateral leaves are similarly reduced in *Phycolepidozia*. (10) The discovery that *Takakia* has a gametophytic complement of 4 or 5 chromosomes, and that in diverse taxa with 8–10 gametophytic chromosomes there are 2 nucleolar organizers (Berrie, 1958b), suggests that *Takakia* is primitive and that there has been widespread paleopolyploidy in hepatics.

The ten fragments of evidence cited above can be likened to additional pillars upon which any edifice of phylogenetic speculation must rest. The phylogenetic speculations of the past (summarized in Schuster, 1966a) are given extra, although admittedly weak, support by these pillars. I remain convinced of the validity of my statement (Schuster, 1966a, p. 257) that "continued absence of adequate fossil evidence is a limiting factor; one can do little more than integrate the few concrete facts, the somewhat less limited circumstantial evidence, and the almost unlimited inferences that can be drawn . . . into a conceptual scheme that satisfies, as nearly as possible, the requirements of factual evidence". In the dozen years since then, however, no data have accumulated that in any significant manner necessitate major revisions in the phylogeny or the phylogenetic principles then adopted. Because of space constraints most of the then available evidence, and the speculations derived, are omitted here; they would serve to further buttress arguments here presented, derived mostly from more recent data. Collectively, I think, they would in part invalidate the pessimistic conclusions of Watson, quoted above, as well as my not very optimistic conclusion, also cited at the beginning of this chapter.

1. Ancestry

Although nothing definitive is known, since the fossil evidence is non-existent at the relevant horizon, it seems likely that the Hepaticae and Musci represent parallel groups which evolved from a common ancestor in which the sporophyte was epiphytic on the gametophyte. This ancestral type presumably had radial symmetry of both gametophyte and sporophyte, gametophytes that probably lacked leaves but were erect in growth (or possessed ramified, erect sectors) and grew by means of a tetrahedral apical cell that cut off derivatives (merophytes) in a spiral sequence. Some of these merophytes, in turn, gave rise to gametangia, each derived from a superficial cell that was cut off, thus giving the exogenous and typically stalked gametangia common to both mosses and hepatics. The aquatic reproduction, an anomaly on land, is a relict of amphibious ancestral types which had already achieved their basic architecture by earliest Devonian times, or earlier. It is assumed that as soon as the archegonium evolved, the potential for embryo retention existed: indeed it is implicit in the structure of the archegonium that the diploid stage would, of necessity, undergo at least its early ontogeny surrounded by gametophytic tissues.

Specifically, the ancestry of the Hepaticae is to be sought, I think, in a type in which four additional traits were already in evidence: (a) the gametophytes were unisexual, as they were in the ancestors of the Musci, but unlike those in all the Tracheophyta; (b) the radial gametophytes produced unicellular outgrowths, slime papillae, whose secretions prevented desiccation of the apical regions and developing gametangia; (c) the sporophytes underwent very rapid reduction and simplification and were soon internalized until the spores were mature; (d) gametangia were produced, in a simple acropetal sequence, from derivatives of the apical cell, and were initially naked, that is, lacked leaf-like protective devices. Such naked gametangia are still found in the Metzgeriales: *Fossombronia*; (Schuster, 1966a, Fig. 50:1), *Takakia*, and some species of *Haplomitrium* (Schuster, 1967, Figs III:6, IV:5–6, and V:4).

In criteria (c) and (d) the Hepaticae early showed major deviations from the Musci; they also evidently did not develop the striking physiological drought resistance found in all early types of Musci (*Sphagnum, Andreaea, Tetraphis* and *Polytrichum*), and they did not develop long-lasting juvenile stages (protonemata). Finally, they may have evolved from a group parallel to the Musci, rather than from an immediately allied ancestral type since they show several attributes that suggest only a very remote affinity to the Musci, such as lack of a columella, the presence of a spore-elater division and of elaters, the presence of oil bodies, a very different ontogeny of gametangia, consistent lack of stomata in the sporophyte. The last criterion is evidently linked with the extremely early internalization of the sporophyte: even if ancestral types of Hepatical agreed with those of Musci in most criteria, including the presence of stomata, the last were lost at a very early date, concurrent with sporophyte internalization. Alternatively, and I think more probably, one can trace back the ancestry of the Hepaticae to an earlier date than that of the Musci, to

an ancestor still sufficiently amphibious that stomata had not yet evolved.
The ancestor deduced for the Hepaticae is thus, of necessity, a very generalized one. It is, furthermore, one from which we can imagine rapid and repeated evolution of the two basic types of gametophyte morphology seen in
the Hepaticae:

(1)   a thallus evolved by planation of the radial system, apparently without
      prior elaboration of leaf-like structures from slime papillae (as seen in
      most of the Metzgeriales, Monocleales, and Marchantiales); and

(2)   a foreshortened, leafy axis, always with elaboration of leaf-like
      appendages (as seen in the Jungermanniidae).

Conditions (1) and (2), as is well known, are connected by numerous transi-
tions. Selection pressures exerted by the diverse ecological niches exploited by
early Hepaticae are visualized as having repeatedly resulted in planation, so that
bilateral, leafy, creeping Jungermanniales are commonplace. By contrast, it
is difficult to visualize selection pressures that would lead to the evolution of
erect, radial types—only extreme competition for space could conceivably
lead to evolution of such a morphology. We must suppose that, until abbreviated, intercalary branches to which gametangia are restricted (e.g. in the
Adelanthaceae) had evolved, the dominant theme in hepatic evolution was
planation and progressively more prostrate and creeping growth, thus bringing
the gametangia closer and closer to ground level. I disagree with the assumption
of Grolle (1969) that this planate condition was the ancestral condition of the
hepatic gametophyte and conclude that the thallus is derivative. Essentially,
all the evidence suggests that the primitive gametophyte was radial, as we
see not only in the isophyllous Jungermanniales (such as *Anthelia*, *Grollea*,
Isophyllaria, or *Lophochaete*) and Calobryales (such as in *Takakia* and primitive
species of *Haplomitrium*). Such radial gametophytes also occur in the majority
of "primitive" living Tracheophyta, such as *Psilotum*, *Stromatopteris* and various
Schizaeaceae, including *Actinostachys* and species of *Schizaea* (Bierhorst, 1953,
1966, 1968a). It is equally true of gametophytes of *Ophioglossum* and of the
more primitive ones of *Lycopodium* (Schuster, 1971a, p. 142). Bierhorst, indeed,
makes the generalization that the "Psilotaceae, Stromatopteridaceae, Gleicheni-
aceae, and Schizaeaceae seem to combine features of a particularly primitive
complex among the major alliances of extant ferns"; and I noted (Schuster,
1971a, p. 142) that it "is exactly in this complex where radial gametophytes
preponderantly occur".

Bierhorst (1971, p. 78) also concluded that "the distribution of axial gametophytes among vascular plants (*Lycopodium*, Ophioglossaceae, Psilotaceae,

Stromatopteridaceae, *Actinostachys*) allows us to postulate [axial morphology] as a possible common denominator for vascular plants in general". He also notes that I (Schuster, 1966a) had already advanced the idea that thalloid hepatics like *Marchantia* may be "a flattened axial type that never went through a leafy stage in its phylogeny".

2. *Time and Place of Origin*

I currently hold the opinion that the divergence between the earliest Tracheophyta and Bryophyta must have occurred at the time (Silurian or earliest Devonian, *c.* 450–390 million years B.P.) when land invasion probably first occurred. Relevant is the existence of a thalloid, Devonian hepatic, *Pallaviciniites devonicus* (Schuster, 1966a, p. 352), a lower Carboniferous moss, *Muscites polytrichaceus*, seemingly allied to the Polytrichales (Frey, 1977), Permian Protosphagnales (Neuburg, 1958, 1960), as well as—by early Mesozoic times—more advanced Jungermanniales (Krassilov, 1973). All of these scattered and diverse fossil finds clearly suggest that the main lines in bryophyte evolution were well established by Paleozoic times and that the divergence between mosses and hepatics had probably occurred by the start of the Devonian. If the assumption is correct that much, if not all, land invasion took place from riverine-estuarine environments, where nutritive-rich and fine grained, water-retentive silts were present (Schuster, 1966a, 1977), then land invasion occurred probably roughly synchronously at sites that were widely separated by inhospitable ocean. Repeated and discontinuous invasion of the land surface would allow initial survival of plants which became adapted to terrestrial life in a variety of ways. This would, in part, explain the now well-documented explosive proliferation of strikingly diverse early land plants, among which we must number several kinds of early bryophyte-like types, many surely long extinct, such as *Sporogonites*.

There is also evidence that, from the Ordovician to the late Silurian the marine invertebrate fauna varied from area to area to a degree comparable to that of today (Whittington and Hughes, 1973). This suggests that during this time continents were not assembled in a single supercontinent, Pangaea, but existed as a number of separate land masses (Smith *et al.*, 1973) and hence that shore lines were about as extensive as they are today. Such an early Paleozoic world would allow independent, early evolution of land types on separate early continents, divided by marine environments.

Although there is at least fragmentary evidence that evolution of bryophyte-like types occurred nearly synchronously with the explosive evolution of an early vascular flora, chiefly of Rhyniales and Zosterophyllales, we have not a shred of evidence as to the time and place of origin of the Anthocerotophyta. Indeed, in the continuing lack of fossil data, all statements as to the time and place of origin of that group must remain pure speculation.

3. Phylogeny: the Linkage Between Life Cycle and Evolution

Any real understanding of evolution within the Hepaticae must involve recognition of the fact that the processes that gave rise to the Hepaticae are exactly those that determined the major subsequent evolutionary patterns within the Hepaticae (Schuster, 1966a). Such processes include the development of gametophytic unisexuality; the evolution of permanently epiphytic sporophytes; the evolution of slime papillae; the evolution of relatively low levels of physiological drought resistance compared to those in mosses, and the early evolution of a reliance on asexual reproduction. Reliance on asexual reproduction is more widespread in the Hepaticae than in the Musci and involves a greater diversity of mechanisms. It is, however, not restricted to bryophytes, as the gemmae of *Psilotum* and the fern *Vittaria* attest. Indeed, *Vittaria* is, in Appalachian populations, only gametophytic and reproduces solely by gemmae. Hence, in effect, under admittedly extreme conditions, ferns may also evolve a sporophyteless, bryophytic means of existence (Emigh and Farrar, 1977).

There is, therefore, no clear distinction between evolution of and within the Hepaticae. Much of the parallelism which befuddles the taxonomist represents simple, inevitable, repeated evolution along parallel or convergent lines.

Evolutionary patterns and directions that ultimately became very important were fixed as soon as the gametophyte emerged onto land. When that happened, selection pressures for bringing gametangia down close to the ground became intense. Equally, and simultaneously, selection pressures for evolution of means to protect gametangia from desiccation also became marked. We therefore find the presumed repeated evolution of leaves (in the Jungermanniales; in certain of the Metzgeriales like *Fossombronia*, some species of *Symphyogyna*, *Noteroclada*, *Phyllothallia*; in *Sphaerocarpos*, Calobryales, and in Treubiales) and the repeated evolution of thalli (in *Pteropsiella*, *Schiffneria*, *Metzgeriopsis*, *Zoopsis* in the Jungermanniales; most Metzgeriales; almost all the Marchantiidae). Both devices were clearly adaptive: with the evolution of leaves, foreshortening became possible since photosynthetic surfaces were increased, and the leaves became protective devices to whose axils the gametangia became restricted. With the evolution of thalli, gametangia came to lie at ground level, and in many cases thallus tissue grew up to surround them (as in *Riccia*, the antheridia of *Verdoornia*, *Pellia*, and of *Blasia*, or the pseudoperianth of *Pallavicinia*).

I would assume that a major factor stimulating the planation of initially leafy gametophytes and the evolution of planate thalloid gametophytes directly from non-leafy ancestors is the absence of a cuticle in almost all the Hepaticae. In the Jungermanniales we find water-repellent cutinized surfaces only in *Lembidium*, *Calypogeia*, and *Douinia*, of genera familiar to me, and none of the Metzgeriales appears to be truly cutinized (Schuster, 1966a, p. 430).

It also seems necessary to assume that planate gametophytes evolved not only repeatedly, and under diverse conditions, but also as a very early response to the harsh environmental conditions that must have characterized land surfaces in Silurian–Devonian times. With very few exceptions, all extant Jungermanniidae exist in forested areas where protected niches abound; the documented inability of the gametophyte to survive drought (Clausen, 1952) is clearly linked. The limited, if repeated, number of ways in which the thallus has evolved in this subclass seems to be linked to a geographical origin of the Jungermanniidae basically different from that of the other subclass of Hepaticae, the Marchantiidae. In the latter there exists only the early Mesozoic fossil *Naiadita* in which a radial gametophyte was preserved. Apart from this, planate and, except for the stenotypic Sphaerocarpales, thalloid gametophytes are the norm. A basic life history that is adapted to long periods of drought through large, durable spores and/or the ability of the gametophyte to tolerate months to years of desiccation, appears to have evolved early in the Marchantiidae, and the earliest known types (Jurassic, Lundblad, 1954 and Triassic, Townrow, 1959) which can definitely be assigned to the Marchantiidae already show the xeromorphic structural adaptations, such as pores and air chambers, that are linked with survival during long, dry periods. Thus the evolution of the two subclasses of the Hepaticae is possibly linked with geographical differences. The Jungermanniidae are basically mesophytic and may have evolved at the edges of the ancient Pacific (Panthalassa), while the Marchantiidae had its evolution conditioned by early adaptation to the physical stress of life in the more continental parts of Pangaea. The relative richness of the Jungermanniidae, especially of isophyllous, triradial, erect types, in the circumpacific basin may reflect the origin of that group in the general oceanic region. By contrast, the paucity of taxa of the Marchantiidae in mesic and, especially, densely forested regions, may reflect the origin and early diversification of that group in regions remote from the margins of the ancient Pacific. This remains a striking feature even of modern types. Thus the relatively rich flora of Jungermanniidae in

Britain and Ireland stands in marked contrast to the paucity of Marchantiidae; by contrast, the Mediterranean area, with distinct alternation of dry and wet periods, is rich in Marchantiidae and has a limited flora of Jungermanniidae.

The life cycles of the two subclasses differ widely. In the Jungermanniidae, short-lived or ephemeral gametophytes are unknown; in the Marchantiidae they are common; in the former, dissemination by small, wind-blown spores of short duration is normal, in the latter exceptional; in the former, asexual gametophytic reproduction by specialized structures is abundant and diverse, in the latter rare and exceptional. Thus, at this level, life history and evolution seem to be intimately linked.

EVOLUTIONARY PATTERNS, PARALLELISMS, AND CONVERGENCES

I assume (as in Schuster, 1977) that there are essentially only two ways in which a gametophyte can become adapted to life on land: evolution from an erect, radial, leafless ancestral type into a thallus or into a foreshortened, leafy gametophyte. All subsequent evolution is secondary modification on these two themes involving such regressions to algal prototypes, as in the clearly neotenic *Protocephalozia*, reduction of the vegetative gametophyte to series of ramified filaments; or in *Phycolepidozia*, reduction to interwoven, ramified, leafless but polyseriate axes. If only two basic kinds of morphological adaptation are possible, then we must accept that during the course of the last 350–400 million years there was repeated evolution of almost identical morphological types. When this has been from roughly similar antecedents we can speak of parallel evolution; when the antecedents were drastically distinct and are, for example, referable to distinct orders or suborders, convergence has clearly occurred. Linked to this is the conservatism of the sporophyte. Hence in trying to work out the phylogeny of the Hepaticae, we face the dual problem of having to contend with repeated convergence and parallelism in the gametophyte and also having very little in the way of evidence of past evolution that can be derived from the sporophyte.

The situation with regard to the sporophyte is not wholly hopeless. Thus the presence of "two-phase" development of epidermal cells of the capsule wall has been used to link phylogenetically the Lepidoziineae and Cephaloziineae (Schuster, 1966a, 1972). The triseriate *Pleurocladopsis*, which is, in aspect, almost like a *Sphagnum* branch, has the same sporophyte anatomy as the Scapanioid *Schistochila* (Schuster, 1971b, 1972; Schuster and Engel, 1977) and, on that account, is best referred to the Schistochilaceae. If we look hard enough, some phylogenetic tracers can be found in the sporophyte, but these are principally at the level of morphogenetic patterns and microanatomy and hence have been inadequately exploited.

Repeated parallel evolution serves, at once, to delight the student of evolution and to confound the taxonomist. Thus *Protocephalozia* and *Phycolepidozia* are not closely allied; the former, mimicking the moss genus *Ephemeropsis*, belongs in the Lepidoziaceae (Schuster, 1972, 1974), the latter, in the Phycolepidoziaceae. Morphological "mimicry" is rampant in the Jungermanniales; thus species of *Jamesoniella*, *Andrewsianthus*, *Gottschelia*, *Mylia*, *Odontoschisma*, and *Solenostoma* (*Jungermannia*) may be so similar in basic gametophytic architecture that students traditionally have trouble in separating them. These parallelisms are inherent in the fact that most hepatics display modifications of the same basic architecture: an axis, derived from segments cut off by a tetrahedral apical cell, with each derivative (merophyte) giving rise to a leaf-like appendage from the anterior (acroscopic) margin, or a thallus. Hence similar or virtually identical modifications, as in leaf form and orientation or in reproductive behaviour, are likely to recur, sometimes several times. This may be true between orders so that, for example, *Dumortiera* and *Monoselenium* (Marchantiales) mimic *Monoclea* (Monocleales); *Metzgeriopsis* (Jungermanniales) may precisely mimic epiphyllous juvenile stages of *Metzgeria* spp. (Metzgeriales) and *Pteropsiella* (Jungermanniales) is superficially identical to *Metzgeria* (Metzgeriales). Such parallelism is even more frequent between more closely related taxa. For example, in the Jungermanniales, *Syzygiella plagiochiloides* Spruce, as the name implies, mimics *Plagiochila*.

Even more striking are the similarities between *Nothostrepta* and *Plagiochila* (Schuster, 1980b) where both genera have strongly decurved shoot apices and compactly spicate androecia, similarly tapered distally. Yet the species of the former had been placed in *Anastrepta* (Grolle, 1961), of the Lophozioideae (Jungermanniaceae) while the latter is the type genus of Plagiochilaceae. A large portion of the phylogenist's energy in recent decades has gone into deciding whether such similarities represent a similar phylogeny, parallelism or convergence.

Although Kashyap (1919) believed that the Hepaticae were on a down-ward slide to oblivion, and waxed almost lyrical about their evolutionary inadequacies, the actual situation is vastly more complex. Kashyap's comments on plants in supposed danger of imminent dissolution derived principally from his study of stenotypic groups (such as the Marchantiales) and, as is evident in his work on the north Indian hepatics (Kashyap, 1929–1932), from his lack of understanding of the leafy Jungermanniales. Some groups, such as Lejeuneaceae, are in the midst of an evolutionary explosion as extensive and intricate as that characterizing their normal hosts, the angiosperms (Schuster, 1966a, p. 307; 1980a). The evolutionary ramifications of an epiphyllous genus like *Cololejeunea* (Schuster, 1963) and the seemingly reticulate and endless patterns of speciation are much as exhibited by various short-lived perennials or annuals such as many Gramineae and the genus *Carex*. And, in *Aphanolejeunea* and *Cololejeunea*, especially in the species that are epiphyllous, we find a similar (although extrinsically imposed) highly abbreviated life-cycle, with very rapid succession of generations. In these taxa reproductive strategies are complex and involved; they involve, normally, close juxtaposition of antheridia and archegonia (paroecism is common, often with female bracts also serving as male bracts; cf. Fig. 9:6 in Schuster, 1966a) and free asexual reproduction by efficiently wind-dispersed cell plates, discoid in form. Freely sexually fertile individuals may produce discoid gemmae, even from male and female bracts. Hence we find the species tend to be "superfertile".

A more rational evaluation would be that certain groups of Hepaticae are as well adapted for survival as the angiosperms. Examples would be most of the Lejeuneaceae, especially the large and complex genera *Lejeunea* and *Cololejeunea*; *Plagiochila*; *Frullania*; *Radula*; and *Porella*. As numerous recent studies by Hattori and Pócs have suggested, species limits in *Frullania* and *Porella* are imprecise and the genetically complex species have as bewildering a structure as those of the most thorny and intractable angiosperm groups.

The study of Hattori (1972) of the *Frullania tamarisci* complex is a good example. Hattori's map (Hattori, 1972, p. 248) shows a fragmented circum-Laurasian range, with clear Pleistocene survival in non-glaciated areas such as Alaska or Siberia in the far north, several centres in Europe, eastern Asia into the Himalayas, eastern and western North America. These relicts were derived from populations that survived south of the glacial boundaries. If this distribution is compared to that of the *Betula papyrifera* complex, we see striking similarities, and there is equally imperfect speciation. The situation is even more complex than portrayed by Hattori, however. *F. tamarisci* subsp. *tamarisci* (supposedly European-north African) and *F. tamarisci* subsp. *asagrayana* (eastern N. America) show intergradation in northeastern North America, and in westernmost Europe (Cornwall, Ireland), suggesting that some gene flow may still occur, presumably via transatlantic spore showers. Indeed, the intraspecific structure of *F. tamarisci*, and its history, with, for example, the separation of European and eastern North American populations by the opening (after 60 million years B.P.) of the North Atlantic, is so highly complex that decades of study would be needed to disentangle the structure of this species adequately.

A large mass of families and genera of the Hepaticae, however, are genus-and/or species-poor and their species may be biotypically depleted. Some orders are very stenotypic such as the Calobryales, including only *Haplomitrium* and *Takakia*, with *c.* 12 species; the Sphaerocarpales, with 3 genera and *c.* 12–15 species; the Monocleales, monogeneric, with 1 or 2 species; and the Treubiales, mono- or digeneric, *c.* 10 species. The reproductive biology of such stenotypes suggests that they are ancient and often (as is clearly the case in *Takakia*, which is reduced to asexually propagating, purely female populations) senescent. Other stenotypic groups, and the phylogenetic significance of stenotypy, are explored in Schuster (1966a, pp. 308–313, 400–401). It is surely not accidental that, apart from one species of *Apotreubia*, all such taxa are unisexual and all, except for a few of *Treubia* and *Riella*, lack gemmae or other specialized methods for asexual reproduction. Such groups, in many cases, give us our best clues to the phytogeography and phylogeny of the Hepaticae. In this essay a perhaps disproportionate amount of space is devoted to them.

PHYLOGENETIC CRITERIA AND LINKAGES

It is axiomatic that in the development of a reasoned and reasonable modern phylogeny, real progress must come principally from evaluation of new criteria, and only secondarily from re-evaluation of previously used criteria. It is equally important to attempt to find linkages between differing criteria, from which phylogenetic patterns and sequences can be derived. The classification outlined on pp. 72–78 is derived from the integration of traditional and modern criteria and the establishment of linkage patterns between them. The following are examples where we see that modern criteria, linked with "traditional" ones, often allow resolution of phylogenetic problems. The following examples serve to indicate the method that has been used in fashioning the phylogeny here adopted.

1. Cellular dimorphism

In all true Jungermaniidae peripheral cells generally bear chloroplasts and oil bodies and no cellular dimorphism occurs within a tissue (such as a leaf); the Marchantiidae (with isolated exceptions where oil cells are lost) show sharp dimorphism between large, chlorophyllose, oil-body free cells and cells that are typically smaller and bear a single oil body unaccompanied by chloroplasts. Because of this criterion, *Monoclea* was placed in the Marchantiidae (Schuster, 1953, 1966a), although it has often been placed in the Metzgeriales (Evans,

1939). Similar dimorphism exists in *Riella*, clearly suggesting that the Sphaerocarpales are members of the Marchantiidae.

2. *Premeiotic Behaviour of Spore Mother Cells*

This criterion suggests that the Marchantiales and Sphaerocarpales, with cell plate formation following meiosis, are allied and may be derived from a type not too dissimilar from the Monocleales, in which Jungermannioid, furrowing-type post-meiotic cytokinesis is still retained. The massive sporophyte plus the lack of a mechanism for preventing the production of more than one sporophyte per gynoecium also suggests that *Monoclea* is primitive compared with the Marchantiales and Sphaerocarpales.

3. *Branching Patterns*

These are relevant at several levels, as the following examples suggest:

In *Haplomitrium* and *Takakia* we find identical types of branching (Schuster, 1967). Firstly, exogenous-intercalary branches are formed by dedifferentiation of mature cortical cells from which are initiated tetrahedral branch initials. These branches arise from mature sectors of the axis and give rise to both permanently leafless axes (horizontal, often branched "stolons" as well as rootlike geotropic axes) and negatively geotropic, eventually leafy "branches". Secondly, sporadic, furcate, "terminal" branches arise in a manner still not fully studied, near the shoot apex. Crandall (1969) claims this last branch type does not exist but I suggest that a comparison of Figs I:3 (*Takakia*) and VI:6 (*Haplomitrium*), in Schuster (1967) shows clear evidence of it. On the basis of branching and other features, I think a good case can be made for considering these two genera as forming suborders of a single order Calobryales.

In both the Porellineae (Porellaceae, Jubulaceae, Lejeuneaceae, Goebeliellaceae) and the Ptilidiineae (Ptilidiaceae, Mastigophoraceae, Chaetophyllopsidaceae), there is an absolute restriction of branching to lateral merophytes, even though broad ventral merophytes are retained. Linked with this are other criteria such as constant formation of perianths, but no other sporophyte protective devices apart from the calyptra; a basically asymmetrically 3(4–5)-lobed leaf, with the dorsal lobe larger and incubously oriented (an exception is found only in leaves of main stems in *Herzogianthus* Schust.) and presence of *Frullania*-type branches, except in some advanced taxa. Taken together, these features suggest that these groups are all allied, and other evidence suggests that the specialized Porellineae probably evolved from the rather unspecialized and malleable Ptilidiineae (Schuster, 1966a, p. 696, 1972). Although the Trichocoleaceae have often been associated with the Ptilidiineae (and were formerly, as in Evans (1939) even placed in the Ptilidiaceae, in part because *Trichocolea* and *Ptilidium* have identical branching patterns) the discovery of ventral branching of the *Acromastigum* type in *Brachygyna*, a monotypic genus of the Trichocoleaceae (Schuster, 1980c), suggests that the Trichocoleaceae are not allied to this sequence but are closer to the Herbertineae, and specifically to the Blepharostomataceae (Pseudolepicoleaceae), as has already been suggested (Schuster, 1959). The fact that both the Blepharostomataceae and all the Trichocoleaceae develop more or less symmetrically quadrifid (rather than asymmetrically trifid) leaves with opposed cilia, and both show tendencies for perianth reduction and coelocaule elaboration, is surely significant. It is equally significant that, when asymmetry develops in the Blepharostomataceae, Trichotemnomaceae, and Trichocoleaceae, as it does in some species of *Temnoma*, *Trichotemnoma*, *Trichocolea*, it is always a reverse type of asymmetry to that found in the Ptilidiineae: the dorsal 1–2 lobes become smaller, and the largest lobe is always the ventral one.

In *Radula* the linkage of exclusively lateral branching with perianth retention and lack of other paragynoecial protective devices apart from the calyptra, might be construed as suggesting a phylogenetic link between the Radulineae and the Ptilidiineae–Porellineae phylogenetic branch. However, in the Rudulineae acroscopic branching is never retained and the branches are uniformly basiscopic; the ventral merophytes never produce rhizoids and there is no evidence that the leaf has been derived from an asymmetrically trifid, Ptilidioid type (the third lobe, or stylus, found in the Jubulaceae and some Lejeuneaceae cannot be distinguished even as a slime papilla). These criteria strongly suggest that the Radulineae have no phylogenetic contacts with the Ptilidiineae. In fact, there is no extant genus to which *Radula* is closely related. Suggestively, even in primitive species such as *R. tenax* Lindb. we already find the ventral merophytes are greatly reduced, with no trace of ventral appendages.

In the preceding cases branching patterns are of major significance in delimiting subordinal groupings and in suggesting affinities. Thus the suborders Lepidolaenineae, Porellineae, Ptilidiineae, Radulineae, and Pleuroziineae have all lost the ability to develop branching from the ventral merophytes; for this, and other reasons they are here placed at the end of the Jungermanniales. This does not automatically mean that the inability to develop postical branches is of subordinal significance. What it does mean is that on three or more occasions postical branching has been lost: in the Radulineae; in the Ptilidiineae–Porellineae–Lepidolaenineae complex of interrelated suborders with the retention (except in advanced Lejeuneaceae) of acroscopic, terminal *Frullania*-type branches; and in the Pleuroziineae, with loss of all terminal branching modes, and the (secondary?) evolution of exclusively lateral-intercalary branching, a mode of branching never seen in the four preceding suborders. It has also been lost in the Perssoniellineae, a suborder with no obvious affinities to any other.

In other cases branching patterns show drastic variation from family to family, from genus to genus, or even from species to species within a single genus. Thus, even though branching patterns can be interpreted as of subordinal significance in the cases already cited, in certain rather primitive families such as the Blepharostomataceae or Lepidoziaceae there is retention of the utmost plasticity in branching. In a phylogenetic context such variability of branching between and within families can perhaps be interpreted as atavistic. Conversely, the evolution of a rigidly defined system of branching, as seen in the preceding suborders, suggests a higher phylogenetic position. In the Blepharostomataceae and Lepidoziaceae the variability of branching patterns is a significant family criterion; differences in branching patterns in these families may represent mere species differences. Thus in the *Pseudocephalozia–Paracromastigum–Bonneria* complex (Lepidoziaceae, subfam. Zoopsidoideae) even environmental conditions may influence the type of branch produced (Schuster, 1965, p. 57). Within that complex we find "progressively increased restriction in branching modes". Initially, in species such as *Paracromastigum granatensis* there are three types of terminal branches (*Frullania*, *Microlepidozia*, and *Acromastigum* types) as well as intercalary, axillary branching. In some populations of *P. densa*, which is perhaps only a phase of *P. bifidum*, we find only intercalary branches, mostly from lateral merophytes. In true *P. bifidum* at least occasional terminal-postical branches of the *Acromastigum* type are still found. In species that can be phylogenetically intercalated between these two extremes, progressive loss of types and numbers of terminal branches is seen. In one species, *P. macrostipa*, branching modes reflect environmental differences between older, central parts of a clone and marginal, pioneer sectors of the clone (Schuster, 1965, p. 58). Central sectors, where plants are nearly erect because of crowding, show only postical branches (intercalary and terminal), whereas creeping axes at the clone margin may still show *Frullania*-type lateral terminal branches. Reasons for such intraclonal variation were examined by Schuster (1965).

4. Wall Pigments

In developing a reasonable phylogeny one must scrupulously avoid a mechanical "present v. absent" type of taxonomy. Criteria which may be of value at the subordinal level in one context may be useless in others, even at the species level. The sporadically scattered ability to develop anthocyanic pigments in the cell walls of the Jungermanniales gives one example. Thus, the Radulineae never contain wall anthocyanins. By contrast, the genera of Lepidolaenineae (*Lepidolaena, Gackstroemia, Lepidogyna, Jubulopsis*) all show such pigments, even though certain species (e.g. *Lepidogyna menziesii*) have lost them. Here the presence or absence of such pigments is of subordinal value. Members of *Lophozia* subgenera *Orthocaulis* and *Barbilophozia* are normally unable to form wall anthocyanins and I regarded their presence in *L. (O.) hamatiloba* Grolle as proof that this could not be an *Orthocaulis*. This species represents a monotypic genus, *Anomacaulis*, properly placed in Jamesonielloideae, a group where reddish pigments are frequent. Yet *Lophozia* subg. *Massula* and *Lophozia* commonly develop wall anthocyanins; here the ability or inability to develop reddish pigments is, at best, of subgeneric significance. Within *Lophozia* it may suggest the relative degree of affinity of subgeneric units; thus *Orthocaulis* and *Barbilophozia* are surely more closely allied to each other than to other subgenera; also *Massula* and *Lophozia* are surely very closely related. Finally, within *Scapania*, ability to form wall anthocyanins may prove at best of varietal or subspecific rank. Thus *Scapania lingulata* Buch, *S. irrigua* (Nees) Dumort, and *S. paludicola* Loesk. K. Müll. all include far northern populations which are able to develop anthocyanin-type pigments while less arctic-boreal populations of these species lack this ability (Schuster, 1974).

These examples could be multiplied indefinitely, but I think I have made my point.

CLASSICAL V. MODERN PHYLOGENIES

Concepts of evolution have gone through three basic stages: (1) In the period immediately after Hofmeister and Darwin, the newly elucidated life-cycles of land plants gave us the classical, very mechanical phylogeny, starting with *Riccia* or the hypothetical *Sphaeroriccia* of Lotsy (1909). Here the dominant theme was the intercalation and subsequent perfection of the sporophyte generation. Overlooked, or neglected, was the intrinsically very simple idea that an epiphytic, or even internally parasitic stage which operates as an organism is unlikely to achieve independence from its host. It is a zoological cliché that parasites have evolved from free-living antecedents and not vice versa.

Recognition of this very simple idea came late; indeed, its significance for the evolution of the bryophytes has hardly ever been analysed, although it is implicit, if not stated, in most modern phylogenies. In any event, these early classifications, which start with thalloid and/or *Riccia*-like organisms with "simple" sporophytes, had a tenacious hold on the imagination and are even reflected in recent texts. Thus in Frye and Clark (1937–1947), Parihar (1959), Bold (1973) and Smith (1955) the Hepaticae uniformly start with *Riccia* and end with *Anthoceros*! This idea is presented with mind-boggling certainty by Frye and Clark (1937, p. 10), whose "philogenetic diagram" deserves to be immortalized (as Fig. 1 here). Since *Riccia* is thalloid and *Sphaerocarpus* is often and erroneously described as thalloid (Smith, 1955; Frye and Clark, 1937–1947), simple logic seemed to dictate that the "thallus" of the ancestral *Riccia* or *Sphaeroriccia* was primitive and that evolution led from thalloid antecedents to leafy groups. For this spurious reason, it is, even today, almost an article of faith to place the mosses after the hepatics.

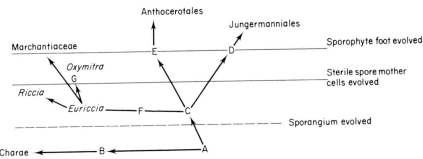

FIG. 1. Hypothetical "Phylogeny" of the Hepaticae of Frye and Clark [Simplified from Frye and Clark, 1937, p. 10][1]

[1] At *A* the "common ancestor" is visualized as a filamentous gametophyte, bearing unseptate rhizoids, and "sperm cells constituting male filaments and giving rise to one sperm each", and an "egg ... covered by sterile branches"—which had a "heavy wall". Meiosis "taking place at once on germination". At *B* is visualized evolution of the charoid habit of branching, and of terminalization of the egg cell. At *C* "development of a dorsiventral habit" accompanied by "closely coherent branches, the haploid body thus becoming the liverwort thallus"; the egg cell loses its "heavy wall" and after fertilization there is "formation of a sphere of diploid cells of which all but the outer layer undergo reduction division". At *D* we find a "tendency of the thallus to remain thin and without intercellular spaces" while at *E* there is a "tendency of the thallus to grow rather thick and to form intercellular air spaces which open ventrally" while at *F* there is a "tendency of the thallus to grow rather thick and to form intercellular air spaces which open dorsally"; at *G* we find "vertical air spaces between columns of cells and partly with air chambers".
    Needless to say, I regard this diagram as almost pure mythology and the ideas reflect a positively medieval approach to phylogeny.

These archaic ideas of progressive sporophyte evolution, derived in part from the antithetic theory of Čelakovský (1874), have not been embraced with enthusiasm by most recent hepaticologists.

(2) In the early decades of this century the renowned text of Wettstein (1903–1908) had a profound effect on subsequent thinking. In essence, although it was not stated in such a bald fashion, Wettstein assumed that mosses and hepatics shared a common ancestor and therefore the nearest thing in the Hepaticae to a radially symmetric moss-like architecture was to be regarded as primitive. Hence Wettstein placed the often radially symmetric Calobryales at the beginning, and ended up with thalloid hepatics as derived groups. The relatively recent discovery of the undoubtedly primitive *Takakia* in which $n = 4$ or 5 gave these ideas a great boost: one could suppose that all other hepatics, in which haploid chromosome numbers of 8, 9, and 10 prevail, were derived by a very early process of polyploidy (Tatuno, 1959), and the presence in various hepatics of paired nucleolar chromosomes has often been cited as evidence for this idea (Berrie, 1958a; Schuster, 1966a).

Even before *Takakia* lent a certain verisimilitude to these ideas, however, Evans (1939), in a now classic paper, carefully developed the idea of progressive and repeated planation of an intitially erect and leafy, triradial gametophyte. He cited as evidence for such general evolutionary progressions the then available evidence from the Jungermanniales, where repeated evolution from leafy to planate and eventually thalloid types could be demonstrated with the end-points being *Zoopsis*, *Schiffneria*, *Pteropsiella* and *Metzgeriopsis*. Evans, however, drove this idea to a final and perhaps insupportable point, entertaining the possibility that, for example, the ventral scales of the Marchantiales were simply modified leaves.

In essence, among many hepaticologists of the first two-thirds of the twentieth century, the idea that all thalloid hepatics could have been derived from leafy ancestral types gradually became fashionable. In some sense, the phylogenies of Cavers (1910–1911) and other, earlier morphologists such as Bower (1908), or Campbell (1895) were reversed: *Riccia* became the end point of all evolution (or, at the last, an end point). A detailed defence of much, but not all, of this philosophy is found in Schuster (1966a).

In this work I rejected the simplistic idea that all leafy types are primitive and gave rise to all thalloid types. "Essentially simultaneous origin of some thallose and 'leafy' gametophytes" is postulated, although it is admitted that some thallose types are obviously recent (Schuster, 1966a, p. 287). Considerable discussion is given to this idea, and it is concluded that, basically, the environmental imperatives obtaining at the time of the transmigration of the ancestral

gametophytes allowed only the two basic solutions outlined above: (*a*) planation of the originally erect, radial, leafless ancestral type and (*b*) foreshortening
of this erect type (with compensatory elaboration of slime papillae to leaf-like
structures, so that an adequate photosynthetic surface is maintained). Indeed,
it is proposed that the "nature of the leaf in the Bryophyta suggests leaves
have evolved several times" (Schuster, 1966a, pp. 505 *et seq.*, 287–288), and that
"the thallose organization has also evolved many times in diverse groups . . . ".

In essence, these ideas constitute a wholly independent, third approach to
liverwort phylogeny and evolution:

(3) Present day reconstructions of liverwort evolution such as those of
Schuster (1966a, 1977) produce a much less neat phylogeny which is much
less cut and dried and not so simplistic when closely examined. I prefer to use
the concept of options (Schuster, 1977): when, at certain points in time an
organism could evolve in two or more ways, there is, logically, nothing to
prevent evolution following both pathways since it tends to be purely exploita-
tive. Thus the transition from an erect-radial (and presumably leafless) ancestral
type to the thallus is one direction for evolution to take; that to the leafy,
erect, Calobryoid gametophyte is an alternative. We must not presume that
one is intrinsically better than the other. Admittedly, in the Hepaticae, which
have less physiological drought resistance than the Musci (Clausen, 1952),
there has been repeated selection pressure toward flattened, prostrate gametophytes. Hence even though in many lines such as *Takakia* and *Haplomitrium*
in the Calobryales and in at least nine of the fourteen suborders of the Jungermanniales (Table I), erect-radial and leafy gametophytes persist, most groups
of hepatics that retain leaves have become secondarily bilateral and usually
dorsiventrally flattened. In the Marchantiidae, where bilateral (and chiefly
thallose) gametophytes prevail, there is reasonable fossil evidence, as offered
by *Naiadita*, that at least in the Sphaerocarpales, there was an isophyllous and
erect ancestral type. This does not mean that all modern Marchantiidae must,
of necessity, have had leafy and erect ancestors!

My 1977 paper provides some documentation of these assertions. I would
like to close this section, however, with an appropriate quotation (Schuster,
1966a, pp. 288–289): "It is, indeed, not inconceivable that the initial reaction
of the gametophyte, on becoming progressively more terrestrial, may have
been toward a prostrate, aplanate form—i.e., toward a thallus . . . however, I
believe this was simply *one of several initial*, different reactions, *by several already
phylogenetically* rather distant groups."

Practically speaking, therefore, there are several important results from such
an approach which allow us to develop the following principles:

(1) We must conceive of a "bush-type" phylogeny—not a "phylogenetic tree"—with all the untidiness this entails, especially when we are forced, using the printed page, to arrange groups in a linear sequence. This "bush-type" phylogeny is evident from p. 71, Fig. 1 (Hypothetical phylogeny of the Hepaticae); here a selected series of salient taxa illustrate the principles espoused. Were additional—or all—taxa included, the "bush" would be so untidy as to defy deciphering. A similar, more complete phylogeny for the Jungermannioles is given in Schuster (1972).

(2) It does not matter whether we start our phylogeny with thalloid or leafy types. I here adhere to a start with the Calobryales, because, as repeatedly emphasized (Schuster, 1966a, 1967, 1971a) the group has many primitive features (e.g., radial symmetry; uniform unisexuality; apparently random dispersal of sex organs in primitive species; lack of any reliable mechanism to prohibit plural sporophyte maturation from a single gynoecium; 4–5 haploid chromosomes of *Takakia*; lack of specialized asexual propagative devices; identity in the initial stages of sex organ ontogeny; and large archegonia).

(3) Since we start with erect, radial types, the Calobryales, in which early attempts at anisophylly admittedly occurred, it is rational to follow the Calobryales with the Jungermanniales. With equal justification, however, one could start with the primitive Fossombroniinae of the Metzgeriales, or with the Treubiales. Any linear sequence has to be arbitrary to some extent; the student should not read into the sequence anything which is not intended.

(4) Since erect radial types in the Marchantiidae are no longer extant (*Naiadita*, the only radial type, is a Rhaetic fossil), and since both leafy types (Sphaerocarpales) and thalloid types (Monocleales, Marchantiales) are clearly relatively advanced, the Marchantiidae on balance, should be placed last in any phylogenetic scheme. The highly reduced sporophytes of all Marchantiidae, apart from *Monoclea*, also dictate such a placement.

(5) Since rigidly "fixed" features are to be regarded as phylogenetically advanced, it follows that organisms that are still variable with regard to the criterion in question must be regarded as *relatively* low phylogenetically. *Haplomitrium* and *Monoclea* lack a well-defined mechanism for preventing multiple sporophyte formation. On that basis, these are primitive genera, and I start the Jungermanniidae therefore with Calobryales, and the Marchantiidae with Monocleales.

(6) The furrowing type of cell division associated with meiosis, as found in all the Jungermanniidae, is probably more primitive than the cell-plate type associated with the Marchantiidae (Pickett-Heaps and Marchant, 1972). Since

the Monocleales, in certain ways, fit well with the Marchantiidae (e.g. the dimorphic cells, with the well-defined oil cells lacking chloroplasts; the androecium; sex organ ontogeny; the positionally dimorphic rhizoids; the 1-stratose capsule wall) yet preserve a Jungermannioid meiosis (Johnson, 1904; Schuster, 1977), it logically follows that the Monocleales should be placed at the base of the Marchantiidae.

(7) Within various orders, sequences are picked from structurally generalized or variable types to those whose morphology is more rigidly defined. Thus, in the Jungermanniales, we start with roughly isophyllous types in which branching is plastic, and from all 3 rows of merophytes as, for example, in *Temnoma*, where 5 distinct branching modes occur (Schuster, 1966a, 1967). Logically, therefore, one can start the Jungermanniales with the Herbertineae, to which *Temnoma* belongs. Furthermore, within this suborder we also find certain probably primitive criteria (stenotypy; bracteolar androecia; a distinct perianth). In the Marchantiales the situation is much more difficult and no one primitive group is left. But one can argue that the Oxymitriaceae–Ricciaceae sequence derives from *Corsinia*-like types (in all of these we find primitive acropetal sex organ development on the leading, "vegetative" thallus). And one can argue that suborders such as the Marchantiineae in which we find sex organs, at least the female, restricted to specialized thallus branches condensed to form receptacles, derive from *Lunularia*-like antecedents. *Lunularia* preserves certain notably primitive attributes such as the receptacle derived from condensation of two pairs of dichotomies, resulting in a deeply 4-lobed structure; the sporophyte seta notably elongating prior to spore discharge, so that the capsule is exserted well beyond receptacular tissue; the regularly 4-valved capsule; the fact that a single receptacle may bear 2–3 or even 4 sporophytes —a feature recalling *Monoclea*!

(8) Other things being approximately equal, preponderantly unisexual taxa are regarded as more primitive than bisexual taxa; taxa lacking asexual reproductive devices are regarded as more primitive than those that rely on these devices to a greater or lesser extent.

These principles apply within groups of the most diverse size. For example, within the Jungermanniales, the Herbertineae are logically placed low down because, within the entire suborder, bisexuality has evolved only twice, in *Lophochaete fryei* (Perss.) Schust. and some phenotypes of *Blepharostoma trichophyllum* and reproduction by specialized asexual structures has evolved only twice, with gemmae in *Blepharostoma* and caducous and fragmenting leaf lobes and/or teeth in *Chaetocolea*. Within the family these same principles apply: in the Lejeuneaceae, the Nipponolejeuneoideae and Ptychanthoideae are fairly primitive since the first lacks asexual reproduction and in the second, some species in a minority of genera, such as *Caudalejeunea* and *Acrolejeunea*, have evolved it; also, bisexuality is infrequent and confined to a relatively few species of the Ptychanthoideae. By contrast, the Lejeuneoideae are higher, with numerous cases of bisexuality, but usually restricted to autoecious spatial orientations and numerous cases of asexual reproduction, involving many structures such as cladia, discoid gemmae or caducous leaves. Finally, the Cololejeuneoideae are placed at the pinnacle of hepatic evolution because of their preponderant bisexuality, with many cases of paroecism; almost all taxa reproduce asexually and epiphylly becomes dominant.

(9) Other things being equal, stenotypic taxa are regarded as more primitive than polytypic ones (Schuster, 1966a, pp. 294–295). However, a group may also be stenotypic because of its recent origin; it, then, simply has had insufficient time to speciate to any extent. In most Hepaticae, however, stenotypy is demonstrably due to the fact that, as with most Gymnospermae, there has been massive extinction. This leads to one other collateral effect of such stenotypy: taxa are widely spaced morphologically. Hence, stenotypy linked with great morphological, and presumably phylogenetic, discontinuity is regarded as indicative of great age.

Using the above examples (many others could be used), we find that this principle gives results that fit well with those derived in other ways. Thus the Herbertineae, as now constituted, include only the Herbertaceae (only *Herberta* and *Triandrophyllum* definitely belong here; the latter has only *c.* 4–5 species, the former has few species, but the taxonomy is so muddled that no intelligent guess is possible), Blepharostomataceae (*Blepharostoma*, with 3 species; *Lophochaete*, with probably 3 species; *Pseudolepicolea*, monotypic; *Grollea*, monotypic; *Archeophylla*, 3 species; *Archeochaete*, 3 species; *Isophyllaria*, monotypic; *Herzogiaria*, monotypic; *Temnoma*, 10–11 species) and Trichotemnomaceae (only *Trichotemnoma*, monotypic). In the Lejeuneaceae, the Nipponolejeuneoideae (only *Nipponolejeunea*, 2 species) and Ptychanthoideae (according to Gradstein (1975, 1978) the 20 genera include a total of 130–190 spp.—thus less than 10 species per genus, on average) are relatively stenotypic. The Cololejeuneoideae include the large genus *Cololejeunea* with at least 12 subgenera and over 250 species!

(10) Other things being not too unequal, each basic group, or suborder, in the Jungermanniales should start with an essentially isophyllous prototype. As Table I makes clear, for the 14 suborders admitted (and for which an outline is given below), it is still possible to find isophyllous to subisophyllous, radial "prototypes" in ten cases. It is assumed that the various modern, aniso-

phyllous to bilateral (to thallose) taxa evolved from extinct taxa similar to these types.

Other principles could be cited and illustrated; the above ten will suffice. It is, however, relevant to show to what extent the isophyllous taxa cited in Column 1 of Table I (the putatively primitive taxa) fulfil some of the requirements of these principles. Examples follow in the sequence given in Column 1: *Eoisotachis* is variable as regards leaf lobe number which ranges from 2–4 on one stem; it shows plastic branching from lateral and ventral merophytes, is unisexual, lacks asexual propagative structures, and is stenotypic (1 or possibly 2 species). *Pleurocladopsis* is morphologically variable (underleaves bi- or unlobed; androecia with or without bracteolar antheridia; plants totally isophyllous to moderately anisophyllous); it is unisexual, lacks asexual propagative structures, and is stenotypic (1 species). *Vetaforma* is excessively variable (antheridia may occur not only in the normal axillary position, but at abaxial bract bases, near the lateral margins of bracts and bracteoles; branching is variable, with *Frullania*- and *Acromastigum*-type terminal branches, lateral, ventral, and intermediate types of intercalary branches; leaf lobe number varies; bracteolar antheridia are retained but the number varies); it is unisexual, lacks asexual devices, and is stenotypic (1 species). *Mastigophora* is variable as regards several relevant criteria (leaf- and underleaf-lobe number is malleable within the species; bracteolar antheridia are retained; the degree of evolution of the perianth is widely variable in *M. flagellifera* (Hook.) Steph.), but rigid and advanced in others (short sexual branches; purely *Frullania*-type branching); it is unisexual and taxa lack asexual propagative structures; it is stenotypic (4 species). *Triandrophyllum* is variable as regards many criteria (branching varies; leaf- and underleaf-lobe numbers range from 2–4; there are bracteolar antheridia); taxa are all unisexual; all lack asexual devices; stenotypy prevails (*c.* 4–5 species; species limits are imprecise and subjective). Other Herbertoid genera (*Blepharostoma*, *Archeophylla*, *Herzogiaria*, *Lophochaete*, *Pseudolepicolea* are all more or less variable in many ways: leaf-lobe numbers vary; branching in some is very plastic, for example *Pseudolepicolea quadrilaciniata* (Sulliv.) Fulf. and Tayl., and they are all stenotypic and (except for *Blepharostoma* and one species of *Lophochaete*) uniformly unisexual. *Tetralophozia* is variable as regards leaf lobe number and also as regards branching modes (Schuster, 1969); taxa are all unisexual and normally lack asexual reproduction (although, very rarely, *T. setiformis* (= *Chandonanthus setiformis*) may have gemmae); the group is stenotypic (2 species). In the Lepidoziineae the isophyllous primitive genera (and subgenera) such as *Neogrollea*, *Isolembidium*, *Hygrolembidium* subg. *Hygrobiellopsis*, *Lepidozia* subg. *Dendrolepidozia* are all stenotypic (1 species

Table I. Morphological progressions in the Jungermanniales (in part after Schuster, 1972)

Primitive, ± isophyllous genus/genera	Anisophyllous and ± advanced intermediate genus/genera	Derivate example(s) (underleaves bilateral, tiny or absent)	Thalloid or semi-thalloid extreme(s)
1. BALANTIOPSIDINEAE: *Eoisotachis* Schust.	*Isotachis* Mitt., *Neesioscyphus* Grolle, *Anisotachis* Schust.	*Hyppoisotachis* Schust., *Balantiopsis* Mitt., *Gyrothyra* Howe	Not evolved
2. PERSSONIELLINEAE: *Pleurocladopsis* Schust.	*Schistochila* Dumort.	*Paraschistochila* Schust., *Perssoniella* Herz.	Not evolved
3. LEPICOLEINEAE: *Vetaforma* Fulf. and Tayl., *Lepicolea* Dumort. p.p.	*Lepicolea* Dumort. p.p.	Not evolved	Not evolved
4. PTILIDIINEAE: *Mastigophora* Nees, p.p.	*Ptilidium* Nees	Not evolved	Not evolved
5. HERBERTINEAE: *Triandrophyllum* Fulf. and Hatch., *Grollea* Schust., *Isophyllaria* Hodgs., *Herbertus* S. F. Gray, etc.	*Temnoma* subg. *Temnoma* Mitt., *Archeophylla* Schust.	Not evolved	Not evolved
6. LEPIDOZIINEAE: *Neogrollea* Hodgs., *Isolembidium* Schust., *Hygrolembidium* Schust.	*Lembidium* Mitt., *Bazzania* S. F. Gray, *Micropterygium* Lindenb.	*Mytilopsis* Spr., *Zoopsidella* Schust.	*Zoopsis* Hook. fil. and Tayl., *Pteropsiella* Spr., *Phycolepidozia*

	Hygrobiella Spr.	Pleuroclada Spr.	Cephalozia Dumort., Jackiella Schiffn.	Schiffneria Steph.
8. ANTHELIINEAE: Anthelia Dumort.		Not evolved	Not evolved	Not evolved
9. GEOCALYCINEAE: Pachyglossa Herz. and Grolle, Lophocolea boveana Mass.		Lophocolea Dumort., p.p. Clasmatocolea Spr., p.p.	Clasmatocolea Spr., p.p.	Not evolved
10. JUNGERMANNIINEAE: Tetralophozia (Schust.) Schljak., Pseudocephaloziella Schust.		Lophozia subg. Orthocaulis (Buch) Schust.	Lophozia subg. Lophozia Dumort.	Not evolved
11. LEPIDOLAENINEAE: Not extant		Lepidolaena Dumort., Lepidogyna Schust.	Not evolved	Not evolved
12. PLEUROZIINEAE: Not extant		Not extant	Pleurozia Dumort., Eopleurozia Schust.	Not evolved
13. RADULINEAE: Not extant		Not extant	Radula Dumort.	Not evolved
14. PORELLINEAE: Not extant		Porella L.	Cololejeunea (Spr.) Schiffn.	Metzgeriopsis Goebel

in each !); *Neogrollea* and *Dendrolepidozia* still retain *Acromastigum*-type branches and branching patterns tend to be variable, especially in *Neogrollea*; none is bisexual and none produces gemmae or other asexual propagules. In the Cephaloziineae *s. str.* only *Hygrobiella* is nearly isophyllous while *Amphicephalozia* (Cephaloziellaceae) is subisophyllous on larger, usually fertile, axes. Both genera are monotypic; both are unisexual; both show considerable variability as regards branching. In the Antheliineae we find only *Anthelia* (2 species, one unisexual, the other bisexual). Branching is plastic (lateral, *Frullania* type; ventral-intercalary; rarely ventral-terminal; Schuster, 1974) ; asexual reproduction is lacking. In Geocalycineae, *Pachyglossa* has only 4 species; all are unisexual, all lack asexual reproduction. It is surely not coincidental that almost all of these taxa (*Eoisotachis, Pleurocladopsis, Vetaforma, Mastigophora, Triandrophyllum* (and other Herbertineae excluding *Blepharostoma*), *Neogrollea, Isolembidium, Hygrolembidium, Anthelia, Pachyglossa*) have rhizoids limited, or virtually so, to fascicles at the underleaf bases. Fasciculate rhizoids are a primitive character (Schuster, 1966a); in this regard only the *Tetralophozia* and *Hygrobiella* elements are advanced. Thus there is abundant evidence that isophylly seems to be linked, in general, with a high incidence of other criteria which may be regarded as relatively primitive.

I have, in the immediately preceding pages, thus tried to give the philosophical foundation for the phylogeny that follows. Although one may argue with bits and pieces of this, I remain convinced that, on the whole, the classification presented on the succeeding pages represents a vast conceptual advance over the only other recent classification for which I have much respect, that of Evans (1939). The leading improvement is that Evans's family Ptilidiaceae is regarded as an artificial group whose 17 genera are now placed into 11 families (Herbertaceae, Antheliaceae, Mastigophoraceae, Cephaloziaceae, Ptilidiaceae, Geocalycaceae, Blepharostomataceae, Isotachidaceae, Lepicoleaceae, Lepidolaenaceae, Trichocoleaceae) that fall into no less than seven suborders (Herbertineae, Antheliineae, Ptilidiineae, Geocalycineae, Balantiopsidineae, Lepicoleineae, Lepidolaenineae). This dramatically illustrates the different results obtained from a truly modern approach. The following classification of the Hepaticae presents my current feelings about interrelationships (and these are illustrated, schematically, in Fig. 2). Some of the morphological progressions which provide evidence for phylogenetic relationships in the Jungermanniales are set out in Table I.

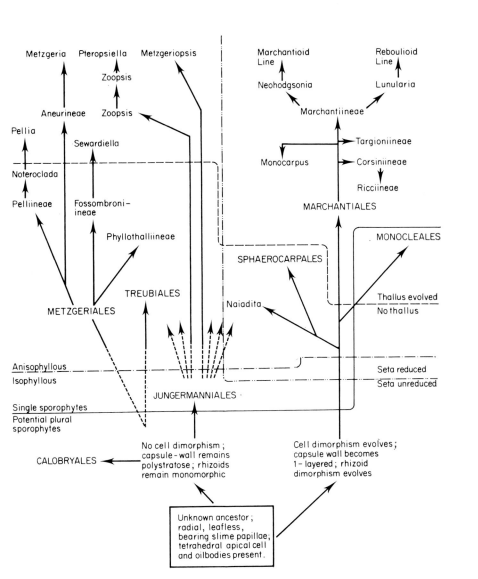

FIG. 2. Presumed phylogeny of the Hepaticae; only phylogenetically important taxa (usually discussed in text) indicated.

A PUTATIVELY PHYLOGENETIC CLASSIFICATION OF THE HEPATICAE

The subjoined classification attempts to arrange the Hepaticae into natural groupings. This system should be compared to the several other modern systems such as those of Verdoorn (1932), Evans (1939), Müller (1939–1940; 1951–1958) and Schuster (1966a). In essence, it is based on a large ensemble of principles, the most important of which have been briefly outlined in the preceding pages. A more elaborate treatment, attempting to justify many of the familial and generic placements, will appear in the Botanical Review. In that paper I use the same linear classification given below as a framework for a series of about 100 annotations which attempt to rationalize various taxonomic and phylogenetic decisions, given here arbitrarily. For convenience of cross-reference, exactly the same sequence is followed in both schemes.

I would like to emphasize that the rank of the group accepted is a matter of indifference to me. Schljakov (1972) has recently elevated most of my suborders to ordinal rank and I have attempted to indicate most of these in the brief synonymy given in the outline classification. The basic reason he does so is that these groups are believed to be fully equivalent to orders now commonly accepted, following Taktajan (1959), in angiosperms. A second reason is that the orders commonly accepted in Musci are, at least in the preponderantly pleurocarpous groups, no more sharply defined than the suborders accepted here. However, if the Schljakov classification is compared with that given here, numerous disagreements as to position of genera (e.g. *Anthelia*) and families, as well as suborders, emerge. This strongly suggests that a conservative attempt at recognition of orders is best retained. For example, the Jungermanniales and Marchantiales, as here (traditionally) broadly delimited are, unquestionably, both readily defined and wholly natural. Both circumscription and affinities of some suborders remain very ambiguous, especially of the Geocalycineae, but also of several other groups.

HEPATICAE (Hepaticopsida)
Subclass Jungermanniidae Schust. (Jungermanniae)
1.   Order Calobryales Campb. ex Hamlin [Haplomitriales Buch ex Schljak.]
        Suborder Takakiineae Schust.:
           Takakiaceae Hatt. and Inoue: *Takakia* Hatt. and Inoue
        Suborder Calobryineae Schust.:
           Haplomitriaceae Dědeček: *Haplomitrium* Nees
2.   Order Jungermanniales [Synonyms: Lophoziales Schljak., Perssoniellales Schljak., Porellales Schljak., Pleuroziales Schljak., Lepidoziales Schljak., Ptilidiales Schljak.]
        Suborder 1. Lepicoleineae Schust.:

Vetaformaceae Fulf. and Tayl.: *Vetaforma* Fulf. and Tayl.

Lepicoleaceae Schust.: *Lepicolea* Dumort.

Suborder 2. Herbertineae Schust.:

Herbertaceae K. Müll.: *Herberta* Gray, *Triandrophyllum* Fulf. and Hatch.

Blepharostomataceae K. Müll. [Pseudolepicoleaceae Fulf. and Tayl.]

    Subfam. Temnomoideae Schust.: *Temnoma* Mitt., *Lophochaete* Schust. *Pseudolepicolea* Fulf. and Tayl., *Archeophylla* Schust., *Archeochaete* Schust., *Isophyllaria* Hodgs., *Herzogiaria* Fulf., *Fulfordia* Hässel

    Subfam. Chaetocoleoideae Schust.: *Chaetocolea* Spr.

    Subfam. Blepharostomatoideae Grolle: *Blepharostoma* (Dumort.) Dumort.

    Subfam. Grolleoideae Solari: *Grollea* Schust.

Trichotemnomaceae Schust.: *Trichotemnoma* Schust.

Trichocoleaceae Nakai: *Trichocolea* Dumort., *Eotrichocolea* Schust., *Brachygyna* Schust.

Suborder 3. Lepidoziineae Schust.:

Lepidoziaceae Limpr.

    Subfam. Lepidozioideae Limpr.: *Lepidozia* Dumort., *Kurzia* v. Mart. [*Microlepidozia* (Spr.) Pears.], *Telaranea* Howe, *Drucella* Hodgs., *Megalembidium* Schust., *Psiloclada* Mitt., *Sprucella* Steph., *Arachniopsis* Spr.

    Subfam. Bazzanioideae Rodw. [Acromastigoideae Grolle]: *Bazzania* S. F. Gray, *Acromastigum* Evs., *Mastigopelma* Mitt.

    Subfam. Neogrolleoideae Schust.: *Neogrollea* Hodgs.

    Subfam. Lembidioideae Schust.: *Isolembidium* Schust., *Chloranthelia* Schust., *Lembidium* Mitt., *Hygrolembidium* Schust.

    Subfam. Zoopsidoideae Schust. [Zoopsoideae]: *Zoopsis* Hook., *Zoopsidella* Schust., *Paracromastigum* Fulf. and Tayl. [incl. *Bonneria* Fulf. and Tayl.], *Pseudocephalozia* Schust., *Pteropsiella* Spr., *Odontoseries* Fulf.

    Subfam. Protocephalozioideae Schust.: *Protocephalozia* Spr.

    Subfam. Micropterygioideae Grolle [Mytilopsidoideae Schust.]: *Micropterygium* Lindenb., *Mytilopsis* Spr.

Phycolepidoziaceae Schust.: *Phycolepidozia* Schust.

Calypogeiaceae (K. Müll.) H. W. Arn.: *Calypogeia* Raddi, *Metacalypogeia* (Hatt.) Inoue

Suborder 4. Cephaloziineae Schust.:

Adelanthaceae (Joerg.) Grolle: *Adelanthus* Mitt., *Wettsteinia* Schiffn., *Calyptrocolea* Schust.

Cephaloziaceae Migula:

    Subfam. Hygrobielloideae (Joerg.) Schust.: *Hygrobiella* Spr.

    Subfam. Cephalozioideae Migula: *Cephalozia* Dumort., *Nowellia* Mitt., *Pleuroclada* Spr., *Metahygrobiella* Schust.

    Subfam. Odontoschismatoideae Buch ex Grolle: *Odontoschisma* Dumort., *Cladopodiella* Buch, *Anomoclada* Spr.

    Subfam. Alobielloideae Schust.: *Alobiella* Spr., *Alobiellopsis* Schust., *Iwatsukia* Kitag. [incl. *Cladomastigum* Fulf.]

    Subfam. Trabelluloideae (Fulf.) Schust.: *Trabacellula* Fulf.

Subfam. Schiffnerioideae Schust.: *Schiffneria* Steph.
Cephaloziellaceae Douin: *Cephaloziella* (Spr.) Schiffn., *Cephaloziopsis* (Spr.) Schiffn., *Amphicephalozia* Schust., *Allisoniella* Hodgs., *Cylindrocolea* Schust., *Cephalojonesia* Grolle, *Kymatocalyx* Herz.
Jackiellaceae Schust.: *Jackiella* Schiffn.
Suborder 5. Antheliineae Schust.
Antheliaceae Schust.: *Anthelia* Dumort.
Suborder 6. Jungermanniineae [subord. Lophoziineae Schljak.]:
Jungermanniaceae Dumort.
  Subfam. Lophozioideae Macv. [incl. Chandonanthoideae Inoue]: *Chandonanthus* Mitt., *Tetralophozia* (Schust.) Schljak., *Lophozia* Dumort. [incl. *Orthocaulis* Buch, *Barbilophozia* Lske., *Massula* K. Müll., *Protolophozia* Schust., *Hypolophozia* Schust., *Leiocolea* K. Müll., *Isopaches* (Buch) Schust., *Obtusifolium* Buch, *Lophozia* Dumort., *Xenolophozia* Schust., all as subg.], *Gymnocolea* Dumort., *Gymnocoleopsis* (Schust.) Schust., *Anastrophyllum* (Spr.) Steph. [incl. subg. *Zantenia* Hatt., *Anastrophyllum* Spr., *Eurylobus* Schust., *Acantholobus* Schust., *Sphenolobus* (Lindb.) Schust., *Schizophyllum* Schust.], *Anastrepta* (Lindb.) Schiffn., *Tritomaria* Schiffn. [incl. subg. *Tritomaria*, *Saccobasis* (Buch) Schust.], *Andrewsianthus* Schust., *Stenorrhipis* Herz., *Pseudocephaloziella* Schust., *Roivainenia* Perss. and Grolle, *Gerhildiella* Grolle, *Sphenolobopsis* Schust. and Kitag.
  Subfam. Jamesonielloideae Inoue: *Jamesoniella* (Spr.) Steph., *Cryptochila* Schust., *Anomacaulis* (Schust.) Schust., *Denotarisia* Grolle, *Cuspidatula* Steph., *Hattoria* Schust.
  Subfam. Nothostreptoideae Schust.: *Nothostrepta* Schust.
  Subfam. Mesoptychoideae Schust.: *Mesoptychia* [Lindb. and Arn.) Evs.
  Subfam. Jungermannioideae Dumort.: *Jungermannia* L. [incl. subg. *Solenostoma*, *Jungermannia*, *Plectocolea*], *Liochlaena* Nees, *Cryptocolea* Schust., *Cryptocoleopsis* Amak., *Horikawaella* Hatt. and Amakawa, *Diplocolea* Amak., *Phragmatocolea* Grolle, *Lophonardia* Schust.
  Subfam. Scaphophylloideae Schust.: *Scaphophyllum* Inoue
  Subfam. Gottschelioideae Schust.: *Gottschelia* Grolle
  Subfam. Mylioideae Grolle: *Mylia* S. F. Gray
  Subfam. Notoscyphoideae Schust.: *Notoscyphus* Mitt.
Gymnomitriaceae Klinggr. [Marsupellaceae Buch]
  Subfam. Gymnomitrioideae Klinggr.: *Acrolophozia* Schust., *Herzogobryum* Grolle, *Marsupella* Dumort., *Gymnomitrion* Corda, *Prasanthus* Lindb. and Arn., *Poeltia* Grolle
  Subfam. Eremonotoideae Schust.: *Eremonotus* Kaal. ex Pears.
  Subfam. Stephanielloideae Schust.: *Stephaniella* Jack
Scapaniaceae Migula:
  Subfam. Scapanioideae Migula: *Scapania* Dumort., *Diplophyllum* Dumort.
  Subfam. Douinioideae Schust.: *Douinia* Buch
Blepharidophyllaceae (Schust.), Schust.: *Blepharidophyllum* Ångstr., *Clandarium* (Grolle) Schust.

Delavayellaceae Schust.: *Delavayella* Steph.

Suborder 7. Geocalycineae Schust. [Lophocoleineae Schljak.]

Geocalycaceae Klinggr. [incl. Lophocoleaceae (Joerg.) Vanden Bergh.].

Subfam. Lophocoleoideae Joerg.: *Lophocolea* Dumort., *Chiloscyphus* Corda, *Pachyglossa* Herz. and Grolle, *Heteroscyphus* Schiffn., *Xenocephalozia* Schust., *Clasmatocolea* Spr., *Tetracymbaliella* Grolle, *Leptophyllopsis* Schust., *Pigafettoa* Massal., *Evansianthus* Schust. and Eng., *Conoscyphus* Mitt. Hepatostolonophora Eng. and Schust.

Subfam. Geocalycoideae Klinggr. [Harpanthoideae C. Jens.]: *Geocalyx* Nees, *Harpanthus* Spr., *Saccogyna* Dumort., *Saccogynidium* Grolle

Subfam. Leptoscyphoideae Schust.: *Leptoscyphus* Mitt. [incl. subg. *Leptoscyphus*, *Anomylia* (Schust.) Schust., *Physoscyphus* Grolle], *Pedinophyllopsis* Schust. and Inoue, *Leptoscyphopsis* Schust., *Platycaulis* Schust.

Plagiochilaceae (Joerg.) K. Müll.

Subfam. Plagiochiloideae Joerg.: *Plagiochila* Dumort., *Pedinophyllum* Lindb., *Chiastacaulon* Carl, *Plagiochilidium* Herz., *Plagiochilion* Hatt., *Acrochila* Schust., *Rhodoplagiochila* Schust.

Subfam. Xenochiloideae Inoue: *Xenochila* Schust.

Subfam. Syzygielloideae Schust.: *Syzygiella* Spr.

Chonecoleaceae Schust.: *Chonecolea* Grolle

Arnelliaceae Nakai [=Southbyaceae K. Müll.]: *Arnellia* Lindb., *Southbya* Spr., *Gongylanthus* Nees

Acrobolbaceae Hodgs. [Marsupidiaceae Schust., Lethocoleaceae S. Arn.]

Subfam. Lethocoleoideae Grolle: *Lethocolea* Mitt., *Goebelobryum* Grolle

Subfam. Acrobolboideae Hodgs.: *Acrobolbus* Nees [incl. subg. *Acrobolbus*, *Xenopsis* Schust., *Marsupellopsis* Schiffn., *Lethocoleopsis* Grolle], *Marsupidium* Mitt., *Tylimanthus* Mitt., *Austrolophozia* Schust.

Suborder 8. Perssoniellineae Schust.

Schistochilaceae Buch

Subfam. Pleurocladopsidoideae Schust.: *Pleurocladopsis* Schust.

Subfam. Schistochiloideae Buch: *Schistochila* Dumort. [incl. subg. *Schistochila*, *Austroschistochila* Schust., *Protoschistochila* Schust., *Pachyschistochila* Schust., *Eoschistochila* Schust. and Eng., *Platyschistochila* Schust. and Eng.], *Paraschistochila* Schust. [incl. subg. *Nothoschistochila* Schust., *Acroschistochila* Schust., *Paraschistochila*]

Perssoniellaceae Schust.: *Perssoniella* Herz.

Suborder 9. Balantiopsidineae Schust.

Balantiopsidaceae Buch

Subfam. Ruizanthoideae Schust.: *Ruizanthus* Schust.

Subfam. Isotachidoideae Schust.: *Isotachis* Mitt. [incl. subg. *Isotachis*, *Hypoisotachis* Schust.], *Eoisotachis* Schust., *Neesioscyphus* Grolle

Subfam. Balantiopsidoideae Nakai: *Balantiopsis* Mitt. [incl. subg. *Balantiopsis*, *Steereocolea* Schust.], *Anisotachis* Schust.

Gyrothyraceae Schust.: *Gyrothyra* Howe

Suborder 10. Pleuroziineae Schust.

Pleuroziaceae K. Müll.: *Pleurozia* Dumort., *Eopleurozia* Schust.
Suborder 11. Radulineae Schust.
    Radulaceae (Dumort.) K. Müll.: *Radula* Dumort.
Suborder 12. Ptilidiineae Schust.
    Mastigophoraceae Schust.: *Mastigophora* Dumort.
    Ptilidiaceae Klinggr.: *Ptilidium* Nees
    Chaetophyllopsidaceae Schust. [=Chaetophyllopsaceae]: *Chaetophyllopsis* Schust., *Herzogianthus* Schust.
Suborder 13. Lepidolaenineae Schust.
    Lepidolaenaceae Nakai
        Subfam. Lepidolaenoideae Nakai: *Lepidolaena* Dumort., *Gackstroemia* Trevis., *Lepidogyna* Schust.
        Subfam. Neotrichocoleoideae (Inoue) Schust.: *Neotrichocolea* Hatt.
        Subfam. Trichocoleopsidoideae Schust.: *Trichocoleopsis* Okamua
    Jubulopsidaceae (Hamlin) Schust.: *Jubulopsis* Schust.
Suborder 14. Porellineae Schust.
    Porellaceae Cavers: *Porella* L., *Ascidiota* Massal., *Macvicaria* Nichols.
    Goebeliellaceae Verd.: *Goebeliella* Steph.
    Jubulaceae Klinggr. [=Frullaniaceae Lorch]
        Subfam. Jubuloideae Klinggr.: *Jubula* Dumort.
        Subfam. Frullanioideae Schust.: *Frullania* Raddi, *Neohattoria* Kamim., *Steerea* Hatt. and Kamim., *Schusterella* Hatt. et al., *Amphijubula* Schust.
    Lejeuneaceae Cavers [Bryopteridaceae Stotler]
        Subfam. Nipponolejeuneoideae Schust.: *Nipponolejeunea* Hatt.
        Subfam. Ptychanthoideae Mizut. ex Schust. [incl. Bryopteridoideae Gradst.]: *Bryopteris* Lindenb., *Brachiolejeunea* (Spr.) Schiffn., *Caudalejeunea* Steph., *Neurolejeunea* (Spr.) Schiffn., *Mastigolejeunea* (Spr.) Schiffn., *Lopholejeunea* (Spr.) Schiffn., *Acrolejeunea* (Spr.) Schiffn., *Schiffneriiolejeunea* Verd., *Marchesinia* S. F. Gray, *Archilejeunea* (Spr.) Schiffn., *Verdoornianthus* Gradst., *Thyisananthus* Lindenb., *Spruceanthus* Verd., *Dicranolejeunea* (Spr.) Schiffn., *Phaeolejeunea* Mizut., *Odontolejeunea* (Spr.) Schiffn., *Stictolejeunea* (Spr.) Schiffn., *Symbiezidium* Trevis., *Blepharolejeunea* S. Arn., *Acanthocoleus* Schust., *Tuzibeanthus* Hatt., *Ptychanthus* Nees
        Subfam. Lejeuneoideae Cavers: *Ceratolejeunea* (Spr.) Schiffn., *Omphalanthus* Lindenb. [incl. subg. *Evansianthus* (Vanden Bergh.) Schust., *Omphalanthus*, *Peltolejeunea* (Spr.) Schust.], *Leucolejeunea* Evs., *Aureolejeunea* Schust., *Amphilejeunea* Schust., *Cheilolejeunea* (Spr.) Schiffn. [incl. subg. *Cheilolejeunea*, *Xenolejeunea* Kachr. and Schust., *Strepsilejeunea* (Spr.) Schust., *Renilejeunea* Schust., *Euosmolejeunea* (Spr.) Schust., *Anomalolejeunea* (Pears.) Schust.], *Hygrolejeunea* (Spr.) Schiffn., *Taxilejeunea* (Spr.) Schiffn., *Trachylejeunea* (Spr.) Schiffn., *Pycnolejeunea* (Spr.) Schiffn., *Lepidolejeunea* Schust. [incl. subg. *Perilejeunea* (Schust. and Kachr.) Schust., *Lepidolejeunea*, *Kingiolejeunea* (Robinson) Schust.], *Lejeunea* [ca. 11 subg.!], *Rectolejeunea*

Evs. [incl. subg. *Chaetolejeunea* (Schust.) Schust., *Rectolejeunea*, *Heterolejeunea* Schust.], *Harpalejeunea* (Spr.) Schiffn., *Drepanolejeunea* (Spr.) Schiffn. [incl. subg. *Drepanolejeunea*, *Kolpolejeunea* Grolle], *Leptolejeunea* (Spr.) Schiffn., *Rhaphidolejeunea* Herz., *Crossotolejeunea* (Spr.) Schiffn., *Pictolejeunea* Grolle, *Echinolejeunea* Schust., *Cyclolejeunea* Evs., *Macrolejeunea* (Spr.) Schiffn., *Cystolejeunea* Evs., *Sphaerolejeunea* Herz., *Cladiantholejeunea* Herz., *Prionolejeunea* (Spr.) Schiffn., *Acantholejeunea* Schust., *Cyrtolejeunea* Evs. [incl. subg. *Cyrtolejeunea*, *Oryzolejeunea* Schust.], *Dactylolejeunea* Schust., *Trachygyna* Schust., *Leiolejeunea* Evs., *Cladolejeunea* Zwickel, *Potamolejeunea* (Spr.) Evs., *Amblyolejeunea* Jovet-Ast, *Placolejeunea* Herz., *Cryptolejeunea* Schust. and Kachr., *Stenolejeunea* Schust., *Ophthalmolejeunea* (Schust.) Schust., *Physantholejeunea* Schust., *Cardiolejeunea* Schust. and Kachroo., *Cephalolejeunea* (Schust. and Kachroo.) Schust., *Echinocolea* Schust.

Subfam. Myriocoleoideae Schust.: *Myriocolea* Spr., *Cladocolea* Schust.

Subfam. Tuyamaelloideae Schust.: *Tuyamaella* Hatt., *Siphonolejeunea* Herz., *Austrolejeunea* (Schust.) Schust., *Haplolejeunea* Grolle, *Nephelolejeunea* Grolle

Subfam. Cololejeuneoideae Herz. ex Grolle [Paradoxae Lacout.]: *Calatholejeunea* Goebel, *Diplasiolejeunea* (Spr.) Schiffn., *Colura* Dumort., *Aphanotropis* Herz., *Myriocoleopsis* Schiffn., *Cololejeunea* (Spr.) Schiffn. [incl. ca. 11 subgenera!], *Aphanolejeunea* Evs.

Subfam. Metzgeriopsidoideae Schust.: *Metzgeriopsis* Goebel

3. Order Treubiales Schljak. [Metzgeriales subord. Treubiineae Schust.]
   Treubiaceae Verd.: *Treubia* Goeb., *Apotreubia* Hatt. *et al.*

4. Order Metzgeriales Schust. emend. Schljak.
   Suborder 1. Fossombroniineae Schust. [Codoniineae]
   Fossombroniaceae Evs.: *Fossombronia* Raddi, *Petalophyllum* Gott., *Sewardiella* Kashyap
   Suborder 2. Phyllothalliineae Schust.
   Phyllothalliaceae Hodgs.: *Phyllothallia* Hodgs.
   Suborder 3. Pelliineae Schust. ex Schljak.
   Pelliaceae Klinggr.: *Noteroclada* Spr., *Pellia* Raddi
   Allisoniaceae (Schust.) Schust. and Inoue: *Allisonia* Herz., *Calycularia* Mitt.
   Makinoaceae Nakai
      Subfam. Makinoioideae Nakai: *Makinoa* Miyaki
      Subfam. Verdoornioideae (Inoue) Schust.: *Verdoornia* Schust.
   Pallaviciniaceae Migula emend. Schust.
      Subfam. Pallavinioideae Migula: *Pallavicinia* Raddi, *Jensenia* Lindb., *Moerckia* Gott., *Hattorianthus* Schust. and Inoue
      Subfam. Symphyogynoideae (Trevis.) Schust.: *Symphyogyna* Nees and Dumort., *Xenothallus* Schust.
      Subfam. Podomitrioideae Schust.: *Podomitrium* Mitt.
   Suborder 4. Blasiineae Schust.
   Treubiitaceae Schust.: *Treubiites* Schust.

Blasiaceae Klinggr.
    Subfam. Blasioideae Klinggr.: *Blasia* L.
    Subfam. Cavicularioideae Inoue: *Cavicularia* Steph.
  Suborder 5. Metzgeriineae Schust. ex Schljak. [Aneurineae]
    Hymenophytaceae Schust.: *Hymenophytum* Dumort.
    Aneuraceae Klinggr.: *Aneura* Dumort., *Riccardia* Raddi, *Cryptothallus* Malmb.
    Metzgeriaceae Klinggr.: *Metzgeria* Raddi, *Apometzgeria* Kuwahara, *Apertithallus* Kuwahara
Subclass Marchantiidae [Marchantiae]
5.   Order Monocleales Schust.
    Monocleaceae: *Monoclea* Hook.
6.   Order Sphaerocarpales Cavers
    Suborder 1. Naiaditineae Schust.: *Naiadita* Buckl.
    Suborder 2. Riellineae Schust.:
      Riellaceae Cavers: *Riella* Mont.
    Suborder 3. Sphaerocarpineae Cavers:
      Sphaerocarpaceae Cavers: *Sphaerocarpus* (Mich.) Boehm., *Geothallus* Campb.
7.   Order Marchantiales Limpr.
    Suborder 1. Corsiniineae Schust. ex Schljak.
      Corsiniaceae Schiffn.: *Corsinia* Raddi, *Cronisia* Berk.
    Suborder 2. Carrpineae Schust. [Monocarpineae]
      Monocarpaceae Carr. [Carrpaceae]: *Monocarpus* Carr. [*Carrpos* Prosk.]
    Suborder 3. Targurioniineae Schust.
      Aitchisoniellaceae Schust.: *Aitchisoniella* Kashyap
      Targioniaceae Endl.: *Targionia* L., *Cyathodium* Kze.
    Suborder 4. Marchantiineae Limpr.
      Lunulariaceae K. Müll.: *Lunularia* Adans.
      Wiesnerellaceae Inoue: *Wiesnerella* Schiffn.
      Conocephalaceae K. Müll.: *Conocephalum* Wigg.
      Aytoniaceae Cavers [Grimaldiaceae, Rebouliaceae]: *Reboulia* Raddi, *Mannia* Corda, *Plagiochasma* Lehm. and Lindenb., *Asterella* Beauv., *Cryptomitrium* Aust.
      Cleveaceae Cavers [Sauteriaceae Evs.]: *Athalamia* Falc., *Sauteria* Nees, *Peltolepis* Lindb.
      Exormothecaceae K. Müll.: *Exormotheca* Mitt., *Stephensoniella* Kashyap
      Marchantiaceae (Bisch.) Endl.
        Subfam. Bucegioideae Schust., subf. n.: *Bucegia* Radian, *Neohodgsonia* Perss.
        Subfam. Marchantioideae (Bisch.) Endl.: *Marchantia* March. f., *Preissia* Corda, *Dumortiera* Nees, *Marchantiopsis* Douin, *Dumortieropsis* Douin
      Monoseleniaceae Inoue: *Monoselenium* Griff.
    Suborder 5. Ricciineae Buch
      Oxymitriaceae K. Müll.: *Oxymitria* Bisch.
      Ricciaceae Dumort.: *Riccia* [Mich.] L., *Ricciocarpus* Corda

REFERENCES

ANDREWS, H. N., Jr. (1960). Notes on Belgian specimens of *Sporogonites*. *Palaeobotanist* 7, 85–89.

BERRIE, G. K. (1958a). The nucleolar chromosome in hepatics, I. *Trans. Br. bryol. Soc.* 3, 422–426.

BERRIE, G. K. (1958b). The nucleolar chromosome in hepatics, II. A phylogenetic speculation. *Trans. Br. bryol. Soc.* 3, 427–429.

BERRIE, G. K. (1962). Australian liverworts, I. *Haplomitrium intermedium* sp. n. *Proc. Linn. Soc. N.S.W.* 87, 191–195.

BIERHORST, D. W. (1953). Structure and development of the gametophyte of *Psilotum nudum*. *Am. J. Bot.* 40, 649–658.

BIERHORST, D. W. (1966). The fleshy, cylindrical, subterranean gametophyte of *Schizaea melanesica*. *Am. J. Bot.* 53, 123–133.

BIERHORST, D. W. (1968a). Observations on *Schizaea* and *Actinostachys* spp., including *A. oligostachys*, sp. nov. *Am. J. Bot.* 55, 87–108.

BIERHORST, D. W. (1968b). On the Stromatopteridaceae (fam. nov.) and the Psilotaceae. *Phytomorphology* 18, 232–268.

BIERHORST, D. W. (1971). "Morphology of Vascular Plants". Macmillan, New York and London.

BOLD, H. (1973). "Morphology of Plants". Third edition. Harper and Row, New York.

BOWER, F. O. (1908). "The Origins of a Land Flora, a Theory Based upon the Facts of Alternation". Macmillan, London and New York.

CAMPBELL, D. H. (1895). "The Structure and Development of the Mosses and Ferns (Archegoniatae)". Macmillan, London and New York.

CAROTHERS, Z. B. and DUCKETT, J. G. (1978). A comparative study of the multilayered structure in developing bryophyte spermatozoids. *Bryophyt. Biblthca* 13, 95–112.

CAVERS, F. (1910–11). The inter-relationships of the bryophytes, I–XI. *New Phytol.* 9, 81–112, 157–186, 196–234, 269–304, 341–353; 10, 1–46, 84–86.

ČELAKOVSKÝ, L. (1874). Über die verschiedenen Formen und die Bedeutung des Generationswechsels der Pflanzen. *Sber. K. böhm. Ges. Wiss.*, Math.-nat. Kl., 1874, 21–61.

CLAUSEN, E. (1952). Hepatics and humidity. *Dansk bot. Ark.* 15(1), 1–80.

CRANDALL, B. A. (1969). Morphology and development of branches in the leafy Hepaticae. *Beih. nov. Hedwigia* 30, 1–261.

EMIGH, V. D. and FARRAR, D. R. (1977). Gemmae: a role in sexual reproduction in the fern genus *Vittaria*. *Science, N.Y.* 198, 297–298.

EVANS, A. W. (1902). The Lejeuneae of the United States and Canada. *Mem. Torrey bot. Club* 8, 113–183.

EVANS, A. W. (1939). The classification of the Hepaticae. *Bot. Rev.* 5, 49–96.

FREY, W. (1977). Neue Vorstellungen über die Verwandtschaftsgruppen und die Stammesgeschichte der Laubmoose. *In* "Beiträge zur Biologie der niederen Pflanzen" (W. Frey, H. Hurka and F. Oberwinkler, eds) pp. 117–136. G. Fischer, Stuttgart.

FRYE, T. C. and CLARK, L. (1937–1947). Hepaticae of North America. *Univ. Wash. Publs Biol.* 6, 1–1018.

GRADSTEIN, S. R. (1975). A taxonomic monograph of the genus *Acrolejeunea* with an arrangement of the genera of Ptychanthoideae. *Bryophyt. Biblthca* **4**, 1–162.

GRADSTEIN, S. R. (1978). Studies on Lejeuneaceae subfam. Ptychanthoideae, IV. *Verdoornianthus*, a new genus from Amazonas, Brazil. *Bryologist* **80**, 606–611.

GROLLE, R. (1961). Notulae hepaticologicae, IV–VI. *Revue bryol. lichén.* **30**, 80–84.

GROLLE, R. (1969). Review of: R. M. Schuster "The Hepaticae and Anthocerotae of North America, 1". *Nova Hedwigia* **16**, 539–542.

HATTORI, S. (1972). *Frullania tamarisci*-complex and the species concept. *J. Hattori bot. Lab.* **35**, 202–251.

JOHNSON, D. S. (1904). The development and relationship of *Monoclea*. *Bot. Gaz.* **38**, 185–205.

JUSSIEU, A. L. de (1789). "Genera Plantarum Secundum Ordines Naturales Disposita". Herissant and Barrois, Paris.

KASHYAP, S. R. (1919). The relationships of liverworts especially in light of some recently discovered Himalayan forms. *Proc. Asiat. Soc. Beng.* N.S. **15**, 152–166.

KASHYAP, S. R. (1929–1932). "Liverworts of the Western Himalayas and the Punjab Plain". Chronica Botanica, New Delhi.

KRASSILOV, V. (1973). Mesozoic bryophytes from the Bureja Basin, far east of the USSR. *Palaeontographica*, Abt. B, **143**, 95–105.

LEMOIGNE, Y. (1968). Observations d'archégones portés par des axes du type *Rhynia gwynne-vaughanii* Kidston et Lang. Existence de gamétophytes vascularisés au Dévonien. *C. r. hebd. Séanc. Acad. Sci., Paris*, Sér. D, **266**, 1655–1657.

LOTSY, J. P. (1909). "Vorträge über botanische Stammesgeschichte". Vol. 2. G. Fischer, Jena.

LUNDBLAD, B. (1954). Contributions to the geological history of the Hepaticae. Fossil Marchantiales from the Rhaetic-Liassic coal mines of Skromberga (Prov. of Scania). *Svensk bot. Tidskr.* **48**, 381–417.

MÄGDEFRAU, K. (1978). Niedere Pflanzen. *In* "Lehrbuch der Botanik" (D. von Denffer, F. Ehrendorfer, K. Mägdefrau and H. Ziegler eds). Thirty-first edition, pp. 542–1012. Gustav Fischer, Stuttgart.

MARTENS, G. von (1870). *Kurzia crenacanthoidea*, eine neue Alge. *Flora, Jena* **28**, 417–418.

MERKER, H. (1959). Analyse der Rhynien und Nachweis des Gametophyten. *Bot. Notiser* **112**, 441–452.

MERKER, H. (1961). Entwurf zur Lebenskreis-Rekonstruktion der Psylophytales nebst phylogenetischem Ausblick. *Bot. Notiser* **114**, 88–102.

MONTAGNE, J. F. C. (1838). Hepaticae. Première Centurie de Plantes cellulaires exotiques ou indigènes nouvelles. *Annls Sci. nat.*, Bot., Sér. 2, **9**, 38–49.

MÜLLER, K. (1939–1940). Lebermoose, Ergänzungsband. *In* "Rabenhorst's Kryptogamen-Flora". Vol. 6. Akademische Verlagsgesellschaft, Leipzig.

MULLER, K. (1951–1958). Die Lebermoose Europas. *In* "Rabenhorst's Kryptogamen-Flora". Third edition, Vol. 6. Akademische Verlagsgesellschaft, Leipzig.

NEUBERG, M. F. (1958). (Palaeozoic mosses from Angarida). *In* "Congreso Geologico Internacional, XXᵃ Sesión, Secc. 7", pp. 97–106.

NEUBERG, M. F. (1960). (Leafy mosses from the Permian deposits of Angarida.) *Trudȳ geol. Inst. Leningr.* **19**, 1–104.

PARIHAR, N. S. (1959). "An Introduction to Embryophyta, I. Bryophyta". Third edition. Central Book Depot, Allahabad.

PICKETT-HEAPS, J. D. and MARCHANT, H. J. (1972). The phylogeny of the green algae: a new proposal. *Cytobios* **6**, 255–264.

SCHLJAKOV, R. N. (1972). On the higher taxa of liverworts (Hepaticae *s. str.*). *Bot. Zh. SSSR* **57**, 496–508.

SCHUSTER, R. M. (1953). Boreal Hepaticae, a manual of the liverworts of Minnesota and adjacent regions. *Am. Midl. Nat.* **49**, 257–648.

SCHUSTER, R. M. (1959). Studies on Hepaticae, I. *Temnoma. Bryologist* **62**, 233–242.

SCHUSTER, R. M. (1963). An annotated synopsis of the genera and subgenera of Lejeuneaceae, I. Introduction; annotated keys to subfamilies and genera. *Beih. nov. Hedwigia* **9**, 1–203.

SCHUSTER, R. M. (1965). Studies on Hepaticae, XXVI. The *Bonneria-Paracromastigum-Pseudocephalozia-Hyalolepidozia-Zoopsis-Pteropsiella* complex and its allies: a phylogenetic study (part I). *Nova Hedwigia* **10**, 19–61.

SCHUSTER, R. M. (1966a). "The Hepaticae and Anthocerotae of North America". Vol. 1. Columbia University Press, New York.

SCHUSTER, R. M. (1966b). Studies on Hepaticae, XXVIII. On *Phycolepidozia*, a new, highly reduced genus of Jungermanniales of questionable affinity. *Bull. Torrey bot. Club* **93**, 437–449.

SCHUSTER, R. M. (1967). Studies on Hepaticae, XV. Calobryales. *Nova Hedwigia* **12**, 3–64 (1966).

SCHUSTER, R. M. (1969). "The Hepaticae and Anthocerotae of North America". Vol. 2. Columbia University Press, New York.

SCHUSTER, R. M. (1971a). Two new antipodal species of *Haplomitrium* (Calobryales). *Bryologist* **74**, 131–143.

SCHUSTER, R. M. (1971b). On the genus *Pleurocladopsis* Schust. (Schistochilaceae). *Bryologist* **74**, 493–495.

SCHUSTER, R. M. (1972). Phylogenetic and taxonomic studies on Jungermanniidae. *J. Hattori bot. Lab.* **36**, 321–405.

SCHUSTER, R. M. (1974). "The Hepaticae and Anthocerotae of North America". Vol. 3. Columbia University Press, New York.

SCHUSTER, R. M. (1977). The evolution and early diversification of the Hepaticae and Anthocerotae. *In* "Beiträge zur Biologie der niederen Pflanzen" (W. Frey, H. Hurka and F. Overwinkler, eds) pp. 107–115. G. Fischer, Stuttgart.

SCHUSTER, R. M. (1980a). "The Hepaticae and Anthocerotae of North America". Vol. 4. Columbia University Press, New York. (In press.)

SCHUSTER, R. M. (1980b). On *Nothostrepta* Schust. *Bryologist*. (In press.)

SCHUSTER, R. M. (1980c). On *Brachygyna* Schust. and the evolution and relationships of the Trichocoleaceae. *Bryologist*. (In press.)

SCHUSTER, R. M. and ENGEL, J. J. (1977). Austral Hepaticae, V. The Schistochilaceae of South America. *J. Hattori bot. Lab.* **42**, 273–423.

SMITH, G. A., BRIDEN, J. C. and DREWRY, G. E. (1973). Phanerozoic world maps. *Spec. Pap. Palaeont.* **12**, 1–42.

SMITH, G. M. (1955). "Cryptogamic Botany". Second edition, Vol. 2. McGraw-Hill, New York.

TAKHTAJAN, A. L. (1959). "Die Evolution der Angiospermae". G. Fischer, Jena.

TATUNO, S. (1959). Chromosomen von *Takakia lepidozioides* und eine Studie zur Evolution der Chromosomen der Bryophyten. *Cytologia* **24**, 138–147.

TOWNROW, J. A. (1959). Two Triassic bryophytes from South Africa. *Jl S. Afr. Bot.* **25**, 1–22.

VERDOORN, F. (1932). Classification of the Hepaticae. *In* "Manual of Bryology" (F. Verdoorn, ed.) pp. 413–432. M. Nijhoff, The Hague.

WATSON, E. V. (1964). "The Structure and Life of Bryophytes". Hutchinson, London.

WETTSTEIN, R. R. von (1903–1908). "Handbuch der systematischen Botanik". Vol. 2. Franz Deuticke, Leipzig und Wien. (Hepaticae on pp. 41–49.)

WHITTINGTON, H. B. and HUGHES, C. P. (1973). Ordovician trilobite distribution and geography. *Spec. Pap. Palaeontol.* **12**, 235–240.

4 | The Genera of the Lejeuneaceae: Past and Present

S. R. GRADSTEIN

Institute of Systematic Botany, Heidelberglaan 2, Utrecht, The Netherlands

Abstract: This paper reviews the historical development of generic concepts in the Lejeuneaceae and the current state of knowledge. Three historical periods are distinguished: (1) 1820–1884, during which many species, but only a few rather artificial genera based solely on perianth and underleaf characters, were described; (2) 1884–1930, in which Spruce and Evans established current generic concepts based on various new characters, and Stephani published numerous new, though often ill-founded species; (3) 1930 to the present day, during which taxonomic work has predominantly been on individual genera and mainly of a regional nature. Many new genera have also been described; these are often monotypic and rarely based on recently discovered taxonomic characters.

Regionally, recent progress has been mainly in Africa; knowledge of tropical American Lejeuneaceae remains poor. The need for more monographic work, frequently stressed in the past, is still felt today. Current generic delimitation—still purely based on gametophyte morphology and anatomy—is weakest in the notoriously difficult subfamily Lejeuneoideae, for which different concepts are employed by various authors. New taxonomic characters are needed to solve the problems involved. Future work should concentrate more on the neglected sporophyte generation, although newly established gametophyte characters should also be taken into account.

INTRODUCTION

The Lejeuneaceae are often considered the most difficult group of Hepaticae as regards the correct delimitation of genera (Schuster and Hattori, 1954). The great morphological diversity displayed by this predominantly tropical family has resulted in the description of over 100 genera, containing more than 1500

Systematics Association Special Volume No. 14, "Bryophyte Systematics", edited by G. C. S. Clarke and J. G. Duckett, 1979, pp. 83–107, Academic Press, London and New York.

described species. Although the number of currently accepted genera varies, at least 70 are accepted even by "lumpers" (Fig. 1).

Traditionally, the genera were classified into two rather artificial groups, the Holostipae and the Schizostipae (Spruce, 1884–1885), based principally on a single character of the gametophyte: the presence of undivided or divided underleaves. This system has now become obsolete and has been replaced by a more natural classification: Ptychanthoideae, Lejeuneoideae and Cololejeuneo-ideae (Mizutani, 1961), based on gametophyte as well as on sporophyte characters. In addition, several smaller subfamilies and tribes were established by Schuster (1963b) and Gradstein (1975). The current subfamily system of Lejeuneaceae is shown in Table I, which also lists all the described genera.

Schuster (1963b) summarized the state of knowledge of Lejeuneaceae up to the late 1950s and discussed the criteria used for generic delimitation. He also introduced an informal taxonomic category, the "genus complex", which proved to be an elegant means for expressing natural relationships among groups of often weakly defined genera. I used this concept to arrange

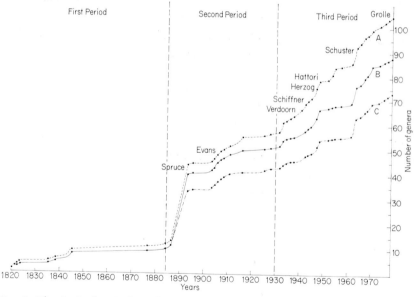

Fig. 1. The rise in the number of genera of the Lejeuneaceae during taxonomic history. Main authors of genera also indicated. A. Described genera (See Table I). B. Genera *currently* accepted by "splitters" (generic names italicized in Table I). C. Genera *currently* accepted by "lumpers". The difference between B and C largely reflects different current concepts of the Lejeuneoideae (p. 92).

TABLE I. The genera of Lejeuneaceae arranged chronologically within the current subfamilies. Monotypic genera are marked with an asterisk. Names currently accepted by authors who apply narrow generic concepts ("splitters") are italicized. These names correspond with line "B" of Fig. 1. Not included are *Bryolejeunea* (= *Bryopteris*), *Dendrolejeunea* (= *Thysananthus*), *Homalolejeunea* (= *Marchesinia*), *Omphalolejeunea* (= *Omphalanthus*), *Platylejeunea* (= *Symbiezidium*), *Ptycholejeunea* (= *Ptychanthus*) and *Thysanolejeunea* (= *Thysananthus*). These names were among the subgenera described by Spruce (1884–1885) and were long regarded as invalid or incorrect at the generic level, although some seem to have been validated as genera by Stephani and Lacouture (Grolle, personal communication)

Bryopteridoideae (Stotler) Gradstein 1975
   Bryopteris (Nees) Lindenberg 1845

Nipponolejeuneoideae Schuster and Kachroo 1963
   Nipponolejeunea Hattori 1944

Ptychanthoideae Mizutani 1961
   Marchesinia S. Gray 1821
   Phragmicoma Dumortier 1822 (= *Marchesinia*)
   Frullanoides Raddi 1823
   Ptychanthus Nees 1838
   Thysananthus Lindenberg 1845
   Ptychocoleus Trevisan 1877 (= *Frullanoides*)
   Symbiezidium Trevisan 1877
   Stictolejeunea (Spruce) Schiffner 1893
   Neurolejeunea (Spruce) Schiffner 1893
   Archilejeunea (Spruce) Schiffner 1893
   Mastigolejeunea (Spruce) Schiffner 1893
   Acrolejeunea (Spruce) Schiffner 1893
   Lopholejeunea (Spruce) Schiffner 1893
   Brachiolejeunea (Spruce) Schiffner 1893

   Dicranolejeunea (Spruce) Schiffner 1893
   Odontolejeunea (Spruce) Schiffner 1893
   Caudalejeunea (Spruce) Schiffner 1893
   Trocholejeunea Schiffner 1932
   Schiffneriolejeunea Verdoorn 1934
   Spruceanthus Verdoorn 1934
   *Heterolejeunea Schiffner 1941 (= *Lopholejeunea*)
   *Tuzibeanthus Hattori 1947
   Phragmilejeunea Schuster 1954 (= *Schiffneriolejeunea*)
   Blepharolejeunea S. Arnell 1962
   Phaeolejeunea Mizutani 1968
   *Acanthocoleus Schuster 1970
   Verdoornianthus Gradstein 1978

Lejeuneoideae Massalongo 1912
   Lejeunea Libert 1820
   Omphalanthus Lindenberg 1845
   Peltolejeunea (Spruce) Schiffner 1893
   Anoplolejeunea (Spruce) Schiffner 1893
   Prionolejeunea (Spruce) Schiffner 1893
   Crossotolejeunea (Spruce) Schiffner 1893

TABLE I—cont.

Harpalejeunea (Spruce) Schiffner 1893
Strepsilejeunea (Spruce) Schiffner 1893
Trachylejeunea (Spruce) Schiffner 1893
Drepanolejeunea (Spruce) Schiffner 1893
Leptolejeunea (Spruce) Schiffner 1893
Ceratolejeunea (Spruce) Schiffner 1893
Taxilejeunea (Spruce) Schiffner 1893
Macrolejeunea (Spruce) Schiffner 1893
Origoniolejeunea (Spruce) Schiffner 1893
Hygrolejeunea (Spruce) Schiffner 1893
Euosmolejeunea (Spruce) Schiffner 1893
Pycnolejeunea (Spruce) Schiffner 1893
Cheilolejeunea (Spruce) Schiffner 1893
*Anomalolejeunea Schiffner 1895
Cyrtolejeunea Evans 1903
Cyclolejeunea Evans 1904
Rectolejeunea Evans 1906
*Cystolejeunea Evans 1906
Leucolejeunea Evans 1906
*Leiolejeunea Evans 1908
Microlejeunea (Spruce) Jack and Stephani 1894
Potamolejeunea (Spruce) Stephani 1916
Stylolejeunea Sim 1926 (= *Lejeunea*)
*Cladolejeunea Zwickel 1933
*Sphaerolejeunea Herzog 1938

Rhaphidolejeunea Herzog 1943
*Nesolejeunea Herzog 1947 (= *Hygrolejeunea*)
*Amblyolejeunea Jovet-Ast 1948
*Placolejeunea Herzog 1948
*Evansiolejeunea Vanden Berghen 1948
*Ciliolejeunea S. Arnell 1953 (= *Lejeunea*)
*Inflatolejeunea S. Arnell 1953 (= *Lejeunea*)
*Cladiantholejeunea Herzog 1954
Stenolejeunea Schuster 1963
*Echinolejeunea Schuster 1963
Echinocolea Schuster 1963
Lepidolejeunea Schuster 1963
*Cardiolejeunea Schuster 1963
*Capillolejeunea S. Arnell 1965
Perilejeunea (Kachroo and Schuster) Robinson 1967
Kingiolejeunea Robinson 1967
Acantholejeunea (Schuster) Schuster 1967
*Dactylolejeunea Schuster 1970
Pictolejeunea Grolle 1977
Aureolejeunea Schuster 1978
*Physantholejeunea Schuster 1978
*Amphilejeunea Schuster 1978

Myriocoleoideae Schuster 1963
*Myriocolea Spruce 1884

Siphonolejeunea Herzog 1942
Tuyamaella Hattori 1947
Austrolejeunea (Schuster) Schuster 1963
*Nephelolejeunea Grolle 1973
*Haplolejeunea Grolle 1975

Cololejeuneoideae Herzog in Grolle 1972
Colura Dumortier 1835
Cololejeunea (Spruce) Schiffner 1893
Diplasiolejeunea (Spruce) Schiffner 1893
Leptocolea (Spruce) Evans 1911 (= Cololejeunea)
Aphanolejeunea Evans 1911

Physocolea (Spruce) Stephani 1916 (= Cololejeunea)
*Calatholejeunea Goebel 1928
Taeniolejeunea Zwickel 1933 (= Cololejeunea)
*Boninoleptocolea Horikawa 1936 (= Cololejeunea)
*Hemilejeunea Schiffner 1941 (= Cololejeunea)
*Myriocoleopsis Schiffner 1942
Campylolejeunea Hattori 1947
*Aphanotropis Herzog 1952
*Schusteria Kachroo 1957 (= Diplasiolejeunea)
*Jovetastella Tixier 1973 (= Cololejeunea)

Metzgeriopsioideae Schuster and Kachroo 1963
*Metzgeriopsis Goebel 1887

the genera of Ptychanthoideae into groups (Gradstein, 1975)—some possibly
more natural than others (Schuster, 1976)—for which nomenclatural status
was not required.

In this paper I will principally focus on the genera and review the historical
development of our concepts and the current state of knowledge. The increase
in number of the genera throughout taxonomic history is illustrated in Fig. 1.
From this picture the main periods of taxonomic research on Lejeuneaceae
become apparent: the first (1820–1884), the second (1884–1930) and the third
period (1930 onwards).

DEVELOPMENT OF TAXONOMIC CONCEPTS

1. First Period: 1820–1884

"Deux espèces d'Hépatiques que j'ai recontrées assez souvent dans mes excursions botaniques aux environs de Malmedy . . . m'ont encore fourni les moyens,
par leur fructification très-singulière, de former un genre nouveau, que . . . j'ai
dédié au savant et modest auteur de la Flore de Spa, M. le docteur Lejeune de
Verviers." (Libert, 1820, p. 6.) Thus was introduced a new genus of liverworts,
Lejeunea Libert, based on *Jungermannia serpyllifolia* Dicks. and a new species,
Lejeunea calcarea Libert, which is now placed in *Cololejeunea*. Though originally
spelt *Lejeunia*, the name was usually written as *Lejeunea* afterwards and has been
conserved in that manner following a proposal by Grolle (1973a). The identity
of Mme Libert's specimen of *Jungermannia serpyllifolia* has been a subject of
considerable confusion (Bonner and Miller, 1960; Schuster, 1963b), but
according to Grolle (1971b) it is the common *Lejeunea cavifolia* (Ehrh.) Lindb.
From my frequent visits to Malmedy in the Belgian Ardennes I can confirm
that this is the only species of *Lejeunea* s. str. in the area, where it is locally
common. The new genus *Lejeunea* was based entirely on sporophyte characters:
the whitish capsule, the fragile, articulate seta and the elaters, which remain
fixed at the valves after capsule dehiscence. For Lejeuneaceae systematics this
was a remarkable beginning; each of the genera described since—more than
one hundred in all—was based entirely on the more obvious generation, the
gametophyte.

Tracing the early history of *Lejeunea*, a remarkably fast expansion of the
genus is seen. While 11 species, the majority European, were assigned to the
genus by Dumortier in 1835, about ten years later the *Synopsis Hepaticarum*
(Gottsche et al., 1844–1847) provided descriptions of almost 300 species, most
of them tropical, and by the end of the first period over 500 species had been
assigned to *Lejeunea* (see the excellent recent catalogue of *Lejeunea* by Bischler

and Lamy, 1978). This rapid expansion coincided with the rise of exotic hepaticology in the 1830s and 1840s, characterized by Schuster (1966) as the "Golden Period" of hepaticology.

In the shadow of *Lejeunea*, smaller genera now assigned to the Lejeuneaceae were established. Among the earliest were *Marchesinia* S. Gray and *Phragmicoma* Dum., proposed almost simultaneously for the oceanic European species *Jungermannia mackayi* Hook., and *Frullanoides* Raddi from Brazil. While *Marchesinia* and *Frullanoides* remained long-forgotten names, *Phragmicoma* grew into an artificial "catch-all" for over 50 species, based solely on the ± flattened perianth and the undivided underleaves. Among the tropical Lejeuneaceae studied by the authors of the *Synopsis* some of the more robust, easier-to-grasp species had also been segregated into separate genera: *Bryopteris* and *Omphalanthus* from tropical America and *Ptychanthus* and *Thysananthus* from tropical Asia. Though also based on characters of the perianth—number of keels and ornamentation—and the underleaves, these were more natural units which have largely been retained in their original conception, except for the separation of *Taxilejeunea* from *Omphalanthus*. Characters of perianth and underleaves were also employed to subdivide *Lejeunea* into sections and subsections, but most of these units were very heterogeneous.

The main achievement of the first period was the description of a great number of tropical species new to science. At that time the taxonomic concepts employed in the Lejeuneaceae were based solely on the study of herbarium specimens.

2. Second Period: 1884–1930

We owe the modern concept of the genus in the Lejeuneaceae to the British hepaticologist Spruce.

> A Yorkshire man by birth and death, Richard Spruce was, in the opinion of some, the most outstanding explorer of South America. An amateur botanist who specialized in bryophytes and whose earliest research dealt with the mosses of England and the Pyrenees, Spruce yearned for field work in tropical forests. An unmarried schoolmaster, he decided, at the age of 32, to travel to South America where he remained, in the Amazon and the Andes, for 15 years. Out of these years of collecting, during which this humble and rather shy man penetrated some of the farthest and most forbidding recesses of the Amazon, came material for a masterpiece of scholarly botany, his *Hepaticae of the Amazon and the Andes of Peru and Ecuador* [Schultes, 1970, p. 270].

This work of Spruce's was published in 1884–1885.

Spruce's contributions to Lejeuneaceae systematics have been reviewed frequently, e.g. by Evans (1902–1912) and Schuster (1963b). He was the first to tackle the group with an intimate knowledge of the species in the field, and was able to apply such field characters as habit, growth form, colour, etc. in conjunction with many other characters. Spruce proceeded to reunite all the genera of Lejeuneaceae distinguished in the *Synopsis* into the single genus *Lejeunea* and redivided it into 37 subgenera. To these he assigned over 250 species, more than half of which were new and based on personal collections. Only one species was excluded and placed in a genus of its own: *Myriocolea irrorata* Spruce. This remarkably specialized rheophyte from the Ecuadorian tributaries of the Amazon has never been found again since Spruce discovered it.

Criteria employed by Spruce to define the subgenera include classical ones such as underleaf shape and perianth, but also many new ones, e.g. branching habit, presence or absence of innovations, ocelli and number of antheridia per bract. His efforts resulted in fairly homogeneous groups, although those placed in the "Schizostipae" (divided underleaves) were often more weakly defined than those in the "Holostipae" (undivided underleaves). Spruce suspected that several of his subgenera, especially among the Holostipae, deserved generic status as well and therefore designed a nomenclature which would facilitate change of rank. Thus he named his subgenera by simply prefixing to the name *Lejeunea* some characteristic term: *Sticto-lejeunea* (leaves dotted by ocelli); *Mastigo-lejeunea* (flagelliform branches); *Cerato-lejeunea* (horned perianths); etc.

The subsequent elevation of the subgenera to generic rank was soon brought about when Schiffner (1893) formally transferred them *en bloc* (except for a few which were transferred later by Lacouture and Stephani). But Bonner *et al.* (1961) argued that Stephani had transferred individual subgenera earlier than 1893. The nomenclatural and taxonomic problems involved have frequently been reviewed, e.g. by Jones (1967), and a renewed investigation of the matter is currently being undertaken by Grolle (e.g. Grolle, 1976).

By the turn of the century over 40 genera were recognized. New ones had been added by Goebel, Schiffner and Stephani: *Metzgeriopsis* Goebel, a remarkable thallose liverwort with *Lejeuneaceous* sexual branches, and *Caudalejeunea* (Steph.) Schiffn. and *Anomalolejeunea* (Spruce) Schiffn., both based again on traditional perianth characters. Schiffner and Stephani also contributed by assigning many species described by the authors of the *Synopsis Hepaticarum* to the Sprucean genera (Stephani, 1890; Schiffner, 1894, 1897).

A major contribution to our understanding of the genera was made by Evans (1902–1912), who studied in detail the West Indian Lejeuneaceae and established several new taxonomic characters. His most important new criterion, reviewed by Greig-Smith (1958), was the position of the hyaline papilla at the lobule apex. It allowed him to sharpen the definition of some groups, e.g.

Archilejeunea, Harpalejeunea, and *Cheilolejeunea,* by splitting off small segregates, and it also served to demonstrate the close relationship between some genera placed far apart by Spruce, e.g. *Euosmolejeunea, Cheilolejeunea* and *Strepsilejeunea.* In some cases, though, his genus concept may have been too narrow, for instance in *Cololejeunea* (Benedix, 1953; Mizutani, 1961). Integrating into taxonomy the achievements of the comparative morphologists Leitgeb and Goebel, Evans also payed considerable attention to branch morphology, leaf segmentation patterns and germination of propagules, and thus paved the way for more recent work (e.g. Fulford, 1956; Crandall, 1969).

The completion of the *Species Hepaticarum* by Stephani (1898–1924), a monumental work considering its scope, is probably the most dramatic event in the history of Lejeuneaceae research. Unfortunately, by describing hundreds of new species, many of them ill-founded or placed in the wrong genus, Stephani obscured the limits of the genera carefully designed by Spruce and Evans. It necessitated a new approach, based largely on revising the *Species Hepaticarum.*

3. Third Period: 1930 onward

The taxonomic revisions of Asiatic Lejeuneaceae by Verdoorn (1934) and Herzog and his students (e.g. Herzog, 1930–1939, 1942, 1943; Eifrig, 1937; Benedix, 1953), mark the beginning of modern Lejeuneaceae research. In this period taxonomic work on Lejeuneaceae has predominantly been on individual genera and is of regional nature, although a few monographic works have also been published. The progress that has resulted from these revisions is reviewed below.

Many new genera were proposed, almost doubling the number recognized, but most of these were monotypic and based on new floristic discoveries or generic transfers of previously misunderstood species. Several of them have been reduced to synonymy again (see Table I). Examples of interesting new discoveries are *Aphanotropis* Herz. (Herzog, 1952) from Borneo and *Myriocoleopsis* Schiffn. from Brazil (Gradstein and Vital, 1975), two rare monotypic taxa found exclusively on rocks in streams. Both show morphological specializations which might be interpreted as adaptations to the aquatic habitat: dimorphic growth (creeping vegetative axes, erect sexual stems), reduced leaf lobules and densely crowded gynoecia. Similar adaptations are seen in the peculiar *Myriocolea* Spruce. This grows in the same habitat as *Myriocoleopsis* and also shares with it a massive stem structure.

Only very few new genera were based on new taxonomic criteria, notably

Phragmilejeunea Schust. (= *Schiffneriolejeunea* Verd., *fide* Gradstein, 1974b), which was primarily characterized by the nature of the oil bodies (Schuster and Hattori, 1954). Other new criteria include stem anatomy, studied by Evans (1935) and Bischler (1961–1967), which has in many cases provided better definitions of known genera and new subdivisions. The hyaline papilla has served to improve the taxonomy of the heterogeneous *Pycnolejeunea* and led to the establishment of *Nipponolejeunea* Hatt. and *Tuyamaella* Hatt., each probably best placed in a distinct subfamily (Kachroo and Schuster, 1961).

The current state of generic delimitation is different in each of the three main subfamilies: Ptychanthoideae, Lejeuneoideae and Cololejeuneoideae. In the Ptychanthoideae circumscriptions established by Spruce and Evans have largely been retained, and only minor improvements have been proposed (Gradstein, 1975). In Cololejeuneoideae the work by Benedix (1953) on Asiatic *Cololejeunea* has led current authors to accept a much broader circumscription for this genus than that employed previously by Evans for tropical America. The subdivision of this large genus is still a major problem, for which different solutions have been proposed by Benedix (1953), Mizutani (1961), Schuster (1963b) and Vanden Berghen (1977).

A somewhat chaotic situation currently exists in the Lejeuneoideae, by far the largest subfamily and considered by some the most difficult group of liverworts as to generic delimitation. The problems involved have been discussed at length by Herzog (1951) and Schuster (1963b). The greatest difficulties seem to be posed by *Lejeunea* s. str. and its allies *Taxilejeunea, Hygrolejeunea, Crossotolejeunea, Microlejeunea* and others, comprising several hundred species. For this group—the *"Lejeunea* complex" *sensu* Schuster—no satisfactory world-wide generic or subgeneric classification exists. For Asiatic *Lejeunea* a very broad concept is currently being employed by Mizutani (e.g. 1970a), who reduced most species of *Taxilejeunea, Hygrolejeunea* and smaller segregates back into *Lejeunea*, whereas for Africa and America the Sprucean segregates are still being retained, though often reluctantly. Spruce himself seems to have been well aware of the weaknesses of his subgenera in this group, as is evident from their long diagnoses (Spruce, 1884–1885). The different generic concepts currently employed for this subfamily are shown in Fig. 1.

(a) *Regional revisions. Asia.* Verdoorn's revision of the entire "Holostipae" for Asia and adjacent areas (Verdoorn, 1934), treating over 500 names in 14 genera, was the first major modern taxonomic revision of Lejeuneaceae. Though some of his genera have proved somewhat heterogeneous, e.g. *Ptychocoleus* (Gradstein, 1975), his treatment nevertheless remains an indispensable identification tool for this group. Herzog and his students provided important studies on large schizostipous genera, e.g. *Drepanolejeunea* (Herzog, 1930–1939), *Leptolejeunea* (Herzog, 1942), *Taxilejeunea* (Eifrig, 1937) and *Cololejeunea* (Benedix, 1953). More recent studies include those by Chen and Wu (1964) on the epiphyllous Lejeuneaceae of China, a revision of *Diplasiolejeunea* (Grolle, 1966b), and the outstanding treatment of Japanese Lejeuneaceae by Mizutani (1961), who has become the present authority for Asiatic Lejeuneaceae. His numerous subsequent taxonomic studies include papers on the Lejeuneaceae of North Borneo (Mizutani, 1966, 1969, 1970a), on Himalayan *Lejeunea* (Mizutani, 1971), on *Harpalejeunea* (Mizutani, 1973) and on critical Stephani species (e.g. Mizutani, 1964, 1967, 1972, 1976). His *Harpalejeunea* study is particularly remarkable because it demonstrated that none of the Asiatic species previously assigned to *Harpalejeunea* belonged there. Nevertheless, seven species could be assigned to that genus, some transferred from other genera, others newly discovered during Mizutani's expedition to Borneo.

Pacific and Australasia. Taxonomic revisions from here are scarce. Apart from the *Colura* and *Acrolejeunea* monographs (Jovet-Ast, 1953; Gradstein, 1975), which include species from these areas, we might mention Verdoorn's Holostipae revision, Miller *et al.* (1963) on the Micronesian atolls, Schuster (1963a) on Australasian groups, and treatments of *Microlejeunea* (Bischler, *et al.*, 1963; Miller *et al.*, 1963, 1967), *Phaeolejeunea* (Mizutani, 1968b), Tuyamaelloideae (Grolle, 1973b) and *Diplasiolejeunea* (Grolle, 1975).

Tropical America. Little revisionary work has been done for the rich neotropical area since the appearance of the *Species Hepaticarum*. Generic revisions include those on *Thysananthus*, *Symbiezidium* and *Ceratolejeunea* by Fulford (1941, 1942, 1945)—the wartime *Ceratolejeunea* paper was based on a limited number of type specimens—and those on *Microlejeunea* by Bischler *et al.* (1963), and on *Leptolejeunea* and the large genus *Drepanolejeunea* by Bischler (1964, 1969). The important review of North American Lejeuneaceae by Schuster (1955–1967) discusses nine different schizostipous genera and provides useful keys to West Indian *Aphanolejeunea* and *Lejeunea* s. str., the latter genus not having been treated by Evans.

Africa. Taxonomic work on African Lejeuneaceae began relatively recently, but our knowledge of the species of this area is already reasonably complete. This is not because of the supposed poverty of this continent, but through the continuing efforts by Vanden Berghen (e.g. 1948, 1951, 1962, 1972, 1977) and Jones (e.g. 1953, 1954, 1957, 1970, 1976), who have prepared many generic revisions as well as notes on individual species. Much is to be learned from the discussions of field characters and environmental modifications by Jones, while as a tool for primary identification Vanden Berghen's flora of Shaba and Zambia (Vanden Berghen, 1972a) is indispensable. For South Africa Arnell (1963) has provided a useful treatment of the family. There has also been much work on the tiny, foliicolous groups, the results of which have been reviewed by Pócs (1978). As in all other areas, the real problems remain in *Lejeunea* and its allies, parts of which were revised recently by Jones (1967, 1968, 1969, 1972).

Since the *Species Hepaticarum* our taxonomic knowledge has become most complete for Africa. The Asiatic Lejeuneaceae have also become reasonably well-known, due principally to the efforts of Verdoorn, Herzog and Mizutani, while the least progress has been made in tropical America, the area where the basis of our present Lejeuneaceae taxonomy had been laid by Spruce and Evans.

(*b*) *Monographs*. "It should be emphasized that the basic need at present is monographic work on the various genera" (Schuster, 1963b, p. 25). For the Lejeuneaceae this is still an understatement. The monograph of the large pantropical genus *Colura* (Jovet-Ast, 1953) was for a long time the only one of its kind for this family. Recently, however, *Rhaphidolejeunea* (Bischler, 1968b), *Tuyamaella* (Tixier, 1973), *Bryopteris* (Stotler and Crandall-Stotler, 1974) and *Acrolejeunea* (Gradstein, 1975), among the larger groups, have also been monographed.

Much can be learned from these studies about the delimitation and distri-bution of species. Today most bryologists are well aware that widespread genera may include species whose areas of distribution cover more than one continent. In the Lejeuneaceae this is particularly true of pantropical groups, and several pluricontinental species are now known, e.g. *Acrolejeunea pycnoclada* (Tayl.) Schiffn., *Cheilolejeunea trifaria* (Reinw., Blume and Nees) Mizutani, *Cololejeunea minutissima* (Sm.) Schiffn., *Colura tenuicornis* (Evans) Steph., *Diplasiolejeunea cavifolia* Steph., *Lejeunea flava* (Sw.) Nees, *Leucolejeunea xan-thocarpa* (Lehm. and Lindenb.) Evans, *Lopholejeunea subfusca* (Nees) Steph. and *Microlejeunea ulicina* (Tayl.) Evans. A single species is often known from different continents by different names, as Pócs (1976) has recently pointed out for Africa and Asia. Jones and Vanden Berghen have repeatedly suggested Afro-american species links in the Lejeuneaceae and relationships between the floras of Nepal and tropical America also exist (Grolle, 1966a).

The known areas of distribution are often wider among the more robust, epiphytic species than in the tiny, foliicolous groups such as *Drepanolejeunea* (Bischler, 1964). However, many foliicolous endemics are based on single specimens (see e.g. Tixier, 1968), and the larger groups have, at best, only been treated regionally, *Colura* excepted. Large areas of the tropics are still under-explored and the smaller or less conspicuous species, including tiny Lejeuneaceae, are often underrepresented in collections, as Touw (1974) has pointed out. Much work remains to be done before the distribution patterns of species in the Lejeuneaceae can be established, and before they can be used to interpret the evolutionary history of the genera and of the family as a whole.

DEVELOPMENT OF NEW CRITERIA

1. Introduction

Mizutani (1961) and Schuster (1963b) have reviewed the taxonomic criteria employed for Lejeuneaceae. Recently, I have discussed newer concepts (Grad-stein, 1975), which include characters of merophyte topography, stem anatomy, branching, oil bodies, trigones, underleaf base anatomy and various sporophyte structures. While the taxonomic relevance of some of these, e.g. stem anatomy and oil bodies, has become firmly established, for other characters this is not the case. A poorly understood character, for instance, is the topography of the merophyte boundaries, introduced by Evans (1935) and employed success-fully to delimit some other taxonomic groups, e.g. *Gottschelia* Grolle and the family Gymnomitriaceae (Grolle, 1971a). In Lejeuneaceae this character has been neglected, although Mizutani (1968a) recently indicated the presence of at least eight different merophyte patterns in Lejeuneaceae.

Sporophytes have not been used for generic delimitation since Libert's (1820) description of *Lejeunea*, although in most other liverwort groups sporo-phyte characters are nowadays considered to be essential for the genus concept (Schuster, 1972; Hässel de Menéndez, 1976; Inoue, 1976). In Lejeuneaceae seta anatomy and the capsule wall have provided important criteria for recognizing subfamilies but at the generic level sound characters derived from the sporo-phyte have not yet been established. This has prompted Schuster (1972) to call for "massive reduction of genera to subgeneric or sectional status" in the Lejeuneaceae, a view with which I cannot agree because sporophytes have carefully been investigated in only very few genera. In at least a third of the genera presently recognized sporophytes remain entirely unknown.

The need for more careful investigation of liverwort sporophytes seems particularly evident in the Lejeuneaceae. Recent studies of the sporophyte in *Bryopteris* (Crandall, 1967) and *Dicranolejeunea* (Stotler and Crandall, 1969) may serve as a model.

In this review I have selected for more detailed discussion two new aspects of comparative morphology and anatomy that seem promising for generic delimitation: the anatomy of the underleaf base anatomy and certain new characters of the branches.

2. Rhizoids and the Anatomy of the Underleaf Base

The Lejeuneaceae always develop their rhizoids at the base of the underleaves. In most genera they are produced in loose fascicles but in foliicolous taxa they form a large, mucilaginous, circular rhizoid disc (Fig. 2b), called "Haftscheibe" (Herzog, 1925), "disc adhesif" (Bischler, 1968a), "secondäre Rhizoidplatte" (Winkler, 1967), or "secondary rhizoid disc" (Gradstein, 1975).

Bischler (1968a) demonstrated that in foliicolous *Drepanolejeunea crucianella* (Taylor) Evans, secondary rhizoid discs are produced only when the stems are tightly appressed to the substrate; aereal stems of the same plant produce fascicled rhizoids. This indicates that the development of these peculiar discs is influenced by the environment. Similarly, the often observed proliferation of the rhizoid apex into a hand-shaped pattern is environmentally induced (Odu and Richards, 1976). Rhizoids originate in the Lejeuneaceae from a small disc of bulging initial cells on the base of the underleaf (Fig. 2a). This disc is usually referred to as "paramphigastrium" following Schiffner (1929), but "primäre Rhizoidplatte" (Winkler, 1967) or "primary rhizoid disc" (Gradstein, 1975) are better terms since they are related to function.

Winkler (1967, 1970) and Bischler (1969) have independently called attention to the taxonomic significance of the anatomy of the underleaf base and its associated rhizoid discs. In transverse section it appears that the underleaf is attached to the stem by large, U-shaped cells (Fig. 2c–e), the so-called "superior central cells" (Gradstein, 1975). These cells are involved in sustaining the primary rhizoid disc and may influence the development of this structure, although morphogenetic evidence for this assumption is still lacking. The number of superior central cells seems to be very constant and taxonomically relevant. While in the tiny Lejeuneoideae there are usually only two, the more robust genera appear to have four superior central cells. Apparently this character is correlated with the number of cortical cells constituting the width of the ventral merophyte and thus with the distinction between the artificial groups Holostipae and Schizostipae. But the number of superior cells seems a much more constant and reliable character; anatomical structures are often stabler than correlated external morphological features.

Transverse sections further show that the underleaf base is bistratose (Fig. 2d) or tristratose (Fig. 2e). The bistratose condition is the more common, but the tristratose underleaf base has been found in *Symbiezidium* and *Odontolejeunea* (Winkler, 1970) and in *Verdoornianthus* (Gradstein, 1978). The polystratose underleaf bases of *Caudalejeunea cristiloba* (Steph.) Gradstein are exceptional (Gradstein, 1974a). In *Odontolejeunea* Winkler (1970) showed that the tristratose underleaf base may become elongated like a stalk, a phenomenon which might be associated with the formation of secondary rhizoid discs in this genus.

The structure of the underleaf base is a systematic character which has been neglected in the Lejeuneaceae and overlooked altogether in *Frullania* and its allies which also produce rhizoids from underleaf bases. Students of both groups should take these interesting structures into account.

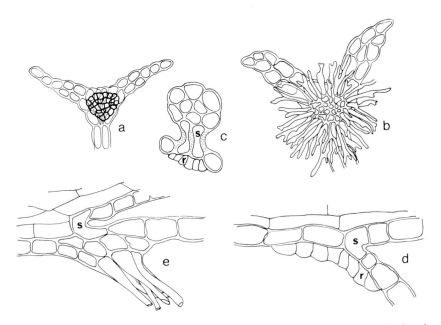

FIG. 2. Rhizoids and underleaf base anatomy in the Lejeuneaceae. a. Underleaf with primary rhizoid disc in *Drepanolejeunea thwaitesiana* (Mitt.) Steph. b. Underleaf with secondary rhizoid disc in *Drepanolejeunea dentistipula* Steph. c. Stem in cross-section, showing two superior central cells subtending the primary rhizoid disc in *Drepanolejeunea dentistipula*. d. Stem in longitudinal section, showing superior central cell and bistratose underleaf base in *Neurolejeunea breutelii* (Gottsche) Evans. e. Stem in longitudinal section, showing tristratose underleaf base and fascicled rhizoids in *Symbiezidium transversale* (Sw.) Trev. (r: rhizoid initial cell; s: superior central cell). a–c are reproduced with permission from Bischler (1969); d–e with permission from Winkler (1970).

3. Branching

Branch morphology is another good illustration of the structural diversity displayed by the Lejeuneaceae. The most important recent work on this subject is by Crandall (1969), who carefully analysed branch structure and development and recognized eleven different branch types in leafy liverworts. A useful key to these branch types, based on external morphology, was published separately (Crandall-Stotler, 1972). Almost simultaneously Mizutani (1970b) published a comprehensive paper on branching in the Lejeuneaceae, based exclusively on external morphology. Mizutani's paper elegantly complements Crandall's morphogenetic data.

Of the 11 branch types distinguished by Crandall, four occur in the Lejeuneaceae: the *Frullania*-type, the *Radula*-type, the *Lejeunea*-type and the *Bryopteris*-type. The *Lejeunea*-type is the most characteristic branch type: while common in the Lejeuneaceae it is not known to exist outside the family. Reports of *Lejeunea*-type branching in the allied Jubulaceae (Schuster, 1970) probably involve the rather similar *Bryopteris*-type branch.

The development of *Lejeunea*-type branches has been the subject of much controversy. The *Lejeunea*-type branch originates from a cortical cell like the allied *Radula*- and *Bryopteris*-type branches (Fulford and Crandall, 1967), but it differs in its basal collar which, according to Crandall (1969), is derived from "leaf brace-cells". These are cells of the decurrent base of a stem leaf associated with the branch. Upon branch formation, the brace-cells are ruptured by the expanding juvenile branch and develop into a lobed collar at the branch base. According to Crandall, leaf brace-cells are found throughout the Lejeuneaceae except in *Bryopteris*, which consequently does not have *Lejeunea*-type branches. This genus—placed in a separate subfamily (Gradstein, 1975) or in a family of its own (Stotler and Crandall-Stotler, 1974)—has instead the *Bryopteris*-type branch, which closely resembles the *Lejeunea*-type branch except for its collar which is practically unlobed and of cortical origin.

Frullania-type branches are probably the most common in liverworts, but in the Lejeuneaceae are restricted to the more robust members of the family: Bryopteridoideae, Nipponolejeuneoideae and Ptychanthoideae. Its presence is sometimes helpful for delimiting genera but in some groups their occurrence seems to be merely the result of luxuriant growth (Jones, 1970; Gradstein, 1975).

The morphology of the first leaf cycle at the base of the *Frullania*-type branches is taxonomically interesting. Many authors have noticed the aberrant and often reduced shape of the first leaf and underleaf. This phenomenon provided the basis for the recognition of three different patterns in the Lejeuneaceae by Mizutani (1970b):

1. *Frullania-Ptychanthus* pattern (Fig. 3a), occurring in various genera of Ptychanthoideae and in *Bryopteris*.
2. *Frullania-Jubula* pattern (Fig. 3b), occurring in e.g. *Stictolejeunea* and in *Nipponolejeunea subalpina* (*Nipponolejeunea* subg. *Mizutania* Inoue).
3. *Frullania-Nipponolejeunea* pattern (Fig. 3c), occurring in *Nipponolejeunea pilifera* (subg. *Nipponolejeunea*).

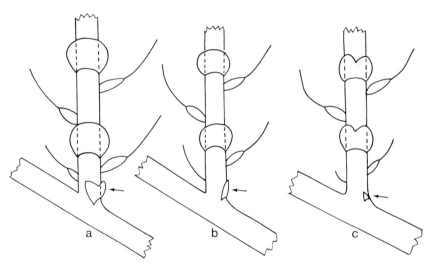

FIG. 3. Schematic representation of *Frullania*-type branch morphology in the Lejeuneaceae, showing the variation of patterns in the first leaf cycle (arrow). a. *Frullania-Ptychanthus* pattern branch. b. *Frullania-Jubula* pattern branch. c. *Frullania-Nipponolejeunea* pattern branch. Modified from Mizutani (1970b).

In branches of the *Frullania-Ptychanthus* pattern the first leafy appendage, which is an underleaf, is large and bilobed. In the *Frullania-Jubula* pattern it is smaller, undivided and scale-like, and in the *Frullania-Nipponolejeunea* pattern it is almost reduced. Similarly, the second leafy appendage, which is the first acroscopic leaf (located on the side of the branch towards the apex of the main stem) is well-differentiated into lobe and lobule in the *Frullania-Ptychanthus* pattern but in the *Frullania-Jubula* pattern it is scale-like and in the *Frullania-Nipponolejeunea* pattern it is almost absent. Further abnormalities are seen in the first underleaf, which is clearly displaced towards the basiscopic side of the branch base (located on the side of the branch towards the base of the stem).

In the *Frullania-Ptychanthus* pattern the insertion line of the bifid first under-leaf extends almost up to the dorsal midline of the branch, which led Mizutani and me (Gradstein, 1975) to believe that it represents an underleaf fused with a modified basiscopic leaf. This viewpoint is hard to correlate, however, with the ordinary sequence of leaf development in young *Frullania*-type branches. As Crandall (1969) pointed out, the first leaf initial cell cut off forms an underleaf, the second an acroscopic leaf, the third a basiscopic leaf, the fourth an under-leaf, etc. If our idea of fusion of underleaf and basiscopic leaf to form the first branch appendage were correct, we would have to assume that the first and second initials had become suppressed in *Frullania-Ptychanthus* pattern branches, or that the segmentation pattern had gone wrong. Neither possibility is supported by our present knowledge of branch ontogeny.

The abnormalities described in the first leaf cycle are, according to Crandall, the result of the slightly oblique orientation of the first leaf initials, which induces displacement and abnormal leaf development. In taxonomy, these abnormalities seem to be significant for distinguishing species, e.g. in *Frullania* (Stotler, 1969; Vanden Berghen, 1976) or for delimiting genera or subgenera of the Lejeuneaceae as Mizutani has shown.

Radula-type branching is for the taxonomy of the Lejeuneaceae probably the most important branching pattern because it forms the subgynoecial innova-tion. Since Spruce, presence or absence of innovations has successfully been employed as a tool for distinguishing genera and subgenera except in some groups of Lejeuneoideae, such as the *Lejeunea* complex (Schuster, 1963b), in which almost all our generic criteria seem to break down.

A new aspect of innovation morphology was described independently by Crandall (1969) and Mizutani (1970b). These authors showed that in the Lejeune-aceae *Radula*-type innovations exhibit two different kinds of spiral leaf-segmen-tation: dextrorse spirals and sinistrorse spirals. While in other branch types the presence of dextrorse or sinistrorse spirals is normally correlated with the side of the stem from which the branch originates, in the *Radula*-type innovation this is not so: a dextrorse or a sinistrorse leaf spiral may originate from either side of the stem.

On this basis two *Radula*-type branching patterns were described by Mizutani (1970b):

1. *Radula-Lejeunea* innovation (Fig. 4a). The first leaf formed is a basiscopic leaf, the second is an underleaf and the third is an acroscopic leaf. Consequently the leaf spiral is dextrorse on innovations originating from the right-hand side of the stem and sinistrorse on innovations on the left-hand side (if viewed from the ventral surface of the plant).

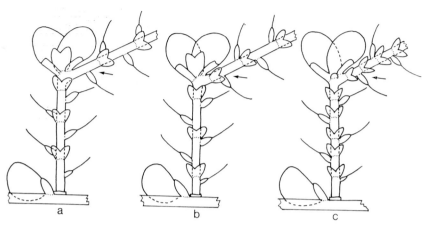

Fig. 4. Schematic representation of the main innovation types in the Lejeuneaceae. a. *Radula-Lejeunea* innovation. b. *Radula-Jubula* innovation. c. Innovation in *Diplasiolejeunea*, showing the pendular leaf segmentation characteristic of *Colo-lejeuneoideae*. Reproduced with permission from Mizutani (1970b).

Mizutani found this type of innovation in *Ptychanthus, Thysananthus, Spruceanthus, Taxilejeunea, Lejeunea* and *Harpalejeunea*, while Crandall illustrated it for *Dicranolejeunea axillaris* (Nees and Mont.) Schiffn.

2. *Radula-Jubula* innovation (Fig. 4b). The first leaf formed is an under-leaf, the second is a basiscopic leaf and the third is an acroscopic leaf. Consequently the leaf spiral is sinistrorse on innovations originating from the right-hand side of the stem and dextrorse on innovations from the left-hand side, the reverse of the pattern of the segmentation in the *Radula-Lejeunea* innovation. The *Radula-Jubula* innovation was observed by Mizutani in *Nipponolejeunea, Brachiolejeunea, Archilejeunea, Neurolejeunea, Leucolejeunea, Ceratolejeunea, Drepanolejeunea* and *Siphono-lejeunea*, while Crandall illustrated it for *Stictolejeunea squamata*.

Distinguishing between the two patterns is surprisingly simple: the *Radula-Lejeunea* innovation starts at its base with a leaf, whereas the *Radula-Jubula* innovation starts with an underleaf. From the evidence presented by Mizutani it seems that in genera or subgenera of the Lejeuneaceae either the one or the other innovation type prevails, and thus an interesting new taxonomic character has come to light. This conclusion was substantiated by Riclef Grolle (personal communication) who has checked this character in a considerable number of taxa.

Species of *Cheilolejeunea* subg. *Euosmolejeunea* with two innovations per gynoecium may exceptionally develop both segmentation types. When this happens, one of the innovations is of the *Radula-Lejeunea* type while the other is of the *Radula-Jubula* type. The two segmentation patterns are of course only seen in plants with ordinary spiral leaf-segmentation. Thus they are lacking in the Cololejeuneoideae, which have pendular instead of spiral leaf-segmentation (Fig. 4c).

ACKNOWLEDGEMENTS

I am much indebted to Dr R. Grolle (Jena) for drawing my attention to the paper by Mizutani on branching in Lejeuneaceae and for his critical reading of the manuscript and to Drs S. Hattori (Nichinan) and H. Inoue (Tokyo) for providing translations of Mizutani's work. Thanks are also due to Mr T. Schipper for the figures.

REFERENCES

ARNELL, S. (1963). "Hepaticae of South Africa". Swedish Natural Research Council, Stockholm.

BENEDIX, E. H. (1953). Idomalayische Cololejeuneen. *Feddes Reprium Beih.* **134**, 1–88.

BISCHLER, H. (1961–1967). Recherches sur l'anatomie de la tige chez les Lejeuneaceae, 1–3. *Revue bryol. lichén.* **30**, 232–252; **33**, 399–458; **34**, 601–675.

BISCHLER, H. (1964). Le genre *Drepanolejeunea* Steph. en Amérique Centrale et Méridionale. *Revue bryol. lichén.* **33**, 15–179.

BISCHLER, H. (1968a). Notes sur l'anatomie des amphigastres et sur le développement du paramphigastre et des rhizoïdes chez *Drepano-*, *Rhaphido-* et *Leptolejeunea*. *Revue bryol. lichén.* **36**, 45–55.

BISCHLER, H. (1968b). Monographie du genre *Rhaphidolejeunea* Herzog. *Revue bryol. lichén.* **36**, 56–104.

BISCHLER, H. (1969). Le genre *Leptolejeunea* (Spruce) Steph. en Amérique. *Nova Hedwigia* **17**, 265–350.

BISCHLER, H. and LAMY, D. (1978). *Jungermanniopsis* to *Lejeunites*. *In* C. E. B. Bonner, "Index Hepaticarum" Vol. 9, 405–745. J. Cramer, Vaduz.

BISCHLER, H., BONNER, C. E. B. and MILLER, H. A. (1963). Studies in Lejeuneaceae, VI. The genus *Microlejeunea* Steph. in Central and South America. *Nova Hedwigia* **5**, 359–411.

BONNER, C. E. B. and MILLER, H. A. (1960). Studies in Lejeuneaceae, I. The typification of *Lejeunea*. *Bryologist* **63**, 217–225.

BONNER, C. E. B., BISCHLER, H. and MILLER, H. A. (1961). Studies in Lejeuneaceae, II. The transition subgenus-genus of Spruce's segregates of *Lejeunea*. *Nova Hedwigia* **3**, 351–359.

CHEN, P.-C. and WU, P.-C. (1964). Study on epiphyllous liverworts of China, I. *Acta phytotax. sin.* **9**, 213–276.

CRANDALL, B. J. (1967). The sporophyte and sporeling of *Bryopteris filicina* (Sw.) Nees. *Bryologist* **70**, 423–431.

CRANDALL, B. J. (1969). Morphology and development of branches in the leafy Hepaticae. *Beih. nov. Hedwigia* **30**, 1–261.

CRANDALL-STOTLER, B. J. (1972). Morphogenetic patterns of branch formation in the leafy Hepaticae—A résumé. *Bryologist* **75**, 381–403.

DUMORTIER, B. C. (1835). "Recueil d'observations sur les Jungermanniacées", Vol. 1. J.-A. Blanquart, Tournay.

EIFRIG, H. (1937). Monografische Studien über die Indomalayischen Arten von *Taxilejeunea*. *Annls. bryol.* **9**, 73–114.

EVANS, A. W. (1902–12). Hepaticae of Puerto Rico, 1–12. *Bull. Torrey bot. Club* **29**, 496–510; **30**, 19–41, 544–563; **31**, 183–226; **32**, 273–290; **33**, 1–25; **34**, 1–34, 533–568; **35**, 155–179; **38**, 251–286; **39**, 209–225.

EVANS, A. W. (1935). The anatomy of the stem in the Lejeuneaceae. *Bull. Torrey bot. Club* **62**, 187–214; 259–280.

FULFORD, M. H. (1941). Studies on American Hepaticae, I. Revision of the genus *Thysananthus*. *Bull. Torrey bot. Club* **68**, 32–42.

FULFORD, M. H. (1942). Studies on American Hepaticae, IV. A revision of the genus *Symbiezidium*. *Lloydia* **5**, 293–304.

FULFORD, M. H. (1945). Studies on American Hepaticae, VI. *Ceratolejeunea*. *Brittonia* **5**, 368–403.

FULFORD, M. H. (1956). The young stages of the leafy Hepaticae. A résumé. *Phytomorphology* **6**, 199–235.

FULFORD, M. H. and CRANDALL, B. (1967). The origin of the *Lejeunea*-type branches in *Brachiolejeunea laxifolia*. *Phytomorphology* **17**, 58–61.

GOTTSCHE, C. M., LINDENBERG, J. B. G. and NEES VON ESENBECK, C. G. (1844–1847). "Synopsis Hepaticarum". Meissner, Hamburg.

GRADSTEIN, S. R. (1974a). Studies on Lejeuneaceae subf. Ptychanthoideae (Hepaticae), II. Two remarkable species of *Caudalejeunea*: *C. grolleana* spec. nov. and *C. cristiloba* (Steph.) comb. nov. *Acta bot. neerl.* **23**, 333–343.

GRADSTEIN, S. R. (1974b). Studies on Lejeuneaceae subfam. Ptychanthoideae, I. Nomenclature and taxonomy of *Ptychocoleus*, *Acrolejeunea* and *Schiffneriolejeunea*. *J. Hattori bot. Lab.* **38**, 327–336.

GRADSTEIN, S. R. (1975). A taxonomic monograph of the genus *Acrolejeunea* with an arrangement of the genera of Ptychanthoideae. *Bryophyt. Biblthca* **4**, 1–162.

GRADSTEIN, S. R. (1978). Studies on Lejeuneaceae subfam. Ptychanthoideae, IV. *Verdoornianthus*, a new genus from Amazonas, Brazil. *Bryologist* **80**, 606–611.

GRADSTEIN, S. R. and VITAL, D. M. (1975). On *Myriocoleopsis* Schiffn. (Lejeuneaceae). *Lindbergia* **3**, 39–45.

GREIG-SMITH, P. (1958). Notes on Lejeuneaceae, III. The occurrence of hyaline papillae. *Trans. Br. bryol. Soc.* **3**, 418–421.

GROLLE, R. (1966a). Die Lebermoose Nepals. *Ergebn. Forschunternehmens Nepal Himalaya* **1**, 262–298.

GROLLE, R. (1966b). Über *Diplasiolejeunea* in Asien. *Feddes Reprium* **73**, 78–89.

GROLLE, R. (1971a). *Jamesoniella* und Verwandte. *Feddes Reprium* **82**, 1–99.

GROLLE, R. (1971b). Miscellanea hepaticologica 111–120. *Trans. Br. bryol. Soc.* **6**, 285–265.

GROLLE, R. (1973a). Nomina conservanda proposita. (366) *Lejeunea* Libert. *Taxon* **22**, 689–690.

GROLLE, R. (1973b). *Nephelolejeunea*—eine neue Gattung der Tuyamaelloideae. *J. Hattori bot. Lab.* **37**, 251–261.

GROLLE, R. (1975). *Diplasiolejeunea* in Australiasien. *Feddes Reprium* **86**, 75–82.

GROLLE, R. (1976). Eine weitere *Siphonolejeunea*—*S. elegantissima* (Steph.) comb. nov. aus Australien. *J. Hattori bot. Lab.* **41**, 405–409.

HÄSSEL DE MENENDEZ, G. G. (1976). Taxonomic problems and progress in the study of the Hepaticae. *J. Hattori bot. Lab.* **41**, 19–36.

HERZOG, T. (1925). Anatomie der Lebermoose. *In* "Handbuch der Pflanzenanatomie" (K. Linsbauer, ed.) Vol. 17 (1), 1–112. Borntraeger, Berlin.

HERZOG, T. (1930–1939). Studien über *Drepanolejeunea*, 1–4. *Annls bryol.* **3**, 126–149; **7**, 57–94; **9**, 115–130; **12**, 98–122.

HERZOG, T. (1942). Revision der Lebermoosgattung *Leptolejeunea* Spr. in der Indomalaya. *Flora, Jena* **135**, 377–434.

HERZOG, T. (1943). *Rhaphidolejeunea* Herz., eine neue Lejeuneaceengattung der Indomalaya. *Mitt. thüring. bot. Ver.* **50**, 100–105.

HERZOG, T. (1951). Kritik des Lejeuneaceensystems. *Feddes Reprium* **54**, 172–184.

HERZOG, T. (1952). *Aphanotropis* Herz., eine neue Gattung der Lejeuneaceae aus Borneo. *Trans. Br. bryol. Soc.* **2**, 62–65.

INOUE, H. (1976). The concept of genus in the Plagiochilaceae. *J. Hattori bot. Lab.* **41**, 13–17.

JONES, E. W. (1953). African hepatics, II. *Leptocolea* with hyaline-margined leaves. *Trans. Br. bryol. Soc.* **2**, 144–157.

JONES, E. W. (1954). African hepatics, VII. The genus *Cheilolejeunea*. *Trans. Br. bryol. Soc.* **2**, 380–392.

JONES, E. W. (1957). African hepatics, XII. Some new or little known Lejeuneaceae. *Trans. Br. bryol. Soc.* **3**, 191–207.

JONES, E. W. (1967). African hepatics, XVIII. *Taxilejeunea* and *Lejeunea* with eplicate perianths. *Trans. Br. bryol. Soc.* **5**, 289–304.

JONES, E. W. (1968). African hepatics, XIX. The *Lejeunea flava* complex. *Trans. Br. bryol. Soc.* **5**, 548–562.

JONES, E. W. (1969). African hepatics, XXI. *Microlejeunea*, *Chaetolejeunea* and *Pleurolejeunea*. *Trans. Br. bryol. Soc.* **5**, 775–789.

JONES, E. W. (1970). African hepatics, XXII. *Dicranolejeunea* and *Marchesinia*. *Trans. Br. bryol. Soc.* **6**, 72–81.

JONES, E. W. (1972). African hepatics, XXIII. Some species of *Lejeunea*. *J. Bryol.* **7**, 23–45.

JONES, E. W. (1976). African hepatics, XXIX. Some new or little-known species and extensions of range. *J. Bryol.* **9**, 43–54.

JOVET-AST, S. (1953). Le genre *Colura*, Hépatiques. Lejeuneaceae, Diplasiae. *Revue bryol. lichén.* **22**, 206–312.

KACHROO, P. and SCHUSTER, R. M. (1961). The genus *Pycnolejeunea* and its affinities to *Cheilolejeunea*, *Euosmolejeunea*, *Nipponolejeunea*, *Tuyamaella*, *Siphonolejeunea* and *Strepsilejeunea*. *J. Linn. Soc., Bot.* **56**, 475–511.

LIBERT, M. (1820). Sur un genre nouveau d'Hépatiques, *Lejeunia*. *Annls gén. Sci. phys. Brux.* **6**, 372–374.

MILLER, H. A., BONNER, C. E. B. and BISCHLER, H. (1963). Studies in Lejeuneaceae,

V. *Microlejeunea* in Pacific Oceania. *Nova Hedwigia* **4**, 551–561.

MILLER, H. A., BONNER, C. E. B. and BISCHLER, H. (1967). Studies in Lejeuneaceae, VIII. *Microlejeunea* in Asia and Australasia. *Nova Hedwigia* **14**, 61–67.

MILLER, H. A., WHITTIER, H. O. and BONNER, C. E. B. (1963). Bryoflora of the atolls of Micronesia. *Beith. nov. Hedwigia.* **11**, 1–89.

MIZUTANI, M. (1961). A revision of Japanese Lejeuneaceae. *J. Hattori bot. Lab.* **24**, 115–302.

MIZUTANI, M. (1964). Studies of little known Asiatic species of Hepaticae in the Stephani herbarium, 1. On some little known southeast Asiatic species of the family Lejeuneaceae. *J. Hattori bot. Lab.* **27**, 139–148.

MIZUTANI, M. (1966). Epiphyllous species of Lejeuneaceae from Sabah (North Borneo). *J. Hattori bot. Lab.* **29**, 153–170.

MIZUTANI, M. (1967). Studies of little known Asiatic species of Hepaticae in the Stephani herbarium, 3. On some little known species of *Cheilolejeunea, Euosmolejeunea* and *Pycnolejeunea. J. Hattori bot. Lab.* **30**, 171–180.

MIZUTANI, M. (1968a). Types of insertion lines of leaves and underleaves to the stem in Lejeuneaceae. *Miscnea bryol. lichen., Nichinan* **5**, 21–22.

MIZUTANI, M. (1968b). Studies of little known Asiatic species of Hepaticae in the Stephani herbarium, 4. *Phaeolejeunea*, a new genus of Lejeuneaceae. *J. Hattori bot. Lab.* **31**, 130–134.

MIZUTANI, M. (1969). Lejeuneaceae subfamily Ptychanthoideae from Sabah (North Borneo). *J. Hattori bot. Lab.* **32**, 129–139.

MIZUTANI, M. (1970a). Lejeuneaceae subfamilies Lejeuneoideae and Cololejeuneoideae from Sabah (North Borneo). *J. Hattori bot. Lab.* **33**, 225–265.

MIZUTANI, M. (1970b). Branching types of Lejeuneaceae. *Miscnea bryol. lichen., Nichinan* **5**, 81–90.

MIZUTANI, M. (1971). *Lejeunea* from the Himalayan region. *J. Hattori bot. Lab.* **34**, 445–457.

MIZUTANI, M. (1972). Studies of little known Asiatic species of Hepaticae in the Stephani herbarium, 7. Some little known species of the subfamily Lejeuneoideae of the Lejeuneaceae. *J. Hattori bot. Lab.* **35**, 399–411.

MIZUTANI, M. (1973). The genus *Harpalejeunea* from Sabah (North Borneo). *J. Hattori bot. Lab.* **37**, 191–203.

MIZUTANI, M. (1976). Studies of little known Asiatic species of Hepaticae in the Stephani herbarium, 9. Some little known species of the family Lejeuneaceae. *J. Hattori bot. Lab.* **40**, 441–446.

ODU, E. and RICHARDS, P. W. (1976). The stimulus to branching of the rhizoid tip in *Lophocolea cuspidata* (Nees) Limpr. *J. Bryol.* **9**, 93–95.

PÓCS, T. (1976). Correlations between the tropical African and Asian bryofloras, I. *J. Hattori bot. Lab.* **41**, 95–106.

PÓCS, T. (1978). Epiphyllous communities and their distribution in East Africa. *Bryophyt. Biblthca* **13**, 681–713.

SCHIFFNER, V. (1893). Hepaticae (Lebermoose). *In* "Die natürlichen Pflanzenfamilien" (A. Engler, ed.). Teil 1, Abt. 3 (1), 3–141. W. Engelmann, Leipzig.

SCHIFFNER, V. (1894). Revision der Gattungen *Bryopteris, Thysananthus, Ptychanthus* und *Phragmicoma* im Herbarium des Berliner Museums. *Hedwigia* **33**, 170–189.

SCHIFFNER, V. (1897). Revision der Gattungen *Omphalanthus* und *Lejeunea* im Herbarium des Berliner Museums. *Bot. Jb.* **23**, 578–600.

SCHIFFNER, V. (1929). Über epiphylle Lebermoose aus Japan nebst einigen Beobachtungen über Rhizoiden, Elateren und Brutkörper. *Annls bryol.* **2**, 87–106.

SCHULTES, R. E. (1970). The history of taxonomic studies in *Hevea*. *Regnum veg.* **7**, 229–294.

SCHUSTER, R. M. (1955–1967). North American Lejeuneaceae, 1–10. *J. Elisha Mitchell Scient. Soc.* **71**, 106–126, 126–148, 218–247; **72**, 87–125, 292–316; **73**, 122–197, 388–443; **78**, 64–68; *J. Hattori bot. Lab.* **25**, 1–80; *J. Elisha Mitchell Scient. Soc.* **81**, 32–50; **83**, 192–229.

SCHUSTER, R. M. (1963a). Studies on antipodal Hepaticae, I. Annotated keys to the genera of antipodal Hepaticae with special reference to New Zealand and Tasmania. *J. Hattori bot. Lab.* **26**, 185–309.

SCHUSTER, R. M. (1963b). An annotated synopsis of the genera and subgenera of Lejeuneaceae, I. Introduction; annotated keys to subfamilies and genera. *Beih. nov. Hedwigia* **9**, 1–203.

SCHUSTER, R. M. (1966). "The Hepaticae and Anthocerotae of North America, I." Columbia University Press, New York.

SCHUSTER, R. M. (1970). Studies on antipodal Hepaticae, III. *Jubulopsis* Schuster, *Neohattoria* Kamimura and *Amphijubula* Schuster. *J. Hattori bot. Lab.* **33**, 266–304.

SCHUSTER, R. M. (1972). Evolving taxonomic concepts in the Hepaticae, with special reference to circum-Pacific taxa. *J. Hattori bot. Lab.* **35**, 169–201.

SCHUSTER, R. M. (1976). Review of: S. R. Gradstein, "A taxonomic monograph of the genus *Acrolejeunea*". *Bryologist* **79**, 380–382.

SCHUSTER, R. M. and HATTORI, S. (1954). The oil-bodies of the Hepaticae, II. The Lejeuneaceae. *J. Hattori bot. Lab.* **11**, 11–86.

SPRUCE, R. (1884–1885). Hepaticae Amazonicae et Andinae. *Trans. Proc. bot. Soc. Edinb.* **15**, 1–590.

STEPHANI, F. (1890). Die Gattung *Lejeunea* im Herbarium Lindenberg. *Hedwigia* **29**, 1–23; 68–99; 133–142.

STEPHANI, F. (1898–1924). "Species Hepaticarum." Georg, Geneva.

STOTLER, R. E. (1969). The genus *Frullania* subgenus *Frullania* in Latin America. *Nova Hedwigia* **18**, 397–555.

STOTLER, R. E. and CRANDALL, B. J. (1969). The sporophyte anatomy of *Dicranolejeunea axillaris*. *Bryologist* **72**, 387–397.

STOTLER, R. E. and CRANDALL-STOTLER, B. J. (1974). A monograph of the genus *Bryopteris* (Swartz) Nees von Esenbeck. *Bryophyt. Biblthca* **3**, 1–159.

TIXIER, P. (1968). *Cololejeunea* de l'Asie du Sud-Est, I. *Leonidentes* et espèces affines. *Revue bryol. lichén.* **36**, 543–594.

TIXIER, P. (1973). Le genre *Tuyamaella* Hatt. (Lejeunéacées). Monographie. *Revue bryol. lichén.* **39**, 221–244.

TOUW, A. (1974). Some notes on taxonomic and floristic research on exotic mosses. *J. Hattori bot. Lab.* **38**, 123–128.

VANDEN BERGHEN, C. (1948). Un nouveau genre d'Hépatiques: *Evansiolejeunea* nov. gen. *Revue bryol. lichén.* **17**, 86–90.

VANDEN BERGHEN, C. (1951). Contribution à l'étude des espèces africaines du genre *Ceratolejeunea* (Spruce) Schiffn. *Bull. Jard. bot. Etat Brux.* **21,** 61–81.

VANDEN BERGHEN, C. (1962). Lejeuneacées épiphylles d'Afrique (Note 1). *Revue bryol. lichén.* **32,** 49–55.

VANDEN BERGHEN, C. (1972a). Hépatiques et Anthocerotées. *In* "Exploration Hydrobiologique du Bassin du Lac Bangweolo et du Luapula" (J. J. Symoens, ed.) Vol. 8 (1), 1–202. Cercle Hydrobiologique, Bruxelles.

VANDEN BERGHEN, C. (1972b). Hépatiques épiphylles récoltées au Burundi par J. Lewalle. *Bull. Jard. bot. natn. Belg.* **42,** 431–494.

VANDEN BERGHEN, C. (1976). Frullaniaceae (Hepaticae) africanae. *Bull. Jard. bot. natn. Belg.* **46,** 1–220.

VANDEN BERGHEN, C. (1977). Hépatiques épiphylles récoltées par J. L. De Sloover au Kivu (Zaïre), au Rwanda et au Burundi. *Bull. Jard. bot. natn. Belg.* **47,** 199–246.

VERDOORN, F. (1934). Studien über asiatische Jubuleae. Die Lejeuneaceae Holostipae der Indomalaya unter Berücksichtung sämtlicher aus Asien, Australien, Neuseeeland und Oceanien angeführten Arten. *Annls bryol. Suppl.* **4,** 40–192.

WINKLER, S. (1967). Die epiphyllen Moose der Nebelwälder von El Salvador, C. A. *Revue bryol. lichén.* **35,** 304–369.

WINKLER, S. (1970). Zur Anatomie der Rhizoidplatten neotropischer, holostiper Lejeuneaceen. *Revue bryol. lichén.* **37,** 47–55.

5 | Taxonomy and Evolution of *Sphagnum*

A. EDDY

*Department of Botany, British Museum (Natural History),
Cromwell Road, London SW7 5BD, Great Britain*

Abstract: Evidence to link *Sphagnum* with the other bryophytes is limited and *Sphagnum* seems to hold a systematically isolated position. On the other hand, evidence from a number of aspects of morphology can be used to suggest relationships within the genus.

Much of the past work on the recognition of infrageneric groups has been done with a strongly European bias and can be misleading when extrapolated to other areas. The apparently sharp distinction between the sections of the genus in Europe are blurred in tropical areas. This, and other features of the taxonomy of *Sphagnum* are described using the example of the basically African *S. capense* group.

The name "*Sphagnum capense* group" denotes a number of taxa which were formerly placed in several different sections of the genus. Although morphologically well defined, the *S. capense* group can be divided into four subgroups by the position of the chlorocysts in the cross-section of the leaf. The subgroups each have a different but overlapping geographical distribution which can be used in conjunction with morphology to suggest evolutionary relationships. Ideas about evolution can then be correlated with evidence of geological history.

Despite the limitations of herbarium-based studies, they can be used to produce hypotheses about evolutionary history and these can be tested by results from other fields.

INTRODUCTION

Any discussion about the evolution of *Sphagnum* must inevitably be limited very largely to evolution within the genus. There is no fossil evidence which might show how the Sphagnopsida is related to other bryophytes (Lacey, 1969); the only links lie in the archegoniate reproduction, spermatozoid ultrastructure

Systematics Association Special Volume No. 14, "Bryophyte Systematics", edited by G. C. S. Clarke and J. G. Duckett, 1979, pp. 109–121, Academic Press, London and New York.

(Duckett and Carothers, this volume, Chapter 17), the mode of growth and the relative prominence of the gametophyte generation. Indeed, much of the morphological evidence emphasizes the isolation of *Sphagnum* from the rest of the Bryophyta.

There is, on the other hand, a great deal of morphological evidence which can be used to deduce relationships between the various species of *Sphagnum* and it is on this aspect of evolution that I shall concentrate here.

Presented with the abundance and diversity of *Sphagnum* in the temperate and subarctic parts of the northern hemisphere, it was inevitable that earlier authors would attempt to sort taxa into infrageneric groups (e.g. Bridel, 1798, 1826; Müller, 1848). In doing this they tended to rely on gross morphology rather than on anatomy. Some of the early groupings now seem bizarre in the light of modern knowledge since they tended to include many convergent lines. Nevertheless, in the latter half of the nineteenth century, largely due to the keen observations of W. P. Schimper (1876) and his contemporaries (e.g. Lindberg, 1862; Schliephacke, 1865), a system was established which has seen little modification up to the present time.

As far as European taxa are concerned, subdivisions of the genus are remarkably clearly defined, and even when the identity of species may be in doubt, the subgenus or section to which it belongs is almost always unequivocal. Had there been less consistency in the gross morphology of the species, these divisions would probably have been promoted to the rank of genus long ago. Each division has a particular characteristic, or combination of features, by which it can be immediately recognized. For example, a plant without cortical fibrils and with adaxially displaced branch-leaf chlorocysts belongs to Section *Acutifolia* while an otherwise similar plant with abaxially displaced chlorocysts is referred to Section *Cuspidata*. This sharp distinction may be an advantage to authors of handbooks and compilers of keys, but it is a source of frustration for the evolutionary taxonomist. It is among the misty regions of ill-defined taxa (at any rank) that the systematist looks for his clues to evolutionary trends. The apparent combination of *"Cuspidata"* and *"Subsecunda"* features in the American *Sphagnum mendocinum* Sull. led Andrews (1937) to merge the two sections. This fusion is not accepted today, but the existence of a species which is apparently intermediate between the two sections might be taken to indicate a possible evolutionary link between them.

When one considers that the genus *Sphagnum* is confined to a relatively narrow range of habitats with no marked discontinuities, and that the majority of the European species have a largely circum boreal geographical range, it is surprising that its sections are so sharply delimited. Even allowing for great antiquity of the genus, to suggest a boreal origin of the larger sections implies a remarkable rate of extinction of intermediate species. It is much more plausible that the boreal flora is derived. In other words, that the boreal zone was, at some time in its geological history, the recipient of sections which had already evolved. Therefore, as is so often the case with other plant groups, the student of evolution must look further south. He may there find refugia for at least some of the intermediate species (of "Pangean" or "Gondwanean" origin) which are absent from the derivative floras of higher latitudes.

The ideal taxonomic study of *Sphagnum* would be a world monograph with all the species properly delimited and arranged systematically according to their evolutionary position. Perhaps the scope of such a project is beyond the capabilities of a single author, if his researches are to be sufficiently penetrating to avoid a "Warnstorfian" proliferation of species on the one hand, or an "Andrewsian" oversimplification on the other. When exotic *Sphagna* are as well understood as the European and North American species, a world coverage will be feasible. The first step towards this ideal is a detailed analysis of the species region by region. These regions should be defined as geographical areas containing the present limits of a characteristic and recognizable flora. As far as *Sphagnum* is concerned, these regions largely coincide with the major continental land masses, with the larger oceans forming the natural range limits, although other physiographic or climatic barriers may operate.

THE SYSTEMATICS OF THE SPHAGNUM CAPENSE GROUP

1. Morphology and Taxonomy

In my own studies, one such region, tropical Asia, has already been examined (Eddy, 1977) and this is currently being followed up by revision of the African *Sphagnum* flora. Some observations on an African species complex may show how the herbarium taxonomist can suggest evolutionary relationships. The chosen complex (Table I) will be referred to as the *Sphagnum capense* group for convenience. This group was chosen because it displays the features of a "pivotal" species, i.e. one which appears to lie on or near the point of divergence of sections. Surprisingly, such pivotal species are apparently rare in the Malaysian flora, although that region contains some widely divergent taxa of presumed antiquity.

Species of the *Sphagnum capense* group are more or less typical of the genus in basic morphology (Figs 1–4). They are relatively delicate plants which in many respects resemble the boreal *S. tenellum*, but with branch-leaf pores more like those of *S. subsecundum* and a perforate, multi-layered hyaloderm.

TABLE I. Species of the *Sphagnum capense* group and the sections in which they were originally placed. The right-hand column indicates the subgroups to which the taxa have been assigned in this paper

Species	Date of publication	Original section	Present subgroup
S. *aloysii-saboudiae* Negri	(1908)	*Subsecunda*	*davidii*
S. *beyrichianum* Warnst.	(1897)	*Subsecunda: Porosa*	*tumidulum*
S. *capense* Hornsch.	(1841)	*Subsecunda: Porosa: Diversiporosa*	*capense*
S. *ceylonicum* Mitt. ex Warnst.	(1879)	*Acutifolia: Rotundata*	*ceylonicum*
S. *chevalieri* Warnst.	(1911)	*Subsecunda: Porosa: Pauciporosa*	*davidii*
S. *davidii* Warnst.	(1905)	*Subsecunda: Porosa: Multiporosa*	*davidii*
S. *ericetorum* Brid.	(1806)	*Cuspidata: Ovalia*	*capense*
S. *goetzianum* Warnst.	(1907)	*Mucronata*	*capense*
S. *islei* Warnst.	(1918)	*Mucronata*	*davidii*
S. *keniae* Dix.	(1918)	*Subsecunda*	*ceylonicum*
S. *kerstenii* Warnst.	(1911)	*Acutifolia: Rotundata*	*ceylonicum*
S. *panduraefolium* C. Muell.	(1887)	*Subsecunda: Porosa: Pauciporosa*	*capense*
S. *perrieri* Thér.	(1922)	*Subsecunda: Multiporosa*	*davidii*
S. *pycnocladulum* C. Muell.	(1887)	*Mucronata*	*capense*
S. *rugegense* Warnst.	(1910)	*Cuspidata: Triangularia*	*capense*
S. *ruwenzorense* Negri	(1908)	*Subsecunda*	*davidii*
S. *tumidulum* Besch.	(1880)	*Mucronata*	*tumidulum*
S. *ugandense* Tayl. and Thomps.	(1955)	*Acutifolia*	*ceylonicum*

Comparatively few names that can be assigned to this group had been published before the mid-nineteenth century. It will come as no surprise that a large proportion of subsequent additions were contributed by Carl Warnstorf (1897, 1905, 1907, 1911). That nearly all these have survived to the present day without being reduced to synonymy is as much a reflection of the neglect of the African flora as it is of the discriminatory powers of earlier authors.

On describing a new species of *Sphagnum* some attempt was usually made to place it into one of the established infrageneric groups. Many of the taxa considered to be new had close relatives in Europe, and their systematic placings were then fairly straightforward. With the remainder, however, assignments seemed to depend largely on whichever character happened to be examined first, or which happened to be at the forefront of the author's mind as he examined the specimen. That Bridel (1826) should place *S. ericetorum* near to the European *S. molluscum* (= *S. tenellum*) was in line with early nineteenth century practice. More surprising is Warnstorf's (1911) much later scattering of species which we now see as part of a single species complex, among four diverse sections. For example, *S. pycnocladulum* was placed in Section *Mucronata*, *S. ceylonicum* in Section *Acutifolia*, *S. ericetorum* in Section *Cuspidata* and *S. capense* in Section *Subsecunda*. Table I gives some idea of the diverse treatment of the *S. capense* group by various authors. All but two of the names listed were included by Warnstorf in his *Sphagnologia Universalis* (1911).

One cannot avoid the impression that most of the species names published at various times by Warnstorf from 1879 onwards were simply transferred, without any kind of serious reappraisal, from the periodicals in which they were first described to the *Sphagnologia* (1911). Where Warnstorf drew comparisons, it was from a thoroughly European standpoint so that tropical species were compared with European species rather than with each other. In this way no significant distribution patterns emerged. No doubt the penetrating eye of LeRoy Andrews would have noticed the discrepancies of Warnstorf's treatment but, apart from Subgenera *Sphagnum* and *Rigida*, he concentrated on the European and North American floras and, by his own admission, baulked somewhat at the *Subsecunda*. Since Warnstorf's time, studies of *Sphagnum* in Africa, e.g. by Renauld and Cardot (1915), Garside (1949) and Taylor and Thompson (1955), were strictly regional and produced little order out of an apparently random scatter of species which were reportedly endemic. Garside (1949) in his studies on the South African species detected the close relationship between *S. pycnocladulum* (Section *Mucronata* according to Warnstorf) and *S. capense* (Section *Subsecunda*) and may well have succeeded in unravelling more of the tangle had he ranged more widely geographically.

Although the *S. capense* group shows a number of features which set it apart from, for example, boreal species of Section *Subsecunda*, it is quite clear that it is to that section that it belongs. Whether the group is worthy of the rank of subsection is debatable. In common with other *Subsecunda*, it has relatively monomorphic branches, serial commissural pores, chlorocysts with thickened walls and oval-sectioned lumina, and relatively isophyllous stem leaves (Figs 1–5). Distinguishing features of the group include the *S. tenellum*-like facies enhanced by the, usually very numerous, regularly arranged, small (c. 1·0–1·3 mm), uniform, very concave, ovate branch leaves. In contrast to other African *Subsecunda*, the leaf apices are usually narrow, often with very little apical resorption, sometimes even approaching the form previously only met with in the Malaysian *S. sericeum* C. Muell. (Eddy, 1977). Branch-leaf pores, except in a few basal leaves of some species, are slightly or considerably larger than in other *Subsecunda* (up to 10·0 μm or more). When present, one of the most conspicuous features is the single or several additional ringed

FIGS 1–4. Some morphological features of the *Sphagnum capense* group. (All drawn from isotypes.)

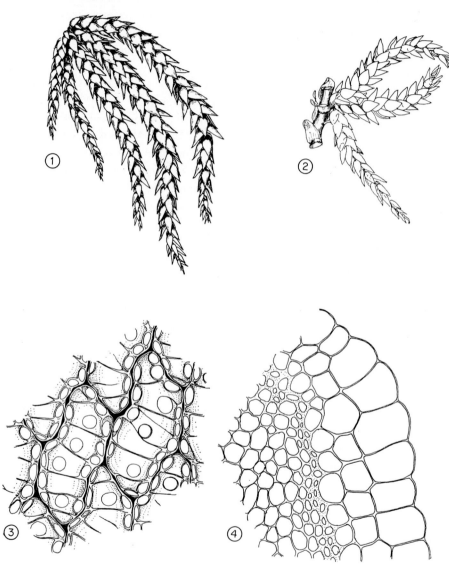

FIG. 1. Branch fascicle of *S. pycnocladulum* showing monomorphic branches and acutely pointed leaves.

FIG. 2. Branch fascicle of *S. capense*.

FIG. 3. Abaxial surface of branch leaf of *S. capense* showing additional free pores.

FIG. 4. Transverse section of stem of *S. davidii* showing well-developed cortex.

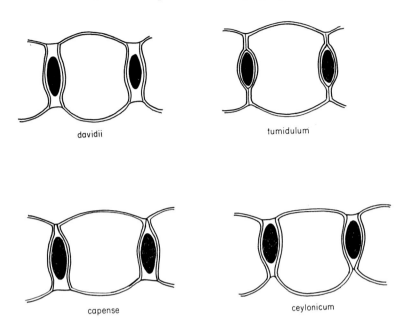

davidii   tumidulum

capense   ceylonicum

Fig. 5. Semi-diagrammatic drawings of the types of leaf section in the *S. capense* group;
all with adaxial side uppermost.

pores over, or near the mid-line of a cell. It was the presence of these in *S. ceylonicum* which first called attention to the close relationship between this species and the African members of the *S. capense* group. The spasmodic appearance of an occasional small, free pore in a *few* subapical leucocysts in the *S. subsecundum* agg. is scarcely comparable to the regular occurrence of such pores in the present group. The stem cortex is (1–)2–3 layered but with the outermost layer markedly more inflated than in other *Subsecunda* of similar stature.

A survey of herbarium material has shown the group to be essentially, but not quite exclusively, African, with its main range on continental Africa and the larger East African islands. There is an outlier in Sri Lanka and a related taxon in South America. On the African mainland the range of the group is confined to regions of relatively high oceanicity, hence it has a somewhat Afro-alpine distribution pattern, descending to moderate altitudes in the Cape region and on the more oceanic islands of Madagascar and Réunion (Fig. 6).

On the basis of several morphological features, the *S. capense* group is more or less well defined and natural, but one of its characters, that of chlorocyst position in the branch leaves, exhibits unusual variation. There can be little doubt

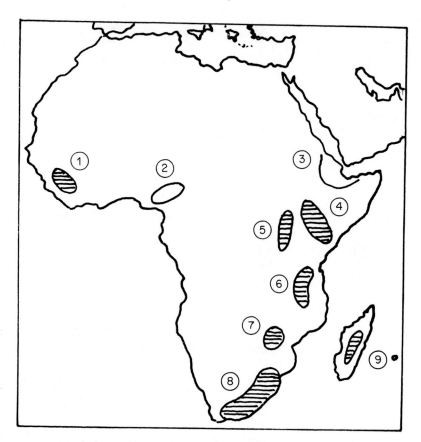

FIG. 6. Map showing the Afro-alpine regions. Shaded areas represent the regions in which species of the *S. capense* group have been collected. The regions are usually labelled as follows: 1. Guinea; 2. Cameroun; 3. Ethiopia; 4. Usambara; 5. Ruwenzori; 6. Mlanje; 7. Chimanimani; 8. Drakensberg; 9. Madagascar.

that this variation was responsible for the varied subgeneric assignments of its component species by different authors. European taxonomists had probably come to rely upon the stability of this feature in familiar boreal taxa and would have been unaware of the dangers inherent in its use in this exotic group.

Although leaf sections show exceptional variation within the *S. capense* group as a whole, they are consistent within its component taxa when these are defined by other characters. The group can be divided into four subgroups on the basis of chlorocyst position (Table I). For convenience, each of these subgroups can be identified with a validly published species name (Fig. 5).

I must stress, however, that the nomenclature of the species is still provisional; a full revision will be the subject of a later paper.

The subgroups are as follows:

1. The *S. davidii* subgroup has median chlorocysts which emerge more or less equally on both sides of the leaf. This form is widespread among members of the section *Subsecunda* and is probably primitive.

2. The *S. capense* subgroup has abaxially displaced chlorocysts which are usually more or less immersed below the adaxial leaf face.

3. The *S. ceylonicum* subgroup has adaxially displaced chlorocysts which may be immersed below the abaxial leaf face.

4. The *S. tumidulum* subgroup has completely immersed chlorocysts which impart a very distinctive appearance to the leaves under the microscope.

In *S. tumidulum* the position of the chlorocysts has been suggested as a reason for separating this species into a separate section. On the African mainland, this type of chlorocyst was only known to occur in the type specimen of *S. beyrichianum* but recent research has shown that the *S. tumidulum* type of chlorocyst can occur spasmodically in *S. pycnocladulum* (*S. capense* subgroup), probably through the suppression of thickening of the posterior cell wall. In *S. pycnocladulum* immersed chlorocysts are rarely consistent throughout a specimen. The feature clearly is less significant here than it is in the Madagascan *S. tumidulum*, but might suggest a relationship.

Within each subgroup there is a degree of variation of characters which in some cases allows further subdivision. In the *S. capense* subgroup, for example, two clearly defined entities can be recognized, one typified by *S. capense* itself and the other by *S. pycnocladulum*. The latter entity approaches the *S. davidii* subgroup in its morphology.

2. Geographical Distribution

Regional representation of African species of *Sphagnum* in the herbaria is very uneven. This is partly because of uneven coverage of the ground by collectors but also partly because of the ecological limitation mentioned above. For a study of distribution, however, a solitary substantiated record can be given as much weight as a large population. In contrast to "mobile" ruderal bryophytes, adventive occurrences of members of the *Sphagnum capense* group are highly unlikely. Allowing for physiographic disjunctions, the ranges of the four chlorocyst-based subgroups show different, but overlapping patterns (Fig. 7) which suggest some evolutionary interpretations. On the African

A. Eddy

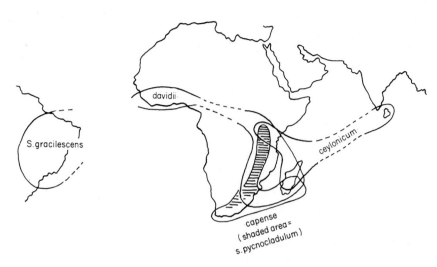

FIG. 7. Map showing the broad distribution patterns of the subgroups of the *S. capense* group. The *S. tumidulum* subgroup is omitted for clarity but *S. gracilescens* has been added to show the link between the floras of south America and west Africa.

mainland, the *S. davidii* subgroup is the most widespread. It is the only subgroup in the complex which has so far been found in west Africa. In the east it extends southwards from Usambara and Ruwenzori to Mozambique and Tanzania (Mlanje) with a slightly divergent outlier in the eastern Drakensberg. It reappears in a somewhat modified form (as *S. perrieri*) in Réunion and Madagascar. The *S. davidii* subgroup is most abundant, and reaches its maximum vigour in growth in the northeast, i.e. in Ruwenzori and Usambara but, perhaps significantly, approaches *S. capense sensu stricto* further south. The *S. ceylonicum* subgroup has a relatively disjunct distribution and is the only one of the four subgroups that crosses the Indian Ocean. It occurs with *S. davidii* in Kenya and Uganda, is apparently rare in Madagascar but relatively frequent in Réunion, and is apparently absent from southeast and south Africa. Some of the mainland specimens are indistinguishable from the plants from Sri Lanka. The range of the *S. capense* subgroup is largely coincident with that of the *S. davidii* subgroup but is more restricted in that it does not occur in the west of the continent, although it does extend further down into the Cape region. The distribution of the *S. capense* subgroup is more complicated than that of the *S. davidii* or *S. ceylonicum* subgroups. Each of the two morphological entities present within the *S. capense* subgroup has a slightly different geographical range. The entity typified by *S. capense* is mainly confined to the Cape region but with variants in Madagascar and Réunion, and that typified by *S. pycnocladulum* has a sparse but widespread distribution from Ruwenzori and Usambara southwards to the Cape. The *S. tumidulum* subgroup is endemic to Madagascar and Réunion.

3. The Evolutionary History of the Sphagnum capense Group

The broad distribution patterns so far presented comprise discrete populations with many disjunctions. Many of the disjunctions are minor, the remnants of more continuous patterns that may have existed perhaps as recently as the Quaternary. The wider disjunctions, between Africa and Sri Lanka or the East African islands, which have provoked some highly original and implausible theories in the past, can now be seen as the relicts of a much earlier Gondwana flora. Many patterns of distribution have been explained by accepting the mobility of continents and it has become fashionable to outline species ranges on "pre-drift" maps (Fig. 8). Can continental drift provide an explanation for the Afro-Asiatic and Afro-Madagascan distributions of some of the subgroups of the *S. capense* group? If one extrapolates back from present distribution patterns, one can visualize that at a time before the separation of Africa, Madagascar and the Indian subcontinent, diversification was taking place within the prototype of the *S. capense* group. The components of this diversification subsequently radiated outwards from a centre which probably lay somewhere near Kenya and Uganda. For a number of reasons, among them the availability of rain-bearing winds from the Tethys, and the higher (about 15°S) latitudinal position of Africa at the time, much of the region is presumed to have been cooler and wetter than at present (Axelrod and Raven, 1978). Thus, the *S. capense* group would presumably have been more abundant and generally more vigorous before the close of the Cretaceous than it is in Africa today. It would be during such favourable conditions that the greatest rate of diversification might occur, encouraged by more local changes in topography or climate.

The sequence of changes which led to the evolution of the present taxa of the *S. capense* group must remain largely conjectural. In my own view the *S. davidii* subgroup approximates most closely to the archetype from which the other subgroups are derived. It is easier to accept the relatively slight shifts in chlorocyst displacement from the *S. davidii* form to the *S. ceylonicum* form on the one hand and the *S. capense* form on the other, than to visualize a sequence from one extreme to the other through the *S. davidii* form. The wider distribution of the *S. davidii* chlorocyst type supports this view.

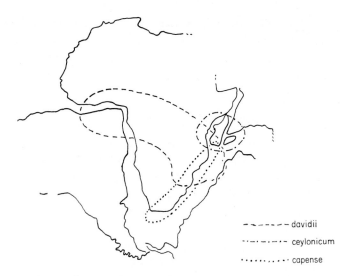

- ------- davidii
-·--·--·- ceylonicum
·········· capense

Fig. 8. Pre-drift map showing the probable distribution patterns of diverging subgroups of the *S. capense* group. At this stage, the *S. tumidulum* morphotype, not shown on this map, was probably confined to Madagascar.

The relationship of the *S. capense* group to the boreal *Subsecunda* is clear enough and is well illustrated by morphotypes within the group itself (especially in juvenile states) and by the existence of intermediate species such as the South American *S. gracilescens* C. Muell. and the Asiatic *S. ovatum* Hamp. *S. gracilescens* is very close in its morphology to the African group and is an interesting phytogeographical link between the floras of tropical west Africa and eastern south America. The divergence of the *S. ceylonicum* and *S. capense* lines could be seen also as the early stages in the evolution of Section *Acutifolia* and the *S. tenellum* group of Section *Cuspidata* respectively. Morphological approaches to Section *Acutifolia* are by no means confined to chlorocyst position, there are numerous other features, such as, for example, the multi-layered cortex or the enlarged pores, which point in that direction. It may not be a coincidence that Section *Acutifolia* is almost absent from the African mainland, for if this section was the end product of the divergence of the *S. capense* group, it may have arisen at some distance from the African continent. Very little immigration of members of Section *Acutifolia* into Africa might have occurred before its geographical isolation.

The Pacific region, especially Indo-Malaya, is usually regarded as the area where angiosperms originated (Smith, 1970). Africa is poor in relict Hepaticopsida (Schuster, personal communication). It was surprising, therefore, to find a group of African *Sphagnum* species with distinctly archetypal features. Perhaps a search will yield other African archetypes among the Bryophyta.

No one is more aware than I of the limitations of herbarium taxonomy. However, I believe that studies such as the one described here can, in addition to their immediate practical aims, produce hypotheses about phylogeny and evolutionary history. Such hypotheses can focus the attention of colleagues in other fields, such as phytochemistry and genetics, on areas where further research could yield valuable results.

REFERENCES

ANDREWS, A. L. (1937). Notes on the Warnstorf *Sphagnum* herbarium, I. *Annls bryol.* **9**, 3–12.

AXELROD, D. I. and RAVEN, P. H. (1978). Late Cretaceous and Tertiary history of Africa. *Monograph. biol.* **31**, 77–130.

BRIDEL, S. E. (1798). "Muscologia Recentiorum." Vol. 2. Ettinger, Gothiae-Parisiis.

BRIDEL, S. E. (1826). "Bryologia Universa." Vol. 1. Barth, Lipsiae.

EDDY, A. (1977). Sphagnales of tropical Asia. *Bull. Br. Mus. nat. Hist., Bot.*, **5**, 357–445.

GARSIDE, S. (1949). *Sphagnum* in South Africa *J. S. Afr. Bot.* **15**, 59–78.

LACEY, W. S. (1969). Fossil bryophytes. *Biol. Rev.* **44**, 189–205.

LINDBERG, S. O. (1862). Torfmossornas byggnad, utbredning och systematiska uppställ-ning. *Öfvers. K. VetenskAkad. Förh. Stockh.* **19**, 113–156.

MÜLLER, C. (1848). "Synopsis Muscorum Frondosorum." Foerstner, Berolini.

RENAULD, F. and CARDOT, J. (1915). Histoire naturelle des plantes. Musci, Sphagnales. *In* "Histoire Physique, Naturelle et Politique de Madagascar" (A. Grandidier, ed.) Vol. 39, pp. 41–54. Imprimerie Nationale, Paris.

SCHIMPER, W. P. (1876). "Synopsis Muscorum Europaeorum". Second edition. E. Schweizerbart, Stuttgart.

SCHLIEPHACKE, K. (1865). Beiträge zur Kenntniss der *Sphagna. Verh. zool.-bot. Ges. Wien* **15**, 383–414.

SMITH, A. C. (1970). "The Pacific as a Key to Flowering Plant History". University of Hawaii, Hawaii.

TAYLOR, J. and THOMPSON, A. (1955). Notes on *Sphagna* from Uganda. *Kew Bull.* **9**, 517–521.

WARNSTORF, C. (1897). Beiträge zur Kenntniss exotischer *Sphagna. Hedwigia* **36**, 145–176.

WARNSTORF, C. (1905). Vier neue exotische *Sphagna. Allg. bot. Zeit.* **1905**, 97–101.

WARNSTORF, C. (1907). Neue europäische und aussereuropäische Torfmoose. *Hedwigia* **47**, 76–124.

WARNSTORF, C. (1911). Sphagnales—Sphagnaceae (Sphagnologia Universalis). *In* "Das Pflanzenreich—Regni Vegetabilis Conspectus" (A. Engler, ed.) Vol. 51. W. Engelmann, Leipzig.

6 | Taxonomy and Phytogeography of Bryophytes in Boreal and Arctic North America

W. C. STEERE

New York Botanical Garden, Bronx, New York 10458, USA

Abstract: The two major problems which confront a bryologist working on the bryophytes of northern North America are (1) ecological and physiological, and (2) phytogeographic.

Most of the North American Arctic has been classified climatically as a polar desert, which is characterized by very low levels of temperature, humidity and available water, as well as by high pH and insolation, all of which affect the physiological reactions of bryophytes adversely. Various combinations of unfavourable ecological factors are reflected by dwarfing and other modifications of the gross morphology of bryophytes, often so much so that they may be unidentifiable. The genus *Bryum* seems to be especially plastic under extreme arctic conditions, as indicated by the dozens of unnecessary new species which have been described from circumpolar regions. Moreover, in *Fissidens arcticus* the degree of development of the specialized leaf border seems directly related to the amount of moisture available.

The phytogeographic problem is simply that a very considerable element of bryophytes, perhaps 15%, occurs at such high latitudes that they have remained little known and hence have been largely overlooked by standard handbooks and manuals.

The geographical distribution of bryophytes in arctic North America is becoming relatively well known because of the increasing tempo of field work. Although further floristic and phytogeographic elements are certain to be recognized eventually, at the moment six elements can be distinguished easily.

(1) The circumboreal bryophyte element of temperate, boreal and arctic climates is by far best known because most of the standard treatments of bryophytes are devoted to it; it comprises 75–80% of the arctic North American bryophyte flora. (2) The circumpolar and arctic-alpine bryophyte element consists largely of relict Tertiary

Systematics Association Special Volume No. 14, "Bryophyte Systematics," edited by G. C. S. Clarke and J. G. Duckett, 1979, pp. 123–157, Academic Press, London and New York.

species which survived in large unglaciated refugia north of Pleistocene glaciation and which now form approximately 15 % of the arctic bryophyte flora. In North America, several species have turned up unexpectedly in unglaciated refugia rather far south of the Arctic Circle in the Rocky Mountains and are therefore considered to be arctic-alpine. Moreover, a new correlation has been discovered between the distribution of circumpolar bryophytes and a narrow unglaciated corridor running far southward from the vast unglaciated regions of northern Alaska and Yukon, not far west of the Mackenzie River. Here, also, several species of circumpolar bryophytes previously thought to be restricted to the Arctic were recently discovered. The occurrence of unglaciated refugia in the Canadian Arctic Archipelago and northern Greenland is still something of a moot question. However, the existence in both areas of so many species normally restricted to unglaciated regions indicates that local refugia must have existed during the Pleistocene. (3) A small but distinct element of otherwise temperate-climate bryophytes also occurs disjunctly in the Arctic. Because of the concentration of such species at Umiat, this element has been termed the "Umiat Syndrome". (4) The desert-and-steppe element consists of relatively few species which are conspicuously disjunct between the deserts of the American Southwest and arctic North America. (5) Disjunct Asiatic species in boreal and arctic North America form a distinctive element. Up to a dozen species which occur particularly in arctic Alaska are otherwise known only from eastern, central and arctic Asia; some of them also extend farther south in western North America. (6) The cosmopolitan element consists of a surprisingly small number of species which are equally widely distributed—and often weedy—in all parts of the world. Eighteen dot-maps illustrate the first five of these six patterns of phytogeographic distribution.

INTRODUCTION

Professor W. B. Schofield has recently completed a splendid new review of the floristic elements of North American mosses, with special attention to endemism and disjunction. When it appears, Schofield's paper will provide the best over-view of North American bryogeography yet available; with his permission, I have used some of his information here, as well as three maps. Published bryogeographical works which I found particularly useful to clarify my ideas are those of Schuster (1966), Schofield (1972) and Brassard (1974). For a review paper of this sort I have not hesitated to quote, paraphrase or even plagiarize my own previous writings on the same subject.

TAXONOMIC PROBLEMS IN BOREAL AND ARCTIC BRYOPHYTES

The taxonomy of boreal and arctic bryophytes at the present time is particularly complicated by two problems, one being ecological and physiological and the other being bryogeographical.

The physiological and morphological reactions of bryophytes to unfavourable or extreme environmental conditions, especially to low levels of temperature, humidity and water availability, as well as high levels of light and pH, are often remarkable, and may result in plants which are so dwarfed or otherwise modified that they are quite unrecognizable even by the specialist. Much of the terrain of arctic North America, especially northern Greenland and the Canadian Arctic Archipelago, is classified by climatologists as a polar desert (Smiley and Zumberge, 1974) because of the very low levels of precipitation, greatly reduced temperatures accompanied by strong winds during the winter—and usually without adequate snow cover for protection of plants—and a generally high alkalinity. Different combinations of these ecological factors impinging on developing gametophytes and sporophytes of mosses and hepatics during the early summer lead to an extraordinarily high degree of variability. The effect of microenvironment on seta length and capsule production, in terms of temperature at and just above soil level at the time of their development, has been discussed by Steere (1954). In the Hepaticae, for example, the arctic forms of *Lophozia* are almost unbelievably plastic in their variations (Schuster, 1969) and in response to the same extreme environmental conditions, a parallel situation has been found in the arctic forms of the genus *Scapania* (Schuster, 1974). In mosses, particularly, great variation occurs in such already variable genera as *Drepanocladus, Bryum, Calliergon* and *Brachythecium*, among others, to such an extent that some collections can be extremely difficult to identify with any known species.

The genus *Bryum* in the Arctic shows the most extreme variation of all and the level of fluctuation in nearly every diagnostic character defies belief! Peristome development seems to be particularly sensitive to environmental conditions, both favourable and unfavourable, and a great variation in structure is a consequence. Largely as the outcome of such remarkable variability under environmental stress, among the 64 species of *Bryum* reported from the collections of the "Fram" Expedition from Ellesmere Island, Bryhn (1906–1907) and Ryan described more than 20 as new species, with a good many more new varieties! Until experimental work on peristome development under controlled conditions of moisture, temperature and light can be carried out, so that some element of predictability can be used to interpret the range of variability in the peristome of any one species in nature, the genus *Bryum* in the high Arctic will remain an almost impenetrable mystery, a veritable morass of superfluous names whose real meaning cannot be determined. At present, in the throes of organizing a multiple-author monographic treatment of all mosses of arctic North America, I am most uncertain of just how to proceed in the matter of this complex and untidy genus. Although Andrews

(1940) revised all the Bryaceae for Grout's *Mosses of North America North of Mexico* (1928–1940), his treatment tended to be so conservative (among American bryologists he was considered to be a "lumper") that many of the species he placed in synonymy will have to be sorted out, re-evaluated and some of them perhaps re-established. To sum up, the genus *Bryum* in the Arctic is probably the ultimate in taxonomic difficulty, particularly since so many collections lack sporophytes. Fortunately for bryologists, no other genus in arctic regions presents the same extreme level of variability and plasticity as *Bryum*, even though many collections of species in other genera are also difficult or nearly impossible to identify with confidence, especially when sterile.

The physiological stresses inherent in the stringent environment of the polar desert and their impact on the anatomical development of bryophytes are also well shown in the genus *Fissidens*. *Fissidens arcticus* Bryhn was described from Ellesmere Island and western Greenland (Bryhn, 1906–1907) as having no differentiated margin at the edge of the leaves, even though the type specimen does show traces of a leaf border (Steere and Brassard, 1974). In Alaska, where higher levels of moisture prevail than in the Canadian Arctic Archipelago or northernmost Greenland, in terms both of humidity and precipitation, the differentiated margin becomes more obvious. The question immediately arises, then, as to whether the presence or absence of a specialized leaf border is only a reflection of environmental conditions, particularly the availability of moisture. The natural corollary to the preceding question is *if* the development of the border is specifically linked to the amount of water available, then why is *F. arcticus* not simply a member of the wide-ranging and variable species, *F. bryoides*? I can report here that Ronald Pursell, the North American specialist on *Fissidens* (personal communication), has adopted this point of view, after examination of a considerable number of collections which had been identified as *F. arcticus*. However, this species would still seem to be one of the best and easiest points of access for experimental work on the effect of different levels of moisture, temperature and light on the degree of development of a differentiated leaf border, under controlled conditions.

The second taxonomic problem is bryogeographic and is more a matter of lack of knowledge than innate taxonomic complexity. When the first botanical explorers were able to visit the arctic tundra vegetation in various parts of the Northern Hemisphere, such as northernmost Norway and Finland, Spitzbergen, Siberia, arctic Canada and Greenland, they discovered many new species of bryophytes which seemed to be endemic to each particular area. Because of the difficulties of travel, only small collections could be brought back, which have existed for decades in some one herbarium and were considered generally to be extraordinarily rare. It was not realized until I began my field work in arctic Alaska in 1951 that all these "endemic" arctic species belong to a widely distributed circumpolar arctic floristic element, since most of them occur in northernmost Alaska (Steere, 1953, 1978), where many of them are not only common but also abundant. We now know that this element consists of at least 100 species of mosses and hepatics (approximately 15% of the total bryophyte flora of the Arctic) which occur at such high latitudes that they have been seen in the field by few bryologists and still exist in too few herbaria. As a result, many of them, except for those species which also occur on high mountains farther south, have been omitted from most of the standard manuals of circumboreal bryophytes, and even from some monographic revisions. However, because of our greater knowledge today concerning the circumarctic floristic element, recent monographs have begun to include the arctic species, notably Mogensen's monograph of *Cinclidium* (1973), Koponen's review of the Canadian Mniaceae (1974) and Vitt's revision of the North American species of *Seligeria* (1976a). The more recent manuals which cover arctic USSR (Abramova *et al.*, 1961) and the acrocarpous mosses of the whole of the USSR (Savich-Lyubitskaya and Smirnova, 1970) do include those circumpolar species which occur in the USSR. Schuster's three volumes (1966–1974) covering the Hepaticae of North America also include his own large collections from northern Ellesmere Island and Greenland. Specimens of many of the high arctic species of bryophytes have also been distributed by Steere *et al.* (1975, 1976) in the two fascicles so far published of *Bryophyta Arctica Exsiccata*.

A brand-new flora, *Mosses of Arctic North America*, to be written by many bryological specialists under the editorship of Gert S. Mogensen (for Greenland), Robert R. Ireland (for Canada) and William C. Steere (for Alaska) has now been organized. Most collaborators have been selected and publication should begin in the foreseeable future. An annotated catalogue of the mosses of arctic Alaska has just appeared (Steere, 1978) and a parallel treatment of the Hepaticae is given by Steere and Inoue (1978). A *Checklist of the Mosses of Canada*, including arctic Yukon and the Northwest Territories, has been completed by a committee of Canadian bryologists (Ireland, Crum, Bird, Brassard, Schofield and Vitt), and has been awaiting publication for two years; it will eventually be published by the National Museums of Canada. Moreover, it will have an advantage over the current and more streamlined *A New List of Mosses of North America North of Mexico* (Crum *et al.*, 1973) in that it will indicate the distribution of each species by province.

To sum up, the high arctic circumpolar species of bryophytes are clear–cut and distinctive and present no special taxonomic problems, apart from being too little known. Moreover, they are beginning to appear in publications and herbaria and will eventually become as well known to bryologists as the very large and pervasive circumboreal floristic element so common at lower latitudes.

THE BRYOGEOGRAPHY OF ARCTIC AND BOREAL NORTH AMERICA

Some years ago (Steere, 1971) I reviewed the history and literature of systematic bryology in all north polar regions. I then emphasized the early expeditions, so that I therefore need to note here only some of the more important publications which have appeared sine 1971. The bryophytes of northernmost Alaska have received major attention very recently, the mosses by Steere (1978) and the hepatics by Steere and Inoue (1978). Schuster and Damsholt (1974) published a substantial treatment of the Hepaticae of arctic West Greenland and Steere (1975) reported on a large collection of bryophytes from the Thule District.

Arctic and subarctic Canada has continued to receive field and herbarium study on a rapidly growing basis, which is fully consistent with the enormous land surface of the Yukon and Northwest Territories. In the Canadian Arctic Archipelago, Brassard (1976) continued his work on northern Ellesmere Island, Vitt (1975) published a major contribution to the muscology of Devon Island and Kuc contributed to our knowledge of several further islands, including Axel Heiberg Island (1973a) and Banks Island (1973b), although largely with relation to their ecology and palaeobotany. Concerning the bryophytes of the Canadian arctic and subarctic mainland, recent contributions to the bryology of the Yukon were made by Bird *et al.* (1977), Douglas and Vitt (1976), Hong and Vitt (1977), Steere and Scotter (1978b, c) and Vitt (1974a, 1976b). With regard to the Northwest Territories during the same span of time, Bird *et al.* (1977) listed the bryophytes of the drainage of the Peel and Mackenzie rivers, Holmen and Scotter (1971) published an important paper on the bryophytes of the Reindeer Preserve, Steere (1977) reported on his collections of bryophytes from Great Bear Lake and Coppermine, and Steere *et al.* (1977) and Steere and Scotter (1978a) reported on the bryology of Nahanni National Park in southwestern Northwest Territories.

To sum up, the bryology of arctic and subarctic North America has in recent years become reasonably well known and our knowledge of it continues to grow at an accelerating tempo, thanks largely to more bryologists becoming interested in arctic bryology and to the increasing scope and opportunity for field work. In fact, for the first time in history we have collections and information adequate for the preparation of a monographic *Moss Flora of Arctic North America*, as mentioned in the previous section.

FLORISTIC AND PHYTOGEOGRAPHIC BRYOPHYTE ELEMENTS

Viewed as a whole, the bryophyte flora of boreal North America consists of several well-marked floristic and bryogeographic elements, most of them strongly disjunct in their distribution. Some of these elements have been difficult to define and to understand well until the last few decades, during which it has become increasingly clear that large areas of arctic North America remained quite free of the continental and cordilleran ice-sheets which almost totally devegetated vast regions farther south during part or all of the Pleistocene (Coulter *et al.*, 1965; Prest, 1969), so that it was available as a refugium—or a series of refugia—for the survival of Tertiary plants, a situation which has had enormous phytogeographic significance through the millennia.

In a region as widely and heavily glaciated by extensive continental and cordilleran ice sheets as northern North America, with the exception of the ice-free areas in the far North, it was only natural for American botanists to concern themselves with the correlations between limits of glaciation and the geographical distribution of plants. Over 40 years ago, I came to the sudden realization that the wide and conspicuously disjunct distribution of *Bryoxiphium norvegicum* (Brid.) Mitt. in North America was rather neatly correlated with the maximum extent of glaciation—and that not a single locality occurred *within* the glacial boundary (Steere, 1937b). This experience was most helpful to me when I began work on arctic American bryophytes, especially in arctic Alaska, much of which escaped glaciation during the Pleistocene and therefore provided refugia for Tertiary plants which appear to have survived there in spite of the great fluctuation and gradual deterioration of climatic conditions over an enormously long period.

One of the most interesting aspects of the bryogeography of northern North America is the existence of a surprisingly large amount of geographic disjunction of various kinds and degrees. Each of the major floristic elements possesses members with strikingly disjunct distributions in North America and elsewhere. Even though the circumboreal element appears today to be quite homogeneous in its distribution, during the Pleistocene it existed over millennia as disjunct populations widely separated by intervening continental glaciation.

Steere (1937a, 1938, 1948, 1953, 1965, 1976a, 1978), Schuster (1966–1974), Schofield (1972), Schofield and Crum (1972) and Brassard (1974) have all made substantial contributions to our understanding of disjunct geographic distributions in northern bryophytes.

To give some idea of the magnitude of the problems we are discussing here, an estimate of the numbers of species involved will be useful. In arctic Alaska, 415 species of mosses are known (Steere, 1978) and 135 species of hepatics (Steere and Inoue, 1978). From the Yukon Territory, 345 species of mosses have been reported (Vitt, 1974a) and 91 species of Hepaticae (Hong and Vitt, 1977). Ireland *et al.* (unpublished checklist) have indicated that 399 species of mosses occur in the Northwest Territories and the number of species of hepatics so far reported can be estimated at approximately 100. For Greenland, the numbers have been estimated by Mogensen (personal communication) as 454 taxa of mosses and 160 of hepatics. Of course, we may be certain that all these numbers will be increased substantially through further field exploration of the regions involved, especially if done by bryologists.

1. The Circumboreal Bryophyte Element of Temperate, Boreal and Arctic North America (Fig. 1)

This element is distributed very widely throughout the temperate, boreal, alpine and arctic regions of North America. It is therefore by far the best known element to the great majority of bryologists of the Northern Hemisphere because it consists of those species of mosses and hepatics which are treated in the standard manuals and handbooks covering the bryophytes of Great Britain, north and central Europe, most of the USSR, mainland China, Japan and much of North America. It is also by far the largest of the North American floristic elements, which, because of its innate complexity, will eventually have to be subdivided into further subgroups or subelements as they are recognized. It contains some 500 species of bryophytes and comprises 75–80% of the known bryological flora of arctic North America, as an extension northward to the tundra regions of those associations and communities of species which are even more common in the forests, swamps, mountains, marshes, bogs and fens of cool-temperate climates considerably further south. This is also, in great measure, the element whose species have colonized the vast areas which were glaciated and devegetated during Pleistocene times (Fig. 1).

Different members of the circumboreal element show remarkable differences in their geographic ranges. Some species are rather narrowly localized by environmental circumstances and special habitat preferences, namely, a specific

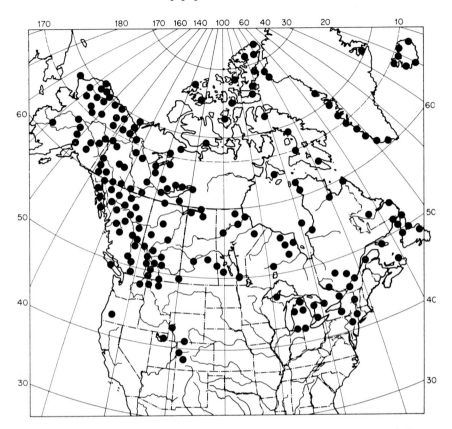

FIG. 1. The North American distribution of *Tomenthypnum nitens* (= *Homalothecium nitens*), a circumboreal species. (From W. B. Schofield, unpublished, with permission.)

type of rock, the ecological conditions of certain types of forest, or a specific climatic zone. On the other hand, many species are extremely wide ranging, and some of them, for example, *Aulacomnium palustre*, *Mnium thomsonii*, *Hypnum lindbergii* and *Fissidens bryoides*, range all the way from Florida and the Gulf Coast of the southern United States northward to the ultimate extent of land. Whether this extremely broad geographic range of one species represents a single genotype or a whole series of populations, each with a different degree of tolerance to cold and other environmental stress factors, only experimental work, especially transplantation experiments, can tell us.

The circumboreal element is today so widely distributed and so omnipresent in northern North America that its historically disjunct distribution is often forgotten. In any event, as improbable as it may seem today in view of the relatively homogeneous distribution of the circumboreal element throughout northern areas at present, it actually consisted over long geological eras of several disjunct populations. With the onset of the Pleistocene ice age perhaps two million years ago, those populations which survived in unglaciated refugia north of the vast continental glaciations became far separated geographically from those populations which migrated southward before the advancing ice sheets and accompanying climatic deterioration, to survice farther south. With the slow melting away of the continental glaciers, this element of bryophytes migrated back into the areas where open terrain was gradually being exposed. Eventually, the entire expanse of open land which had once been covered to great depths by ice was recolonized to such an extent that the present range of most species appears to be homogeneous (Fig. 1).

When methodologies become sufficiently refined, it will be rewarding to make a comparative study of the presumably stable relict arctic populations which have persisted in the same refugia for an enormous period of time and those populations of the same species which have been relatively mobile in their migration southward and their return to the north over areas that had been glaciated. It seems safe to predict the discovery of subtle differences in physiological adaptations to various factors of the environment, in ecological tolerances, in degree of fertility, in a tendency toward weediness or aggressiveness in competition and even in chromosomal behaviour and number.

Some measure of the length of time needed for the migration of boreal mosses is given in a very thoughtful paper by Crum (1972), who pointed out that 245 species are now present in the "Great Lakes Forest" of the northern part of the Lower Peninsula of Michigan. Crum emphasized, however, that this region was completely covered by continental glaciers until approximately 10 000 years ago, after which it was inundated successively by two great freshwater seas produced by melting glaciers, Lake Algonquin and Lake Nipissing, as late as 2500 years ago.

2. The Circumpolar Arctic and Arctic-alpine Bryophyte Element (Figs 2–10)

Several of the many major arctic expeditions of the eighteenth and nineteenth centuries returned to their home base with a large collection of plants from which new species of mosses and hepatics were occasionally described. Because the new species from one expedition rarely overlapped with or duplicated those described from other expeditions, the idea very naturally arose and was rather widely accepted (Herzog, 1926) that several distinct centres of apparent

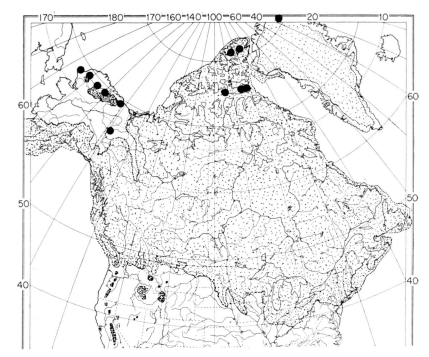

Fig. 2. The North Anerican distribution of *Seligeria polaris* Berggr., an arctic circum-polar species. (From Vitt, 1976a, with permission.) The stippled areas of this and several subsequent maps indicate maximum Pleistocene glaciation during the Wisconsin stage.

endemism existed for arctic bryophytes, such as the Lena and Yenisei rivers of Asiatic USSR, the islands of the Canadian Arctic Archipelago, Spitzbergen and several areas of Greenland. However, during my first field season in arctic Alaska in 1951, I discovered that most of the species considered to be endemic to relatively small areas in Spitzbergen, Siberia, Ellesmere Island and Greenland also occurred in the vast tundra areas south of Point Barrow. By the end of the 1951 field season, it had become evident to me that the "endemic" species of earlier bryologists were not endemic at all to any individual area, but that all such species must belong to a single high-arctic circumpolar floristic element. Equally stimulating, it then became possible to predict that any arctic bryo-phyte apparently restricted to a single locality or region would in all probability be discovered eventually in all or most of the major regions of the Arctic. Since I proposed this hypothesis (Steere, 1953), greatly accelerated field work throughout the Arctic has provided adequate evidence to substantiate the

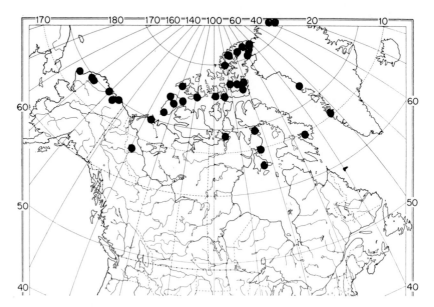

Fig. 3. The North American distribution of *Voitia hyperborea* Grev. and Arnott, an arctic circumpolar species.

existence of the circumpolar arctic floristic element. Brassard (1974) has discussed the origins and evolution of this element at some length and with considerable insight. About 100 species of bryophytes belong to this element in arctic North America (about 15% of the total bryophyte flora) and many more are certain to be added to it as careful field work by bryologists continues (Figs 2–10).

A corollary of the recognition of an arctic circumpolar element consisting of species which are not closely related to those of the boreal regions farther south, and which are probably Tertiary relicts, is that the presence of such an element must be closely correlated with the lack of extensive continental glaciation in much of Alaska and the Yukon, as elsewhere in arctic North America. These great unglaciated regions therefore served as refugia for Tertiary plants. As I said in proposing the arctic circumpolar element (Steere, 1953), "It would not require too great a stretch of the imagination to postulate that we are dealing with the last remnants of a widely distributed Tertiary or inter-glacial flora that originated in some far distant area, probably in the southern hemisphere, and that has now been restricted to the arctic regions by the destructive activities of continental glaciations at lower latitudes, as well as

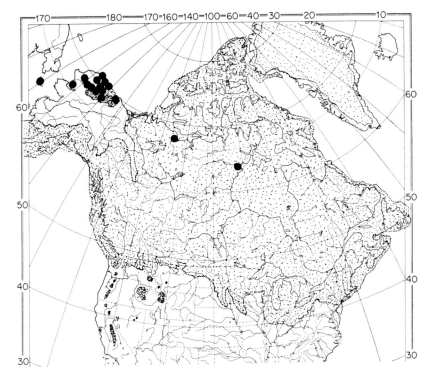

F‍ɪɢ. 4.   The North American distribution of *Pseudolepicolea fryei* (Perss.) Grolle and Ando,
an arctic hepatic originally described from St Lawrence Island; it also occurs
in arctic Siberia.

by the success in competition of more rapidly speciating genera and of younger
and more aggressive species there."

Two excellent and thought-provoking maps are now available to illustrate
the remarkable small amount of continental glaciation which occurred in
Alaska. Coulter *et al.* (1965) produced a map showing the maximum glaciation
through the whole duration of the Pleistocene, and Prest (1969) mapped the
maximum extent of ice in all of North America, but only during the Wisconsin
stage, the last major advance of ice sheets during the Pleistocene. These maps
both show graphically the relatively small amount of continental glaciation in
northern Alaska and the Yukon, compared with the large amount of land surface
that remained free of ice. This situation comes as something of a shock to many
biologists who had assumed that those areas must have been covered to a very
great depth by a massive ice-cap, just as central Greenland and Antarctica are

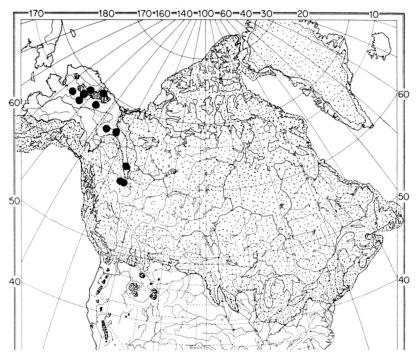

FIG. 5.   The total known distribution of *Andreaeobryum macrosporum* Steere and B. Murray,
an arctic American species apparently restricted to unglaciated refugia.

today—and just as Alaska was shown to be in most early maps of the Ice Age,
based on assumption only.

In addition, it has become increasingly evident that certain areas long considered to have been completely glaciated have actually escaped glaciation,
within otherwise glaciated regions, and were thus able to serve as local refugia
for plants. Many members of the circumpolar arctic element occur only within
the Arctic Circle, whereas others, although more frequent in arctic areas,
also extend southward where high altitude or cold climates provide appropriate
environmental conditions, as arctic-alpine species (Gams, 1955; Holmen, 1955,
1960; Steere, 1953, 1965, 1976a, 1978). Several species which had previously
been thought to be restricted to the Arctic have now turned up rather unexpectedly in the northern Rocky Mountains in isolated areas which appear to have
escaped Pleistocene glaciation, or at least some phases of it (Packer and Vitt,
1974), for example, *Didymodon johansenii* (Williams) Crum (Crum, 1965,
1969), *Bryobrittonia longipes* (Mitt.) Horton (Vitt, 1974b; Horton, 1978) (Fig.

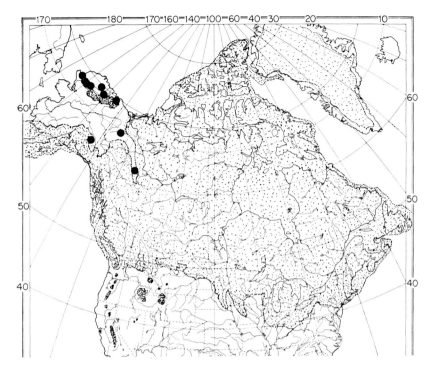

Fig. 6. The total known distribution of *Lejeunea alaskana* (Schust. and Steere) Inoue and Steere, an arctic hepatic which exists outside the Arctic Circle only in unglaciated refugia.

10), *Oreas martiana* (Hopp. and Hornsch.) Brid. (Weber, 1960, 1973; Schofield, 1972) (Fig. 8) and *Hypnum procerrimum* (= *Ctenidium*) (Bird, 1968; Schofield, 1972).

In addition to occasional local areas in the Rocky Mountains as far south as Alberta and British Columbia which apparently escaped Pleistocene glaciation or some phases of it, there existed east of the Rockies, just west of the Mackenzie River, a long and narrow ice-free corridor which ran from the enormous unglaciated regions of Alaska and the Yukon southward to the southernmost reaches of the Northwest Territories and probably still farther south (Prest, 1969)—and perhaps as far south as northern British Columbia (Figs 5 and 10). This ice-free corridor was apparently caused by the failure of the Cordilleran ice sheets moving eastward out of the Rocky Mountains to meet and join, over a broad front, the vast Keewatin continental glaciers which flowed in all directions from the Canadian Shield. A section of the South

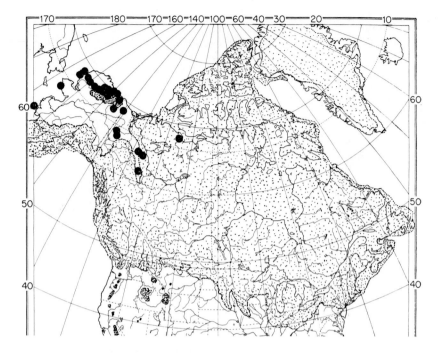

FIG. 7. The North American distribution of *Radula prolifera* Arn., an arctic hepatic which generally occurs in unglaciated refugia; it also occurs in arctic Siberia, where it was first discovered.

Nahanni River, in southwestern Mackenzie District of the Northwest Territories, lies within the southern part of this ancient corridor. In an excellent illustration of the clear correlation between the circumpolar arctic element of bryophytes and unglaciated areas, it was recently discovered that the unglaciated parts of Nahanni National Park, toward the southern end of an unbroken migratory pathway from the unglaciated areas of northern Alaska and the Yukon, have an abundance of the arctic element (Steere *et al.*, 1977; Steere and Scotter, 1978a) (Figs 5–7). From further north, in the unglaciated parts of the Mackenzie Mountains of the Northwest Territories, Brassard (1972) reported two species of the circumpolar element, and Hong and Vitt (1977), Bird *et al.* (1977), Vitt (1976b) and Vitt and Horton (1978) made parallel discoveries of the same element in the unglaciated parts of the Yukon Territory.

Prest's map showing the maximum extent of ice-cover in North America during the Wisconsin stage of the Pleistocene (Prest, 1969) indicates that the

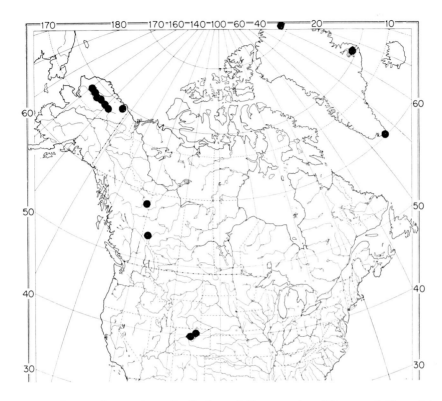

Fɪɢ. 8. The North American distribution of *Oreas martiana* (Hopp. and Hornsch.) Brid., an arctic-alpine species, which also occurs in Europe and Asia. (After Steere, 1978, modified.)

Canadian Archipelago, with the exception of Banks Island, was totally glaciated. However, the geological observations of England and Bradley (1978) and the bryological work of Schuster *et al.* (1959), Brassard (1971, 1976) and Schofield (1972) strongly suggest that some unglaciated refugia must have been available, because of clear evidence provided by glacial moraines, ice thickness, and the survival of so many relict species characteristic of ice-free areas (Figs 2, 10, 11 and 13). Schuster and Damsholt (1974) have presented parallel evidence for the survival and persistence of relict Tertiary species in unglaciated regions of West Greenland.

The recently recognized, clear-cut correlation between lack of glaciation and the distribution of bryophytes becomes obvious when maps showing the total known distribution of the species under consideration are compared

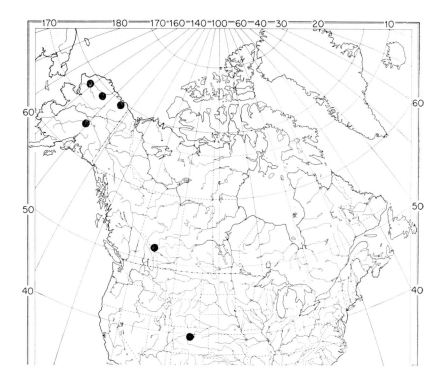

FIG. 9. The North American distribution of *Voitia nivalis* Hornsch., an arctic-alpine species, which also occurs in Europe and Asia. (From Steere, 1978.)

carefully with the maximum extent of Pleistocene glaciation in North America (Figs 4–6).

Continued search will doubtless yield much more information on the occurrence of other members of the same element considerably further south in North America than we would have anticipated originally, just as some of the otherwise wholly arctic species in Europe persist in the Alps. Also, many typically arctic American bryophytes occur in the "tundra zone" in the immediate vicinity of Lake Superior (Steere, 1937a, 1938; Schuster, 1966–1974) and Great Bear Lake (Steere, 1977), where the very low prevailing temperature of a great volume of water in the deep lakes drastically affects the littoral vegetation and its constitution.

An important phytogeographic question which has not yet been satisfactorily answered is how so many species from a considerably warmer Tertiary era could have survived until today, even in the large refugia available to them,

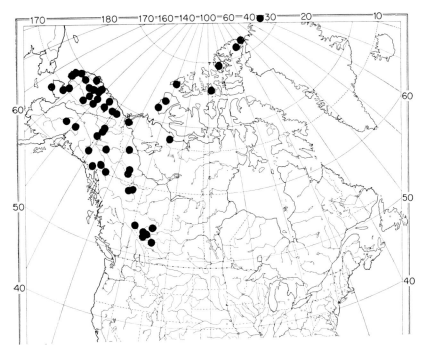

Fig. 10. The North American distribution of *Bryobrittonia longipes* (Mitt.) Horton, an arctic-alpine species which also occurs in Siberia. (From Horton, 1978, with permission.)

because of the dramatic changes which occurred in their environment, particularly the presumably much colder climatic regime of the Pleistocene and the consequent disappearance of trees from a previously forested region. Several of the relict species, particularly those in such genera as *Radula*, *Frullania* and *Lejeunea*, are normally epiphytic on the trunks of trees and yet they have survived very well in the treeless tundras of arctic North America. I have suggested (Steere, 1953) that the rather xeric tundra tussocks of *Carex*, *Eriophorum* and other sedges, as well as tussock-forming grasses, in and upon which these small plants grow today, resemble tree trunks ecologically in that all moisture available to them must come from fog, mist, rain, dew, seepage water, capillary action and other external sources. Like the hepatics, the mosses have also obviously become adapted to microhabitats and microenvironments in which they could survive, in spite of climatic deterioration over a long period. Of course, as I have pointed out elsewhere (Steere, 1976a), bryophytes have a remarkable ability to withstand low temperatures as well as drought and high temperatures. It seems, in fact, that temperature by itself is not necessarily a limiting or restrictive factor in the distribution of northern mosses and hepatics, although the indirect effects of temperature on available water may affect the environment drastically.

Many biologists have thought that plants of temperate species generally become smaller as they extend northward into boreal and arctic climates, and that the arctic flora is therefore simply a reduced and dwarfed temperate flora. However, although such a belief is based on a certain amount of accurate ecological observation with regard to the ubiquitous circumboreal element of bryophytes, it does not apply at all to the relict arctic floristic element, in which most species are as large and some of them conspicuously larger than their congeners which occur further south in much more favourable environments— a few examples are *Aulacomnium acuminatum* (Lindb. and Arn.) Par., *Cinclidium latifolium* Lindb., *C. subrotundum* Lindb., *Schistidium holmenianum* Steere and Brassard, *Tortella arctica* (Arnell) Crundw. and Nyh. and *Trichostomum arcticum* Kaal., each of which is the largest species known in its respective genus; many parallel examples exist in the Hepaticae (Steere and Inoue, 1978). In spite of their larger size, these relict arctic species are not necessarily polyploids; in fact, polyploidy among bryophytes is no more prevalent among arctic species than among those in temperate climates (Steere, 1954; Inoue, 1976).

As already noted in passing, some members of the arctic floristic element seem to be related to tropical or subtropical species of the Southern Hemisphere (Steere, 1953), which would suggest a very ancient origin and a very long period of isolation in northern regions of Tertiary species. These anomalous phytogeographic and evolutionary relationships have been commented upon also by Schuster (1974), when he established a new monotypic subsection for the arctic hepatic, *Scapania simmonsii* Bryhn and Kaal. Persson (1952) made the same point with reference to the moss, *Trichostomum cuspidatissimum* (*T. arcticum*), which he characterized as a "southern" species.

Those species of mosses and hepatics considered to comprise the circumpolar arctic element were listed by Steere (1965, 1976a) and Schofield (1972).

3. *Temperate Bryophytes Which Occur Disjunctly in the Arctic; the "Umiat Syndrome"* (Figs 11 and 12)

What I have termed the "Umiat Syndrome" (Steere, 1965) consists of a small but reasonably homogeneous group of disjunct species of bryophytes which are far more characteristic of cool temperate climates much farther south and which would never have been predicted by bryologists to survive

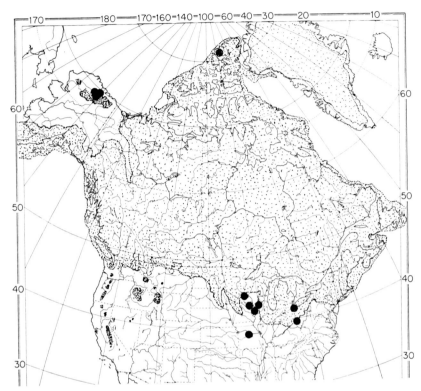

FIG. 11. The North American distribution of *Seligeria pusilla*, a temperate species which has apparently persisted in the Arctic since preglacial times; an example of the so-called "Umiat Syndrome". (From Vitt, 1976a, with permission.)

in the Arctic, yet nearly all of them occur at Umiat in arctic Alaska, hence the name. Brassard (1974) has observed a parallel survival of temperate species in what are presumably unglaciated refugia in Ellesmere Island (Fig. 11). These disjunct species were evidently stranded in arctic America at the onset of the Pleistocene ice age, at or toward the end of a much warmer period and of much more widely distributed temperate floras, and yet have been able to survive even under the much more rigorous climatic conditions and the changes in habitats caused by the disappearance of the forest. Some species are separated from the closest known populations southward by great distances indeed (Steere, 1965) (Figs 11 and 12), whereas others are not. Many of this group of disjunct species are characteristic of the upper Mississippi River basin of the north central United States as well as farther east, as in the St Lawrence River Valley and the Great Lakes region (Crum, 1976) (Figs 11 and 12), although

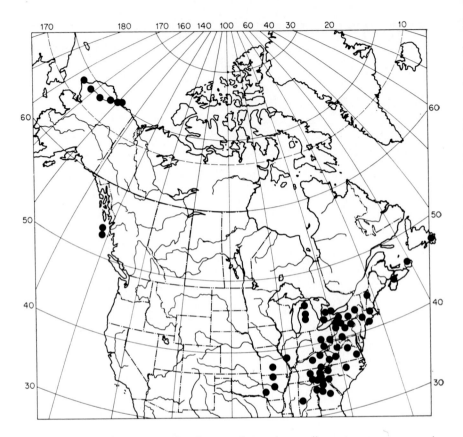

FIG. 12. The North American distribution of *Ctenidium molluscum*, a temperate species which has apparently persisted in the Arctic as a preglacial relict. The West Coast localities indicated are the unglaciated Queen Charlotte Islands. (From Schofield, unpublished, with permission.)

some species have western and southern affinities. A few members of this disjunct element occur on soil or humus but most grow directly on south-facing cliffs and ledges of calcareous sandstone or under the projecting base of large rock fragments.

The 15 or so species of bryophytes which comprise the temperate disjunct element have been listed earlier by Steere (1965, 1976a).

4. *The Desert-and-steppe Bryophyte Element in Arctic America* (Figs 13 and 14)

It may come as something of a shock even to bryologists to learn that several

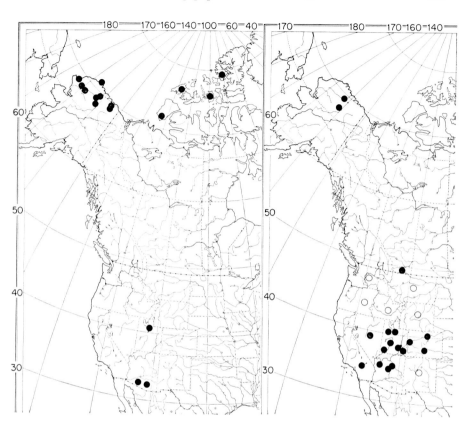

FIG. 13. The conspicuously disjunct North American distribution of *Pterygoneurum lamellatum* (left) and the total known distribution of *Tortula bistratosa* Flow. (right) showing the same kind of disjunction. The open circles represent reports for a state only, without a specific locality. (From Steere, 1978.)

species of mosses earlier known only from the southwestern and western deserts and steppes of temperate North America have also been discovered recently in the polar desert of arctic America. Like its southern counterpart, the polar desert has a very low level of precipitation and very low atmospheric humidity; moreover, both share low winter temperatures and relatively high summer temperatures, and the soil of both types of desert has a rather uniformly high level of alkalinity.

In arctic North America bryophytes of this element grow primarily on, or are restricted to, the fine-grained silt which is extruded in the form of "frost-boils" under the stress of the early winter freeze-up, and such silt extrusions

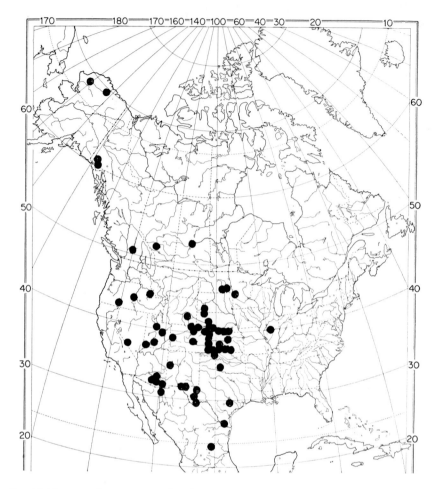

FIG. 14. The North American distribution of *Pterygoneurum subsessile* (Brid.) Jur.,
    showing its disjunct occurrence in arctic Alaska. (After Steere and Iwatsuki,
    1974.)

are found throughout the high Arctic. However, the curious and undoubtedly
ancient ecological specialization on silt probably developed during the warm
and dry climate of pre-Pleistocene or interglacial steppes and subdeserts and
not in the cold and dry one of the Arctic. In North America, *Pterygoneurum
lamellatum* was known until recently only from Arizona and Utah, when I
realized that what I had described as *P. arcticum* Steere from Alaska (Steere,
1959) was probably better placed in *P. lamellatum*, at least for the moment, in

spite of the remarkable and seemingly improbable disjunction involved (Steere, 1976b). It has also been reported from Ellesmere Island (Brassard, 1976) and from elsewhere in the Canadian Arctic Archipelago by Kuc (1973a) (Fig. 13). *Pterygoneurum subsessile* (Brid.) Jur. has a wide distribution in the drier parts of western and central North America, ranging from northern Mexico to the Yukon (Fig. 14), so that its recent discovery in northernmost Alaska by Steere and Iwatsuki (1974) does not represent nearly as great a disjunction as does the distribution of *P. lamellatum*. Like *Pterygoneurum subsessile*, *Stegonia latifolia* (Schwaegr.) Broth. has a very wide distribution in the drier parts of North America, especially in the Rocky Mountains, as far south as Nevada, northward to Alaska, the Yukon and Northwest Territories and Greenland. Gams (1934) considered this to be a characteristic steppe species is Asia and Europe. *Stegonia pilifera* (Brid.) Crum, Steere and Anderson has much the same distribution pattern as *S. latifolia*, as well as the same ecological preference for calcareous silt. First described from Utah, *Tortula bistratosa* Flow. is now known to be widespread almost throughout the arid American Southwest, extending northward in the dryer areas to Oregon, Washington and Alberta, it was only very recently reported from arctic Alaska (Steere, 1978) where it was collected in two localities by Iwatsuki in 1974. *Tortula bistratosa* is closest in its pattern of geographical distribution in North America (Fig. 13) to *Pterygoneurum lamellatum*, of the five species so far discovered in this unique element; it is very close in both ecology and morphology to *T. desertorum* Broth., and perhaps identical with it.

The extension of bryophytes from the more southern steppes and deserts into the polar desert of North America raises the interesting question of whether the converse could also occur. As an example, *Fossombronia alaskana* Steere and Inoue is very widely distributed in Alaska (Steere and Inoue, 1974) and had recently been reported from Greenland by Mogensen and Brassard (1978). Since all the other species of *Fossombronia* occur in much warmer climates, it is not beyond the realm of possibility that *F. alaskana* may unexpectedly turn up in the American Southwest, even though it at present appears to be endemic to the arctic circumpolar element.

5. Disjunct Asiatic Bryophytes in Boreal and Arctic North America (Figs 15–18)

Several species of both mosses and hepatics which have been generally considered to be restricted to Asia, particularly eastern Asia, have been discovered in arctic North America, especially in its western region, northernmost Alaska. *Claopodium pellucinerve* (Mitt.) Best, originally described from Japan, is widespread in the mountains of western North America from Alaska (including

FIG. 15. The total known North American distribution of *Ascidiota blepharophylla* Mass., a hepatic known elsewhere only in Central China.

FIG. 16. The total known North American distribution of *Gollania turgens* (C. Muell.) Ando, known elsewhere only in Central China. (From Steere, 1978.)

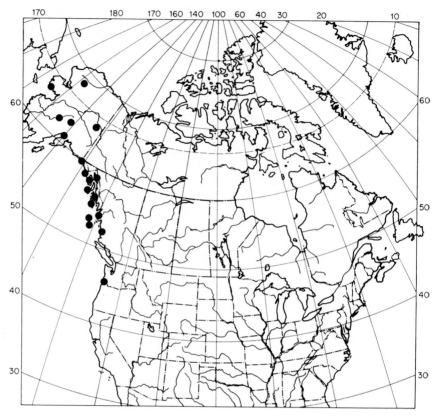

Fig. 17. The North American distribution of *Habrodon leucotrichus* (Mitt.) Perss., known otherwise only from Japan and elsewhere in East Asia. (Schofield, unpublished, with permission.)

the Arctic) and the Yukon, southward in the Rocky Mountains through British Columbia to Arizona, New Mexico and Mexico. On the Asiatic side it occurs in arctic USSR, the Himalayas, China, Korea and Japan (Noguchi, 1964). *Gollania turgens* (C. Muell.) Ando was described originally from Shensi and western Kansu provinces in central China, to which it remained endemic for a half century. However, in 1951 it was rediscovered in two localities in Alaska, and first reported from North America from Kantishna as *G. densepinnata* Dix. (Sherrard, 1957). I also collected it in arctic Alaska at Driftwood Creek in 1951 and it is now known from more arctic localities (Steere, 1978). Vitt and Horton (1978) very recently reported it from an unglaciated area in the Yukon, a remarkable range extension for this essentially Asiatic species (Fig.

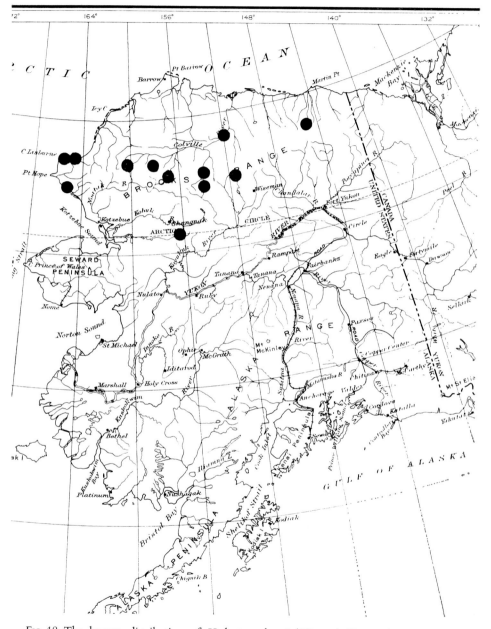

Fig. 18. The known distribution of *Herbertus sakuraii* (Warnst.) Hatt. subsp. *arcticus* Inoue and Steere. The subspecies *sakuraii* occurs in East Asia, especially Japan, and in British Columbia.

16). *Habrodon leucotrichus* (Mitt.) Pers., a common corticolous moss of the alpine forests of Japan, was first reported from Alaska by Persson (1946). Since then, it has been collected in several other localities in Alaska, where it appears to have a wide range (Steere, 1978); it has also been found in western British Columbia and on Saddle Mountain, Oregon (Schofield, unpublished data) (Fig. 17). It has been reported from arctic North America from only one locality, at Walker Lake on the south slope of the Brooks Range of Alaska (Sherrard, 1955). *Herzogiella adscendens* (Lindb.) Iwats. and Schof. (Iwatsuki and Schofield, 1973) is a wide-spread disjunct between eastern Asia and western North America. In Asia it is most common in Japan, and in North America in Alaska and southward along the Pacific Coast, where it reaches the mountains of British Columbia. *Myuroclada maximowiczii* (Borszcz.) Steere and Schof., a widely distributed species in eastern and northern Asia, was reported from Alaska by Steere and Schofield (1956) and is now known to be widespread there (Steere, 1978).

In the Hepaticae, too, one finds several species with a conspicuously disjunct distribution between eastern Asia and boreal and arctic North America. *Ascidiota blepharophylla* Mass. has very nearly the same history and present distribution pattern as *Gollania turgens* (Figs 15, 16). It was described originally from central China and rediscovered in 1951 at Driftwood Creek in arctic Alaska (Steere and Schuster, 1960). It is now known only from two further stations in northernmost Alaska (Steere and Inoue, 1978). *Herbertus sakuraii* (Warnst.) Hatt. subsp. *arcticus* Inoue and Steere is so far endemic to arctic Alaska (Fig. 18). whereas the subspecies *sakuraii* is much more widespread. Originally described from Japan, it is now known from the Himalayas and British Columbia, as a conspicuous disjunct (Steere and Inoue, 1978). *Lejeunea alaskana* (Schust. and Steere) Inoue and Steere was originally described from arctic Alaska (Schuster and Steere, 1958, as *Hygrolejeunea*), where it is widespread, and was recently reported from farther east in the same unglaciated area, in Yukon (Hong and Vitt, 1977) and from the southwestern part of the Northwest Territories (Steere and Scotter, 1978a) (Fig. 6). Although this species has not actually been discovered in Eastern Asia, its closest relatives are in Japan (Steere and Inoue, 1978); it has no close relatives at all in North America. Several other species, such as *Calycularia laxa* Lindb. and Arn., *Pseudolepicolea fryei* (Perss.) Grolle and Ando (Fig. 4) and *Radula prolifera* Arn. (Fig. 7) may be considered as Asiatic-North American disjuncts in a technical sense, since they also occur in arctic Siberia. However, they are better considered simply as members of the arctic circumpolar element which is restricted to high latitudes in each continent and region.

6. The Cosmopolitan Bryophyte Element

Many biologists are surprised at the small number of weedy bryophytes that really exist as most textbooks of botany give the impression that, because of the enormous numbers of spores which they produce, bryophytes can easily populate any available habitat. In fact, only a very small number of species are able to live up to this expectation; the great majority of species are strictly limited in their total present distribution by environmental, climatic and historical-geological events and factors.

In most areas of the earth, we have come to think of the weedy species of bryophytes which are truly cosmopolitan as being carried about by man and following human activities such as fires, and as being true weeds in artificially disturbed habitats. However, in many localities in arctic America, *Marchantia polymorpha*, *Bryum argenteum*, *Ceratodon purpureus*, *Funaria hygrometrica* and *Leptobryum pyriforme*, the most frequent of all weedy bryophytes and the most widely distributed over the globe, are both common and abundant in areas which have never been disturbed by human activities, but where the disruption of the habitats has been caused by other animals and by such natural events as erosion and ice action (Steere, 1978).

In summary, we may be sure that still further floristic elements exist among the bryophytes of arctic North America, and that they will be recognized and defined as their distribution patterns become better known. Moreover, some of the floristic elements described here will eventually be subdivided as they are studied in greater depth, and anomalous subgroups can be segregated out.

REFERENCES

ABRAMOVA, A. L., SAVICH-LYUBITSKAYA, L. I. and SMIRNOVA, Z. N. (1961). "Manual of Leafy Mosses of Arctic USSR." Akad. Nauk SSSR, Moscow-Leningrad. (In Russian.)

ANDREWS, A. L. (1940). *Bryum. In* "Moss Flora of North America North of Mexico" (A. J. Grout, ed.) Vol. II, pp. 211–240. Newfane, Vermont.

BIRD, C. D. (1968). New or otherwise interesting mosses from Alberta. *Bryologist* **71**, 358–361.

BIRD, C. D., SCOTTER, G. W., Steere, W. C. and MARSH, A. H. (1977). Bryophytes from the area drained by the Peel and Mackenzie rivers, Yukon and Northwest Territories, Canada. *Can. J. Bot.* **55**, 2879–2918.

BRASSARD, G. R. (1971). The mosses of northern Ellesmere Island, Arctic Canada. I. Ecology and phytogeography, with an analysis for the Queen Elizabeth Islands. II. Annotated list of the taxa. *Bryologist* **74**, 233–311.

BRASSARD, G. R. (1972). Mosses from the Mackenzie Mountains, Northwest Territories. *Arctic* **25**, 308.

Brassard, G. R. (1974). The evolution of arctic bryophytes. *J. Hattori bot. Lab.* **38**, 39–48.
Brassard, G. R. (1976). The mosses of northern Ellesmere Island, Arctic Canada. III. New or additional records. *Bryologist* **79**, 480–487.
Bryhn, N. (1906–1907). Bryophyta in itinere polari norvagorum secundo collecta. In "Report of the Second Norwegian Arctic Expedition in the 'Fram', 1898–1902". Pp. 1–260. Videnskabs-Selskabet, Kristiania.
Coulter, H. W., Hopkins, D. M., Karlstrom, T. N. V., Péwé, T. L., Wahrhaftig, C. and Williams, J. R. (1965). "Map Showing Extent of Glaciations in Alaska." Misc. Geol. Investig. Map I–415. U.S. Geological Survey, Washington, D.C.
Crum, H. A. (1965). *Barbula johansenii*, an arctic disjunct in the Canadian Rocky Mountains. *Bryologist* **68**, 344–345.
Crum, H. A. (1969). A reconsideration of the relationship of *Barbula johansenii* (Musci) *Can. Fld Nat.* **83**, 156–157.
Crum, H. A. (1972). The geographic origins of the mosses of North America's eastern deciduous forest. *J. Hattori bot. Lab.* **35**, 269–298.
Crum, H. A. (1976). "Mosses of the Great Lakes Forest." Revised edition. University of Michigan, Ann Arbor.
Crum, H. A., Steere, W. C. and Anderson, L. E. (1973). A new list of mosses of North America North of Mexico. *Bryologist* **76**, 85–130.
Douglas, G. W. and Vitt, D. H. (1976). Moss-lichen flora of St. Elias-Kluane Ranges, southwestern Yukon. *Bryologist* **79**, 437–456.
England, J. and Bradley, R. S. (1978). Past glacial activity in the Canadian High Arctic. *Science, N.Y.* **200**, 265–270.
Gams, H. (1934). Beiträge zur Kenntnis der Steppenmoose. *Annls bryol.* **7**, 37–56.
Gams, H. (1955). Zur Arealgeschichte der arktischen und arktischoreophytischen Moose. *Feddes Reprium Spec. nov. Veg.* **58**, 80–92.
Grout, A. J. (1928–1940). "Moss Flora of North America North of Mexico." Newfane, Vermont.
Herzog, T. (1926). "Geographie der Moose." Fischer, Jena.
Holmen, K. (1955). Notes on the bryophyte vegetation of Peary Land, North Greenland. *Mitt. thüring. bot. Ges.* **1**, 96–106.
Holmen, K. (1960). The mosses of Peary Land, North Greenland. A list of the species collected between Victoria Fjord and Danmark Fjord. *Meddr Grønland* **163** (2), 1–96.
Holmen, K. and Scotter, G. W. (1971). Mosses of the Reindeer Preserve, Northwest Territories, Canada. *Lindbergia* **1**, 34–56.
Hong, W. S. and Vitt, D. H. (1977). Hepaticae of the Yukon Territory, Canada. *Bryologist* **80**, 461–569.
Horton, D. G. (1978). *Bryobrittonia longipes*, an earlier name for *B. pellucida* (Encalyptaceae, Musci). *Brittonia* **30**, 16–20.
Inoue, H. (1976). Chromosome studies in some arctic hepatics. *Bull. natn. Sci. Mus., Tokyo*, Ser. B, Bot., 2, 39–46.
Iwatsuki, Z. and Schofield, W. B. (1973). The taxonomic position of *Campylium adscendens*. *J. Hattori bot. Lab.* **37**, 609–615.
Koponen, T. (1974). A guide to the Mniaceae in Canada. *Lindbergia* **2**, 160–184.
Kuc, M. (1973a). Bryogeography of Expedition Area, Axel Heiberg Island, N.W.T., Canada. *Bryophyt. Biblthca* **2**, 1–120.

Kuc, M. (1973b). Addition to the arctic moss flora. VI. Moss-flora of Masik River Valley (Banks Island) and its relationship with plant formations and the postglacial history. *Revue bryol. lichén.* **39**, 253–264.

Mogensen, G. S. (1973). A revision of the moss genus *Cinclidium* Sw. (Mniaceae Mitt.). *Lindbergia* **2**, 49–80.

Mogensen, G. S. and Brassard, G. R. (1978). *Fossombronia alaskana* found in Greenland. *Bryologist* **81**, 155.

Noguchi, A. (1964). A revision of the genus *Claopodium. J. Hattori bot. Lab.* **27**, 20–46.

Packer, J. G. and Vitt, D. H. (1974). Mountain Park: a plant refugium in the Canadian Rocky Mountains. *Can. J. Bot.* **52**, 1393–1409.

Persson, H. (1946). The genus *Habrodon* discovered in North America. *Svensk. bot. Tidskr.* **40**, 317–324.

Persson, H. (1952). Critical or otherwise interesting bryophytes from Alaska–Yukon. *Bryologist* **55**, 1–25, 88–116.

Prest, V. K. (1969). "Retreat of Wisconsin and Recent Ice in North America. Speculative Ice-marginal Positions During Recession of Last Ice-sheet Complex." Map 1257A. Geol. Survey of Canada, Ottawa.

Savich-lyubitzkaya, L. I. and Smirnova, Z. (1970). "The Handbook of Mosses of the USSR." Akad. Nauk. SSSR, Leningrad. (In Russian.)

Schofield, W. B. (1972). Bryology in arctic and boreal North America and Greenland. *Can. J. Bot.* **50**, 1111–1133.

Schofield, W. B. and Crum, H. A. (1972). Disjunctions in bryophytes. *Ann. Mo. Bot. Gdn* **59**, 174–202.

Schuster, R. M. (1966–1974). "The Hepaticae and Anthocerotae of North America East of the Hundredth Meridian." Columbia University Press, New York.

Schuster, R. M. and Damsholt, K. (1974). The Hepaticae of West Greenland from ca. 66°N to 72°N. *Meddr Grønland* **199**(1), 1–373.

Schuster, R. M. and Steere, W. C. (1958). *Hygrolejeunea alaskana* sp. n., a critical endemic of northern Alaska. *Bull. Torrey bot. Club* **85**, 188–196.

Schuster, R. M., Steere, W. C. and Thomson, J. W. (1959). The terrestrial cryptogams of northern Ellesmere Island. *Bull. natn. Mus. Can.* **164**, 1–132.

Sherrard, E. M. (1955). Bryophytes of Alaska. I. Some mosses from the southern slopes of the Brooks Range. *Bryologist* **58**, 225–236.

Sherrard, E. M. (1957). Bryophytes of Alaska. II. Additions to the mosses and hepatics of the Mt. McKinley region. *Bryologist* **60**, 310–326.

Smiley, T. L. and Zumberge, J. H. (eds) (1974). "Polar Deserts and Modern Man." Univ. Arizona Press, Tucson.

Steere, W. C. (1937a). Critical bryophytes from the Keweenaw Peninsula, Michigan. *Rhodora* **39**, 1–14, 33–46.

Steere, W. C. (1937b). *Bryoxiphium norvegicum,* the sword moss, as a preglacial and interglacial relic. *Ecology* **18**, 346–358.

Steere, W. C. (1938). Critical bryophytes from the Keweenaw Peninsula, Michigan. II. *Annls bryol.* **11**, 145–152.

Steere, W. C. (1948). Musci. *In* "Botany of the Canadian Eastern Arctic. Part II. Thallophyta and Bryophyta" (N. Polunin, ed.) *Bull. natn. Mus. Can.* **97**, 370–490. (Dated "1947", but published April, 1948.)

STEERE, W. C. (1953). On the geographical distribution of arctic bryophytes. *Stanford Univ. Publs* Biol. Sci. **11**, 30–47.

STEERE, W. C. (1954). Chromosome number and behaviour in arctic mosses. *Bot. Gaz.* **116**, 93–133.

STEERE, W. C. (1959). *Pterygoneurum arcticum*, a new species from northern Alaska. *Bryologist* **62**, 215–221.

STEERE, W. C. (1965). The boreal bryophyte flora as affected by Quaternary glaciation. *In* "The Quaternary of the United States" (H. E. Wright, Jr and D. G. Frey, eds) pp. 485–495. Princeton Univ. Press, Princeton.

STEERE, W. C. (1971). A review of arctic bryology. *Bryologist* **74**, 428–441.

STEERE, W. C. (1975). Mosses and hepatics from the Thule District, northwestern Greenland. *Am. Midl. Nat.* **94**, 326–347.

STEERE, W. C. (1976a). Ecology, phytogeography and floristics of Arctic Alaskan bryophytes. *J. Hattori bot. Lab.* **41**, 47–72.

STEERE, W. C. (1976b). Identity of *Pterygoneurum arcticum* with *P. lamellatum*. *Bryologist* **79**, 221–222.

STEERE, W. C. (1977). Bryophytes from Great Bear Lake and Coppermine, Northwest Territories, Canada. *J. Hattori bot. Lab.* **42**, 425–465.

STEERE, W. C. (1978). The mosses of Arctic Alaska. *Bryophyt. Biblthca* **14**, 1–508.

STEERE, W. C. and BRASSARD, G. R. (1974). The systematic position and geographical distribution of *Fissidens arcticus*. *Bryologist* **77**, 195–202.

STEERE, W. C. and HOLMEN, K. A. (1975). "Bryophyta Arctica Exsiccata. Fasciculus 1 (No. 1–50)." New York Botanical Garden, New York and Botanical Museum, Copenhagen.

STEERE, W. C. and INOUE, H. (1974) *Fossombronia alaskana*, a new hepatic from arctic Alaska. *Bryologist* **77**, 63–71.

STEERE, W. C. and INOUE, H. (1978). The Hepaticae of Arctic Alaska. *J. Hattori bot. Lab.* **44**, 251–345.

STEERE, W. C. and IWATSUKI, Z. (1974). The discovery of *Pterygoneurum subsessile* (Brid.) Jur. in arctic Alaska. *J. Hattori bot. Lab.* **38**, 463–473.

STEERE, W. C. and SCHOFIELD, W. B. (1956). *Myuroclada*, a genus new to North America. *Bryologist* **59**, 1–5.

STEERE, W. C. and SCHUSTER, R. M. (1960). The hepatic genus *Ascidiota* Massalongo new to North America. *Bull. Torrey bot. Club* **87**, 209–215.

STEERE, W. C. and SCOTTER, G. W. (1978a). Additional bryophytes from Nahanni National Park and vicinity, Northwest Territories, Canada. *Can. J. Bot.* **56**, 234–244.

STEERE, W. C. and SCOTTER, G. W. (1978b). Bryophytes of the northern Yukon Territory, Canada, collected by A. J. Sharp and others. *Brittonia* **30**, 271–288.

STEERE, W. C. and SCOTTER, G. W. (1978c). Bryophytes from the southeastern Yukon Territory, Canada. *Brittonia* **30**, 395–403.

STEERE, W. C., HOLMEN, K. A. and MOGENSEN, G. S. (1976). "Bryophyta Arctica Exsiccata. Fasciculus II (No. 51–100)." New York Botanical Garden, New York and Botanical Museum, Copenhagen.

STEERE, W. C., SCOTTER, G. W. and HOLMEN, K. (1977). Bryophytes of Nahanni National Park and vicinity, Northwest Territories, Canada. *Can. J. Bot.* **55**, 1741–1767.

VITT, D. H. (1974a). The mosses reported for the Yukon Territory. *J. Hattori bot. Lab.* **38**, 299–312.

VITT, D. H. (1974b). The distribution of *Bryobrittonia pellucida* Williams (Musci). *Arctic* **27**, 237–241.

VITT, D. H. (1975). A key and annotated synopsis of the mosses of the northern lowlands of Devon Island, N.W.T., Canada. *Can. J. Bot.* **53**, 2158–2197.

VITT, D. H. (1976a). The genus *Seligeria* in North America. *Lindbergia* **3**, 241–275.

VITT, D. H. (1976b). Mosses new to the Yukon from the Ogilvie Mountains, *Bryologist* **79**, 501–506.

VITT, D. H. and HORTON, D. G. (1978). Bryophytes new to the Yukon. *Bryologist* **81**, 167–168.

WEBER, W. A. (1960). A second American record for *Oreas martiana*, from Colorado. *Bryologist* **63**, 241–244.

WEBER, W. A. (1973). Guide to the mosses of Colorado. Keys and ecological notes based on field and herbarium studies. *Occ. Pap. Inst. arct. alp. Res. Univ. Colorado* **6**, 1–48.

7 | A Historical Review of Japanese Bryology

T. KOPONEN

*Botanical Museum, University of Helsinki,
Unioninkatu 44, SF-00170 Helsinki 17, Finland*

Abstract: Research into the bryophyte flora of Japan was strictly limited during the period of the "closed country" policy which ended in 1854. Most of the early collections were made by Europeans, and European bryologists were responsible for naming them. This work was of mixed quality and has since in many cases been corrected by the Japanese themselves. Japanese bryologists first became active at the end of the nineteenth century but competent monographic work was not produced until about 1930 for the hepatics and as recently as 1947 for the mosses. Nowadays, however, the Japanese bryoflora is one of the best known; there are some 500 hepatics and 1400 mosses recorded. The work of Japanese bryologists has now extended to cover the whole of eastern and southeastern Asia.

A SHORT HISTORY OF JAPANESE BRYOLOGY

1. Bryology During the "Closed Country" Policy

The history of Japanese bryology is very closely connected with the history of Japan itself. Many westerners are not familiar with Japanese history so it is necessary to outline the relevant points before showing how bryology fits into it.

The first westerners who came into contact with the Japanese were Portuguese (Schmid, 1942; Rudolph, 1974). Three Portuguese travelling in a Chinese junk were shipwrecked on Tanegashima in South Japan in 1542 or 1543. Others soon followed, and in 1549 Francis Xavier, a famous Jesuit missionary, entered the country to teach Christianity to the Japanese. In addition to the Portuguese, the Spanish and the Dutch established trading relations with Japan. Difficulties

---

Systematics Association Special Volume No. 14, "Bryophyte Systematics", edited by G. C. S. Clarke and J. G. Duckett, 1979, pp. 159–183, Academic Press, London and New York.

arose, however, with the Jesuits. The Japanese became aware of the militant attitude of the missionaries, and feared the growth of colonialism in the Far East. Finally, Christianity was completely forbidden and the foreigners were expelled. Japan then entered the *sakoku* period of the so-called "closed country" policy. The missionaries did not leave willingly but only after violent persecution; the Catholic Church records over 3000 martyrdoms of both missionaries and Japanese converts in Japan at that time.

Japan was, therefore, almost completely isolated and no westerners except for a small number of Dutch were allowed into the country. Japanese were forbidden to leave their country, and if they did they were not permitted to return. The construction of seagoing ships was banned. Even the import of foreign books was not allowed, and foreigners were prohibited from studying the Japanese language. In 1825 an order was issued that any foreign vessel, no matter what its intentions, must be driven away from Japanese shores. The firearms, musket and cannon, which the Portuguese had brought and which the Japanese soon learned to manufacture and use, helped to keep the seclusion complete. The closed country policy lasted some 200 years from 1639 to 1854.

Under conditions like these, exploration, including bryological exploration, was not very effective. During the late eighteenth and early nineteenth centuries, while explorers and plant collectors from France, Britain and other nations were travelling widely in Africa, the Americas and the Pacific area, Japan remained practically a virgin land for botany. However, as mentioned above, the Japanese allowed one exception to their strict rules. A small number of Dutch Protestants, who did not bring missionaries with them, were allowed to maintain a small so-called "factory" on the small island of Deshima in Nagasaki. Only a dozen Europeans and some Javanese servants were allowed to stay there at any one time. They were kept under such very close surveillance that they were effectively prisoners. They were always cared for by European physicians, some of whom were not only medical men but scientists and naturalists as well. Through their activity the first specimens of bryophytes collected in Japan reached Europe. The earliest botanists visiting Japan were A. Cleyer (in Japan 1682–1687), G. Meister (1682–1687) and E. Kämpfer (1690–1692) (Schmid, 1942). They paid little attention to mosses, although at least some specimens older than Thunberg's collections exist (Mitten, 1891, p. 189).

Linnaeus's student Carl Peter Thunberg (1743–1828) stayed in Japan for 15 months in 1775–1776. His trip is well documented by his own reports (1795), and by Stearn (1958), Schmid (1942) and Rudolf (1974). Since Thunberg was Swedish he studied the Dutch language for three years in Cape Town before his trip and entered Japan as a Dutch physician. He collected mainly vascular plants, and only five mosses and four hepatics are mentioned in his *Flora Japonica* (1784; see also Takaki, 1973; Iwatsuki, 1976). The small number of specimens is easily understood when the conditions of collecting are remembered (Rudolf, 1974).

The most important early collections of Japanese bryophytes were made by Philipp Franz v. Siebold (1796–1866), who stayed in Japan from 1823 to 1830 and 1859 to 1862, and some of his contemporaries (Schmid, 1942). Siebold was mainly a vascular plant taxonomist, and his bryophyte collections went to Dozy and Molkenboer (1844, 1845–1854, 1854) and to Sande Lacoste (1857, 1864, 1866–1867) for determination (Asahina, 1959; Kitagawa, 1971–1973). In addition to Siebold, C. J. Textor, Pompe van Meerdervoort, and Itō Keisuke are listed as collectors by Sande Lacoste.

In addition to the Dutch, or persons masquerading as Dutch, there was another important naturalist who entered Japan during the 200 years of seclusion policy (Lensen, 1959). He was Lieutenant Adam Laxmann (1766–1803), the son of Erik Laxmann, originally a Finnish correspondent of Linnaeus. A number of Japanese fishermen had been stranded in the Aleutians, and the Russians sent them to St Petersburg (now Leningrad) via Siberia. An expedition was arranged to return the two surviving Japanese. Adam Laxmann was the leader of the expedition, and while trying to negotiate trade privileges for the Russians, he collected plants and other material in 1792 in Nemuro, Hokkaido. His collections, now in the British Museum, may contain bryophytes, but there is no record of this.

2. The Period of European Domination

On 3 July 1853, Commodore Perry and his squadron of four ships were anchored off the city of Uraga, on the western side of the Bay of Yedo (now the Bay of Tokyo). Perry had a message from the President of the United States requesting a treaty to the effect that Japan should open her borders and allow free trading. The treaty was signed in Kanagawa on 31 March 1854 and the long period of seclusion began to crumble.

However, the restless times which followed in Japan hindered free travel and the inner parts of Japan were especially inaccessible to foreigners. Permission was still needed to collect plants. A good example is the visit of the Russian botanist Maximovicz (Bretschneider, 1898) who stayed in Hakodate in 1860–1861 and was allowed to travel only about 30 km from the city. Nevertheless, several expeditions including botanists visited Japan and many diplomats also collected specimens, including mosses. Missionaries were again allowed to enter Japan. The following list (Table I) gives the most significant foreign collectors and the dates of their visits to Japan, if known. Foreign bryologists who named and published their specimens are also cited in the table, as are sources of additional information. The list is preliminary and therefore still incomplete.

TABLE I. Collectors of Japanese bryophytes during the second half of the nineteenth century, with a list of the bryologists who named their specimens

Collectors, with dates of visit (if known)	Bryologists who named the specimens and sources of further information
Anckarkrona	Brotherus (1899)
Bisset, J., 1879, 1887–1888	Mitten (1891), Brotherus (1899), Evans (1905)
Brauns, D., 1881	Reimers (1937)
Buerger	Mitten (1891)
Charen, P. J., 1893	Warnstorf (1907, 1911)
Dampax (and F. F. Gaultiers), 1902	Paris (1904)
Dickins, F. W. ("Dickens", "Dikkens"), 1876	Mitten (1891), Bescherelle (1893), Salmon (1900)
Faurie, P. U. J., 1873–1913	Bescherelle (1893, 1894, 1898, 1899), Cardot (1897, 1907, 1908, 1909, 1911, 1912), Brotherus (1899, 1920), Warnstorf (1900, 1904, 1907, 1911), Stephani (1898–1925), Paris (1902, 1904), Hagen (1906), cf. Sayre (1975)
Ferrié, B. J., 1880	Renauld and Cardot (1904), Thériot (1907, 1908, 1909), Brotherus (1920), cf. Sayre (1975)
Ford, C., 1890	Bescherelle (1893), Salmon (1900)
Harin	Brotherus (1899)
Henon	Duby (1876), Mitten (1891)
Hilgendorf, F. W., 1874	Brotherus (1899), Reimers (1937)
Kjellman, F. R., 1879 (Vega expedition)	Geheeb (1881)
Maiden, J. H., 1885	Salmon (1900)
Maries, C., 1877–1878	Mitten (1891)
Maximovicz, C., 1860–1864	Lindberg (1873), cf. Bretschneider (1898)
Mayr, H., 1890	Bescherelle (1893), Brotherus (1899), Salmon (1900)
Moseley, H. N., 1875 (The Challenger expedition)	Mitten (1891), Salmon (1900), cf. Sayre (1975)
Nordenskiöld, A. E., 1879 (Vega expedition)	Bescherelle (1900), Stephani (1900)

Oldham, R., 1861–1862	Mitten (1865), Salmon (1900), cf. Bretschneider (1898), Sayre (1975)
Perry expedition, 1853, 1854	Sullivant (1857), cf. Bretschneider (1898), Iwatsuki (1966), Sayre (1975)
Piotrowski, 1888	Bescherelle (1893)
Rein, 1874, 1877	Brotherus (1899)
Roux, L., 1888	Bescherelle (1893)
Savatier, P. A. L., 1866–1871, 1873–1876	Husnot (1883), Salmon (1900), Warnstorf (1911), Cardot (1912), cf. Sayre (1975)
Schaal, F.	Jaeger and Sauerbeck (1876–1879), Mitten (1891), Bescherelle (1893)
Schottmüller, O., 1861	Reimers (1937)
Stone, J.-B., 1891	Dixon (1931)
Thomson	Mitten (1891)
Wichura, M. E., 1860–1861 (Preussische Expedition nach Ostasien)	Bescherelle (1898), Brotherus (1899, 1920), cf. Reimers (1931), Sayre (1975)
Wilford, C., 1857, 1859	Mitten (1891), Salmon (1900)
Woods	Mitten (1891)
Wright, C., 1855 (Ringgold and Rodgers expedition)	Sullivant and Lesquereux (1859), Bescherelle (1893), cf. Bretschneider (1898), Iwatsuki (1966), Sayre (1975)

The list suggests that Japanese bryology was, at that time, largely dominated by European bryologists. The collections were named and published by those who had primary access to the collections or secondary access through interest in special groups. The Americans Sullivant and Lesquereux (1857, 1859) studied specimens collected by early American expeditions, but later American bryologists hardly played any role in research on the Japanese bryoflora. Indeed, such American bryologists as J. H. Holzinger forwarded specimens sent to him by Japanese bryologists to Cardot (Cardot, 1907) and specimens reaching E. G. Britton in the New York Botanical Garden were sent to V. F. Brotherus. William Mitten had access to the specimens collected by British explorers and collectors. His *An Enumeration of all the Species of Musci and Hepaticae Recorded from Japan* (Mitten, 1891) was the second compilation of the Japanese bryophyte flora, following Miquel's (1866) *Prolusio Florae Japonicae*. The title of Mitten's paper is not in close accord with the text; in addition to Japanese records, it deals with collections from China and India. One of the collectors listed, Dr Maingau ("A number of specimens without labels collected by Dr Maingau"), never visited Japan (Bretschneider, 1898; van Steenis-Kruseman, 1950), his specimens, among them *Octoblepharum albidum* Hedw. (Iwatsuki and Noguchi, 1973), originated from China.

The impact of the French bryologists Emile Bescherelle, Jules Cardot, Marie
Thériot and Jean Paris, on the bryology of Japan is remarkable. This is partly
due to the fact that they received specimens collected by the most successful
collector, French missionary Père Faurie (1847–1915) (Fournier, 1932), who
stayed in Japan for about 40 years and collected all over the country, as well
as in Korea, Taiwan, Sakhalin and Hawaii (Horikawa, 1949). But in addition,
the leading bryologists of the time were in close contact through a frequent
exchange of letters and specimens (Paris, 1910). To illustrate this, Table II
lists the letters received by V. F. Brotherus from other bryologists who published on the Japanese flora (these letters deal also with the floras of other parts
of the world).

TABLE II. Correspondence to V. F. Brotherus from bryologists who worked on Japanese
bryophytes

Correspondent	Years of correspondence	Number of letters
Bescherelle, E.	1876–1900	30
Cardot, J.	1883–1923	96
Dixon, H. N.	1905–1928	45
Evans, A. W.	1893–1924	5
Geheeb, A.	1870–1909	87
Hagen, I.	1900–1915	35
Mitten, W.	1901–1906	8
Paris, J. E. G. N.	1892–1911	223
Reimers, H.	1923	1
Renauld, F.	1878–1909	77
Salmon, E. S.	1898–1902	13
Stephani, F.	1888–1909	35
Warnstorf, C. F.	1883–1913	51

Warnstorf seems finally to have received most of the *Sphagna*, and Stephani
dominated the field of hepaticology. In addition to unidentified specimens,
others bearing unpublished manuscript names were frequently distributed by
such bryologists as Brotherus. Colleagues often published these names (Paris,
1904) or sometimes replaced Brotherus's names with others (Cardot, 1911).
The number of taxa described by these European bryologists is very high. This
is partly due to the fact that they usually had a narrow species concept and
partly to ignorance of the phytogeography of Japan. Species described earlier,
e.g. from the Himalayas (Mitten, 1859), were repeatedly described as new
from Japan. Inaccurate citation, transcription and transliteration of Japanese

place names into European languages has always caused difficulties, for instance, Stephani often reported specimens and new species for Japan on the basis of Hawaiian collections (Hattori, 1949; Mizutani, 1961). It is very fortunate that Carl Müller never became interested in the Japanese bryoflora!

3. Early Japanese Bryologists

The first Japanese known to collect mosses for scientific purposes was Itō Keisuke (1803–1901), a student of Siebold's and the translator of Thunberg's *Flora Japonica* into Japanese (Itō, 1887; Rudolf, 1974). His name is recorded as a collector by Dozy and Molkenboer (1845–1854), Sande Lacoste (1866–1867) and Mitten (1891). However, a list of species in Itō's own herbarium (Miquel, 1866) includes no bryophytes.

The Japanese who became interested in bryophytes at the end of the nine-teenth century faced the same problem as Americans had faced decades earlier and botanists of developing nations face today: most of the collections and, original type specimens were abroad, mainly in Europe. The only way to begin was to ask for help from foreign bryologists. V. F. Brotherus became one of the most frequently used helpers of early Japanese bryologists. He himself evidently contacted Japanese botanists (Ando, 1974), but his fame was also spread by bryologists' letters: "I have heard from one Miyake, a Japanese scholar, who is now studying in a university of Germany, that you are one of the most celebrated and most enthusiastic investigators on mosses . . . " wrote H. Nakani-shiki to Brotherus in 1903, then asking him to study his specimens. Brotherus was known for his rapid answers and identifications, and numerous Japanese began to send him their material. Table III lists Brotherus's Japanese correspond-ents. Some of these letters have been published by H. Inoue (1972) and Ando (1974, 1975, 1976).

Brotherus's Japanese correspondents sent not only their own specimens to him, but frequently also those of their students and friends. The total number of native Japanese collectors represented in Brotherus's herbarium and reliquiae (Koponen, 1976) is at least 107. There are many thousand specimens, most of which have never been published. Brotherus's (1899) paper is the first to be based largely on specimens collected by Japanese botanists, mainly Kingo Miyabe and his colleagues. Since Brotherus dealt only with mosses, those Japanese interested in hepatics were advised to contact Stephani, Schiffner or Evans. As Table III shows, the most active early bryologists were Eikichi Iishiba, Shutai Okamura, Kyuichi Sakurai and Hisahiko Sasaoka. All of them dealt chiefly with mosses.

TABLE III. Early Japanese bryologists who corresponded with V. F. Brotherus. Number of publications according to Merrill and Walker (1938), Walker (1960) and Mizushima (1963). Those who dealt mainly with hepatics are marked with an asterisk

Name	Years of correspondence	Number of letters	Number of bryological publications	The year of the first publication
Fujii, K.	1905–1911	4	—	—
Gono, M.	1908–1910	4	—	—
Iishiba (Ihsiba), E.	1906–1921	17	27	1907
*Inoue, T.	1905	2	11	1894
(= Yoshinaga, T.)				
Kono, G.	—	—	2	1906
Miyabe, K.	1891–1894	3	—	—
*Miyake, K.	—	—	3	1898
Nakanishiki, K.	1903–1908	3	5	1905
Okamura, S.	1907–1915	13	c. 50	1908
Sakurai, K.	1910–1926	32	c. 50	1931
Sasaoka, H.	1915–1929	10	c. 25	1910
*Sawada, K.	—	—	2	1909
Toyama, R.	—	—	7	1935
Tsuchiga, Y.	1928	1	—	—
*Tsuge, C.	—	—	4	1887
Vematsu, E.	1907–1917	8	—	—
Yasuda, A.	1913–1918	6	4	1915
Yatabe, R.	1891	1	—	—

Many of the first papers of the Japanese bryologists were merely lists of bryophytes which had been identified by foreign specialists, or reproductions of foreign papers (compare, for instance, Brotherus, 1906, and Okamura, 1908). However, Okamura (1877–1947) was rather successful as a bryologist, and his bryological work culminated in monographic works (1915–1916). He also wrote the part dealing with the bryophytes for Makino's *Illustrated Flora of Japan* (1940). The *Cryptogamae Japonicae Iconibus Illustrae* (Matsumura and Miyoshi, 1899–1901) was one of the first comprehensive books dealing with cryptogams and it was followed shortly afterwards by the first manual of Japanese mosses by Iishiba, Uematsu and Katō (1912). Iishiba (1873–1936) went on to write another more complete manual (1929), which enumerated 1222 species of mosses in Japan, Korea and Formosa, and, in 1932, a text book about mosses giving keys to families, a synopsis of genera, and an index of vernacular names. In both of these works he validated numerous manuscript names of Brotherus and other European bryologists by giving a brief Japanese description for each species. He also published a checklist of Japanese hepatics and a key to the families and genera of Japanese hepatics (Iishiba, 1930, 1940). Toyama's series *Spicilegium Muscologiae Asiae Orientalis* (1935, 1937a, b, c, 1938) should also be mentioned.

K. Sakurai (1889–1963) was the dominant muscologist in Japan from the early forties to the early sixties. He was a professional physician and evidently the first Japanese bryologist to go abroad. He attended a course of medical training in Berlin during the winter of 1925–1926 and used his spare time to study mosses in the Berlin–Dahlem Botanical Garden where he contacted Dr Hermann Reimers. Reimers and Sakurai (1931) wrote one of the earliest critical papers dealing with the Japanese bryoflora. It is a tragedy of Japanese muscology that Sakurai did not follow Reimer's lead, but turned to a rather careless splitting. In later years he never contacted Reimers or other professional bryologists, and described a large number of species without paying enough attention to comparisons with related species. For instance, of the nine species of *Mnium* described by him, only one, *Plagiomnium tezukae* (Sak.) T. Kop. is valid today. This is despite the fact that Reimers's student Kabiersch (1936) was at the same time undertaking very competent monographic research on Mniaceae and related families. Sakurai, Sasaoka and some other Japanese also sent their specimens to H. N. Dixon, who was another careless worker when foreign floras were in question. Dixon (1942) wrote: "I have received specimens of Japanese mosses for determination from H. Sasaoka, an indefatigable collector. They contain a large number of undescribed species. . . ." Of the 21 new Japanese species described in that paper, only a single one is accepted today, and even it, *Gollania cochlearifolium* Broth. ex Sak., is based on a doubtful label name by Brotherus, validated before Dixon by Sakurai (1939). Sakurai summarized his studies in *Muscologia Japonica* (1954), which gives a systematic enumeration with literature citations, synonymy, Japanese names, and distribution in Japan. It also includes some keys in Japanese, and is illustrated. An estimation of the quality of Japanese bryology at that time was given by Verdoorn (1938).

In conclusion, the earliest Japanese muscologists seem to have followed the worst of the European traditions. Because of this, modern studies of mosses were delayed several decades by comparison with those of hepatics.

4. The Period of Monographing

The Japanese hepaticologists began to publish independently more than ten years before the muscologists, and serious monographic research also began earlier in the field of hepaticology. I believe that a single person, the late Professor Yoshiwo Horikawa (1902–1976), deserves the credit for this (cf. Suzuki, 1977; Ando, 1977). Horikawa was a strong enough personality to begin to work independently of European traditions. His early monographs (1929–1933, 1934, 1934–1936) are pioneer work and the beginning of serious taxonomic bryology in Japan. In his time, Hiroshima University became the most important centre of bryological research in Japan (Ando, 1975). Many of the Japanese bryologists active today are Horikawa's students or his students' students.

The major stimulus for bryology in Japan was the establishment of the Hattori Botanical Laboratory in 1946 and the foundation of the *Journal of Hattori Botanical Laboratory* (Hattori and Iwatsuki, 1973). An estimation of the significance of these events was given by Schuster (1975).

The *Journal of the Hattori Botanical Laboratory* has been, since its foundation, the major journal for publishing monographs of Japanese mosses and hepatics. Today the hepatic flora is thoroughly known; there is a monograph or revision for each major family and genus of hepatics. Also most of the moss genera and families have been monographed or revised during the last thirty years, and there are several monographs in preparation.

In 1972, the Bryological Society of Japan was established and a new journal *The Proceedings of the Bryological Society of Japan* was founded. The Society is large with about 400 members. The interest in bryology amongst the Japanese is demonstrated by the fact that Iwatsuki and Mizutani's *Coloured Illustrations of Bryophytes of Japan* (1972) has run to a fifth edition (1978), and the publisher has already sold more than 13 000 copies.

5. Bryological Exploration by the Japanese outside Japan

For historical reasons Japanese bryologists have been active in the study not only of their own flora but of the floras of neighbouring countries. Taiwan was under Japanese rule from 1895–1945, and most of its bryoflora has been studied and published by Japanese workers, especially Horikawa and Noguchi (Wang, 1970). The Japanese ruled Korea from 1910–1945, and parts of Manchuria from 1905–1945. Papers dealing with these areas (e.g. Noguchi, 1954; Osada, 1958; Hong and Ando, 1959) are not as numerous as those for Taiwan, but there are many records for these areas in the Japanese literature.

Once the Japanese bryoflora became better known, interest in studying other areas naturally rose. There have been many Japanese expeditions to other Asiatic countries with bryologists as participants, and bryophytes have also been collected by vascular plant botanists. Papers have been published on the floras of the Himalayas (Noguchi, 1964a, 1964b, 1966, 1971; Noguchi and Iwatsuki, 1975; Noguchi *et al.*, 1966; Hattori, 1966, 1968, 1971, 1975), Thailand (Horikawa and Ando, 1964; Noguchi, 1972, 1973a), New Guinea (Takaki, 1975), Ceylon (Noguchi, 1973b) and Borneo (Noguchi and Iwatsuki, 1972; Iwatsuki and Noguchi, 1975). The co-operative Hattori–Tennessee bryological expeditions (Iwatsuki, 1963, 1964; Sharp, 1965) deserve a special mention in this connection. These expeditions have returned with hundreds of thousands of bryophyte specimens, and reports of the identification of the materials are frequently published. Still more remote floras have also been studied such as those of the Antarctic (Horikawa and Ando, 1961) and Chile (Seki, 1974). Japanese workers, however, still mainly deal with east and south-east Asia, although a few tackle worldwide monographs, for example H. Ando is working on *Hypnum*, H. Ochi on *Bryum* and H. Inoue on *Plagiochila*.

A GUIDE TO THE JAPANESE BRYOLOGICAL LITERATURE

In the historical review given above I have avoided giving many references, since the bibliographies of eastern Asiatic botany (Merrill and Walker, 1938; Walker, 1960) cover the period up to 1958. The paper by H. Inoue (1972) should also be consulted. These bibliographies include, in addition to the authors and titles, general, regional and systematic indices in which biographic and other special topics are easily found; those articles which are entirely in Japanese are briefly reviewed. These bibliographies also contain a bibliography of every early Japanese bryologist. The bryological literature after 1958 is listed in the standard bryological journals, and the bibliographies published in *Flora Malesiana Bulletin* also cover Japan. Reviews of the literature published in Japan are published regularly in *Miscellanea Bryologica et Lichenologica*, and similar lists appear in the *Proceedings of the Bryological Society of Japan* (e.g. Iwatsuki, 1975, 1976).

There is no complete modern moss or hepatic flora of Japan, and the old floras mentioned previously are out of print. Iwatsuki and Mizutani's *Coloured Illustrations of Bryophytes of Japan* (1972) figures most of the common bryophytes, and Noguchi's *Handbook of Japanese Mosses* (1976) is useful for the muscologists. H. Inoue's *Illustrations of Japanese Hepaticae* (1973, 1976) will be a major manual once all the planned five volumes are completed. The above books are all in Japanese, although Inoue's contains annotations in English, and bryologists unable to read that language must consult the original monographs. Muscologists have an easy task thanks to Iwatsuki and Noguchi's (1973) checklist of Japanese mosses which is one of the most useful reference books for the whole of eastern Asia. The index lists the names of all genera and species of mosses described or reported from Japan, based on the literature available to the end of 1972. Each valid epithet is followed by author citations, literature references to illustrations published in Japan, distributional area in Japan, and Japanese name. Taxonomic and nomenclatural synonyms are given. The index also lists the botanical journals in which most of Japanese bryology is published, and the floras and manuals on Japanese mosses. A similar list for the hepatics is under preparation by Hattori and Mizutani. As mentioned above, the bibliographies contain a systematic section and useful lists of references of older literature dealing with hepatics are in Horikawa (1929) and Hattori (1952b).

In this connection several exsiccatae published in Japan are worth mentioning. *Hepaticae Japonicae Exsiccatae* have been published since 1946 by Hattori and, later, jointly by Hattori and Mizutani (Mizutani and Hattori, 1976). Twenty-one series and 1050 specimens have appeared so far. *Musci Japonici Exsiccati* by Hattori, Noguchi and Hattori, and Noguchi and Iwatsuki (Iwatsuki and Noguchi, 1976), which began in 1947, has 30 series and 1500 specimens. Iwatsuki and Mizutani began a new series under the name *Bryophyta Exsiccata* in 1977 made in 500 sets and sold at a moderate price. H. Inoue has been publishing *Bryophyta Selecta Exsiccata*, which also contains specimens from outside Japan, since 1970. Eight fascicles and 400 specimens have been already distributed. All of these exsiccates are invaluable for anyone who needs to identify a bryophyte collected in Japan.

PHYTOGEOGRAPHY

Bryogeography has been closely co-ordinated with the taxonomy of bryophytes since Schimper's (1860) and Brotherus's (1884) works on the zonal distribution of bryophytes. This applies also to Japan which, because of its mountainous terrain, is an ideal place for such studies. Hiroshima University and Horikawa's name must again be mentioned in this connection. There are in the world few schools which possess a comparable tradition in both sociological phytogeography and in the taxonomy of bryophytes. Japanese phytosociologists follow the strictly nomenclatural Braun–Blanquet method in their research and describe the bryophyte associations along with lichens and higher vegetation. This means that there are few papers which deal solely with the bryophyte associations (Nakanishi, 1962; Kobayashi, 1971). Iwatsuki and Hattori's long series on the epiphytic cryptogams (e.g. 1970), Nagano's (1969) paper on the bryophytes on limestone, Ando and Taoda's (1967) and Nakanishi and Suzuki's (1977) works on urban mosses are remarkable examples. There are also numerous articles dealing with Japanese moss gardens (e.g. Iwatsuki and Kodama, 1961; Ando, 1971; I. Ishikawa, 1973, 1974).

The mapping of bryophyte distribution has been popular among Japanese bryologists who often give not only the ordinary distribution map, but also the vertical range which, in many cases, is much more informative. The three dimensional method of mapping plant distributions developed by Horikawa (Horikawa, 1955, 1972; Ando, 1977) is unique among the various mapping projects. Mizushima (1960), Mizutani (1961), Osada (1966), and Watanabe (1972) among others have arranged Japanese bryophytes into distributional groups, which were originally applied to vascular plants (Hara and Kanai, 1958–1959).

Another topic in the field of phytogeography which has received much attention from bryologists is the relationship of the Japanese bryoflora to that of other continents, especially North America. This has been discussed by many Japanese and foreign bryologists (Hattori, 1951, 1952a; Iwatsuki, 1958, 1972; Schofield, 1965; Sharp and Iwatsuki, 1965; Iwatsuki and Sharp, 1967, 1968; Abramova and Abramov, 1969; Steere, 1969). In 1972, in connection with the 25th anniversary of the Hattori Botanical Laboratory, a symposium on "Distributional Patterns and Speciation of Bryophytes in the Circum–Pacific Regions" was held in Tokyo (Steere and Inoue, 1972).

FLORA

According to the checklists of Mizutani and Hattori (1969) and Iwatsuki and Noguchi (1973) the bryoflora of Japan consists of 501 species of hepatics and 1571 species of mosses. Since 1972, however, about 100 species of moss have been reduced to synonymy, and there are still others in the same category (Z. Iwatsuki, personal communication). The number of moss species may actually be about 1400. This large number can be explained by the large variety of habitats suitable for bryophytes within subtropical, meridional, temperate, oroboreal, and orohemiarctic bioclimatic vegetation zones (Hämet-Ahti *et al.*, 1976), and by the geological history of Japan. Japan was never completely glaciated, so the flora was able to persist through all the glacial periods. Nearly all the circumpolar boreal, and many of the arctic circumpolar bryophytes, grow in Japan. The only scarce habitat in Japan is extensive tracts of wet boreal fen and bog. For this reason *Drepanocladus* and similar genera are poorly represented. Remarkably, some of these boreal plants occur in Japan in more southerly vegetation zones than they do in the area of their main occurrence, and are clearly relics. Thus, *Cyrtomnium hymenophyloides* (Hüb.) T. Kop., which is considered to be a high arctic species, occurs in Japan in orotemperate zones.

One of the main questions posed by the Japanese flora is whether the byrophytes occurring there are conspecific with those in North America and Europe. In general, the boreal ones are the same, but the temperate and more southern species are frequently different. This is well demonstrated by the family Mniaceae (Koponen, 1973, 1978).

Although the study of the Japanese bryoflora has been pursued for more than a century, continued collecting still produces novelties. The finding of *Takakia* in Japan is a good example (Hattori *et al.*, 1968). Monographic research still adds species new to science (Koponen, 1971; Kanda, 1975, 1977). Although Japan is an island, there is less endemism in the flora than has previously been thought. Sakurai (1954), for example, classified 49·4% of Japanese mosses as endemic, and Herzog (1926) also thought the proportion to be high. Taking again the example of the Mniaceae, in 1974 it seemed likely that three or four species were endemic to Japan (Koponen, 1974). Within three years, all of these except *Pseudobryum speciosum* (Mitt.) T. Kop. were also found to occur in neighbouring areas (Koponen, 1977). Further study of floras in the Soviet Far East, such as that by Koponen *et al.* (1978) and in continental China will no doubt reduce still further the number of supposed Japanese endemic species.

MODERN BRYOLOGY IN JAPAN

1. Cytology

There are some early papers which deal with the karyotypes of Japanese bryophytes (Miyake, 1905; M. Ishikawa, 1916; Shimotomai and Kimura, 1932; Tatuno, 1933, 1941; Kurita, 1937) but the principal work in this field has been carried out since 1959 by S. Tatuno, K. Yano, S. Inoue, M. Segawa and K. Ono (for references see Fritsch, 1972; Steere, 1972). The Japanese often work on a genus or family in co-operation with taxonomists (e.g. S. Inoue and Iwatsuki, 1976), which gives the best results from the taxonomic point of view. Thus the karyotypes of 28% of Japanese bryophytes are already known (H. Inoue, personal communication).

2. Ultrastructure

Japanese bryologists were among the first electron microscopists, both scanning (Shin, 1963; Miyoshi, 1969a, 1969b) and transmission (Shin, 1963; Shin and Muroya, 1965; Miyoshi, 1969a). Since then many papers have been published reporting work using these techniques (Kamimura, 1970, 1971, 1972; Miyoshi, 1973a; Hirohama 1973, 1975; Saito and Hirohama, 1974).

3. Chemistry

At present there are two very productive groups of chemists studying bryophytes, one at Hiroshima University (Ando, 1975) and the other at Tokushima-Bunri University (for references see Suire, 1975; Huneck, 1977). Both groups deal with hepatics, and work in close co-operation with taxonomists (Matsuo et al., 1978; Asakawa and Takemoto, 1978).

4. Pollution Studies

The many papers by Taoda (1972, 1973a, 1976, 1977) can be mentioned under this heading. Especially interesting are his attempts to construct a "bryometer" to measure air pollution in big cities. Nakamura (1976) and Yokobori (1978) have also worked in this field.

5. Numerical Methods

Seki's (1968) revision of the family *Sematophyllaceae* is so far the only example of the application of numerical classification to higher systematic categories in bryophytes.

ACKNOWLEDGEMENTS

I wish to express my sincere thanks to Drs Hattori Sinske and Iwatsuki Zennosuke (Nichinan) and Inoue Hiroshi (Tokyo) and to Professors Daniel H. Norris (Arcata, California) and Ando Hisatsugu (Hiroshima) for their comments on the manuscript.

REFERENCES

ABRAMOVA, A. L. and ABRAMOV, I. I. (1969). Eastern-Asiatic affinities of the Caucasian bryoflora. *J. Hattori bot. Lab.* **32,** 151–154.

ANDO, H. (1971). Les jardins de mousses au Japon. *Bull. Soc. Amateurs Jard. alp.* **5,** 290–294.

ANDO, H. (1974, 1975, 1976). Old letters from Japanese botanists to Prof. V. F. Brotherus, 1–3. *Proc. bryol. Soc. Japan* **1**, 88–89, 110–111, 145–147.

ANDO, H. (1975). Bryology at Hiroshima University, Japan. *Taxon* **24**, 255–256.

ANDO, H. (1977). History of the distributional studies by Prof. Y. Horikawa and the publication of his "Atlas of the Japanese Flora". *Hikobia* **8**, 22–33.

ANDO, H. and TAODA, H. (1967). Bryophytes and their ecology in Hiroshima City. *Hikobia* **5**, 46–68.

ASAHINA, Y. (1959). Lichen and bryophyte specimens collected by Siebold and his contemporaries in Japan. *Bull. natn. Sci. Mus. Tokyo* **4**, 374–387.

ASAKAWA, Y. and TAKEMOTO, T. (1978). Chemical constituents of *Trichocolea*, *Plagiochila* and *Porella*. *Bryophyt. Biblthca* **13**, 335–353.

BESCHERELLE, M. E. (1893). Nouveaux documents pour la flore bryologique du Japon. *Annls. Sci. nat.*, Sér. 7, Bot., **18**, 320–393.

BESCHERELLE, M. E. (1894). Enumération des Hépatiques récoltées par M. l'abbé Faurie au Japon et déterminées par M. Stephani. *Revue bryol.* **21**, 25–27.

BESCHERELLE, M. E. (1898). Bryologiae japonicae supplementum I. (Fin). *J. Bot. Paris* **12**, 280–300.

BESCHERELLE, M. E. (1899). Bryologiae japonicae supplementum. I. and II. Pleurocarpi (1). *J. Bot. Paris* **13**, 37–45.

BESCHERELLE, M. E. (1900). Liste des muscinées récoltées au Japon par M. le Professeur A. E. Nordenskiöld au cours du voyage de la Vega, autour de L'Asie en 1878–1879. *Öfvers. K. VetenskAkad. Förh. Stockh.* **57**, 289–296.

BRETSCHNEIDER, E. V. (1898). "History of European Botanical Discoveries in China." Sampson and Low, London.

BROTHERUS, V. F. (1884). "Etudes sur la Distribution des Mousses au Caucase." Thèse de doctorat. Univ. Helsingfors.

BROTHERUS, V. F. (1899). Neue Beiträge zur Moosflora Japans. *Hedwigia* **38**, 204–247.

BROTHERUS, V. F. (1906). *Orthomniopsis* and *Okamuraea*, zwei neue Laubmoosgattungen aus Japan. *Ofvers. finska VetenskSoc. Förh.* **49**(10), 1–4.

BROTHERUS, V. F. (1920). Musci novi japonici. *Öfvers. finska VetenskSoc. Förh.* **62**, Avd. A, no. 9, 1–55.

CARDOT, J. (1897). Fontinales japonaises. *Revue bryol.* **24**, 33–36.

CARDOT, J. (1907). Mousses nouvelles du Japon et de Corée. *Bull. Herb. Boissier*, Sér. 2, **7**, 709–717.

CARDOT, J. (1908) Mousses nouvelles du Japon et de Corée. *Bull. Herb. Boissier*, Sér. 2, **8**, 331–336.

CARDOT, J. (1909). Mousses nouvelles du Japon et de Corée. *Bull. Soc. bot. Genève*, Sér. 2, **1**, 120–132.

CARDOT, J. (1911). Mousses nouvelles de Japon et de Corée. *Bull. Soc. bot. Genève*, Sér. 2, **3**, 275–294.

CARDOT, J. (1912). Mousses nouvelles de Japon et de Corée. *Bull. Soc. bot. Genève*, Sér. 2, **4**, 378–387.

DIXON, H. N. (1931). Contributions to Japanese bryology. *Revue bryol.* N.S. **4**, 153–169.

DIXON, H. N. (1942). Some new Japanese mosses. *Revue bryol. lichén.* **13**, 10–19.

DOZY, F. and MOLKENBOER, J. H. (1844). Musci frondosi ex Archipelago Indico et Japonia. *Annls Sci. nat.*, Sér. 3, Bot., 2, 297–316.

Dozy, F. and Molkenboer, J. H. (1845–1854). "Musci frondosi inediti archipelagi indici, sive descriptio et adumbratio muscorum frondosorum in insulis Java, Borneo, Sumatra, Celebes, Amboina, nec non in Japonia nuper detectorum minusve cognitorum." Hazenberg, Lugduni-Batavorum.

Dozy, F. and Molkenboer, J. H. (1854). Musci frondosi. *In* "Plantae Junghuhnianae" (F. A. W. Miguel, ed.) pp. 312–341. Synthoff, Lugduni-Batavorum.

Duby, J.-E. (1876). Choix de mousses exotiques nouvelles ou mal connues. *Mém. Soc. Phys. Hist. nat. Genève* **26**, 1–14.

Evans, A. W. (1905). A remarkable *Ptilidium* from Japan. *Revue bryol.* **32**, 57–60.

Fournier, P. (1932). Voyages et découvertes scientifiques des missionaires naturalistes français. A travers le monde pendant cinq siècles XVe siècle. *Encycl. biol.* **10**, 1–285.

Fritsch, R. (1972). Chromosomenzahlen der Bryophyten. Eine Übersicht und Diskussion ihres Aussagewertes für das System. *Wiss. Z. Friedrich-Schiller-Univ. Jena*, Math. Naturwiss. Reihe, **21**, 839–944.

Geheeb, A. (1881). Bryologische fragmente, I. *Flora, Jena* **64**, 289–297.

Hagen, I. (1906). Mélanges bryologiques. *Revue bryol.* **33**, 49–54.

Hämet-Ahti, L., Ahti, T. and Koponen, T. (1974). A scheme of vegetation zones for Japan and adjacent regions. *Annls bot. fenn.* **11**, 59–88.

Hara, H. and Kanai, H. (1958–1959). "Distribution maps of flowering plants in Japan." Inoue Book Co., Tokyo.

Hattori, S. (1949). Short review of the Japanese species of *Plagiochila. J. Jap. Bot.* **24**, 149–154.

Hattori, S. (1951). On the distribution of Hepaticae of Shikoku and Kiushiu (southern Japan). *Bryologist* **54**, 103–118.

Hattori, S. (1952a). *Ptilidium californicum* and other nearctic liverworts in Japan. *Bryologist* **55**, 147–149.

Hattori, S. (1952b). Hepaticae of Shikoku and Kyushu, southern Japan, 1. *J. Hattori bot. Lab.* **7**, 38–61.

Hattori, S. (1966). Anthocerotae and Hepaticae. *In* "The Flora of Eastern Himalaya" (H. Hara, ed.) pp. 501–536. University of Tokyo Press, Tokyo.

Hattori, S. (1968). Résultats des expéditions scientifiques genèvoises au Népal en 1952 et 1954 (partie botanique) 20. Hepaticae. *Candollea* **23**, 275–285.

Hattori, S. (1971). Hepaticae. *In* "The Flora of Eastern Himalaya, Second Report" (H. Hara, ed.) Bull. Univ. Mus. Tokyo. Vol. 2, 222–240.

Hattori, S. (1975). Anthocerotae and Hepaticae. *In* "The Flora of Eastern Himalaya. Third Report" (H. Ohashi, ed.) Bull. Univ. Mus. Tokyo. Vol. 8, 206–242.

Hattori, S. and Iwatsuki, Z. (1973). Foundation and activities of the Hattori Botanical Laboratory, Japan. *Taxon* **22**, 167–168.

Hattori, S., Sharp, A. J., Mizutani, M. and Iwatsuki, Z. (1968). *Takakia ceratophylla* and *T. lepidozioides* of Pacific North America and a short history of the genus. *Miscnea bryol. lichen, Nichinan.* **4**, 137–149.

Herzog, T. (1926). "Geographie der Moose." Gustav Fischer, Jena.

Hirohama, T. (1973). Scanning electron microscopic observations on the spores or Japanese mosses, I. *The Cell* **5**(6), 25–27.

Hirohama, T. (1975). Spore morphology of bryophytes observed by scanning electron microscopy, I. Dicranaceae. *Bull. natn. Sci. Mus., Tokyo.*, Ser. B, 2, 61–72.

HONG, W. S. and ANDO, H. (1959). An enumeration of mosses recorded from Korea, with some new additions to the Korean flora. *Theses Catholic med. Coll., Seoul* **3**, 371–395.

HORIKAWA, Y. (1929–1933). Studies on the Hepaticae of Japan I–III, *Sci. Rep. Tōhoku Univ.* **4**, 37–72; 395–429. **5**, 623–650. IV–VIII, *J. Sci. Hiroshima Univ.* **1**, 13–35; 55–76; 77–94; 121–134; 197–205.

HORIKAWA, Y. (1934). Monographia hepaticarum australi-japonicarum. *J. Sci. Hiroshima Univ.*, Ser. B, Div. 2, **2**, 101–325.

HORIKAWA, Y. (1934–1936). Symbolae florae bryophytae orientali-asiae. I–X. *Bot. Mag. Tokyo* **48**, 452–462; 599–609; 708–719. **49**, 49–59; 211–221; 588–595; 671–678. **50**, 201–206; 380–385; 556–561.

HORIKAWA, Y. (1949). Records of U. Faurie's botanical trips in Japan and neighbouring areas. *Bull. biol. Soc. Hiroshima Univ.* **12**, 1(1), 30–33.

HORIKAWA, Y. (1955). Distributional studies of bryophytes in Japan and adjacent regions. *Contr. Phytotax. Geobot. Lab. Hiroshima Univ.*, N. Ser., **27**, 1–152.

HORIKAWA, Y. (1972). "Atlas of the Japanese Flora, an Introduction to Plant Sociology of East Asia." Gakken Co. Ltd., Tokyo.

HORIKAWA, Y. and ANDO, H. (1961). Mosses of the Ongul Islands collected during the 1957–1960 Japanese Antarctic Research Expedition. (Appendix: A historical review of Antarctic bryology). *Hikobia* **2**, 160–178.

HORIKAWA, Y. and ANDO, H. (1964). Contributions to the moss flora of Thailand. *Nature Life S.E. Asia* **3**, 1–44.

HUNECK, S. (1977). Neue Ergebnisse zur Chemie der Moose, eine Übersicht. Teil 5. *J. Hattori bot. Lab.* **43**, 1–30.

HUSNOT, T. (1833). *Eustichia savatieri* Husn. *Revue bryol.* **10**, 85–86.

IISHIBA (IHSIBA), E. (1907). A list of Musci and Hepaticae from Sendai. *Bot. Mag. Tokyo* **21**, (44)–(45).

IISHIBA, E. (1929). "Nihon-san senrui sōsetsu (A manual of the mosses of Japan)." Tokyo.

IISHIBA, E. (1930). "Species Hepaticarum Nipponicarum." Sendai (mimeographed).

IISHIBA, E. (1932). "Nihon-san senrui no bunrui (Classification of mosses of Japan)."

IISHIBA, E. (1940). (A key to the families and genera of Japanese Hepatics). *Acta phytotax. geobot.* **9**, 217–226.

IISHIBA, E., UEMATSU, E. and KATŌ, T. (1912). "Futsu Nihon senrui zusetsu (Illustrations of common Japanese mosses)."

INOUE, H. (1972). History of the bryology in Japan 1–2. *Shizengaku to Hakubutsukan* **39**, 158–170; 193–207.

INOUE, H. (1973–1976). "Illustrations of Japanese Hepaticae." Tsukiji Shokan Co., Tokyo.

INOUE, S. and IWATSUKI, Z. (1976). A cytotaxonomic study of the genus *Rhizogonium* Brid. (Musci). *J. Hattori bot. Lab.* **41**, 389–403.

INOUE, T. (1894). Hepaticae of Tosa. *Bot. Mag. Tokyo* **8**, (291)–(293).

ISHIKAWA, I. (1973, 1974). Bryophyta in Japanese gardens. 1, 2. *Hikobia* **6**, 272–283; **7**, 65–78.

ISHIKAWA, M. (1916). A list of the number of chromosomes. *Bot. Mag. Tokyo* **30**, 404–448.

ITŌ, T. (1887). On the history of botany in Japan. *J. Bot. Lond.* **25**, 225–229.

IWATSUKI, Z. (1958). Correlations between the moss floras of Japan and of the southern Appalachians. *J. Hattori bot. Lab.* **20**, 304–352.

IWATSUKI, Z. (1963–1964). An account of the bryological research expedition to North Borneo by the Hattori Botanical Laboratory (1–4). *Miscnea bryol. lichen, Nichinan.* **3**, 50–56; 73–76; 87–90; 97–100.

IWATSUKI, Z. (1966). Critical re-examination of the Asiatic mosses reported by Sullivant and Lesquereux in 1857 and 1859. *J. Hattori bot. Lab.* **29**, 53–69.

IWATSUKI, Z. (1972). Distribution of bryophytes common to Japan and the United States. *In* "Floristics and Paleofloristics of Asia and Eastern North America" (A. Graham, ed.) pp. 107–137. Elsevier, Amsterdam.

IWATSUKI, Z. (1975, 1976). Bryological literature published in Japan in 1974 (1–2). *Proc. bryol. Soc. Japan* **1**, 138–140; 152–153.

IWATSUKI, Z. (1976). On two Japanese mosses reported by C. P. Thunberg. *Proc. bryol. Soc. Japan* **1**, 142–143.

IWATSUKI, Z. and HATTORI, S. (1970). Studies on the epiphytic moss flora of Japan, 19. The epiphytic bryophyte communities in the broadleaved evergreen forests on the Tsushima Islands, Kyushu. *Mem. natn. Sci. Mus., Tokyo* **3**, 365–374.

IWATSUKI, Z. and KODAMA, T. (1961). Mosses in Japanese gardens. *Econ. Bot.* **15**, 164–169.

IWATSUKI, Z. and MIZUTANI, M. (1972). "Coloured illustrations of bryophytes of Japan." Hoikusha Publishing Co., Ltd., Osaka.

IWATSUKI, Z. and NOGUCHI, A. (1973). Index muscorum japonicarum. *J. Hattori bot. Lab.* **37**, 299–418.

IWATSUKI, Z. and NOGUCHI, A. (1975). Mosses of North Borneo, II. *J. Hattori bot. Lab.* **39**, 315–333.

IWATSUKI, Z. and NOGUCHI, A. (1976). Musci japonici exsiccati, ser. 1–25. Alphabetical list of species. *J. Hattori bot. Lab.* **40**, 291–325.

IWATSUKI, Z. and SHARP, A. J. (1967). The bryogeographical relationships between eastern Asia and North America, I. *J. Hattori bot. Lab.* **30**, 152–170.

IWATSUKI, Z. and SHARP, A. J. (1968). The bryogeographical relationships between eastern Asia and North America, II. *J. Hattori bot. Lab.* **31**, 55–58.

JAEGER, A. and SAUERBECK, F. (1876–1879). "Genera et species muscorum systematice disposita seu adumbratio florae muscorum totius orbis terrarum. Vol. 2". St Gall. naturwiss. Gesellshaft. San Gall.

KABIERSCH, W. (1936). Studien über die ostasiatischen Arten einiger Laubmoosfamilien (Mniaceae–Bartramiaceae). *Hedwigia* **76**, 1–94.

KAMIMURA, M. (1970). Electron microscopic studies on the fine structure of spore walls in Frullaniaceae (Hepaticae). I. *Bull. Kochi-Gakuen Junior Coll.* **1**, 43–50.

KAMIMURA, M. (1971). Studies on the fine structures of spore walls in Frullaniaceae (Hepaticae), II. *Miscnea bryol. lichen., Nichinan* **5**, 187–190.

KAMIMURA, M. (1972). Scanning electron microscope studies on the fine structure of the spore walls of a North American Hepaticae, *Leucolejeunea unciloba*. *Bull. Kochi-Gakuen Junior Coll.* **3**, 5–7.

KANDA, H. (1975, 1977). A revision of the family Amblystegiaceae of Japan. I, II. *J. Sci. Hiroshima Univ.*, Ser. B, Div. 2, Bot., **15**, 201–276; **16**, 47–119.

KITAGAWA, N. (1971–1973). The collection of Japanese Hepaticae by Siebold and his contemporaries, 1–2. *Miscnea bryol. lichen., Nichinan.* **5**, 171–172; **6**, 101–103.

KOBAYASHI, K. (1971). Phytosociological studies on the scrub of dwarf pine (*Pinus pumila*) in Japan. *J. Sci. Hiroshima Univ.*, Ser. B, Div. 2, Bot., **14**, 1–52.

KONO, G. (1906). On two new species of Muscineae. *Bot. Mag. Tokyo* **20**, (79)–(82).

KOPONEN, T. (1971). A report on *Rhizomnium* (Mniaceae) in Japan. *J. Hattori bot. Lab.* **34**, 365–390.

KOPONEN, T. (1973). *Rhizomnium* (Mniaceae) in North America. *Annls bot. fenn.* **10**, 1–28.

KOPONEN, T. (1974). A preliminary report on the Mniaceae in Japan, II. *Hikobia* **7**, 1–20.

KOPONEN, T. (1976). Bryology at the Botanical Museum, University of Helsinki, Finland. *Taxon* **25**, 365–374.

KOPONEN, T. (1977). Miscellaneous notes on Mniaceae (Bryophyta). II. *Annls bot. fenn.* **14**, 62–64.

KOPONEN, T. (1978). On the taxonomy and phytogeography of *Mnium* Hedw. s. str. (Musci, Mniaceae). *Abstr. bot.* **5**, Suppl. 3, 63–73.

KOPONEN, T., JÄRVINEN, I. and ISOVIITA, P. (1978). Bryophytes from the Soviet far east, mainly the Khbarovsk Territory. *Annls bot. fenn.* **15**, 107–121.

KURITA, M. (1937). Geschlechtschromosomen und Chromosomenzahlen bei einigen Laubmoosen. *Z. indukt. Abstamm. u. VererbLehre.* **74**, 24–29.

LENSEN, G. A. (1959). "The Russian Push Toward Japan. Russo-Japanese Relations, 1697–1875." Princeton University Press, Princeton.

LINDBERG, S. O. (1873). Contributio ad floram cryptogamam Asiae boreali-orientalis. *Acts Soc. scient. fenn.* **10**, 223–280.

MATSUMURA, J. and MIYOSHI, M. (1899–1901). "Cryptogamae Japonicae Iconibus Illustratae; or Figures with Brief Descriptions and Remarks on the Musci, Hepaticae, Lichenes, Fungi, and Algae of Japan." Keigyōsha and Co., Tokyo.

MATSUO, A., NAKAYAMA, M., HAYASHI, S., SEKI, T. and AMAKAWA, T. (1978). A comparative study of the diterpenoids from several species of the genus *Jungermannia*. *Bryophyt. Biblthca.* **13**, 321–328.

MERRILL, E. D. and WALKER, E. H. (1938). "A Bibliography of Eastern Asiatic Botany." Arnold Arboretum, Jamaica Plain.

MIQUEL, F. A. W. (1866). "Prolusio Florae Iaponicae." Van der Post, Amstelodami.

MITTEN, W. (1859). Musci Indiae orientalis. An enumeration of the mosses of the East Indies. *J. Proc. Linn. Soc. Bot. Suppl.* **1**, 1–171.

MITTEN, W. (1865). On some species of Musci and Hepaticae additional to the floras of Japan and the coast of China. *J. Proc. Linn. Soc.* **8**, 148–158.

MITTEN, W. (1891). An enumeration of all the species of Musci and Hepaticae recorded from Japan. *Trans. Linn. Soc. London,* Ser. 2, **3**, 153–206.

MIYAKE, K. (1898). A new genus of Hepaticae in Japan. *Bot. Mag. Tokyo* **12**, (85)–(86).

MIYAKE, K. (1905). On the centrosome of Hepaticae. *Bot. Mag. Tokyo* **29**, 98–101.

MIYOSHI, N. (1969a). Light- and electron-microscopic studies of spores in the Musci, 1. *Andreaea rupestris* var. *fauriei, Buxbaumia aphylla, Pogonatum sphaerothecium* and *Polytrichum commune. Hikobia* **5**, 172–177.

MIYOSHI, N. (1969b). Light- and electron-microscopic studies of spores in the Musci, 2. Spores of *Schistostega pennata* and *Hedwigia ciliata. J. Jap. Bot.* **44**, 295–299.

MIYOSHI, N. (1973a). Scanning electron microscopic observation of spores in the Musci. *Jap. J. Palynol.* **12**, 13–18.

MIYOSHI, N. (1973b). Spore wall structure of *Dumortiera hirsuta* (Hepaticae). *Hikobia* 6, 251–252.

MIZUSHIMA, U. (1960). Japanese Entodontaceae. *J. Hattori bot. Lab.* 22, 91–158.

MIZUSHIMA, U. (1963). Dr Kyuichi Sakurai (1890–1963). *Miscnea bryol. lichen., Nichinan.* 3, 57–59.

MIZUTANI, M. (1961). A revision of Japanese Lejeuneaceae. *J. Hattori bot. Lab.* 24, 115–302.

MIZUTANI, M. and HATTORI, S. (1969). Check list of Japanese Hepaticae and Anthocerotae. *Miscnea bryol. lichen., Nichinan.* 5, 33–43.

MIZUTANI, M. and HATTORI, S. (1976). Hepaticae japonicae exsiccatae ser. 1–20. Alphabetical list of species. *J. Hattori bot. Lab.* 40, 327–353.

NAGANO, I. (1969). Comparative studies of moss vegetations developing on the limestone, chert, and other rocks lying adjacent to each other in the Chichibu Mountain area, central Japan. *J. Hattori bot. Lab.* 32, 155–203.

NAKAMURA, T. (1976). Bryophytes as indicator of urbanization. *Proc. bryol. Soc. Japan* 1, 178–182.

NAKANISHI, K. and SUZUKI, H. (1977). A phytosociological study of bryophyte communities developed on stone walls and concrete blocks in Hiroshima City. *Hikobia* 8, 197–211.

NAKANISHI, S. (1962). The epiphytic communities of beech forest in Japan. *Bull. Fac. Educ. Kobe Univ.* 27, 141–220.

NAKANISHIKI, K. (1905). Mosses found in Kosheki-yama, Tosa Province. *Bot. Mag. Tokyo* 19, (145)–(146).

NOGUCHI, A. (1954). Notulae bryologicae, V. A list of mosses from Manchuria and North Korea. *J. Hattori bot. Lab.* 12, 27–33.

NOGUCHI, A. (1964a). Résultats des expéditions scientifiques genèvoises au Népal en 1952 et 1954 (partie botanique), 19. Musci. *Candollea* 19, 167–189.

NOGUCHI, A. (1964b). A list of mosses from Kashmir and Pakistan. *Candollea* 19, 191–198.

NOGUCHI, A. (1966). Musci. *In* "The flora of Eastern Himalaya" (H. Hara, ed.) pp. 537–591. The University of Tokyo Press. Tokyo.

NOGUCHI, A. (1971). Musci. *In* "The Flora of Eastern Himalaya, Second Report" (H. Hara, ed.) Bull. Univ. Mus. Tokyo. Vol. 2, 241–258.

NOGUCHI, A. (1972). Mosses of Thailand. *Lindbergia* 1, 169–183.

NOGUCHI, A. (1973a). Contributions to the bryology of Thailand. *J. Hattori bot. Lab.* 37, 235–250.

NOGUCHI, A. (1973b). Mosses of Ceylon collected by Dr Hiroshi Inoue. *Bull. natn. Sci. Mus., Tokyo* 16, 305–316.

NOGUCHI, A. (1976). "Handbook of Japanese Mosses." Hokuryukan, Tokyo.

NOGUCHI, A. and IWATSUKI, Z. (1972). Mosses of North Borneo. 1. *J. Hattori bot. Lab.* 36, 455–486.

NOGUCHI, A. and IWATSUKI, Z. (1975). Musci. *In* "Flora of Eastern Himalaya, Third Report." Bull. Univ. Mus. Tokyo. Vol. 8, 243–282.

NOGUCHI, A., TAKAKI, N. and INOUE, H. (1966). Bryophytes collected by Dr K. Yoda in eastern Nepal. *Bull. natn. Sci. Mus., Tokyo* 9, 359–386.

OKAMURA, S. (1908). Contributions to the study of Japanese Bryophyta. I. On two new genera of Musci. *Bot. Mag. Tokyo* 22, 29–31; 41–43.

OKAMURA, S. (1915–1916). Contributiones novae and floram bryophyton japonicam. 1–2. *J. Coll. Sci. Univ. Tokyo* **36** (7), 1–51; **38** (4), 1–100.

OKAMURA, S. (1940). Bryophyta. *In* "An Illustrated Flora of Nippon. Nippon shokubutsu dzukan" (T. Makino, ed.) pp. 977–1002. Hokuryukan Co., Tokyo.

OSADA, T. (1958). An additional list of mosses from North Korea. *J. Hattori bot. Lab.* **19**, 60–66.

OSADA, T. (1966). Japanese Polytrichaceae, II. The genera *Polytrichum, Oligotrichum, Bartramiopsis* and *Atrichum*, and phytogeography. *J. Hattori bot. Lab.* **29**, 1–52.

PARIS, E. G. (1902). Musci Japonici a R. P. Faurie anno 1900 lecti. *Bull. Herb. Boissier,* Sér. 2, **11**, 918–933; **12**, 988–993.

PARIS, E. G. (1904). Quelques nouvelles pleurocarpes japonaises et tonkinoises. *Revue bryol.* **31**, 56–65; 93–95.

PARIS, E. G. (1910). "Quatrième liste des mousses et hépatiques offertes en échange par M. Le Général Paris, 1–8."

REIMERS, H. (1931). Beiträge zur Moosflora Chinas, I. *Hedwigia* **71**, 1–77.

REIMERS, H. and SAKURAI, K. (1931). Beiträge zur Moosflora Japans, I. *Bot. Jb.* **64**, 537–560.

RENAULD, F. and CARDOT, J. (1904). Musci exotici novi vel minus cogniti, X. *Bull. Soc. r. Bot. Belg.* **41**, 7–122.

RUDOLF, R. C. (1974). Thunberg in Japan and his Flora Japonica in Japanese. *Monumenta Nippon.* **29**, 163–179.

SAITO, K. and HIROHAMA, T. (1974). A comparative study of the spores of taxa in the Pottiaceae by use of the scanning electron microscope. *J. Hattori bot. Lab.* **38**, 475–488.

SAKURAI, K. (1939). Beobachtungen über Japanische Moosflora, XVIII. *Bot. Mag. Tokyo* **53**, 59–67.

SAKURAI, K. (1954). "Nippon no senrui. Muscologia Japonica." Iwanami Shoten, Tokyo.

SALMON, E. S. (1900). On some mosses from China and Japan. *J. Linn. Soc. Bot.* **34**, 449–474.

SANDE LACOSTE, C. M. van der (1857). Synopsis Hepaticarum Javanicarum, adjectis quibusdam speciabus Hepaticarum novis extra-Javanicis. *Verh. K. Akad. Wet. Amst.* **5**, 1–112.

SANDE LACOSTE, C. M. van der (1864). Hepaticae. Jungermannieae Archipelagi Indici, adiectis quibusdam speciebus japonicis. *Annls Mus. Bot. Lug. -Bat.* **1**, 287–312.

SANDE LACOSTE, C. M. van der (1866–1867). Musci frondosi; Musci et Hepaticae. *In* "Prolusio flora Iaponicae (F. A. W. Miguel, ed.) pp. 180–188; 373. Van der Post, Amstelodami.

SASAOKA, H. (1910). List of mosses for Etchu Province. *Bot. Mag. Tokyo* **24**, 197–198.

SAWADA, K. (1909). *Frullania* in Japan. *Bot. Mag. Tokyo* **23**, 178–183.

SAYRE, G. (1975). Cryptogamae exsiccatae—An annotated bibliography of exsiccatae of Algae, Lichenes, Hepaticae, and Musci, V. Unpublished exsiccatae, I. Collectors. *Mem. N.Y. bot. Gdn* **19**, 277–432.

SCHIMPER, W. P. (1860). "Synopsis muscorum europaeorum." Schweizerbart, Stuttgart.

SCHMID, G. (1942). Über Ph. Fr. v. Siebolds Reise nach Japan. Mit Briefen aus den Jahren 1822 bis 1827. *Bot. Arch. Berlin* **43**, 487–536.

SCHOFIELD, W. B. (1965). Correlations between the moss floras of Japan and British Columbia, Canada. *J. Hattori bot. Lab.* **28**, 17–42.

SCHUSTER, R. M. (1975). Foreword. *In* "Selected Bryological Papers by Dr Sinske Hattori Published Between 1940–1951" (Committee for the Celebration of Dr Sinske Hattori's Sixtieth Birthday, ed.) pp. iii–iv. Hattori Bot. Lab., Nichinan.

SEKI, T. (1968). A revision of the family Sematophyllaceae of Japan with special reference to a statistical demarcation of the family. *J. Sci. Hiroshima Univ.*, Ser. B., Div. 2, Bot., **12** (1), 1–80.

SEKI, T. (1974). A moss flora of Provincia de Aisén, Chile. Results of the Second Scientific Expedition to Patagonia by Hokkaido and Hiroshima Universities, 1967. *J. Sci. Hiroshima Univ.*, Ser. B, Div. 2, Bot., **15** (1), 9–101.

SHARP, A. J. (1965). The co-operative Hattori-Tennessee Bryogeographic expedition, 1. Taiwan. *Miscnea bryol. lichen., Nichinan* **3**, 179–180.

SHARP, A. J. and IWATSUKI, Z. (1965). A preliminary statement concerning mosses common to Japan and Mexico. *Ann. Mo. bot. Gdn* **52**, 452–456.

SHIMOTOMAI, N. and KIMURA, K. (1932). Geschlechtschromosomen bei *Pogonatum inflexum* Lindb. und Chromosomenzahlen bei einigen anderen Laubmoosen. *Bot. Mag. Tokyo* **46**, 385–391.

SHIN, T. (1963). Electron-microscopical observations of the peristome and spores of mosses, 1. *Hikobia* **3**, 202.

SHIN, T. and MUROYA, T. (1965). Electron-microscopical observations of the peristome and spores of mosses, 2. *Hikobia* **4**, 178–180.

STEARN, W. T. (1958). Botanical exploration to the time of Linnaeus. *Proc. Linn. Soc. Lond.* **169**, 172–196.

STEENIS-KRUSEMAN, M. J. van (1950). Malaysian plant collectors and collections. *In* "Flora Malesiana Ser. I, Spermatophyta" Vol. 1, (C. G. G. J. van Steenis, ed.) 1–639. Noordhoff-Kolff N.V., Djakarta.

STEERE, W. C. (1969). Asiatic elements in the bryophyte flora of western North America. *Bryologist* **72**, 507–512.

STEERE, W. C. (1972). Chromosome numbers in bryophytes. *J. Hattori bot. Lab.* **35**, 99–125.

STEERE, W. C. and INOUE, H. (1972). Distributional patterns and speciation of bryophytes in the circum-Pacific regions: Introduction. *J. Hattori bot. Lab.* **35**, 1–2.

STEPHANI, F. (1898–1924). "Species Hepaticarum" Vols 1–6. Herbier Boissier, Genève.

STEPHANI, F. (1900). Liste des muscinées récoltées an Japon par M. le Professeur A. E. Nordenskiöld au cours du voyage de la Vega, autour de l'Asie en 1878–79. Hepaticae. *Öfvers. K. VetenskAkad. Förh. Stock.* **57**, 295.

SUIRE, C. (1975). Les données actuelles sur la chimie des bryophytes. *Revue bryol. lichén.* **41**, 105–256.

SULLIVANT, W. S. (1857). Musci and Hepaticae. *In* "List of Dried Plants Collected in Japan, by S. Wells Williams Esq. and Dr James Morrow" (A. Gray, ed.) p. 330. U.S. House Executive Documents, Washington.

SULLIVANT, W. S. and LESQUEREUX, L. (1859). Characters of some new Musci collected by Wright in the North Pacific Exploring Expedition, under the command of Captain John Rogers. *Proc. Am. Acad. Arts Sci.* **4**, 275–282.

SUZUKI, H. (1977). Yoshiwo Horikawa (1902–1976). *Hikobia* **8**, 1–21.

TAKAKI, N. (1973). The earliest publication in Europe on Japanese bryophytes. *Miscnea bryol lichen., Nichinan* **6**, 120–121.

TAKAKI, N. (1975). Some noteworthy mosses of Papua New Guinea. *In* "Reports on the Cryptogams in Papua New Guinea" (Y. Otani, ed.) pp. 87–97. Natn. Sci. Mus., Tokyo.

TAODA, H. (1972). Mapping of atmospheric pollution in Tokyo based on epiphytic bryophytes. *Jap. J. Ecol.* **22,** 125–133.

TAODA, H. (1973a). The effect of air pollution on bryophytes. *Proc. bryol. Soc. Japan* **1,** 45.

TAODA, H. (1973b). Bryo-meter, an instrument for measuring the phytotoxic air pollution. *Hikobia* **6,** 224–228.

TAODA, H. (1973c). Effect of air pollution on bryophytes, I. SO_2 tolerance of bryophytes. *Hikobia* **6,** 238–250.

TAODA, H. (1976). Bryophytes as indicators of air pollution. *In* "Science for Better Environment, Proceedings of the International Congress on the Human Environment" (Kyoto 1975), pp. 291–301. Asahi Evening News, Tokyo.

TAODA, H. (1977). Bryophytes in the urban ecosystems. *In* "Tokyo Project. Interdisciplinary Studies of Urban Ecosystems in the Metropolis of Tokyo" (M. Numata, ed.) pp. 99–117. Tokyo.

TATUNO, S. (1933). Geschlechtschromosomen bei einigen Lebermoosen, I. *J. Sci. Hirosima Univ.,* Ser. B2, **1–2,** 165–182.

TATUNO, S. (1941). Zytologische Untersuchungen über die Lebermoose von Japan. *J. Sci. Hiroshima Univ.,* Ser. B2, **4,** 73–187.

THERIOT, M. I. (1907). Diagnoses d'espèces nouvelles. *Monde Pl.,* Sér 2, **9,** 21–22.

THERIOT, M. I. (1908). Diagnoses d'espèces et de variétés nouvelles de mousses (5e article). *Bull. Acad. int. Géogr. bot.* **18,** 250–254.

THERIOT, M. I. (1909). Diagnoses d'espèces et de variétés nouvelles de mousses (6e article). *Bull. Acad. int. Géogr. bot.* **19,** 17–24.

THUNBERG, C. P. (1784). "Flora japonica." Müller, Lipsiae.

THUNBERG, C. P. (1795). "Travels in Europe, Africa, and Asia made between the years 1770 and 1779." Third edition, Vols 1–4. F. and C. Rivington, London.

TOYAMA, R. (1935, 1937a–c, 1938). Spicilegium Muscologiae Asiae Orientalis I–V. *Acta phytotax. geobot.* **4,** 213–219; **6,** 42–45, 101–107, 169–178; **7,** 102–111.

TSUGE, C. (1887). *Marchantia polymorpha* L. *Bot. Mag. Tokyo* **1,** 151–154.

VERDOORN, F. (1938). On some recent contributions to Japanese bryology. *Ann. Bryol.* **10,** 1–2.

WALKER, E. H. (1960). "A Bibliography of Eastern Asiatic Botany. Supplement I." American Institute of Biological Sciences, Washington.

WANG, C. K. (1970). "Phytogeography of the Mosses of Formosa." Tunghai Univ., Taichung.

WARNSTORF, C. (1900). Neue Beiträge zur Kenntniss europäischer und exotischer *Sphagnum*formen. *Hedwigia* **39,** 100–110.

WARNSTORF, C. (1904). Neue europäische und exotische Moose. *Beih. bot. Zbl.* **16,** 237–252.

WARNSTORF, C. (1907). Neue europäische und aussereuropäische Torfmoose. *Hedwigia* **47,** 76–124.

WARNSTORF, C. (1911). Sphagnales-Sphagnaceae (Sphagnologia universalis). *In* "Das Pflanzenreich. Regni Vegetabilis Conspectus" (A. Engler, ed.) **51.** W. Engelmann, Leipzig.

WATANABE, R. (1972). A revision of the family Thuidiaceae in Japan and adjacent areas. *J. Hattori bot. Lab.* **36**, 171–320.

YASUDA, A. (1915). Eine neue Art von *Bartramia. Bot. Mag. Tokyo* **29**, 23–24.

YOKOBORI, M. (1978). Measuring of phytotoxic air pollution based upon response of bryophytes using a filtered-air growth chamber. *Jap. J. Ecol.* **28**, 17–23.

8 | The Systematics of Tropical Mosses

G. C. G. ARGENT

Royal Botanic Gardens, Edinburgh EH3 5LR,
Great Britain

Abstract: An account is given of the special problems involved in studying the systematics of tropical mosses. These problems are demonstrated using the example of the African members of the families Pterobryaceae and Meteoriaceae.

The history of systematic studies in temperate and tropical regions has been very different. For example, in the temperate areas, very large genera were maintained until rational decisions about how they should be split up could be taken. In the tropics, on the other hand, both species and genera have been allowed to proliferate almost unchecked. It is therefore very important that monographic studies covering areas as wide as possible should be undertaken.

By comparison with north-temperate species, tropical mosses are poorly represented in herbaria and a high proportion of specimens are misidentified. In addition, herbaria are usually sited in countries remote from the tropics where the species grow. It is thus difficult to make direct comparisons between field and herbarium material and to incorporate knowledge of phenotypic plasticity in herbarium-based monographic revisions.

At the present state of knowledge the taxonomy of tropical mosses is an essentially practical subject and a number of suggestions are made to encourage its advancement.

As with flowering plants, the modern scientific study of mosses, including tropical mosses, started in Europe and was for a long time dominated by the preconceived idea of some European botanists that the tropics were wildly exotic and almost astronomically rich in species. This was the era at the close of the nineteenth century and opening of the twentieth century when, as Touw (1974) says, "industrious bryologists almost poured out novelties".

Systematics Association Special Volume No. 14, "Bryophyte Systematics", edited by G. C. S. Clarke and J. G. Duckett, 1979, pp. 185–193, Academic Press, London and New York.

Collections were commonly brought back by non-bryologists, often non-botanists; collection data were frequently lacking, descriptions brief, illustrations rare and thousands of invalid species were described. Thus compare Touw's (1974) estimate of 7000 "good" species of moss globally with the 18 000 valid species in *Index Muscorum* (R. E. Magill, personal communication) or the approximately 80 000 names (Crosby, personal communication) from a rough page count estimate of *Index Muscorum* (van der Wijk *et al.*, 1959–1969).

The tropics are indeed rich in moss species but strictly in comparison with the numbers of species in temperate and arctic areas and not with the number of stars in the sky. The tropics offer a range of different habitats which is probably broader than that of the temperate and arctic regions combined. Within the tropical zone, climates range from hot and humid at sea-level, through mist forest to, in places, a permanent snow line. Because large areas of the Gondwana land mass have migrated northwards and are now situated in the tropics, the tropical zone is the meeting place for the southern Gondwana flora and the northern Laurasian flora. When this rich variety of genomes is set in the innumerable islands and mountain valleys of areas like south-eastern Asia, there is enormous potential for building up isolated populations which may eventually result in new species. Africa, however, which receives disproportionate attention in this paper, fared less well than other parts, probably because of its early isolation from Gondwanaland and its harsher extremes of climate. Africa demonstrates the fact that tropical moss floras, even of moist areas, are not always rich in species.

So to practical problems and difficulties. I shall use as examples experiences encountered when working on a taxonomic revision of African Pterobryaceae and Meteoriaceae (Argent, 1973, and unpublished data) since they are by no means exceptional. Some of the problems, although obvious and common, are no less real. The first problem is finding material; I was lucky in having the excellently documented collections of Professor Richards as the basis for the study but private collections apart, one quickly moves to national herbaria. It is, however, impossible to search the often vast amounts of unsorted material on short visits to foreign herbaria or to request others to do this on one's behalf, and it is usually impossible or impracticable to borrow it.

Another problem is in the delimitation of families. A frustrating aspect of moss taxonomy is that the families, though often easily recognizable to a trained eye, almost defy meaningful description when read by the intelligent non-expert, and the difficulty of keying them may be judged by the very few family keys attempted in the literature. I am confident that I am not the only bryologist who has sat down with a totally unplaced plant and thumbed through the pages of Brotherus (1924–1925) in the hope of making a match—a not wholly scientific procedure. A surprising number of specimens of the two families studied from national herbaria were misidentified as to family. This was partially remedied because the two families were being revised simultaneously, but the misidentified Pterobryaceae and Meteoriaceae specimens in other families are lost, at least until the next revision, which may be a very long time. It is interesting that when I tabulated some differences between these two families using African material in an attempt to assist others in placing specimens correctly, During (1977) found that these differences were invalid in southeast Asia.

Delimitation of genera raised problems because it was impossible to monograph the families completely. One was obliged to force species into existing genera rather than to rearrange groups with which one was only partly familiar and thus create new nomenclatural combinations of dubious value. One cannot help comparing the development of the taxonomy of European mosses with that of the tropical species. In Europe, very large genera such as *Neckera* and *Hypnum* were maintained until most of their European species were well known, and they were then split up. Contrast this with the early division of African Pterobryaceae into small genera, often containing only a single species, leaving the generic concepts to warp and break under the strain of subsequently discovered species with varying characters. The result is a chaos of nomenclatural combinations. For example, *Jaegerina scariosa* (C. Muell.) C. Muell., a common, distinctive and widespread species has been recorded under five different genera (Argent, 1973). Smith (1978), writing about the taxonomy of British mosses, comments that the broad generic concept of Dixon (1924) is totally out of place in the advanced state of present-day British moss taxonomy. Nevertheless, the concept was historically advantageous in helping to minimize nomenclatural combinations based on inadequate knowledge. I do not think I would be alone in appealing to bryologists not to publish new genera without very careful checking.

Many difficulties arise at the species level. Touw (1974) says that "much more collecting should be done by bryologists themselves". This is a view with which most bryologists would agree, but it needs to be repeated in the hope that it may be read by people controlling finance, who may otherwise feel that bryophytes can be picked up by flowering plant collectors incidentally. This practice leads to a high percentage of unnamable collections, or at least collections which can only be named after a great deal of difficulty. A short field trip in West Africa enabled me to comprehend more easily patterns of phenotypic plasticity at the species level which once realized can often be spotted in related taxa. As an illustration I shall describe the variation in *Squami-dium biforme* (Hamp.) Broth. Two of the three species of mainland Africa were reduced to varieties of the earliest name, and for the sake of uniformity a further variety *chlorothrix* Argent was proposed to cover a previously over-looked variant. These varieties, apart from *S. biforme* var. *rehmannii* (C. Muell.) Argent, are probably growth forms with little if any genetic basis. Thus, *Squamidium* produces two distinctly different shoot units characterized by different leaf shape, which I designated "stems" and "branches"; it also grows in a variety of habitats. In exposed conditions the main stem creeps, firmly attached to the substrate, and gives rise to erect "determinate" branches with larger leaves. In moister conditions the "stems" creep along the substrate and then hang pendent with whorls of determinate "branches". Under the wettest conditions the determinate "branches" grow on to become "stems". They maintain their distinct "branch" leaf-form in the basal portion but lose it further on (Argent, 1973, Fig. 3). Once this pattern of variation was realized it became extremely useful in interpreting that of other genera. *Pilotrichella* is similar although its branches have, perhaps, more resistance to becoming indeterminate and reverting to stem form. In *Floribundaria* it provided the only reliable vegetative character to separate *F. floribunda* (Doz. and Molk.) Fleisch. from what was then regarded as *F. vaginans* (Welw. and Dub.) Broth. Both these species at first sight show little relation to the whorled nodal architecture of *Squamidium* or *Pilotrichella* but by using leaves just beyond the bases of branches, leaf characters which are constantly different could be found.

An understanding of phenotypic plasticity is essential for good taxonomy. Ideally, growth chamber or transplant experiments would supply the most reliable data but we have little enough of this even for north-temperate moss floras. Intelligent field observations by bryologists can immensely improve taxonomy in this respect and the days should now be past when three pheno-typic variants could be published as new species alongside records of the species now considered to encompass them, as happened in African *Papillaria* (Müller, 1890). In the tropics one must be watchful of variation patterns which are not encountered in temperate bryophytes, particularly in what are essentially tropical families. Thus Edwards (personal communication) attributes the differ-ence in the scale of reduction of species names to synonymy in the two African families he studied to the fact that the variation of the one, Polytrichaceae (a reduction from 17 to 15) was well understood by European bryologists familiar with this family, while the other, Calymperaceae (a reduction from 60 to 15), an essentially tropical family, was not. Another example which might be given is the problem of anisomorphous sexes in several Pterobryaceae.

In some *Jaegerina* and *Calyptothecium* species male plants are distinctly smaller than females and could easily be mistaken for different species. Thus sexuality is a crucial character in the key to *Jaegerina* (Argent, 1973). This may not be an exclusively tropical problem but it is not one I have come across in temperate species.

These examples serve as a warning to temperate bryologists familiar with temperate variation patterns that things can be different in the tropics. A good deal of tropical taxonomy will certainly continue to be done in temperate regions by bryologists who either do not want field experience or, more often, do not get opportunities to do it.

I have already commented on the generic problems which arise from revising the species from a limited area and not undertaking a complete monograph. Ideally, species should also be examined over their total range but this is not always easy when the plant may be recorded under different generic and specific names in different parts of its range. During a study of limited scope one is bound to be haunted by related taxa which one never had the chance to investigate.

In the case of *Floribundaria floribunda* I was fortunately able to examine good material with capsules from both Asia and Africa. *Floribundaria cameruniae* (Dus.) C. Muell. was published without any reference to *F. floribunda*. Fleischer (1904–1923) commented on the fact that *F. cameruniae* was inseparable from *F. floribunda* in the vegetative state but thought it distinct from the Asian plant because the drawing of the exostome by Dusén (1895) showed no basal membrane. Potier de la Varde (1933–1936) examined one capsule and agreed with Fleischer's assumption, confirming that there was no basal membrane on the exostome. On examination of three collections of *F. cameruniae* with several capsules, however, I found that no basal membrane shows when the teeth are examined *in situ*. They are inserted a short distance below the mouth of the capsule and if the rim is dissected away the teeth are found to be standing on a distinct basal membrane. I examined four specimens of *F. floribunda* from Asia with capsules and of these, three had the teeth very shallowly inserted near the rim of the capsule; they tended to have a persistent annulus which often remained at least partly covering the basal membrane. The teeth of the fourth specimen, however, were at least as deeply inserted under the rim of the capsule as those in the African material, and had no basal membrane showing above the rim. In view of the great variability of the capsules of both African and Asian material and the absence of other diagnostic characters I reduced the African *F. cameruniae* to synonomy with *F. floribunda* and thus extended the range of this species from southeast Asia across to West Africa.

It is comforting that Pócs (personal communication), who has field experience of both Africa and Asia, agrees with me.

This example illustrates the relative ease with which a little careful checking can change the taxonomic picture when good material is available. I had seven fruiting gatherings to study compared with the solitary capsule which led Potier de la Varde to conclude that the African and Asian plants are distinct. This underlines the need for more good collections from all parts of the tropics and again points to the need for bryologists themselves to make collections. I was less fortunate in the case of *Floribundaria vaginans*. Happy to reduce four other species into this, I failed to look at specimens labelled *Meteoriopsis*, a genus not then recorded for Africa. I only subsequently realized the true affinities of this species when I became familiar with *Meteoriopsis reclinata* (Mitt.) Fleisch. in Papua New Guinea.

The amateur in the tropics has somewhat different problems. There are excellent opportunities for field work and observation but almost invariably very limited library and herbarium facilities. Both Greene (1976) and Touw (1974) have commented on the problems created when names of doubtful validity are provided to meet demand. But field workers must have names to use as handles, even if they must be changed later. Misidentifications must be avoided, but a name is not "wrong" in this sense because it later proves to be synonymous with another. Such names can be corrected and although it is tiresome when names change, it is better than having to wait indefinitely for a name. I agree with van Steenis when he advocates that taxonomists should prepare accounts with which biologists who are not specialists can name their own plants but they should not spend the whole time on this. I disagree, however, with his statement (van Steenis, 1965) that "In the tropics . . . floristic botany should definitely be discouraged in the present state of knowledge. Eagerness to publish lists and enumerations should be suppressed . . ." Systematics must advance on several fronts; both floristics and provisional lists have an important part to play, along with monographs, in building up the understanding of a group of plants.

As to limited library facilities, the era of xerox reprints has made it possible to gather together a very scattered and specialized literature at fairly moderate cost, and I think the impact of this on taxonomy is enormous. Working in the tropics with this literature must be very disappointing however as so many descriptions are, in the words of Jones (1954) on hepatics, but equally applicable to mosses, "so brief that they are quite inadequate to define the species, and often indeed are insufficient to indicate the genus as now understood to which the plant should be assigned". Good illustrations are the next best thing to accurately named herbarium material when trying to identify a plant and they are a good deal quicker to use than herbarium specimens.

The view that tropical herbaria are bound to fall into neglect whereas north-temperate ones are invulnerable cannot be sustained. Fortunately, the present day economic value of plants (mainly the woody ones) means that many tropical countries run good local herbaria often attached to forestry departments. It is a matter of courtesy, even if not a statutory obligation, to deposit a set of the specimens one has collected locally, and such specimens will stimulate local interest. More should be done to pressurize timber companies to provide funds so that the less immediately valuable plants in the forests they are destroying are at least recorded.

The herbarium worker in temperate countries is in the best position to pursue taxonomy since he has his own institution's specimens to work on and is usually able to borrow specimens from other institutions—something workers in many tropical countries cannot do. But he very easily loses touch with his working public—the people who use his keys or monographs. Universities have a great advantage here in that both keys and plants may be set before students. With minds unfettered by over-familiarity, students are usually only too quick to point out where jargon obscures meaning or size differences are unquantified, so that taxonomic accounts can be made considerably more useful. There is still a tendency amongst some taxonomists to endow their work with mystic obscurity, lending them the false importance of necessitating plants to pass through their hands for naming. Neither is taxonomy well served by someone who waits to publish until sure of every last detail. A taxonomist who publishes nothing may be almost as bad as one who publishes a lot of ill-founded species.

Without wishing to encourage the publication of invalid species, I feel that we have the means, if *Index Muscorum* is kept up to date, of keeping order in the mire of indigestible names. It would help if type descriptions had to be accompanied by scale drawings—a more universal language than Latin. This would be easier than trying to prevent the publication of new names. Possibly it would help to submit reprints to a central governing body so that names could not be missed from the index supplements. I would also like to see moss floras produced in loose-leaf fashion in the way some palynologists produce fossil pollen descriptions (e.g. Kremp *et al.*, 1957), in whatever order was convenient.

As has been stressed before by Crum and Steere (1957) and Richards and Clear (1967), monographic treatments of genera and families are the best way of removing the difficulties confronting anyone wishing to study tropical mosses. In addition, more numerous well-preserved and documented collections are badly needed, particularly from areas where the natural habitat is being destroyed at such a rate that species are threatened with extinction. But little advance is to be made if new collections merely accumulate in large, unsorted masses as is the case in some herbaria, and the greatest need is simply for more taxonomists. Bryophytes should not be neglected merely because they are of no economic value. They are an intrinsically interesting group and have an important contribution to make to solving problems of evolution and phytogeography.

REFERENCES

ARGENT, G. C. G. (1973). A taxonomic study of African Pterobryaceae and Meteoriaceae, I. Pterobryaceae. *J. Bryol.* **7**, 353–378.

BROTHERUS, V. F. (1924–1925). Musci (Laubmoose). *In* "Die natürlichen Pflanzenfamilien" (A. Engler, ed.). Second edition, Vol. 10, 129–478; Vol. 11, 1–542. W. Engelmann, Leipzig.

CRUM, H. A. and STEERE, W. C. (1957). The mosses of Porto Rico and the Virgin Islands. *Scient. Surv. P. Rico* **7**, 395–599.

DIXON, H. N. (1924). "The Student's Handbook of British Mosses." Third edition. Sumfield, Eastbourne.

DURING, H. J. (1977). A taxonomical revision of the Garovaglioideae (Pterobryaceae, Musci). *Bryophyt. Biblthca* **12**, 1–244.

DUSÉN, P. (1895). New and some little known mosses from the west coast of Africa. *K. svenska VetenskAkad. Handl.* **28** (2), 1–56.

FLEISCHER, M. (1904–1923). "Die Musci der Flora von Buitenzorg." E. J. Brill, Leiden.

GREENE, S. W. (1976). Are we satisfied with the rate at which bryophyte taxonomy is developing? *J. Hattori bot. Lab.* **41**, 1–6.

JONES, E. W. (1954). The task of the tropical hepaticologist. *In* "Huitième Congrès International de Botanique. Rapports et Communications aux Sections 14, 15 et 16" pp. 53–54. Paris.

KREMP, G. O. W., AMES, H. T. and GREBE, H. (1957). "Catalog of Fossil Spores and Pollen" Vol. I. College of Mineral Industries, Pennsylvania State University, Pennsylvania.

MÜLLER, K. (1890). Die Moose von vier Kilimandscharo-expeditionen. *Flora, Jena* **73**, 465–499.

POTIER DE LA VARDE, R. (1933–1936). Mousses du Gabon. *Mém. Soc. natn. Sci. nat. math. Cherbourg* **42**, 1–270.

RICHARDS, P. W. and CLEAR, I. D. (1967). Notes on African mosses, III. *Campylopus* and *Microcampylopus. Trans. Br. bryol. Soc.* **5**, 305–315.

SMITH, A. J. E. (1978). "The Moss Flora of Britain and Ireland." Cambridge University Press, Cambridge.

STEENIS, C. G. G. J. van (1965). Nature and purpose of botanical classification from the standpoint of the producer, particularly with respect to tropical plants. *Bull. bot. Surv. India* **7**, 8–14.

Touw, A. (1974). Some notes on taxonomic and floristic research on exotic mosses. *J. Hattori bot. Lab.* **38**, 123–128.

Wijk, R. van der, Margadant, W. D. and Florschütz, P. A. (1959–1969). "Index Muscorum." International Association of Plant Taxonomists, Utrecht.

.

9 | Towards an Experimental Approach to Bryophyte Taxonomy

A. J. E. SMITH

*School of Plant Biology, University College of North Wales,
Bangor, Gwynedd LL57 2UW, Great Britain*

Abstract: The present poor state of bryophyte biosystematics compared with that of flowering plants is discussed, as is the value of the biosystematic approach. Examples from the small number of experimental studies that have been made are given. The use of cultivation experiments under uniform and varied conditions in assessing inter- and intraspecific variability and the status of taxa, especially in relation to certain species of *Lophocolea*, *Bryum* and *Pohlia*, is considered. Approaches to the study of genecological variation in *Marchantia polymorpha*, *Lophocolea heterophylla* and two *Pohlia* species are evaluated. The importance of mitotic and meiotic chromosome studies is assessed. In view of the existence of natural hybrids the feasibility of using hybridization in assessing relationships and in determining the genetic basis of species differences is commented on. The present lack of information on breeding systems is deplored and the necessity for work on this topic before the processes of micro-evolution in bryophytes can be fully understood is stressed. It is concluded that bryophytes show evolutionary potential comparable to that of the angiosperms and that there is no reason why information on the evolutionary processes in bryophytes, equivalent to that now available concerning angiosperms, should not be built up.

INTRODUCTION

Biosystematic studies on higher plants over the last 40 or 50 years have produced a vast amount of data on the origins of species and the nature of variation within them. It has proved possible to publish books on the subject such as those by Manton (1950) and Stebbins (1950, 1971), quite apart from reviews

Systematics Association Special Volume No. 14, "Bryophyte Systematics", edited by G. C. S. Clarke and J. G. Duckett, 1979, pp. 195–206, Academic Press, London and New York.

and symposium reports on special aspects of biosystematics. With bryophytes the situation is very different and the small amount of experimental taxonomic information available can be summarized in relatively short review papers such as those of Anderson (1964), Smith (1978b), Steere (1958) and Stotler (1976). It is clear from a survey of the literature that there have been very few workers in the field of bryophyte biosystematics. To illustrate this point, of the taxonomic papers published in *The Bryologist, Journal of Bryology* and *Lindbergia* since 1970, five were of a biosystematic nature while 92 were descriptive.

Davis and Heywood (1963) say that the knowledge of the world's flora is extremely uneven and its classification can be considered in four overlapping groups:

1. *The pioneer or exploratory phase* in which the flora is known only from a limited number of herbarium specimens.

2. *The consolidation phase* in which data concerning morphological variation, frequency and distribution, based on herbarium and field studies, are accumulated.

3. *The biosystematic phase* in which chromosome studies, breeding and cultivation experiments are carried out, i.e. there is an emphasis on micro-evolution and variation.

4. *The encyclopaedic phase* where data of both orthodox and experimental types are incorporated in the production of monographs.

Insofar as bryology is concerned, the present state of knowledge of most of the world's flora is very much in the first, exploratory phase (hence the emphasis on descriptive taxonomy). It is only in Europe, parts of North America and Japan that bryophyte taxonomy has passed into the second phase. Here there is sufficient information available for a start to be made on experimental studies of micro-evolution and variation. Yet, even in Britain, where there has probably been a more intensive study of the bryophyte flora over a longer period of time than anywhere else in the world, the amount of experimental work done is negligible. The only taxa that have been studied to any depth using experimental techniques in Britain are the *Bryum bicolor* agg. (Smith and Whitehouse, 1978), four *Dicranum* species (Briggs, 1965), two *Drepanocladus* species (Lodge, 1960), *Campylopus* (Corley, 1976), *Pohlia* section *Pohliella* (Lewis and Smith, 1977) and *Lophocolea* (Steel, 1978). Outside Britain there have been studies on *Brachythecium rutabulum* and *B. rivulare* (Wigh, 1975), *Polytrichum alpestre* (Longton, 1974), *Weissia* s. l. (Anderson and Lemmon, 1972, 1974) and *Lophocolea heterophylla* (Hatcher, 1967). It is high time that more experimental rather than descriptive work is carried out in those countries where there is a detailed knowledge of the bryophyte flora.

THE BIOSYSTEMATIC APPROACH

Biosystematics may be defined in a variety of ways depending upon whether the approach is genetic or taxonomic. In the former it is concerned only with micro-evolutionary phenomena and it is the population that is the basic unit; in the latter it is a modernized type of systematics which takes into account the genetic and evolutionary relationships of taxa and in which the species is the basic unit. Unfortunately, some categories defined by experiment do not correspond to orthodox taxonomic categories, and some types of variation are not amenable to orthodox taxonomic treatment. Attempts have been made to define evolutionary units as they exist in nature but this has led to the creation of a plethora of experimental categories. From a practical point of view, if experimental categories cannot be expressed in terms of variety, subspecies or species then they are of little taxonomic value even though they may be of evolutionary interest.

The situation in *Atrichum undulatum* illustrates this point admirably. There are three cytotypes within the species, one haploid, one diploid and the third triploid. Biologically, because there can be no gene exchange between them, they are separate evolutionary units. Lazarenko and Lesnyak (1977) say these should be treated as cryptic sibling species. They are, however, indistinguishable morphologically unless the chromosomes are counted, which is not a practical proposition. It is, in terms of common sense, best to regard *Atrichum undulatum* as a single species containing three cytotypes. Newton (1968a) took this approach with *Tortula muralis* in which there is a series of chromosome races that intergrade. On the other hand, if there are morphological differences between micro-evolutionary units, then they should be recognized taxonomically, as in the instance of *Bryum pseudotriquetrum* or *Mnium marginatum* where the dioecious and monoecious varieties are haploid and diploid respectively.

The techniques that have traditionally been used in experimental taxonomy are cultivation and breeding experiments and chromosome studies. More recent techniques include isozyme studies for assessing genetic variability within and between species and phytochemistry for investigating phylogenetic relationships (Suire, this volume, Chapter 19). Experimental results are often subjected to sophisticated statistical treatment. Such techniques cannot make up for inconclusive experimental results, but are able to distinguish between those deductions based purely on intuition and those which are justified by the available evidence.

1. Cultivation Experiments

Much can be achieved by the use of cultivation experiments. It is often argued that the cultivation of bryophytes is difficult, yet perusal of the literature reveals that they have frequently been cultured for experimental purposes, especially by physiologists and morphogeneticists. Cultures can be maintained for considerable periods of time. Corley (1976) grew pot cultures of *Campylopus* species for two to three years, Lewis and Smith (1977) grew bulbil-bearing *Pohlia* species for two years; Whitehouse (personal communication) has grown *Bryum bicolor* in pots for up to 15 years. Admittedly it is difficult to keep cultures free of weeds, especially if they are grown in enclosed containers. Thus, Steel (1978) was only able to grow *Lophocolea* species in glass containers for up to three months before they were overcome by contaminants. Axenic cultures on agar only grow well for about three months because of deterioration of the medium. Frequently, however, three months is adequate time to obtain satisfactory results whether using axenic cultures as with *Pohlia annotina* agg. (Lewis and Smith, 1977) or non-axenic ones as with *Lophocolea* (Steel, 1978) and members of the Mniaceae (Wigh, 1972).

From work that has been carried out using both uniform and variable cultural conditions it is evident that valuable information may be obtained concerning variability within species and the constancy, or otherwise, of differences between related taxa, this latter with particular reference to the status of varieties and the validity of key taxonomic characters.

(a) *Experimental study of the status of taxa.* There has been, and sometimes still is, a tendency amongst bryologists to describe anything deviating from the norm as a variety, without providing any experimental evidence. With some species the number of varieties described is remarkable: *Bryum capillare*, 50 varieties; *Ceratodon purpureus*, 39; *Ctenidium molluscum*, 20; *Dicranum scoparium*, 24; *Hypnum cupressiforme*, 39; *Schistidium apocarpum*, 35; *Sphagnum subsecundum*, 82; (all from van der Wijk *et al.*, 1959–1969). This is clearly an absurd situation yet these intraspecific taxa cannot be eliminated indiscriminately as at least some may have a genetic basis.

Some idea of the status of a particular variety may be obtained by the examination of a large number of gatherings although herbaria may well contain selected specimens which will give a biased result as was found by Smith and Hill (1975) in studies on herbarium material of *Ulota*. Results from the study of herbarium specimens may also be controversial because of the selection of the characters used. In studies of the *Sphagnum subsecundum* agg. Hill (1975) used quantitative characters and concluded that *S. inundatum* was a variety of *S. auriculatum*. Eddy (1977), on the other hand, used qualitative characters and concluded that *S. inundatum* was a subspecies of *S. subsecundum*. This sort of situation can only be resolved by the cultivation of numerous gatherings over a period of time. That this can be done with *Sphagna* was demonstrated by Agnew (1958) in experimental studies on *Sphagnum recurvum*.

The variations in *Bryum bicolor* and in *Tortula subulata* provide examples of the value of cultivation experiments before a morphological variant is accepted as an intraspecific taxon. It was reported by Rilstone (1950) that in the west of Britain there was a form of *Bryum bicolor* which differed from usual plants in being more densely tufted and glossy and in having smaller bulbils. In culture, however, it proved impossible to draw a line between this and the typical form (Whitehouse, personal communication). In contrast, plants of *Tortula subulata* var. *graeffii* grown for three years in mixture with var. *subulata* remained morphologically distinct, indicating that there were good grounds for maintaining it as a separate taxon (Crundwell, 1956).

Work on *Dicranum* and *Pohlia* indicates the types of variation that may be found within and between species. Briggs (1965) worked on four species of *Dicranum*, *D. bonjeanii*, *D. fuscescens*, *D. majus* and *D. scoparium*, which he showed to be specifically distinct. Dixon (1924) recognized a total of eight varieties of these species. Briggs showed by cultivation experiments that whilst distinct genotypes are likely to exist, they show such phenotypic plasticity that they cannot be distinguished except by experiment and therefore cannot be given taxonomic recognition. Lewis and Smith (1977) studied the seven species of *Pohlia* in Britain that regularly produce axillary propagules. They found that within each of five of the species the gemmae remained morphologically very similar regardless of the environment. These species thus show neither genetic nor environmental variation. In a sixth species, *P. drummondii* there was considerable morphological variation in the gemmae in field populations but under uniform cultural conditions the gemmae of these populations became indistinguishable, indicating that there was environmentally induced variation but no genetic variation in the species. In *Pohlia proligera*, the seventh species, it was shown by cultivation experiments that variation in gemma form between wild populations had both an environmental and also a genetic basis.

These results indicate the necessity of experiment in assessing the status of taxa, especially in taxonomically difficult groups. It is also clear that conclusions which are valid for one species cannot be extended to related species. Each must be studied individually.

Difficulties are often experienced in identifying bryophytes because characters such as the shape and size of parts do not always correspond with descriptions in floras. This may be because certain characters show greater variation than was thought, or because poor discriminatory characters have been selected. Thus Macvicar (1926) says that in *Lophocolea cuspidata* the size of the leaf cells is 33–42 μm whilst in *L. bidentata* it is 28–35 μm. Steel (1978) showed that when grown under a range of environmental conditions there was such an overlap in cell size between the two species that this character could not be used to distinguish them. Bulbil morphology is regarded as an important taxonomic character in certain species of *Pohlia*. Lewis and Smith (1977) showed that variation between *P. proligera* and the plant traditionally referred to as *P. annotina* was continuous and that, on the basis of bulbil morphology, they could only be regarded as a single species. They also showed that the plant called, in Britain, *P. rothii* was merely a lowland environmental variant of *P. drummondii*. Conversely, cultivation experiments have shown that British material of *Bryum bicolor* represents four distinct species which can be recognized by the morphology of their bulbils and leaves (Smith and Whitehouse, 1978).

(*b*) *Genecological differentiation*. An aspect of micro-evolution that is of con- siderable interest and which can be investigated by cultivation is genecological differentiation. A simple but striking example of this phenomenon is found in *Marchantia polymorpha*. Briggs (1972) showed that populations of this plant from central Glasgow were much more tolerant of lead than were country populations. This indicated that there had been selection in relatively recent times for lead tolerance in populations of *Marchantia polymorpha* in polluted areas.

A more detailed study, providing evidence of genetic differentiation between different populations, was carried out by Hatcher (1967) on *Lophocolea heter- ophylla*. He raised five genetically distinct clones of *L. heterophylla* from spores and cultured them for four to five months under a series of different environ- mental conditions. The responses of each of the clones to the different conditions varied markedly, some doing well in a particular environment, others doing badly or dying under the same conditions. He was thus able to demonstrate that not only did morphological differences between populations have a genetic basis but that there were genetically determined physiological differences as well, a classic case of ecological differentiation.

Clarke and Greene (1971) showed by cultivation experiments that there were differences in reproductive behaviour between populations both of *Pohlia cruda* and *P. nutans* from South Georgia and Great Britain indicating that there had been selection for reproductive cycles related to the environmental conditions under which the populations occurred.

Experiments of this type provide evidence of the action of natural selection and may help to explain the degree of variability that may occur in some species.

2. *Chromosome Studies*

A review of work on bryophyte chromosomes is given by Newton in this volume (Chapter 10) and will not be repeated here. Cytological investigations play an important part in experimental systematics and some aspects deserve to be emphasized.

In nearly all cases where cytological studies have been used for taxonomic ends, the chromosomes, whether meiotic or mitotic, have been used as an additional morphological character. As vehicles of heredity which are not subject to environmental modification, chromosomes are of considerably greater taxonomic significance than other morphological characters. Except in the case of polyploidy, however, they have provided little information on micro-evolution.

As Newton points out, modern cytological techniques which allow more accurate comparisons of karyotypes are likely to give a great deal of new information. Another open field for study is the investigation of meiosis in hybrid bryophytes, both natural and artificially produced. The degree of chromosome pairing in hybrids gives an indication of the closeness of relationship and in the case of polyploids may be used to determine parentage. The only published study to date on this subject is that by Anderson and Lemmon (1972) on natural hybrids between *Astomum* and *Weissia*. In two sets of hybrids meiosis was normal but subsequent germination of the spores was very low. In view of the normal meiosis it was concluded that the species of *Astomum* and *Weissia* are closely related. In my opinion, this calls into doubt the validity of the generic separation.

3. *Breeding Experiments*

(*a*) *Assessment of relationship.* Breeding experiments may be used as a test of affinity and for assessing the genetic basis of sporophyte differences between species. In flowering plant hybrids the normality of meiosis and the viability of pollen in the F_1 and the viability of hybrid seed are used as measures of relationship; the higher the degree of normality, the closer the two species concerned are. In bryophytes the form of hybrid sporophytes, the nature of meiosis in such sporophytes and the fertility of the spores should also provide a measure of relationship. There are numerous reports of natural hybrids in the *Ditrich-aceae*, *Funariaceae* and *Weissia* s.l. and occasional reports in other genera such as *Campylopus* (Corley, 1969), *Bryum* and *Ulota* (Smith, 1978a). It is very likely that hybridization occurs elsewhere but where there are no obvious differences between the sporophytes of the species concerned these will escape notice. This argument is supported by the occurrence of capsules containing abnormal spores or in which meiosis is abnormal, both suggesting hybridity. That artificial hybrids can be raised is indicated by the work of Wettstein (1928, 1932) on the Funariaceae and of Burgeff (1943) on *Marchantia*. There is no reason why crossing experiments should not be carried out between dioecious species or autoecious species where male shoots can be removed. It may be more difficult with synoecious species if they are self compatible.

Anderson and Lemmon (1972), in discussing hybridization in mosses, stated that in all recorded crosses one of the parents is a cleistocarp, neither reciprocal crosses nor F_2s have been found in nature and there is no evidence of intro-gression. That it was possible to make such statements as recently as 1972, indicates the paucity of knowledge about hybridization in bryophytes, a state of affairs that urgently needs rectifying. There are now a few examples of hybridization in which both parents are operculate and reciprocal hybrids have been reported in *Weissia* (Smith, 1978a, p. 275). That F_2s and introgressants have not been found may be due to difficulty in detection. It may well be that variability in some species of genera such as *Brachythecium*, *Hypnum* and *Trichostomum* may be due to hybridization or introgression, as also may the existence of the occasional indeterminate bryophyte. In the *Fissidens viridulus* complex a number of doubtful species have been described, apparently based upon an arbitrary selection of characters such as nature of leaf border, leaf cell size and distribution of gametangia. The random way these characters may be assorted could be due to genetic segregation following hybridization or to introgression. Problems such as these warrant further study.

(*b*) *Genetic basis of character differences.* In genera in which there are differences between the sporophytes of various species it should prove possible to assess the genetic basis of these differences if spores from hybrid sporophytes can be raised and used to produce F_2 sporophytes. It would, for example, be interesting to know the genetic basis of immersed/exserted capsules and operculate/inoper-culate capsules in the genus *Weissia*. It would be much more difficult to deter-mine the nature of gametophyte differences since the gametophyte is functionally haploid. The only possible approach would be to produce aposporous diploid gametophytes from F_1 hybrid sporophytes. This would provide an indication of whether two states of a character are under simple Mendelian control. In mosses the situation is further complicated by the fact that about 80% of species are polyploid (Smith, 1978b) and it is futile to speculate on the genetics of species differences before experimental results are available.

4. Breeding Systems

For a proper understanding of genetic variability within species it is essential to know something of the breeding systems of the plants concerned. Unfortunately almost nothing is known of breeding systems in bryophytes although there is a little observational and experimental data on gene flow distances (Anderson and Lemmon, 1974; Clayton-Greene *et al.*, 1977). Whilst dioecious species are obviously outbreeding it is stated by some authors (Gemmell, 1950; Lewis, 1961; Iwatsuki, 1972) that monoecious species are obligatorily inbreeding. This latter assumption is questionable. Lazarenko and Lesnyak (1972) reported that the monoecious species *Desmatodon cernuus* is self-incompatible in monospore culture. That the great majority of reported natural hybrids involve monoecious species is indicative of at least some degree of outbreeding. Newton (1968b) pointed out that the percentage of meiotic irregularities of the type associated with structural hybridity or genetic heterozygosity found in monoecious species was highly indicative of at least some degree of outbreeding.

Anderson and Lemmon (1974) in observations on *Weissia controversa*, which is autoecious, point out that only about 10% of female plants had male branches and that it must therefore be assumed that 90% of capsules arise by cross fertilization.

It is clear that before any generalizations about breeding systems and variation, such as those of Gemmell (1950) can be made, a much more thorough investigation of breeding systems is required. The great diversity within many bryophyte genera and families (e.g. *Bryum, Fissidens, Scapania,* Amblystegiaceae, Lejeuneaceae) and the adaptive success of the group as a whole suggests genetic variability and hence an efficient outbreeding system or systems.

CONCLUSION

Within the limits of their structure, bryophytes are a highly successful group of plants. It has been argued that they show less evolutionary potential than angiosperms (Gemmell, 1950; Anderson, 1963; Crum, 1972). Even on the basis of the limited amount of data on their biology that are at present available this is very questionable. They show as wide a range of adaptation to habitat as the angiosperms; they possess a similar range of types of life cycle, from short-lived ephemerals to long-lived perennials and many are able to withstand long periods of unfavourable environmental conditions either by producing resistant spores or vegetative propagules or by physiological drought resistance. To gain at least some insight into the reasons for their success and their evolutionary potential, much more work must be done on genetic variability, breeding systems and the mechanisms involved in speciation and diversification. In view of the detailed knowledge that has accumulated about angiosperm evolution there seems no reason why a similar body of knowledge should not be built up about bryophytes. This will not advance knowledge on phylogeny; until a much improved fossil record is built up or some new techniques evolved, bryophyte phylogeny will remain a matter of speculation. Biosystematics will, however, provide information on the mechanisms of evolution and on phylogeny at the specific and possibly the generic levels.

REFERENCES

AGNEW, S. (1958). "A Study in the Experimental Taxonomy of some British *Sphagna* (Section *Cuspidata*) with Observations on their Ecology." Ph.D. thesis, Univ. Wales.

ANDERSON, L. E. (1963). Modern species concepts: Mosses. *Bryologist* **66**, 107–119.

ANDERSON, L. E. (1964). Biosystematic evaluations in the Musci. *Phytomorphology* **14**, 27–51.

ANDERSON, L. E. and LEMMON, B. E. (1972). Cytological studies of natural intergeneric hybrids and their parental species in the moss genera *Astomum* and *Weissia*. *Ann. Mo. bot. Gdn* **59**, 382–416.

ANDERSON, L. E. and LEMMON, B. E. (1974). Gene flow distances in the moss, *Weissia controversa* Hedw. *J. Hattori bot. Lab.* **38**, 67–90.

BRIGGS, D. (1965). Experimental taxonomy of some British species of the genus *Dicranum*. *New Phytol.* **64**, 366–386.

BRIGGS, D. (1972). Heavy metal tolerance in bryophytes. *J. Bryol.* **7**, 149.

BURGEFF, H. (1943). "Genetische Studien an *Marchantia*." G. Fischer, Jena.

CLARKE, G. C. S. and GREENE, S. W. (1971). Reproductive performance of two species of *Pohlia* from temperate and sub-Antarctic stations under controlled environmental conditions. *Trans. Br. bryol. Soc.* **6**, 278–295.

CLAYTON-GREENE, K. A., GREEN, T. G. A. and STAPLES, B. (1977). Studies of *Dawsonia superba*. I. Antherozoid dispersal. *Bryologist* **80**, 439–444.

CORLEY, M. F. V. (1969). Hybrid sporophytes in *Campylopus*. *Trans. Br. bryol. Soc.* **5**, 829.

CORLEY, M. F. V. (1976). The taxonomy of *Campylopus pyriformis* (Schultz) Brid. and related species. *J. Bryol.* **9**, 193–212.

CRUM, H. (1972). The geographic origins of the mosses of North America's eastern deciduous forest. *J. Hattori bot. Lab.* **35**, 269–298.

CRUNDWELL, A. C. (1956). Further notes on *Tortula subulata* var. *graeffii*. *Trans. Br. bryol. Soc.* **3**, 125–126.

DAVIS, P. H. and HEYWOOD, V. H. (1963). "Principles of Angiosperm Taxonomy." Oliver and Boyd, Edinburgh and London.

DIXON, H. N. (1924). "The Student's Handbook of British Mosses." Third edition. Sumfield, Eastbourne.

EDDY, A. (1977). *Sphagnum subsecundum* agg. in Britain. *J. Bryol.* **9**, 309–319.

GEMMELL, A. R. (1950). Studies in the Bryophyta, I. The influence of sexual mechanism on varietal production and distribution of British Musci. *New Phytol.* **49**, 64–71.

HATCHER, R. E. (1967). Experimental studies of variation in Hepaticae, I. Induced variation in *Lophocolea heterophylla*. *Brittonia* **19**, 178–201.

HILL, M. O. (1975). *Sphagnum subsecundum* Nees and *S. auriculatum* Schimp. in Britain. *J. Bryol.* **8**, 435–441.

IWATSUKI, Z. (1972). Geographical isolation and speciation of bryophytes in some islands of Eastern Asia. *J. Hattori bot. Lab.* **35**, 126–141.

LAZARENKO, A. S. and LESNYAK, E. N. (1972). A comparative study of two moss sibling species—*Desmatodon cernuus* (Hüb.) BSG—*D. ucrainicus* Laz. *Zh. obshch. Biol.* **33**, 657–667.

LAZARENKO, A. S. and LESNYAK, E. N. (1977). On chromosome races of the moss *Atrichum undulatum* (Hedw.) Brid. in the west of the U.S.S.R. *Ukr. bot. Zh.* **34**, 383–388.

LEWIS, K. and SMITH, A. J. E. (1977). Studies on some bulbiliferous species of *Pohlia* section *Pohliella*, I. Experimental investigations. *J. Bryol.* **9**, 539–556.

LEWIS, K. R. (1961). The genetics of bryophytes. *Trans. Br. bryol. Soc.* **4**, 111–130.

LODGE, E. (1960). Studies of variation in British material of *Drepanocladus fluitans* and *Drepanocladus exannulatus*, II. An experimental study of the variation. *Svensk. bot. Tidskr.* **54**, 387–393.

LONGTON, R. E. (1974). Genecological differentiation in bryophytes. *J. Hattori bot. Lab.* **38**, 49–65.

MACVICAR, S. M. (1926). "The Student's Handbook of British Hepatics." Second edition. Sumfield, Eastbourne.

MANTON, I. (1950). "Problems of Cytology and Evolution in the Pteridophyta." Cambridge University Press, Cambridge.

NEWTON, M. E. (1968a). Cyto-taxonomy of *Tortula muralis* Hedw. in Britain. *Trans. Br. bryol. Soc.* **5**, 523–535.

NEWTON, M. E. (1968b). "Cytology of British Bryophytes." Ph.D. Thesis, Univ. Wales.

RILSTONE, F. (1950). *Bryum atropurpureum* W. and M. var. *gracilentum* Tayl. *Trans. Br. bryol. Soc.* **1**, 369–370.

SMITH, A. J. E. (1978a). "The Moss Flora of Britain and Ireland." Cambridge University Press, Cambridge.

SMITH, A. J. E. (1978b). Cytogenetics, biosystematics and evolution in the Bryophyta. *Adv. bot. Res.* **6**, 195–276.

SMITH, A. J. E. and Hill, M. O. (1975). A taxonomic investigation of *Ulota bruchii* Hornsch. ex Brid., *U. cripsa* (Hedw.) Brid. and *U. crispula* Brid., I. European material. *J. Bryol.* **8**, 423–433.

SMITH, A. J. E. and WHITEHOUSE, H. L. K. (1978). An account of the British species of the *Bryum bicolor* complex including *B. dunense* sp. nov. *J. Bryol.* **10**, 29–47.

STEBBINS, G. L. (1950). "Variation and Evolution in Plants." Columbia University Press, New York.

STEBBINS, G. L. (1971). "Chromosomal Evolution in Higher Plants." Edward Arnold, London.

STEEL, D. T. (1978). The taxonomy of *Lophocolea bidentata* (L.) Dum. and *L. cuspidata* (Nees) Limpr. *J. Bryol.* **10**, 49–59.

STEERE, W. C. (1958). Evolution and speciation in mosses. *Am. Nat.* **92**, 5–20.

STOTLER, R. E. (1976). The biosystematic approach in the study of the Hepaticae. *J. Hattori bot. Lab.* **41**, 37–46.

WETTSTEIN, F. von (1928). Morphologie und Physiologie des Formwechsels der Moose auf genetische Grundlage. *Biblthca genet.* **10**, 1–216.

WETTSTEIN, F. von (1932). Genetik. *In* "Manual of Bryology" (F. Verdoorn, ed.), pp. 232–272. M. Nijhoff, The Hague.

WIGH, K. (1972). Cytotaxonomical and modification studies in some Scandinavian mosses. *Lindbergia* **1**, 130–152.

WIGH, K. (1975). Scandinavian species of the genus *Brachythecium* (Bryophyta), I. Modifications and biometric studies in the *B. rutabulum-B. rivulare* complex. *Bot. Notiser* **128**, 463–475.

WIJK, R. van der, MARGADANT, W. D. and FLORSCHÜTZ, P. H. (1959–1969). "Index Muscorum." International Association for Plant Taxonomy, Utrecht.

10 | Chromosome Morphology and Bryophyte Systematics

M. E. NEWTON

Departments of Botany and Zoology, University of Manchester*

Abstract: Bryophyte cyto-taxonomy based on chromosome morphology has hitherto relied on measurement and description of m-chromosomes, heterochromatin distribution, morphological and structural sex-chromosomes and accessory chromosomes. Of these, the accurate definition of chromosome proportions and of the location of heterochromatin are identified as fundamental to further systematic studies. Possible ways of improving technique are considered. In particular, the potential use of Giemsa C-banding is assessed with reference to the literature and to new work on *Rhizomnium pseudopunctatum* and *Atrichum crispum*. As a means of distinguishing between the morphologically similar chromosomes of many liverworts and mosses, Giemsa C-banding is likely to prove invaluable but it also provides a useful tool for attempting to discriminate between different kinds of heterochromatin in bryophytes. A knowledge of heterochromatin types is essential to a better understanding of sex-chromosomes and accessory chromosomes but it is difficult to apply to bryophytes the term "facultative heterochromatin", as defined for diploid organisms. Q- and N-banding techniques may also prove helpful means of furthering bryophyte cyto-taxonomy.

INTRODUCTION

Initial work on bryophyte chromosomes around the turn of the century (Farmer, 1895; van Hook, 1900; Davis, 1901; Beer, 1903; Moore, 1905) formed only part of a more general interest in cell division itself. For that reason, and also because the precise nature of chromosome number and

* Address for correspondence: Shaw Bank, 143 Mottram Old Road, Stalybridge, Cheshire, SK15 2SZ, Great Britain

Systematics Association Special Volume No. 14, "Bryophyte Systematics", edited by G. C. S. Clarke and J. G. Duckett, 1979, pp. 207–229, Academic Press, London and New York.

morphology had not then been realized, comparatively little attention was paid to chromosome description. However, the situation was remedied to the extent that Allen (1917, 1919) was able to describe one chromosome which differed according to the sex of *Sphaerocarpos donnellii* Aust., whereas Heitz (1928a, b) could go further to recognize internal variation within and between individual *Pellia* chromosomes and thus to identify heterochromatin. Both observations, amply substantiated by subsequent workers, are particularly noteworthy because they are fundamental to so much of our present knowledge of chromosome morphology and hence to bryophyte cyto-taxonomy.

The type of cytological data accumulated in the 50 or 60 years following the work of Heitz and Allen has depended on whether the source was mitosis or meiosis. The majority of liverwort cytologists have concentrated almost exclusively on mitotic studies, thereby obtaining information about chromosome structure as well as number. Meiosis reveals fewer structural details, yet, with relatively few exceptions, it has been the usual choice in mosses. Useful summaries have been provided by Berrie (1960), Anderson (1962), Schuster (1966) and Fritsch (1972) for liverworts and by Wylie (1957), Anderson (1962), Fritsch (1972) and Steere (1972) for mosses. In general, it would appear that liverwort evolution has occurred without associated changes in chromosome number, whereas mosses show evidence of extensive polyploidy and aneuploidy. The kind and extent of any correlated structural rearrangements during the course of evolution are difficult to judge, although a number of hypotheses have been put forward (Mehra and Khanna, 1961; Berrie, 1963; Maekawa, 1963; Tatuno and Nakano, 1970). They rely heavily on the distribution of certain types of chromosomes which are characteristic of, and perhaps in some cases unique to, bryophytes and which it is the purpose of this paper to consider.

KARYOTYPE ANALYSIS

1. Measurement

Many bryophyte taxa have been described or defined wholly or in part on the basis of chromosome number (Lorbeer, 1934; Lefèbvre, 1968). The practice is not, however, without its risks. For instance, structural rearrangements including deletions, insertions, translocations, inversions and Robertsonian fusions or fissions could, during the course of evolution, transform numerically identical karyotypes into ones that differed in chromosome number; the reverse might also be expected to occur. Moreover, it was for this reason that Yano (1957a) suggested that some evidence of polyploidy in mosses may be overlooked. Uniformity of chromosome number in liverworts and in certain families and genera of mosses has therefore resulted in cyto-taxonomic characters being sought in morphological variation within the karyotype.

Detailed description of karyotypes is usually concerned with mitotic metaphase chromosomes, where interpretation is complicated by neither the presence of chiasmata nor the extreme condensation associated with meiosis. The major criteria are length, centromere position, differential staining and, in certain chromosomes, the location of secondary constrictions. As far as is known at present, each bryophyte chromosome has a single localized centromere. Although the centromere does not appear as a constriction in some species and genera (Lewis, 1958; Newton, 1971, 1977a), its position is readily identified at the onset of anaphase. The only exception is the largest chromosome of *Pleurozium schreberi* in which Vaarama (1954) reported two alternative sites of centromere activity.

Japanese cytologists, as the chief workers in this field (S. Inoue, 1964, 1971; Segawa, 1965a, b, c; H. Inoue, 1967, 1968; Ono, 1970a, b, c), but also Indian bryologists (Chatterjee and Gangulee, 1970), have developed a system of nomenclature which in its best form allows for the concise yet accurate presentation of data. Thus, length is quoted as a percentage of the total length of the entire complement and centromere position and other markers are also defined. Following Tatuno (1941) and Yano (1957a, b, c), it became customary to describe metacentric or slightly sub-metacentric chromosomes as *V* chromosomes, whereas *J* denoted those with distinctly submedian or subterminal centromeres. In addition, the letter *I* has sometimes been used to distinguish telocentric chromosomes. In this its basic form, the classification is not only subjective but also tends to obscure significant differences between karyotypes (Ramsay, 1969). Bearing in mind the possible effects of structural rearrangement, it cannot even be assumed, contrary to the view of Ramsay (1969), that indentically ranked and similarly shaped chromosomes in related species are necessarily homologous (Rees, 1972).

Later refinements, as used by Segawa (1965a, b, c) and S. Inoue (1973), eliminated the subjective element. They calculated the length of the shorter arm as a percentage (form % or *F* %) of the total and thus provided a precise description of centromere position. Unfortunately, *F* % has been translated back into *V* and *J* terminology (Segawa, 1965a; S. Inoue and Iwatsuki, 1976), the categories being used to designate ranges of more than and less than 33·3%, respectively, or in some cases (Ono, 1970a, b, c) more or less than 40%. In addition, chromosomes less than half the length of the longest member of the complement have been distinguished by small letters *v* and *j* (Tsutsumi *et al.*,

1973; S. Inoue and Iwatsuki, 1976). Presented in formulae, any similarity in the proportion of chromosome types has been taken to imply homology, not only within liverworts (Tatuno, 1941, 1959; Berrie, 1963; S. Inoue, 1973) and mosses (Mehra and Khanna, 1961; S. Inoue and Uchino, 1969) but also between the two groups (Yano, 1960; Tatuno and Nakano, 1970). Such widely drawn comparisons have been criticized for assuming the monophyletic nature of the Bryophyta and also because they disregard the possibility of countless small structural changes (Schuster, 1966; Smith and Newton, 1968; Ramsay, 1974). In the ox and the goat, for instance, chromosome proportions and numbers are identical but are associated with a 10% difference in the amount of DNA (Bostock and Sumner, 1978).

Despite these difficulties, the work of Iverson (1963), who found that $F \%$ varied between species of *Frullania*, clearly indicated an important role for chromosome measurement in bryophyte taxonomy. Wigh (1972) pointed out significantly different relative chromosome lengths in *Pseudobryum cinclidioides* from North America (Lowry, 1948), Japan (Ono, 1970b) and Sweden (Wigh, 1972) and it is possible that this variation may also prove to be taxonomically significant. What is not certain is the extent to which these deviations depend on inadequate technique. Sybenga (1959), for instance, found that squash techniques such as are used for almost every bryophyte study can cause serious distortion, both of total length and of $F \%$. Moreover, chromosomes such as the smallest in British *Odontoschisma sphagni* (Newton, 1973) attain maximum condensation before the eight larger chromosomes (Fig. 1). Reproducible results from stages prior to anaphase must depend, therefore, on a degree of precision that is impossible to ensure among cells which do not divide synchronously. Anaphase can be recognized more accurately and should be preferred for karyotype analysis (Newton, 1977a). It is still subject to distortion during slide preparation, but large samples help to reduce the problem.

Perhaps the most promising means of improving the precision of bryophyte karyotype analysis is suggested by the work of Solari and Counce (1977) and Moses (1977) on the synaptonemal complexes formed during meiosis in mammals. These have the advantage of combining an extended form with recognizable centromeres. It is conceivable that in those bryophyte chromosomes which have considerable amounts of heterochromatin, differential rates of condensation in euchromatin and heterochromatin may destroy the observed correlation between relative length of synaptonemal complexes and mitotic metaphase chromosomes but the system deserves investigation.

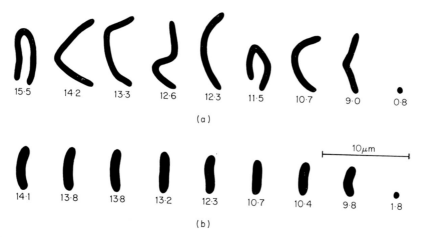

(a)

(b)

FIG. 1. Karyotypes from single thallus of *Odontoschisma sphagni*, showing relative lengths of chromosomes derived from photographs. (a) late prophase; (b) metaphase.

2. *m-Chromosomes*

Micro- or m-chromosomes are widespread in bryophytes and were first described in liverworts (Heitz, 1927), one conspicuously small chromosome being characteristic of many haploid complements of $n = 8$ $(7+1m)$ and $n = 9$ $(8+1m)$. Although proof of homology in the two main groups of bryophytes is lacking, the term was applied to mosses by Vaarama (1953) and Bryan (1955). Wylie (1957) defined limits of one tenth the size of the next larger member of the complement for m-chromosomes in mosses but later work suggests that the limit is too low for recording accurately the frequency with which marked discontinuity of variation occurs at a point between the smallest chromosome and the rest of the complement (Newton, 1971). A size up to half that of the next larger chromosome has therefore been applied to mosses and liverworts alike (Newton, 1971, 1973, 1975, 1977a) but there is no general agreement on appropriate limits (Smith and Newton, 1966). Indeed, Berrie (1960, 1963) reviewed the subject and found that the m-chromosome recognized by Tatuno (1959) in *Takakia lepidozioides* Hattori and Inoue was almost two-thirds the length of the longest. As a result, Berrie (1960) recognized three categories of m-chromosomes in liverworts and subdivision for mosses has been similarly advocated, for example by Mehra and Khanna (1961), Smith and Newton (1968) and Vysotska (1975).

Following the discovery of a small chromosome that was considered to be the y-chromosome (see sex-chromosomes below) of *Pellia endiviifolia* (Heitz, 1928a), m-chromosomes came to be associated with sex-determination, although this cannot be accepted as universally true. They have also been equated with the small heterochromatic h-chromosomes (see below) reported by Tatuno (1941) in liverworts and by Yano (1957a, b, c) in mosses as well as by many subsequent authors. In fact, it was only because he was unable to detect heteropycnosis (differential staining associated with heterochromatin) in *Atrichum pallidum* Ren. and Card. that Sharma (1963) described the smallest and largest chromosomes as m- and M-chromosomes, respectively. As mentioned previously (Newton, 1975), the term "h-chromosome" has now generally supplanted "m-chromosome" in liverwort cytology, with the result that a chromosome is so described if it is only fractionally the smallest member of the complement. Relative size in mosses remains, however, the main criterion for recognizing m-chromosomes, although the meiotic properties of light staining and precocious segregation of half-bivalents are considered by some to be equally diagnostic (Ramsay, 1964, 1969; Bryan, 1973; Snider, 1973; Wigh, 1973). It is possible, however, that early disjunction may be no more than a function of their small size (Smith and Newton, 1968; Wigh, 1973; Ramsay, 1974), particularly as the vast majority of the m-chromosomes so far known in mosses display no other meiotic abnormalities and are quite regular members of the complement. Nevertheless, a few have been found to behave irregularly during meiosis and it is possible that some may be supernumerary (see accessory chromosomes below) as suggested, for example, by Vaarama (1953, 1968), Steere (1954), Lewis (1961), Mehra and Khanna (1961), Anderson (1974) and Vysotska (1975).

3. Heterochromatin

(*a*) *Discovery*. Certain parts of *Pellia* chromosomes were found by Heitz (1928b) to be more condensed than most during prophase of mitosis and to remain condensed during telophase and the ensuing interphase. Their constituent material, which he called heterochromatin as opposed to euchromatin, stains intensely with acetic-orcein, aceto-carmine or Feulgen during prophase and interphase when euchromatin is dispersed and weakly staining. The distinction is not confined to *Pellia* and the relative distribution of the two types of chromatin has been seized upon as a possible source of cyto-taxonomic data.

(*b*) *H- and h-chromosomes*. On the basis of an extensive range of liverwort studies, Tatuno (1941) defined a large and a small heterochromatic chromosome, H- and h-chromosomes respectively, each of which was universally present in a haploid gametophytic complement. Yano (1957a, b, c) extended the terminology to mosses and identified the H-chromosome as V shaped with a secondary constriction. Subsequent records in the literature confirm the widespread occurrence of H- and h-chromosomes to the extent that one of each is generally taken as an indication of basic haploidy or at least of the absence of recent polyploidy (Anderson, 1964; Yano, 1967; Wigh, 1975). Heterochromatin is not, however, confined to these two chromosomes. Tatuno and Ono (1966) and Ono (1970a, b, c, 1972), for instance, recognized H_1-, H_2- and h-chromosomes within a complement. Heterochromatic chromosomes are frequent in bryophytes (Berrie, 1960, 1963; Anderson, 1964; Segawa, 1965a, b, c; Ramsay, 1966; Ono, 1970a, b, c) but there are difficulties associated with their designation as H- or h-chromosomes (Berrie, 1963), and Newton (1977a, b) proposed the abandonment of the terms.

(c) Constitutive and facultative heterochromatin. For the bryophytes, heterochromatin provides the most powerful cyto-taxonomic tool we have. Brown (1966) distinguished between facultative and constitutive heterochromatin on the basis of their distribution in diploid organisms. Relative characteristics were discussed at length by Comings (1972a, b), Comings and Mattoccia (1972), Cattanach (1975) and Bostock and Sumner (1978) but a point of particular interest in the present context is the fact that facultative heterochromatin appears to be a behavioural phenomenon which affects only one of the two homologues in any given cell of higher organisms, whereas constitutive heterochromatin is a structural feature of both homologues. Essentially, the current view is that constitutive heterochromatin is composed of highly repeated short sequences of DNA. Its component nucleotides vary between chromosomes and species (Bostock and Christie, 1975; Bostock and Sumner, 1978) and the timing of its synthesis is equally variable (Newton, 1977a). In the liverworts studied so far, synthesis is completed rather earlier than is that of euchromatin (Tatuno *et al.*, 1970; Tatuno *et al.*, 1971; Masubuchi, 1971, 1973, 1974). Facultative heterochromatin, however, is a condensed expression of nucleotide sequences found in the euchromatin, differing from the latter in the proportion of non-histone proteins (Maclean, 1976). At least in higher organisms, facultative heterochromatin replicates late (Bostock and Sumner, 1978).

The only opportunity to compare these two types of heterochromatin in bryophytes is presented by the sporophyte (Berrie, 1974), yet it is from the gametophyte that systematic characters are largely drawn. To distinguish between the two is, however, fundamentally important because constitutive heterochromatin, by virtue of its permanence, and its facultative counterpart are not equally significant for taxonomy.

(d) *Giemsa C-banding*. With certain limitations, Giemsa C-banding techniques provide a means of recognizing constitutive heterochromatin throughout the cell cycle, including metaphase and anaphase (Fig. 4), when differential condensation does not occur. Ever since the nomenclature of chromosome bands was formally agreed (Hamerton *et al.*, 1972), C-bands have been identified with constitutive heterochromatin. Fox *et al.* (1973) claimed that the correlation is invalid because telomeric C-bands in small chromosomes of the locust, *Schistocerca gregaria* (Forskål), represent regions that are not heterochromatic but photographs of acetic-orcein preparations (Fox, 1973) are inconclusive on this point. Alternatively, Bostock and Sumner (1978) were of the opinion that not all of the constitutive heterochromatin is necessarily C-banded, quoting as evidence the work of Gallagher *et al.* (1973) who found that only certain parts of totally heterochromatic chromosomes were C-banded in the grasshopper, *Myrmeleotettix maculatus* (Thunb.). However, Gallagher *et al.* themselves pointed out that repetitive DNA had not been localized to particular regions of the chromosomes concerned and Dover (1976) and Rees *et al.* (1977), for example, have implicated facultative heterochromatin in the phenotype of B-chromosomes. Although these unresolved points must be borne in mind, they do not detract from the potential value of Giemsa C-banding techniques in bryophyte systematics.

For example, it is clear that the distribution of darkly staining C-bands along the chromosomes of species of *Pellia* (Newton, 1977a) is in very close agreement with the distribution of heterochromatin as recognized by Heitz (1928b). All nine chromosomes of *Pellia neesiana*, five of *P. epiphylla* and four of *P. endiviifolia* are individually recognizable by their C-band patterns and recognition of the remainder is made easier by reducing the number in which relative length must be considered. Because C-band patterns remain distinctive during prophase, nucleolar organizing chromosomes are also easier to identify. Evidence of chromosome rearrangement has been detected in *P. neesiana* but perhaps the most promising aspect of the work comes from the existence of very pronounced inter-specific differences between the C-banded karyotypes of *Pellia* species and of *Riccardia pinguis* and *Cryptothallus mirabilis* (Fig. 4). Though the systematics of these species is not in dispute, the evidence augurs well for the taxonomic application of Giemsa C-banding to critical groups of bryophytes.

(e) *The example of Rhizomnium pseudopunctatum.* Preliminary studies in a single gathering of R. *pseudopunctatum* from Holden Clough, near Ashton–under–Lyne, Lancashire, are indicative of the kind of investigations likely to prove rewarding. Despite showing little morphological variation throughout its range (Koponen, 1972), two chromosome numbers are known for this species, in addition to an unpublished count of $n = 7$ by Anderson, quoted by Mehra and Khanna (1961). Thus, $n = 14$ (12+2m) has been found in North America (Lowry, 1948; Bowers, 1968) and Britain (Newton, 1971), whereas reports from elsewhere in Europe (Heitz, 1942; Vaarama, 1950; Bowers, 1968) are of $n = 13$ (12+1m). The specimen reported here had a gametophytic number of $n = 12 + 1m$ (Fig. 2a). Heitz (1942) suggested that hybridization between species with $n = 6$ and $n = 7$ had been involved in the ancestry of R. *pseudopunctatum* but other authors have considered autopolyploidy of an $n = 7$ (6+1m) species followed by a reduction in the number of m-chromosomes as more likely. Lowry (1948), for instance, found seven pairs of chromosomes, the relative lengths of which were very similar to those of individual chromosomes of $n = 7$ R. *punctatum* but Bowers (1968) considered the similarity to $n = 7$ R. *andrewsianum* (Steere) Kop. was greater. Both suggestions imply that structural rearrangement of the karyotype following an initial chromosome doubling was considered to be slight. If evolutionary change in the amount and distribution of constitutive heterochromatin had been equally conservative, as it is

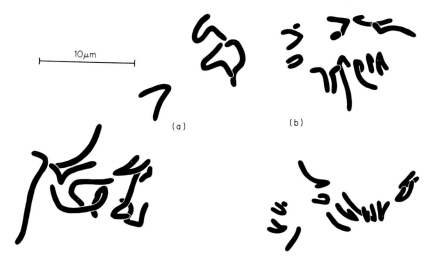

FIG. 2. Gametophytic chromosome configurations of *Rhizomnium pseudopunctatum*, $n = 12 + 1m$. (a) metaphase; (b) anaphase

believed to have been in *Allium* (El-Gadi and Elkington, 1975) for example, seven (or 6 + 1) pairs of similarly C-banded chromosomes might be expected. Work is continuing in this direction but it is already quite clear that only one chromosome of the specimen under investigation included a substantial C-band (Fig. 5). The band was situated distally and occupied more than half of the shorter arm of a distinctly submetacentric or acrocentric chromosome which was not the longest of the complement. This, and karyotype analysis on the basis of anaphase configurations (Fig. 2b, Table I), deserves consideration in the context of Ono's (1970b) work with *R. punctatum*.

TABLE I. Relative chromosome length and $F\%$ in British *Rhizomnium pseudopunctatum* (present paper) compared with Japanese *R. punctatum* (Ono, 1970b)

Rhizomnium pseudopunctatum			*Rhizomnium punctatum*				
Ch. pair	Relative length	$F\%$	Ch.	Relative length		$F\%$	
				♂	♀	♂	♀
1	23·9	24·4	1	20·2	20·5	39	38
2	18·7	27·0	2	16·6	16·3	50	50
3	16·6	48·7	3	15·9	16·3	7	7
4	14·3	44·8	4	15·9	15·6	50	50
5	13·4	38·2	5	14·8	14·5	49	49
6	10·3	32·7	6	13·4	14·5	22	20
7	2·8	?	7	3·2	3·2	44	44

The $F\%$ of the heavily C-banded chromosome indicates that it belongs to pair 2. This is in very close agreement with the discovery that the largest block of heterochromatin in Japanese *R. punctatum* occurred in chromosome 2 (Ono, 1972). That chromosome, however, was metacentric (Table I) as was the H_1-chromosome of all 18 species of *Mnium sensu lato* examined by Ono. If it is related to the C-banded chromosome of British *R. pseudopunctatum*, substantial rearrangement must have occurred. Alternatively, it is possible that the latter may be related to chromosome 3 of Japanese *R. punctatum*, which was also partially heterochromatic. This chromosome was of a similar length to chromosome 2, but was acrocentric. Whichever of these suggested relationships is correct, the large quantities of heterochromatin expected to occur in one or both of a pair of chromosomes were not detected in British *R. pseudopunctatum*. This heterochromatin may have been lost, or dispersed as smaller segments throughout the karyotype of the gathering, or it may not have been involved in the cytotype's evolution. Alternatively, it may have been present but failed to respond to Giemsa C-banding techniques. Answers to these questions would provide a sounder understanding of the evolution and taxonomy of the Mniaceae.

4. Sex Chromosomes

(a) *Morphological.* A large chromosome in female *Sphaerocarpos donnellii* was shown by Allen (1917, 1919) to be homologous with a small one in male plants. He called them X and Y chromosomes, respectively. Chromosomes which differ in size according to sex have since been described in about 60 species of hepatics (Segawa, 1965a), and are widely accepted as morphological sex chromosomes. It has also been claimed that they occur in mosses, since bivalents composed of unequal homologues are not infrequently seen (e.g. Steere *et al.*, 1954; Ramsay, 1966; Smith and Newton, 1968). Ramsay (1966) followed the meiotic segregation of such bivalents in species of *Macromitrium* and identified the larger component with female gametophytes of *M. archeri* Mitt. *in* Hook. However, she recognized that comparison with male tissue was desirable. Thus, while proof is lacking, the evidence indicates that morphological sex chromosomes will be found in mosses, although it is illogical to assume that a dimorphic bivalent must represent a pair of such chromosomes. Smith and Newton (1968), for example, drew attention to dimorphic bivalents in monoecious mosses and John and Hewitt (1966) demonstrated the cytological basis of a dimorphism which was quite unrelated to sex determination in the grasshopper, *Chorthippus parallelus* Zetterstedt.

There are also instances of sex-specific chromosomes which differ in number as well as size. They are known only in the subgenus *Galeiloba* of *Frullania*, there being two X chromosomes in the female complement compared with a single Y in male plants (Berrie, 1960; Lewis, 1961; Iverson, 1963). It is interesting in this context that H. Inoue (1974) has examined three Asian species of *Plagiochila* in which female plants ($n = 8$) are without one of the chromosomes of the male complement ($n = 9$).

(b) *Structural.* Following the work of Tatuno and Segawa (1955), a second type of sex-specific chromosome has been recognized in mosses (Ono, 1970a, b, c) and in liverworts (Segawa, 1965a, b, c). These are structural sex chromosomes which differ solely in the amount and distribution of heterochromatin, the Y of male gametophytes possessing more than the X of female plants. Their use as potential marker chromosomes in bryophyte systematics is obvious, particularly as a means of studying the origins of putative autodiploids. However, their validity has been questioned (Berrie, 1974) on the grounds that different quantities of heterochromatin may be no more than phenotypic variation depending on whether the chromosome is in the environment of a male or female genome.

(c) *The example of* Atrichum crispum. The feasibility of taxonomic exploitation of structural sex chromosomes has been considered in the light of results from a gathering of *A. crispum* from Devil's Bridge, northwest of Tintwistle, Cheshire. This species occurs in North America and Europe (Nyholm, 1971) but is known in Britain only as male plants (Smith, 1967), which reduced the cytological problems of dealing with a dioecious species. In common with North American material (Lowry, 1954), a gametophytic chromosome number of $n = 7$ was observed in the British specimen, the differential condensation of prophase chromosomes being exceptionally clear in orcein preparations (Fig. 6). The distribution of heterochromatin is summarized in Table II; in chromosome C

TABLE II. Gametophytic karyotype data for *Atrichum crispum* derived from propionic-orcein stained prophase and Giemsa C-banded anaphase configurations. Since centromere position is obscure during prophase, heterochromatin is described in relation to chromosome halves, irrespective of whether these refer to more or less than a single arm

	Prophase			Anaphase		
Ch.	Relative length	Recognizable heterochromatin	Ch.	Relative length	$F\%$	Prominent C-bands
A	21·9	— : —	1	19·9	48·8	distal and subterminal in one arm
B	17·5	distal : —	2	18·8	45·7	—
C	15·7	$\pm$ total : distal	3	16·5	47·9	—
D	12·4	distal : —	4	12·3	30·2	median in shorter arm
E	12·0	subterminal : distal	5	11·4	28·6	—
F	11·5	— : —	6	11·1	20·8	—
G	8·9	distal : distal	7	10·0	11·6	—

it coincided with that of structural sex-chromosomes of other species of *Atrichum* described by Ono (1970c) and Tatuno and Kise (1970). This is one of five types defined by Ono (1970c) on the basis of heterochromatin distribution and is the Polla type, which he found to be invariably the largest member of the complement. It is metacentric with one totally heterochromatic arm, the other being distally and proximally heterochromatic in male plants but only distally heterochromatic in female plants. The distinction has not been attempted in British material in view of the obscurity of centromeres during prophase. The h-chromosomes of Japanese species possessed two unequal distal blocks of heterochromatin exactly like the pattern seen in chromosome G of *A. crispum*. Ono (1970c) also found a substantial terminal block of heterochromatin in a medium-sized H_2-chromosome of this genus. Chromosomes B and D were of this type in *A. crispum*, but there was also subterminal heterochromatin in chromosome E. Thus, there are striking resemblances between Japanese and British specimens, but also considerable differences which remain puzzling in the context of Giemsa C-banding.

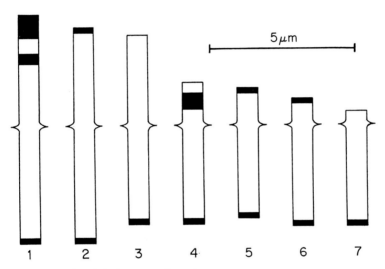

FIG. 3. Giemsa C-banded idiogram of *Atrichum crispum*

Apart from minute telomeric C-bands, only chromosomes 1 and 4 possessed substantial C-bands (Figs 3 and 7). The largest chromosome had two in the distal half of one arm and would appear to correspond to chromosome C of prophase. Discrepancies in relative length are likely to result from differential rates of condensation in heterochromatin and euchromatin such as Ho and Kasha (1974) found in *Medicago sativa* L. Similarly, chromosomes 4 and E appear to coincide. Attention may be drawn to the absence of extensive C-bands in the smallest chromosome, which, by analogy with Japanese species, could be identified as an h-chromosome. It follows that not all the heterochromatin visible in Fig. 6 was responsive to the Giemsa C-banding technique

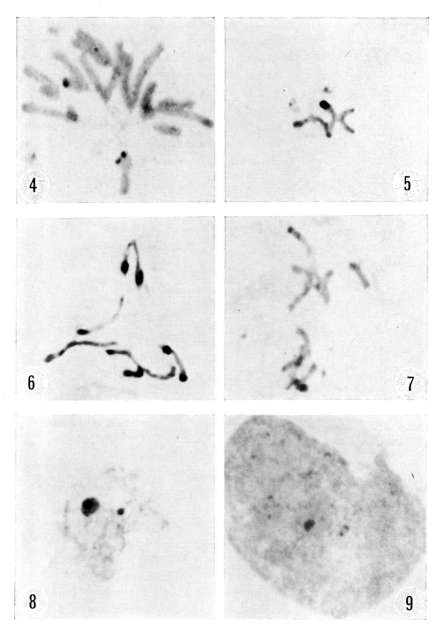

Figs 4, 5, 6, 7, 8, 9

and this was confirmed in interphase nuclei. Only two conspicuous blocks, the larger in association with the nucleolus, were revealed by Giemsa C-banding (Fig. 8). Their size and location were those of two heterochromatin bodies seen alone or with others in orcein preparations, though the one discrete from the nucleolus sometimes appeared to be compound (Fig. 9). Additional heterochromatin bodies occurred in some nuclei but failed to respond to the Giemsa C-banding technique. Hence, either there was a significant amount of facultative heterochromatin, or the technique failed to detect all the constitutive heterochromatin. If the former is true, it has serious implications for karyotype analysis carried out with a view to taxonomic application. Chromosome 1 is particularly pertinent because of its connotations of sex determination and its similarity in different species and genera (Ono, 1970c; Tatuno and Kise, 1970).

5. Accessory Chromosomes

The possibility that accessory chromosomes occur in bryophytes has been confused by the use of that term to include m-chromosomes which appear to be regular members of the complement (Steere *et al.*, 1954; Anderson and Lemmon, 1974). However, the subject has been widely discussed for bryophytes (Steere, 1954; Anderson, 1964; Vaarama, 1968; Wigh, 1973) and the term as used here refers only to supernumerary or B-chromosomes. Vysotska (1975) designated them B_m-chromosomes to distinguish them from A_m-chromosomes which are a constant feature of the species in which they occur.

As Wigh (1973) pointed out, evidence of accessory chromosomes in mosses has been drawn either from mitotic or meiotic studies but not from a consideration of both. This is obviously a serious difficulty in the way of identifying

---

FIG. 4. *Cryptothallus mirabilis*, $n = 10$. One pole of Giemsa C-banded gametophytic anaphase. $\times 2350$

FIG. 5. *Rhizomnium pseudopunctatum*. Giemsa C-banded gametophytic chromosomes at the onset of anaphase. $\times 2350$

FIG. 6. *Atrichum crispum*, $n = 7$. Gametophytic prophase stained with propionic-orcein, showing differential condensation. $\times 2350$

FIG. 7. *Atrichum crispum*. Giemsa C-banded gametophytic mitosis at the onset of anaphase. $\times 2350$.

FIG. 8. *Atrichum crispum*. Giemsa C-banded gametophytic interphase, showing the larger heterochromatin body associated with the nucleolus. $\times 2350$.

FIG. 9. *Atrichum crispum*. Gametophytic interphase stained with propionic-orcein, showing a small compound heterochromatin body and a larger one associated with the nucleolus. $\times 2350$.

chromosomes whose diagnostic characters include irregular segregation at mitosis and/or meiosis, resulting in numerical variation in and between populations, individuals, tissues or cells (Battaglia, 1964; Muntzing, 1974; Jones, 1975).

Evidence of meiotic irregularity is persuasive (Vaarama, 1949, 1950, 1953, 1968; Vysotska, 1975) but without intensive mitotic studies is inconclusive. Wigh (1973), on the other hand, obtained impressive data on intra-specific variation in chromosome number within the Brachytheciaceae, from which he concluded that accessory chromosomes were present in two species of *Brachythecium* and two of *Homalothecium*, as did Nyholm and Wigh (1973) for *Rhynchostegium megapolitanum* and S. Inoue (1968, 1972) for *Pohlia proligera* and *Lesquereuxia robusta* Lindb. Similarly, Anderson and Lemmon (1974) recognized m-chromosomes of *Weissia controversa* as supernumerary, despite the fact that they underwent normal meiosis and maintained a constant number within a colony. However, because of the limitations of karyotype analysis discussed above, these data are open to other interpretation. For instance, it is possible to envisage situations in which different numbers of chromosomes represented precisely the same genetical information and quantity of DNA, but in different linkage groups. Correlated differences between phenotypes could evolve but are not an essential part of the thesis, which would make it particularly attractive in such a taxonomically problematic genus as *Brachythecium*. Thus, it is difficult to decide whether accessory chromosomes occur in bryophytes but the means for examining the point are now available.

CONCLUSION

Stotler (1976) was not unduly pessimistic in recognizing the unpromising nature of liverwort cyto-taxonomy. Techniques up to that time had proved inadequate to differentiate clearly between apparently similar karyotypes. To a more limited extent, the same was also true of mosses (Crundwell, 1970). Steere (1972), for instance, in mentioning three of the greatest achievements in bryophyte cytology, named discoveries of half a century earlier. Work in the interim has built up the worldwide knowledge required for the essential transformation of the subject by modern cytological methods. Three points deserve particular attention.

(1) Relative lengths and $F \%$ of chromosomes are invaluable to karyotype description but standardization of technique is required. Unless inhibitors of spindle formation are used, mitotic anaphase provides the most reliable source of fully condensed chromosomes for measurement, since it eliminates the uncertainty of distinguishing between incompletely condensed late prophase

chromosomes and those of metaphase (Fig. 1). However, there is room for further refinement, perhaps involving the use of synaptonemal complexes.

(2) Vaarama (1976) expressed the opinion that the description of heterochromatin distribution is a particularly important part of karyotype analysis. Such a view is clearly upheld by subsequent discoveries, but there is an urgent need to discover exactly what it is that bryologists are calling heterochromatin.

The role that constitutive heterochromatin and Giemsa C-banding can be expected to play in bryophyte systematics has been demonstrated (Newton, 1977a, b and this paper). A significant contribution can also be predicted for facultative heterochromatin with respect to accessory chromosomes and to sex-specific chromosomes. It must be stressed, however, that the current definition of facultative heterochromatin is inadequate when haploid gametophytes of bryophytes are considered. Material appearing as facultative heterochromatin in one of the two homologues of a sporophyte need not represent an identical segment of a similarly inactivated chromosome in the gametophyte. For instance, there is evidence to suggest that heterochromatin other than Giemsa C-banding material does occur in moss gametophytes (Newton, 1977b, and this paper) but it cannot be compared with its homologue as the definition of facultative heterochromatin requires (Bostock and Sumner, 1978). As a basis for further study, it might be suggested that genes specific for the development, as opposed to determination, of a female gametophyte could appear as facultative heterochromatin in male plants and vice versa. Berrie's (1974) discovery of equal amounts of heterochromatin *sensu lato* in a pair of homologues in *Plagiochila praemorsa* Stephani sporophytes compared with unequal amounts in male and female gametophytes of the same genus (Segawa, 1965a, b) is relevant to the whole question of structural sex chromosomes and would justify consideration in this light. It is also possible that the search for accessory chromosomes could be facilitated by the knowledge that at least some supernumerary chromosomes of higher organisms appear to be facultatively heterochromatic (Dover, 1976; Rees *et al.*, 1977).

(3) The cyto-taxonomic application of Giemsa C-banding to bryophytes is promising. Some of the other banding techniques reviewed by Hsu (1973), Schnedl (1974), Dutrillaux and Lejeune (1975) and Vosa (1975) may also be rewarding. G-bands have not yet been produced in plant material, possibly due to more pronounced condensation of plant than animal chromosomes (Greilhuber, 1977). Q-banding techniques, however, have been successfully applied to higher plants (Vosa, 1976) and more exact information about the location of nucleolar organizers might also be usefully sought in bryophytes by N-banding (Funaki, *et al.*, 1975).

ACKNOWLEDGEMENTS

I should like to thank Dr A. J. E. Smith for various discussions on bryophyte sex chromosomes, Dr D. I. Southern for the use of cytological facilities, and the Royal Society for a scientific grant-in-aid.

REFERENCES

ALLEN, C. E. (1917). A chromosome difference correlated with sex differences in *Sphaerocarpos*. *Science, N.Y.* **46,** 466–467.

ALLEN, C. E. (1919). The basis of sex inheritance in *Sphaerocarpos*. *Proc. Am. phil. Soc.* **58,** 289–316.

ANDERSON, L. E. (1962). Chromosome numbers: Bryophytes. *In* "Growth" (P. L. Altman and D. S. Dittmer, eds) pp. 45–57. Federation of American Societies for Experimental Biology, Washington, D.C.

ANDERSON, L. E. (1964). Biosytematic evaluations in the Musci. *Phytomorphology* **14,** 27–51.

ANDERSON, L. E. (1974). 25 years of botany: Bryology, 1947–1972. *Ann. Mo. bot. Gdn* **61,** 56–85.

ANDERSON, L. E. and LEMMON, B. E. (1974). Gene flow distances in the moss, *Weissia controversa* Hedw. *J. Hattori bot. Lab.* **38,** 67–90.

BATTAGLIA, E. (1964). Cytogenetics of B-chromosomes. *Caryologia* **17,** 245–299.

BEER, R. (1903). The chromosomes of *Funaria hygrometrica*. *New Phytol.* **2,** 166.

BERRIE, G. K. (1960). The chromosome numbers of liverworts (Hepaticae and Anthocerotae). *Trans. Br. bryol. Soc.* **3,** 688–705.

BERRIE, G. K. (1963). Cytology and phylogeny of liverworts. *Evolution, Lawrence, Kan.* **17,** 347–357.

BERRIE, G. K. (1974). Sex chromosomes of *Plagiochila praemorsa* Stephani and the status of structural sex chromosomes in Hepatics. *Bull. Soc. bot. Fr., Colloq. Bryol.,* **121,** 129–135.

BOSTOCK, C. J. and CHRISTIE, S. (1975). Chromosomes of a cell line of *Dipodomys panamintinus* (Kangaroo rat). A banding and autoradiographic study. *Chromosoma* **51,** 25–34.

BOSTOCK, C. J. and SUMNER, A. T. (1978). "The Eukaryotic Chromosome." North-Holland, Amsterdam, New York and Oxford.

BOWERS, M. C. (1968). A cytotaxonomic study of the genus *Mnium* in Colorado. *Revue bryol. lichén.* **36,** 167–202.

BROWN, S. W. (1966). Heterochromatin. *Science, N.Y.* **151,** 417–425.

BRYAN, V. S. (1955). Chromosome studies in the genus *Sphagnum*. *Bryologist* **58,** 16–39.

BRYAN, V. S. (1973). Chromosome studies on mosses from Austria, Czechoslovakia and other parts of central Europe. *Öst. bot. Z.* **121,** 187–226.

CATTANACH, B. M. (1975). Control of chromosome inactivation. *A. Rev. Genet.* **9,** 1–18.

CHATTERJEE, A. K. and GANGULEE, H. C. (1970). Cytological studies on the mosses of Eastern India, VI. Karyotypes. *Nucleus, Calcutta* **13,** 118–125.

COMINGS, D. E. (1972a). Replicative heterogeneity of mammalian DNA. *Expl Cell Res.* **71,** 106–112.

COMINGS, D. E. (1972b). The structure and function of chromatin. *Adv. Hum. Genet.*, *N.Y.* **3**, 237–431.

COMINGS, D. E. and MATTOCCIA, E. (1972). DNA of mammalian and avian heterochromatin. *Expl Cell Res.* **71**, 113–131.

CRUNDWELL, A. C. (1970). Infraspecific categories in Bryophyta. *Biol. J. Linn. Soc.* **2**, 221–224.

DAVIS, B. M. (1901). Nuclear studies on *Pellia*. *Ann. Bot.* **15**, 147–180.

DOVER, G. A. (1976). Molecular properties of B-chromosome DNA: A clue to its origin in the Triticinae and Orthoptera. *In* "Current Chromosome Research" (K. Jones and P. E. Brandham, eds) pp. 67–75. North-Holland, Amsterdam, New York and Oxford.

DUTRILLAUX, B. and LEJEUNE, J. (1975). New techniques in the study of human chromosomes: Methods and applications. *Adv. Hum. Genet.*, *N.Y.* **5**, 119–156.

EL-GADI, A. and ELKINGTON, T. T. (1975). Comparison of the Giemsa C-band karyotypes and the relationships of *Allium cepa*, *A. fistulosum* and *A. galanthum*. *Chromosoma* **51**, 19–23.

FARMER, J. B. (1895). On spore-formation and nuclear division in the Hepaticae. *Ann. Bot.* **9**, 469–523.

FOX, D. P. (1973). The control of chiasma distribution in the locust, *Schistocerca gregaria* (Forskål). *Chromosoma* **43**, 289–328.

FOX, D. P., CARTER, K. C. and HEWITT, G. M. (1973). Giemsa banding and chiasma distribution in the desert locust. *Heredity, Lond.* **31**, 272–276.

FRITSCH, R. (1972). Chromosomenzahlen der Bryophyten, eine Übersicht und Diskussion ihres Aussagewertes für das System. *Wiss. Z. Friedrich Schiller-Univ. Jena*, Math. Nat. Reihe, **21**, 839–944.

FUNAKI, K., MATSUI, S. and SASAKI, M. (1975). Location of nucleolar organizers in animal and plant chromosomes by means of an improved N-banding technique. *Chromosoma* **49**, 357–370.

GALLAGHER, A., HEWITT, G. and GIBSON, I. (1973). Differential Giemsa staining of heterochromatic B-chromosomes in *Myrmeleotettix maculatus* (Thunb.) (Orthoptera: Acrididae). *Chromosoma* **40**, 167–172.

GREILHUBER, J. (1977). Why plant chromosomes do not show G-bands. *Theor. Appl. Genet.* **50**, 121–124.

HAMERTON, J. L., JACOBS, P. A. and KLINGER, H. P. eds. (1972). Paris Conference (1971): Standardization in human cytogenetics. *Cytogenetics* **11**, 313–362.

HEITZ, E. (1927). Ueber multiple und aberrante Chromosomenzahlen. *Abh. naturw. Ver. Hamburg* **21**(3–4), 47–58.

HEITZ, E. (1928a). Der bilaterale Bau der Geschlechtschromosomen und Autosomen bei *Pellia fabbroniana*, *P. epiphylla* und einigen anderen Jungermanniaceen. *Planta* **5**, 725–768.

HEITZ, E. (1928b). Das Heterochromatin der Moose I. *Jb. wiss. Bot.* **69**, 762–818.

HEITZ, E. (1942). Über die Beziehung zwischen Polyploidie und Gemischtgeschlechtlichkeit bei Moosen. *Arch. Klaus-Stift. VererbForsch* **17**, 444–448.

HO, K. M. and KASHA, K. J. (1974). Differential chromosome contraction at the pachytene stage of meiosis in Alfalfa (*Medicago sativa* L.). *Chromosoma* **45**, 163–172.

HSU, T. C. (1973). Longitudinal differentiation of chromosomes. *A. Rev. Genet.* **7**, 153–176.

INOUE, H. (1967). Chromosome studies on some Japanese liverworts. *Bot. Mag., Tokyo* **80**, 172–175.

INOUE, H. (1968). Chromosome numbers of some Malayan and Taiwan liverworts. *Bull. natn. Sci. Mus. Tokyo* **11**, 397–403.

INOUE, H. (1974). Some taxonomic problems in the genus *Plagiochila. J. Hattori bot. Lab.* **38**, 105–109.

INOUE, S. (1964). Karyological studies on mosses, I. *Bot. Mag., Tokyo* **77**, 412–417.

INOUE, S. (1968). B-chromosomes in two moss species. *Miscnea bryol. lichen., Nichinan* **4**, 167–169.

INOUE, S. (1971). Karyological studies on Mosses, VII. Karyotypes of fourteen species in Sematophyllaceae, Hypnaceae and Hylocomiaceae. *Bot. Mag., Tokyo* **84**, 247–255.

INOUE, S. (1972). B-chromosomes in *Lesquereuxia robusta* Lindb. *J. Byrol.* **7**, 150.

INOUE, S. (1973). Karyological studies on *Takakia ceratophylla* and *T. lepidozioides. J. Hattori bot. Lab.* **37**, 275–286.

INOUE, S. and IWATSUKI, Z. (1976). A cytotaxonomic study of the genus *Rhizogonium* Brid. (Musci). *J. Hattori bot. Lab.* **41**, 389–403.

INOUE, S. and UCHINO, A. (1969). Karyological studies on mosses, VI. Karyotypes of fourteen species including some species with the intraspecific euploid and aneuploid. *Bot. Mag., Tokyo* **82**, 359–367.

IVERSON, G. B. (1963). Karyotype evolution in the leafy liverwort genus *Frullania. J. Hattori bot. Lab.* **26**, 119–170.

JOHN, B. and HEWITT, G. M. (1966). A polymorphism for heterochromatic supernumerary segments in *Chorthippus parallelus. Chromosoma* **18**, 254–271.

JONES, R. N. (1975). B-chromosome systems in flowering plants and animal species. *Int. Rev. Cytol.* **40**, 1–100.

KOPONEN, T. (1972). Speciation on the Miniaceae. *J. Hattori bot. Lab.* **35**, 142–154.

LEFEBVRE, J. (1968). Clé de détermination des Plagiotheciaceae de Belgique. *Buxbaumia* **21**, 79–85.

LEWIS, K. R. (1958). Chromosome structure and organization in *Pellia epiphylla. Phyton. B. Aires* **11**, 29–37.

LEWIS, K. R. (1961). The genetics of bryophytes. *Trans. Br. bryol. Soc.* **4**, 111–130.

LORBEER, G. (1934). Die Zytologie der Lebermoose mit besonderer Berücksichtigung allgemeiner Chromosomenfragen. I. *Jb. wiss. Bot.* **80**, 567–818.

LOWRY, R. J. (1948). A cytotaxonomic study of the genus *Mnium. Mem. Torrey bot. Club* **20**(2), 1–42.

LOWRY, R. J. (1954). Chromosome numbers and relationships in the genus *Atrichum* in North America. *Am. J. Bot.* **41**, 410–414.

MACLEAN, N. (1976). "Control of Gene Expression." Academic Press, London and New York.

MAEKAWA, F. (1963). Reduction in chromosomes and major polyploidy: their bearing on plant evolution. *J. Fac. Sci. Tokyo Univ.*, sect. 3, Bot., **8**, 377–398.

MASUBUCHI, M. (1971). Early DNA synthesis of heterochromatin and replication of Y-chromosome in *Pellia neesiana. Bot. Mag., Tokyo* **84**, 24–29.

MASUBUCHI, M. (1973). Evidence of early replicating DNA in heterochromatin. *Bot. Mag., Tokyo* **86**, 319–322.

MASUBUCHI, M. (1974). Early replicating DNA in heterochromatin of *Plagiochila ovalifolia* (liverworts). *Bot. Mag., Tokyo* **87**, 229–235.

MEHRA, P. N. and KHANNA, K. N. (1961). Recent cytological investigations in mosses. *Res. Bull. Panjab Univ.* N.S., Sci., **12**, 1–29.

MOORE, A. C. (1905). Sporogenesis in *Pallavicinia*. *Bot. Gaz.* **40**, 81–96.

MOSES, M. J. (1977). Microspreading and the synaptonemal complex in cytogenetic studies. *In* "Chromosomes Today", Vol. 6 (A. de la Chapelle and M. Sorsa, eds) pp. 71–82. North-Holland, Amsterdam, New York and Oxford.

MÜNTZING, A. (1974). Accessory chromosomes. *A. Rev. Genet.* **8**, 243–266.

NEWTON, M. E. (1971). Chromosome studies in some British and Irish bryophytes. *Trans. Br. bryol. Soc.* **6**, 244–257.

NEWTON, M. E. (1973). Chromosome studies in some British and Irish bryophytes, II. *J. Bryol.* **7**, 379–398.

NEWTON, M. E. (1975). Chromosome studies in some British bryophytes. *J. Bryol.* **8**, 365–382.

NEWTON, M. E. (1977a). Heterochromatin as a cyto-taxonomic character in liverworts: *Pellia*, *Riccardia* and *Cryptothallus*. *J. Bryol.* **9**, 327–342.

NEWTON, M. E. (1977b). Chromosomal relationships of heterochromatin bodies in a moss, *Dicranum tauricum* Sapehin. *J. Bryol.* **9**, 557–564.

NYHOLM, E. (1971). Studies in the genus *Atrichum* P. Beauv. A short survey of the genus and the species. *Lindbergia* **1**, 1–33.

NYHOLM, E. and WIGH, K. (1973). Cytotaxonomical studies in some Turkish mosses. *Lindbergia* **2**, 105–113.

ONO, K. (1970a). Karyological studies on Mniaceae and Polytrichaceae, with special reference to the structural sex-chromosomes I. *J. Sci. Hiroshima Univ.*, ser. B, div. 2, **13**, 91–105.

ONO, K. (1970b). Karyological studies on Mniaceae and Polytrichaceae, with special reference to the structural sex-chromosomes II. *J. Sci. Hiroshima Univ.*, ser. B, div. 2, **13**, 107–166.

ONO, K. (1970c). Karyological studies on Mniaceae and Polytrichaceae, with special reference to the structural sex-chromosomes III. *J. Sci. Hiroshima Univ.*, ser. B, div. 2, **13**, 167–221.

ONO, K. (1972). On the chromosomes of *Dendroligotrichum dendroides* (Hedw.) Broth. from Patagonia. *J. Jap. Bot.* **47**, 38–43.

RAMSAY, H. P. (1964). The chromosomes of *Dawsonia*. *Bryologist* **67**, 153–162.

RAMSAY, H. P. (1966). Sex chromosomes in *Macromitrium*. *Bryologist* **69**, 293–311.

RAMSAY, H. P. (1969). Cytological studies on some mosses from the British Isles. *Bot. J. Linn. Soc.* **62**, 85–121.

RAMSAY, H. P. (1974). Cytological studies of Australian mosses. *Aust. J. Bot.* **22**, 293–348.

REES, H. (1972). DNA in higher plants. *In* "Evolution of Genetic Systems" (H. H. Smith, ed.) pp. 394–418. Gordon and Breach, New York, London and Paris.

REES, H., TEOH, S. B. and JONES, L. M. (1977). Heterochromatisation and the possibility of gene inactivation in B chromosomes of *Picea glauca*. *Heredity* **38**, 272 (Abstr.).

SCHNEDL, W. (1974). Banding patterns in chromosomes. *In* "Aspects of Nuclear Structure and Function" (G. H. Bourne, J. F. Danielli and K. W. Jeons, eds) pp. 237–272. *Int. Rev. Cytol.* suppl. 4.

SCHUSTER, R. M. (1966). "The Hepaticae and Anthocerotae of North America east of the hundredth Meridian." Vol. 1. Columbia University, New York and London.

SEGAWA, M. (1965a). Karyological studies in liverworts, with special reference to structural sex chromosomes. I. *J. Sci. Hiroshima Univ.*, ser. B, div. 2, **10**, 69–80.

SEGAWA, M. (1965b). Karyological studies in liverworts, with special reference to structural sex chromosomes. II. *J. Sci. Hiroshima Univ.*, ser. B, div. 2, **10**, 81–148.

SEGAWA, M. (1965c). Karyological studies in liverworts, with special reference to structural sex chromosomes. III. *J. Sci. Hiroshima Univ.*, ser. B, div. 2, **10**, 149–178.

SHARMA, P. D. (1963). Cytology of some Himalayan Polytrichaceae. *Caryologia* **16**, 111–120.

SMITH, A. J. E. (1967). Distribution map of *Atrichum crispum* (James) Sull. in Britain. *Trans. Br. bryol. Soc.* **5**, 362.

SMITH, A. J. E. and NEWTON, M. E. (1966). Chromosome studies on some British and Irish mosses. I. *Trans. Br. bryol. Soc.* **5**, 117–130.

SMITH, A. J. E. and NEWTON, M. E. (1968). Chromosome studies on some British and Irish mosses. III. *Trans. Br. bryol. Soc.* **5**, 463–522.

SNIDER, J. A. (1973). Chromosome studies of some mosses of the Douglas Lake region, III. *Mich. Bot.* **12**, 107–117.

SOLARI, A. J. and COUNCE, S. J. (1977). Synaptonemal complex karyotyping in *Melanoplus differentialis*. *J. Cell Sci.* **26**, 229–250.

STEERE, W. C. (1954). Chromosome number and behavior in Arctic mosses. *Bot. Gaz.* **116**, 93–133.

STEERE, W. C. (1972). Chromosome numbers in bryophytes. *J. Hattori bot. Lab.* **35**, 99–125.

STEERE, W. C., ANDERSON, L. E. and BRYAN, V. S. (1954). Chromosome studies on Californian mosses. *Mem. Torrey bot. Club* **20**(4), 1–75.

STOTLER, R. E. (1976). The biosystematic approach in the study of the Hepaticae. *J. Hattori bot. Lab.* **41**, 37–46.

SYBENGA, J. (1959). Some sources of error in the determination of chromosome length. *Chromosoma* **10**, 355–364.

TATUNO, S. (1941). Zytologische Untersuchungen über die Lebermoose von Japan. *J. Sci. Hiroshima Univ.*, ser. B, div. 2, **4**, 73–187.

TATUNO, S. (1959). Chromosomen von *Takakia lepidozioides* und eine Studie zur Evolution der Chromosomen der Bryophyten. *Cytologia* **24**, 138–147.

TATUNO, S. and KISE, Y. (1970). Cytological studies on *Atrichum*, with special reference to structural sex chromosomes and sexuality. *Bot. Mag., Tokyo* **83**, 163–172.

TATUNO, S. and NAKANO, M. (1970). Karyological studies on Japanese mosses. I. *Bot. Mag., Tokyo* **83**, 109–118.

TATUNO, S. and ONO, K. (1966). Zytologische Untersuchungen über die Arten von Mniaceae aus Japan. *J. Hattori bot. Lab.* **29**, 79–95.

TATUNO, S. and SEGAWA, M. (1955). Uber strukturelle Geschlechtschromosomen bei *Mnium maximowiczii* und Nukleolinuschromosomen bei einigen Bryophyten. *J. Sci. Hiroshima Univ.*, ser. B, div. 2, **7**, 1–10.

TATUNO, S., TANAKA, R. and MASUBUCHI, M. (1970). Early DNA synthesis in the X-chromosome of *Pellia neesiana*. *Cytologia* **35**, 220–226.

TATUNO, S., TANAKA, R. and YONEZAWA, Y. (1971). H^3-thymidine autoradiographic study on the heteropycnosis and DNA synthesis in *Calobryum rotundifolium* ($n = 9$). *Miscnea bryol. lichen., Nichinan* **5**, 133–134.

TSUTSUMI, S., TAGUCHI, M. and INOUE, S. (1973). Karyological studies on Swedish mosses. *Miscnea bryol. lichen., Nichinan* **6**, 82–84.

VAARAMA, A. (1949). Meiosis in moss-species of the family Grimmiaceae. *Port. Acta Biol.* sér. A, R. B. Goldschmidt vol., 47–78.

VAARAMA, A. (1950). Studies on chromosome numbers and certain meiotic features of several Finnish moss species. *Bot. Notiser* **1950**, 239–256.

VAARAMA, A. (1953). Chromosome fragmentation and accessory chromosomes in *Orthotrichum tenellum*. *Hereditas* **39**, 305–316.

VAARAMA, A. (1954). Cytological observations on *Pleurozium schreberi* with special reference to centromere evolution. *Suomal. eläin-ja kasvit. Seur. van. kasvit. Julk.* **28** (1), 1–59.

VAARAMA, A. (1968). Structurally and functionally deviating chromosome types in Bryophyta. *Nucleus, Calcutta* supp. vol., 285–294.

VAARAMA, A. (1976). The cytotaxonomic approach to the study of bryophytes. *J. Hattori bot. Lab.* **41**, 7–12.

VAN HOOK, J. M. (1900). Notes on the division of the cell and nucleus in liverworts. *Bot. Gaz.* **30**, 394–399.

VOSA, C. G. (1975). The use of Giemsa and other staining techniques in karyotype analysis. *Curr. Adv. Plant Sci.* **6**, 495–510.

VOSA, C. G. (1976). Heterochromatic banding patterns in *Allium*, II. Heterochromatin variation in species of the *paniculatum* group. *Chromosoma* **57**, 119–133.

VYSOTSKA, E. I. (1975). New data on chromosome numbers of Bryopsida in the Ukraine. *Ukr. bot. Zh.* **32**, 498–503.

WIGH, K. (1972). Cytotaxonomical and modification studies in some Scandinavian mosses. *Lindbergia* **1**, 130–152.

WIGH, K. (1973). Accessory chromosomes in some mosses. *Hereditas* **74**, 211–224.

WIGH, K. (1975). Scandinavian species of the genus *Brachythecium* (Bryophyta) II. Morphology, taxonomy and cytology in the *B. rutabulum—B. rivulare* complex. *Bot. Notiser* **128**, 476–496.

WYLIE, A. P. (1957). The chromosome numbers of mosses. *Trans. Br. bryol. Soc.* **3**, 260–278.

YANO, K. (1957a). Cytological studies on Japanese mosses. I. Fissidentales, Dicranales, Grimmiales, Eubryales. *Kyoiku Kag.* **6**, 1–31.

YANO, K. (1957b). Cytological studies on Japanese mosses. II. Hypnobryales. *Mem. Takada Branch, Niigata Univ.* **1**, 85–127.

YANO, K. (1957c). Cytological studies on Japanese mosses. III. Isobryales, Polytrichinales. *Mem. Takada Branch, Niigata Univ.* **1**, 129–159.

YANO, K. (1960). On the chromosomes in some mosses. XIII. Chromosomes in seven *Brotherella* species. *J. Hattori Bot. Lab.* **23**, 93–98.

YANO, K. (1967). On the chromosomes in some mosses. XV. Chromosomes in six species of *Acanthocladium, Acroporium, Ctenidium, Drepanocladus* and *Bryhnia*. *Jap. J. Genet.* **42**, 83–88.

11 | Spore Morphology and Bryophyte Systematics

G. C. S. CLARKE

British Museum (Natural History), Cromwell Road, London SW7 5BD, Great Britain

Abstract: Various features of the morphology of bryophyte spores have been used as taxonomic characters, but the most easily observed features such as shape and size are often the ones which are most affected by environmental influences. Characters such as the construction of the spore wall and its ornamentation are less affected in this way and are therefore more reliable in taxonomic work, although they are more difficult to observe. Scanning and transmission electron microscopy have enabled much more accurate and detailed observations to be made and preliminary results suggest that there is a great deal of information relevant to systematics to be derived from spores, although coverage of the whole range of bryophytes is as yet poor.

A review of the literature shows that while many studies have concentrated on variations in spore morphology at the species level, the greatest potential use of spores in elucidating taxonomy and phylogeny is at the level of the genus and above.

INTRODUCTION

Over the last 25 years or so pollen morphology has been found to give significant insights into the systematics of flowering plants. There is now a general appreciation of the importance of pollen morphology and also of what kind of information is likely to be learned from its study. While it would be wrong to suggest that systematic investigations of flowering plants are supplemented as a matter of routine by observations on pollen, this kind of study is regularly undertaken. One could not say the same of spore data in bryophyte systematics. Investigations of a few genera of mosses and hepatics traditionally involve

Systematics Association Special Volume No. 14, "Bryophyte Systematics", edited by G. C. S. Clarke and J. G. Duckett, 1979, pp. 231–250, Academic Press, London and New York.

a study of spore morphology (*Fossombronia*, *Riccia*, and *Encalypta*, would be examples) but they are very much in the minority. In other cases perfunctory descriptions of spores were often given but rarely used. Why is this the case? Is there some reason to suppose that bryophyte spore morphology is intrinsically less useful to systematic studies than pollen morphology? Or are we merely ignoring characters which could be of great use but have never been adequately taken into account?

Pioneers such as Leitgeb (1884) and Roth (1904–1905) provided a great deal of basic data which has since been added to by others such as Knox (1939), Jonas (1952), McClymont (1955) and Erdtman (1957, 1965). In recent years new techniques have become available, notably scanning and transmission electron microscopy, and a number of workers have used them in the study of spores. It therefore seems a suitable time to review past work on bryophyte spores and their relevance to systematics, and to assess the likely importance of spore morphology to future investigations. This review concentrates solely on morphology, and basically that of the spore wall. There is information of other kinds to be drawn from spores, particularly in studies on development (Neidhart, this volume, Chapter 12), and on the physiology of germination.

THE TAXONOMIC CHARACTERS OF BRYOPHYTE SPORES

The potential of bryophyte spores in taxonomic studies is governed by the number of recognizable characters they possess. This depends on their structure, and structure can be divided into several different aspects: shape, size, wall ornamentation, wall layering, presence and nature of apertures, and polarity. Each of these categories includes features which may vary from taxon to taxon and can therefore be used in taxonomy. I shall not attempt to give a catalogue of all the known states of these features, but rather to review each in turn and assess their suitability as taxonomic characters.

1. Shape

This is one of the most obvious attributes of a spore, but one that is of limited usefulness. The shape of a spore depends upon its condition; many spores are thin-walled structures which collapse easily. A mature spore in a moist environment where its cytoplasm is fully turgid takes on a different shape from a spore which has dried out and shrivelled. The variations of shape that this kind of phenomenon can induce are considerable and must be discounted (Reitsma, 1969). When comparing spores it is important to examine them in a comparable state or the differences one sees are liable to be of no consequence. There are two important factors which affect shape, first the spatial relationship of the spores to one another in the tetrad (see p. 236) and second, the character- istic tendency in some advanced taxa, such as the Lejeuneaceae, Frullaniaceae or *Ptychanthus* (Ono, 1966) among the hepatics and Dicnemonaceae, *Mulleriella* or *Drummondia* (Vitt and Hamilton, 1974) among the mosses, to have spores which undergo the first stages of germination before they are released from the sporophyte. Such spores are multicellular at maturity and often take on elongated shapes rather than the more or less spherical shape typical of the majority of bryophytes.

As a taxonomic character, shape is of significance only at the broadest level. Variations in shape between spores seem either to be the result of transitory phenomena such as those mentioned above or else of major differences in development which probably imply considerable phylogenetic differentiation (Knox, 1939; Jovet-Ast, 1954; Inoue, 1960; Miyoshi, 1966; Ono, 1966).

2. Size

Size is another very obvious attribute of a spore, and one that has traditionally been observed. Unlike shape, size has proved to be of considerable use in bryophyte taxonomy. There is often more variation between related taxa, and this variation is open to statistical treatment. Most bryophyte spores are more or less spherical and measure between 10 μm and 50 μm in diameter although there are some conspicuous exceptions to this (Erdtman, 1957, 1965). Jovet-Ast (1954) has pointed out that those mosses that have large spores are often cleistocarpous (e.g. *Acaulon*, *Phascum*, *Lorentziella*, *Ephemerum*), or lack a peristome (e.g. *Gigaspermum*). This suggests a close relationship between the size of spores and their method of distribution. Environmental factors which affect the turgidity of the spores are again important here, but a number of other phenomena are equally important as recent studies have shown. For example, some species produce spores of two different sizes, not only from the same sporangium, but as a result of a single meiosis (Vitt, 1968; Ramsay, this volume, Chapter 13). Vitt called this system anisospory (as opposed to the heterospory of vascular plants). The occurrence of anisospory is itself a character of taxonomic importance; it occurs in *Macromitrium* and *Schlotheimia*, both in the Orthotrichaceae (Ernst-Schwarzenbach, 1939). This phenomenon has to be distinguished from the far more common case of the presence of both viable and aborted spores in the same capsule which also leads to variations in size. Mogensen (1978) has produced an interesting study of this in the genus *Cinclidium* where he has found that a large proportion of spores regularly abort and fail to swell to full size, apparently as a result of lethal genetic factors. More mundane phenomena like bad weather have been shown to produce a similar effect in pollen grains (Stanley and Linskens, 1974) and may do the same in bryophyte spores. Mogensen also points out that during their maturation *Cinclidium* spores do not increase in volume at a steady rate but enlarge in a series of steps with relatively inactive periods between them. This emphasizes the importance of measuring only mature spores.

3. Wall Ornamentation

This is often a more significant character than shape or size since it is not affected by alterations in the turgidity of the spores. Environmental factors can influence ornamentation; Taylor *et al.* (1972) found, for example, that spores of *Asterella tenera* (Mitt.) Schust. grown in axenic culture were different from those collected in the field. But this example involves growth in artificial conditions and I know of no parallel case for plants growing in the wild.

Certain genera have spores with prominent, easily observed ornamentation and these have received considerable attention in floras, but it is probably fair to say that many spores with less prominent ornamentation have been unduly neglected. There are several reasons for this. One is that ornamentation is often rather irregular and difficult to describe accurately. Another is that preparation techniques have very often been inadequate and not enough time has been spent in making sure that the spores are clean and in a suitable state for examination. In some cases, as Schuster (1966) pointed out, careless microscopy has led to inaccurate observations. The study of McClymont (1955) showed the importance of observing ornamentation in fully mature spores. He studied spores from a range of *Bruchia* species and could recognize two distinct types of ornamentation. But immature spores had intermediate ornamentation patterns which could have obscured the distinction. Hirohama (1977a) made a similar observation with *Orthotrichum* species.

Recent results with the scanning electron microscope (SEM) have shown that there is often another order of ornamentation present above that seen in the light microscope in both Musci and Hepaticae. Dickson (1969, 1973), for example, has shown that the spores of British species of *Polytrichum*, which all look similar in the light microscope, can be distinguished in the SEM by their ornamentation. In a similar way, Denizot (1971) showed that *Lunularia* spores, which appear smooth in the light microscope, are clearly ornamented at the higher magnifications of the SEM. She also found that other genera such as *Conocephalum, Dumortiera, Reboulia* and *Targionia* all had more complex ornamentation than had been suspected from light microscope observations.

Since the introduction of the SEM there have been several publications which demonstrate the range of variation present in bryophyte spores. Taylor, Hollingsworth and Bigelow (1974), for example, illustrate a wide range of hepatic spores and Hirohama (1976, 1977a, b, 1978) has covered a number of moss families.

The transmission electron microscope (TEM) is also of great potential value here, used either with carbon replicas (Shin, 1963), with sectioned material (McClymont and Larson 1964; Gullvåg, 1966; Sorsa, 1976) or in the silhouette technique (Lewinsky, 1974a, b, 1977) where spores are placed directly on the EM grid. Lewinsky (1974a) was able to show that spores of a range of species in the Plagiotheciaceae all looked similar at a magnification of 6000 in the SEM, but that at × 30 000 in the TEM all the species could be differentiated.

It would be wrong, however, to give the impression that electron microscopy is the only way to make useful observations. There is still a great deal to be discovered within the limits of resolution of the light microscope (Zigliara, 1971–1972; Boros and Járai-Komlódi, 1975; Schuster and Engel, 1977), but careful microscopy and suitable preparation techniques are essential.

4. Wall Layering

For this the TEM comes into its own and the light microscope can only give rudimentary data. The failure of the light microscope to be of much use here is easily explained; the walls of bryophyte spores are often very thin and the various layers of which they are formed are at or below the limits of optical resolution (Heckman, 1970; Sorsa, 1976). Erdtman, one of the pioneers in the study of wall layering in pollen grains, gave diagrams from light microscope observations of wall sections in many bryophyte genera (Erdtman, 1957), but in practically every case even he was unsure of his interpretations.

With the extra resolving power of the TEM, a great deal of new data has come to light, especially in the Hepaticae, but also in the Musci. Bryophyte spores have walls which are composed of a series of layers in much the same way as those of pteridophyte spores, and are comparable in a number of respects with pollen grains (Gullvåg, 1966–1967; Neidhart, this volume, Chapter 12). The innermost layer, which seems to correspond to the intine of pollen grains, is relatively structureless, but the outer layers (exine and perine) are much more complex. Not only do these layers carry the ornamentation with its variable morphology, but the way they are built up also varies from taxon to taxon.

Some of the most detailed work on bryophyte spore walls has been carried out in the Marchantiales and Sphaerocarpales by Denizot. This work has been summarized by Denizot (1976) and by Neidhart (this volume, Chapter 12).

Heckman (1970) studied wall structure in a number of species of the Jungermanniales. She found that the walls are laid down round a series of lamellae and that the nature and distribution of the lamellae varied considerably from species to species. In *Ptilidium pulcherrimum*, sheet-like lamellae form a continuous layer round the spore and the papillate ornamentation results from folds in this sheet. Sheet lamellae are absent in *Lophocolea heterophylla* but there are numerous small "slip lamellae" which lie perpendicular to the cell surface on which they form small papillae. *Diplophyllum apiculatum* (Evans) Steph. has both sheet and slip lamellae. Sorsa (1976), on the other hand, reported that the Musci are quite different; they have no lamellae, and sporopollenin is laid down in the walls in a granular fashion. Heckman (1970) reports that *Anthoceros* is like Musci rather than Hepaticae in this respect, but this needs confirmation.

There is still a considerable amount of basic research to be done on wall layering. There are many families which have never been examined in this way and we have a great deal to learn about the homology of the layers between different taxa.

5. Polarity

The process of meiosis results in the formation of four spores which, at least to start with, are associated in a tetrad. Spores have polarity because of the way they are arranged with respect to one another in the tetrad. One part (the proximal pole) of each spore points towards the centre while the opposite, distal pole is not in contact with the other spores. Mature spores sometimes retain this polarity in their morphology.

The tetrad is usually arranged in a tetrahedral shape like a triangular pyramid. There is also the possibility of a flat tetrad where all the spores are in a single plane.

As the spores are pressed together in a tetrahedral tetrad they take on a characteristic shape with a convex distal face and three more or less flat proximal faces which tend to come together in a point. There is evidence from studies of development (Neidhart, this volume, Chapter 12) that the resistant, sculptured parts of the spore wall are laid down shortly after meiosis. At that time the spores may either still be associated in the tetrad or they may have separated.

If they are pressed together when the outer layers of the wall are laid down, the morphology of the spores reflects it, sometimes in shape and sometimes in ornamentation. There is often a Y-shaped trilete mark which represents the edges of the different facets of the proximal face of the spore as in *Sphagnum* (Terasmae, 1955; Tallis, 1962). There may also be some kind of distinction between the poles in their surface ornamentation; the elements may be differently shaped, or have differences in their distribution. Taylor *et al.* (1972) have given an example of this in *Asterella tenera*.

On the other hand, if the spores are freed from the tetrad before the exine ornamentation is laid down there may be no remaining evidence of polarity since the ornamentation is distributed evenly all over the spore. In a similar way, mature spores often lose the shape they had in the tetrad and become more or less spherical.

A feature of polarity which has only been noted since the introduction of electron microscopy is the thickening of the intine on one side of the spore. This has been found in *Fissidens* (Mueller, 1974). *Bruchia* and *Ceratodon* are similar, and in these genera the exine is thinnest where the intine is thickest (Larson, 1964; Valanne, 1966). So spores which have no apparent polarity when examined in the usual way, may in fact have a distinct internal polarity. Sorsa (1976) found that this was the case in the majority of the Eubryales he studied.

In some exceptional cases the spores may be shed from the sporophyte and dispersed as tetrads rather than as individuals. This has been reported for *Andreaea* species (Erdtman, 1957, 1965) and for *Sphaerocarpos donnellii* Aust. (Doyle, 1975). It is a completely distinct phenomenon from the multicellular spores described above.

6. Apertures

This is an area where the literature shows a need for basic research. Some bryophyte spores have distinct apertures, but many do not (Erdtman, 1957, 1965; Miyoshi, 1966; Boros and Járai-Komlódi, 1975). Where apertures occur they are often irregularly-shaped areas of the spore wall in which the exine is thin or absent as in the Pottiales and Hookeriales or they might be trilete structures as in the Marchantiales, Sphagnales or Andreales (Erdtman, 1965). Occasionally (e.g. *Tetraplodon* and *Splachnum*) the apertures may be single, straight furrows (Erdtman, 1965). Erdtman has interpreted these as two of the branches of a trilete aperture which are aligned with one another. It would be interesting to know how this type of aperture links with the arrangement of the spores in the tetrad.

If spore morphology is to be used effectively in taxonomic studies, there must be an adequate and generally agreed system of terminology. Many features of spores are difficult to describe, particularly those which concern three-dimensional shapes, or subtle differences between patterns of ornamentation.

A number of glossaries of terms suitable for describing spore morphology have been compiled, notably those of Erdtman (1952, 1967), McClymont (1955) and Miyoshi (1966). Reitsma (1970) in a glossary of pollen terminology gave many terms which are equally applicable to spores and he reviewed much of the hotly-debated literature on the subject.

The ornamentation patterns of many spores are irregular and a verbal system of terminology is unlikely to be adequate to describe them. In such cases photographs are essential; light micrographs may give all that is needed but SEM prints are much simpler to interpret and compare.

Many of the terms used for describing pollen and pteridophyte spores can be used for bryophyte spores, but some, notably those for wall layers (e.g. intine, exine, perine) imply homologies that are not always appropriate. Neidhart (this volume, Chapter 12) discusses this problem.

PREPARATION TECHNIQUES

Experience with pollen grains has shown the advantages of using a standard preparation technique by which the grains can be cleaned, stained and expanded to a repeatable extent. For pollen, the technique which is normally used is acetolysis (Erdtman, 1960), a process developed for pollen analysis but which has proved very useful in studies of the morphology of pollen walls. The outer layers of the walls of both pollen and spores are composed of sporopollenin, a highly resistant carotenoid polymer (Brooks and Shaw, 1978) and it is these layers that give the majority of characters used in taxonomic studies. Acetolysis involves heating spores in a mixture of concentrated sulphuric acid and acetic anhydride; it dissolves away extraneous material and leaves only those structures composed of sporopollenin. The sporopollenin is stained brown at the same time. Dried specimens from the herbarium are often shrivelled so Reitsma (1969) introduced a process by which pollen or spores could be expanded using a photographic wetting agent before acetolysis. This ensures that they are in a suitable state for examination of wall structures.

The use of acetolysis not only cleans the surface of spore walls and removes the cytoplasm they surround, it also goes some way to preventing the swelling of the spores by the osmotic effect of aqueous mounting media such as glycerine or glycerine jelly.

For SEM work, untreated spores are often taken directly from herbarium specimens (Taylor *et al.*, 1974; Saito and Hirohama, 1974), and this can give satisfactory results. There is, nevertheless, always the possibility that untreated spores may be coated with extraneous material which could obscure, or give a false impression of wall ornamentation. Also, spores containing cytoplasm can be swollen to an unnatural shape when the electron beam of the SEM raises their temperature and expands their contents. Both these problems are overcome by acetolysis.

TEM observations of spores require a range of different techniques which are chosen according to the aim of the study. Sorsa (1976) lists many of them and they will not be repeated here.

THE USES OF DATA ON SPORE MORPHOLOGY

1. At the Species Level

Many early descriptions of bryophytes included some details of spore morphology, largely to record the range of variation present rather than to make any special use of the data. Among general floras, that of Roth (1904–1905) stands out for its many accurate descriptions of spores, although earlier examples could be cited. It was during this exploratory phase of the study of bryophyte anatomy that the existence of many of those taxa which have unusually large or prominently ornamented spores came to light. For other taxa, however, where the spores appeared to have nothing to offer the taxonomist, descriptions omitted any mention of spores or passed over them in a most perfunctory way.

There have been a number of publications which catalogue the morphology of a range of spores so that species preserved in fossil, especially Quaternary, deposits can be identified. Knox (1939), Jonas (1952), Erdtman (1957, 1965) and more recently Boros and Járai-Komlódi (1975) have given a large number of illustrations and descriptions based on observations in the light microscope. None of these works includes any kind of key to the spores, so their practical use is limited and one must riffle through the plates to find a match for an unknown spore. Very few attempts have been made at keys for spore identification. Tallis's (1962) key to British *Sphagnum* species is one of the few examples. Probably because of this, lists of species recorded from Quaternary deposits rarely include bryophytes, even though there are species with recognizable spores which could be expected to occur (Dickson, 1973).

For the systematist, this kind of work is of limited use; it mainly serves to uncover variation which might prove a useful source of data for further study. Pollen analysts are more or less limited to light microscope work for technical reasons and their publications therefore include only a limited amount of detail.

In taxonomic studies one of the most frequent uses of spore morphology has been similar in approach to that of the pollen analysts—to distinguish species from one another. There are a number of genera where spores are regularly used like this—*Fossombronia* (e.g. Stephani, 1900; Ladyzhenskaya and Zenkova 1955; Ladyzhenskaya, 1970) and *Riccia* (Lindenberg, 1836; Macvicar, 1912; Müller, 1954; Jovet-Ast, 1971–1972) are obvious examples. The majority of genera where spores are regularly examined are in the Hepaticae although there are also moss examples such as *Pottia* (Chamberlain, 1978; Rejment-Grochowska, 1978) or *Encalypta* (Schimper, 1876; Mårtensson, 1956). All these genera are exceptional in their spore morphology—their spores are either larger than average or they have more prominent ornamentation, or both. In many of these cases the species are difficult to separate on their gametophyte anatomy and spores give more reliable or convenient characters. In the hepatics, at least, spores are structures which are often less plastic in phenotype than the body of the gametophyte plant.

Exceptionally, a single species may contain more than one type of spore. Anisosporous species are an example of this (Vitt, 1968; Ramsay, this volume, Chapter 13). Recently Bischler (1978) has found that *Plagiochasma eximium* (Schiffn.) Steph. has spores with two kinds of ornamentation, one irregularly reticulate with shallow muri and another with deep, regular reticulations. The significance of this has not yet been explained.

In the majority of cases where more easily studied characters are sufficient to distinguish species, spores have little to offer in routine identification. There has been a general acceptance that, especially among the Bryales and Jungermanniales, there is little variation in spore morphology which can be used for this kind of study. Knox (1939) went so far as to say that moss spores are "monotonously uniform . . . it becomes at once evident that spore structure would be of very limited taxonomic value".

Spore morphology can, however, be of considerable use in the original delimitation of species by confirming suspected distinctions found by using other characters. For example, *Targionia hypophylla* and *T. lorbeeriana* K. Müll. can be separated on the morphology of their gametophytes (Jovet-Ast and Zigliara, 1966), but it is doubtful whether they would have been accepted as distinct were it not for the unequivocal differences between their spores (Müller, 1942; Zigliara, 1971–1972).

I would therefore disagree with McClymont's comment that "there is every reason to advocate the extension of the use of spores in identification" (McClymont, 1955) but would support him when he goes on to advocate their use in specific delimitation.

2. Recognition of Taxa above the Species Level

Until recently, the main preoccupation of bryophyte taxonomists was the basic work of recognition and description of species. For many areas of the world this is still the case. There has been comparatively little original thought about supra-specific taxa although there are, of course, notable exceptions to this as the reviews in this volume by Schuster (Chapter 3) and Miller (Chapter 2) make clear. In the last few years, however, there has been a growing concern with taxa of higher rank in the bryophytes, a concern which has, for example, stimulated this symposium.

It is not surprising that research into the systematic importance of spore morphology has followed a similar path. There is little early work in which spores have been used to suggest generic or family relationships, but a considerable amount has appeared in recent years. It is no coincidence that this change in interest has taken place at the same time as new techniques have been developed.

The particular advantage of spore morphology in recognizing higher ranking taxa lies in the fact that spores are less variable than many other parts of the plant. Because there are more bryophyte species than different types of spore, the species are automatically grouped into taxa in which the spores are similar. Recognition of these groups may not necessarily be as simple as this statement implies since there is the possibility that unrelated taxa may have evolved superficially similar spores by a process of convergence. For example, multicellular spores are found in unrelated genera such as *Conocephalum* and *Dendroceros* (Miyoshi, 1966). Convergence can, in many cases, be recognized by a more complete study of morphology and should not lead to a misinterpretation of relationships. That the examples given above are convergent can be demonstrated by studies of the ultrastructure of the spores in question or by the structure of the plants themselves. This problem is, of course, a general one which is by no means exclusive to spores.

(a) *Hepaticae.* As pointed out above, much of the work so far published on bryophyte spores has concentrated on the Hepaticae rather than the Musci, and particularly, on the more primitive orders which often have larger spores than the more advanced orders (Schuster, 1966). Stephani (1900), for example, used spore morphology to divide the genus *Fossombronia* into three groups and these groups were themselves subdivided on spore characters. In a more recent example, Grolle (1967) showed that spore features could be used, amongst others, to separate the genera *Lepidolaena* Dum. and *Gackstroemia* Trev. Similarly, Schuster (1960) cited the exceptionally large spores of the genera *Chaetophyllopsis* Schust. and *Herzogianthus* Schust. as one character to justify setting up a new family, the Chaetophyllopsidaceae, to incorporate them. Many other examples could be given.

Schuster and Engel (1977) went a stage further than this in their paper on the Schistochilaceae of South America. They found that "divergences in spore–exine structure are phenomenal" in the family and were able to suggest which spore types were primitive and which advanced. Using this kind of information, hypotheses about relationships between taxa can be derived and tested against data from other sources.

There have been several attempts to survey the spores of the Hepaticae in a more general way. Inoue (1960), in a paper mainly concerned with the germination of spores rather than their morphology, showed that the Marchantiales included three major spore types: apolar, cryptopolar and polar. Ono (1966), considering the Jungermanniales, found that the spores of the 67 species he studied could be divided into eight major types which differed in shape, size and ornamentation. Ono confirmed the findings of earlier workers such as Knox (1939) that the majority of species fell into a single type which he named the *Cephalozia* type. He also made some general comments on the use of his results in systematics. Miyoshi (1966) went much further in this respect in his monograph on the spores of Japanese hepatics. Miyoshi recognized five basic spore types which are more broadly defined than those of Ono and which probably have more systematic significance since they are constructed on more fundamental characters such as polarity and the nature of the tetrad scar. Within these spores types, Miyoshi suggested which members had primitive or advanced characters and how they might be related to one another. Schuster (1966), in a general introduction to the study of hepatic spores and elaters also included some comments on their phylogenetic significance. He used data on spore ornamentation and, particularly, size to suggest how, during the course of evolution, spores were discharged and distributed in a progressively more effective way.

None of the works cited so far included ultrastructural data. One of the first to do this was that of Heckman (1970) who found considerable differences in the details of spore wall construction in various species of the Jungermanniales. Heckman's results have been described above (p. 236). They show that there is considerable potential for further study, but the species she chose to examine did not give a very broad representation of the Hepaticae and we are still ignorant of spore ultrastructure in many families. This is, however, a most promising field for research.

(*b*) *Musci.* There is little early work on the Musci which makes use of spores to elucidate taxa of higher rank. One of the first to do this was McClymont (1955) who made a detailed study of spores on the genus *Bruchia*. McClymont had originally set out to see whether *Bruchia* species could be identified from their spores. He was frustrated in this, but discovered that the genus comprised two different but related spore types. He suggested that six species which had been supposed to belong to the genus but had spores of a quite different type should be excluded from *Bruchia*. He also concluded that *Bruchia* is phylogenetically intermediate between *Trematodon* and *Archidium*, a theory for which he found some support from other characters.

More recently, Saito and Hirohama (1974) have used the SEM to study spores of the Pottiaceae. They found that variations in spore morphology did not correlate with other morphological features or with accepted taxonomic divisions. Spores from species in the same genus, such as *Trichostomum platyphyllum* (Broth.) Chen and *T. crispulum*, sometimes had very different ornamentation patterns, while some species of the genera *Trichostomum*, *Weissia*, *Hymenostomum* and *Pseudosymblepharis* had spores which were apparently identical. Saito and Hirohama did not draw many conclusions about the taxonomic implications of their results but they suggested that, from the species they had seen, there was no justification for separating the genera *Astomum* and *Hymenostomum* from *Weissia*. It is interesting that Smith (this volume, p. 201) drew the same conclusion from evidence of an entirely different kind.

Lewinsky (1974b), in a study of the European species of *Tortula* had similar difficulties in interpreting her results. All the species have papillate ornamentation but there is some variation in the size and density of the papillae. Using these data, Lewinsky divided the species into a number of groups, but she found that these groups did not coincide with the sections of the genus. Lewinsky (1977) had no more success in correlating spore morphology with taxonomic divisions in the genus *Orthotrichum*. When the Orthotrichaceae is taken as a whole, however, Hirohama (1977b) found that the ornamentation of the spore surface is characteristic for each genus. Hirohama drew up a sequence of genera in the family according to whether, in his opinion, the spores were primitive or advanced. He also (Hirohama, 1977a) deduced relationships of this kind in the Bartramiaceae and suggested, from the spore evidence, that the Grimmiaceae has a close affinity with the Dicranaceae (Hirohama, 1978).

The work of Sorsa and Koponen (1973) on the Mniaceae showed that although the morphology of spores in the various genera of the family is not particularly variable, it gave confirmatory evidence for the division of the family into tribes. Sorsa (1976) extended this study to cover a number of families related to the Mniaceae and found considerable differences between the families in both wall structure and ornamentation. The TEM was especially useful in this since it could be used to examine a range of characters beyond the resolution of other microscopes.

Two recent studies on the spores of *Encalypta* species have emphasized the importance of interpreting results correctly and also of assessing the spore data in conjunction with data from other sources. Járai-Komlódi and Orbán (1975) examined the spores of nine European species while Vitt and Hamilton (1974) examined eight species from North America. All the species from North America occur also in Europe and we can therefore compare the results of the two investigations directly. Table I sets out the opinions of both pairs

TABLE I. *Encalypta* species grouped by their spore types according to two different systems

Species	Járai-Komlódi and Orbán (1975)	Vitt and Hamilton (1974)
E. ciliata	1	2
E. rhapdocarpa	2	2
E. vulgaris	2	2
E. brevicolla Bruch	2	1
E. affinis (Hedw. fil.) Web. and Mohr	2	1
E. longicollis Bruch	2	3
E. alpina	3	2
E. procera Bruch	3	1
E. streptocarpa	3	—

of authors on how spore morphology suggests the species should be grouped. In both cases three groups are proposed, but there the resemblance between the systems ends; the composition of the groups in both systems could not be more different. The only way to discriminate between the two is by reference to other characters. Járai-Komlódi and Orbán have not included data of this kind, but Vitt and Hamilton show that their system is very closely paralleled by one proposed on the basis of peristome characters. In a similar way, Koponen (1978) correlated the morphology of spores in the Splachnaceae with the biology of spore dispersal and used this correlation to deduce which species were advanced.

Another example which shows how spore data can be used in conjunction with data from other fields is given by Smith (1974). Before studying the spores, Smith had proposed "major generic realignments" of the Polytrichaceae using a range of characters. When he tested these realignments by an investigation of spore morphology he found that the spores bore out his decisions in a very accurate way; each of Smith's "realigned" genera was consistent in the ornamentation of its spores.

CONCLUSIONS

From this review several general conclusions stand out. Spore morphology has not so far played a significant role in studies of bryophyte systematics and there are two main reasons for this. In the first place the literature shows that there has been little understanding of how the various features of spore morphology can be used as taxonomic characters. Descriptions of spores have tended to concentrate on those characters which are most obvious and easy to describe, but these characters more influenced by environmental factors than some others which require more painstaking observation. Linked with this has been a general lack of appreciation of the importance of careful preparation of specimens and of high standards of microscopy, whether optical or electronic.

The other reason that spore morphology has been neglected lies in the type of study for which it has often been used. Only in exceptional cases are spores of use to the taxonomist at the level of species. Their potential contribution to systematics is in the recognition of higher ranking taxa such as genera or, especially, families. Most studies which include observations of spores have been directed at the specific level and spores have proved largely irrelevant to them. This has led to a general view that spores have little to offer the taxonomist and they have often been overlooked, even in those cases where they could have been of use.

We now have sufficient experience in the study of spores to be able to understand much more clearly which characters can give significant evidence for systematics and how observations of them should be used. As a rule, those features, such as shape and size, which are most prominent, are only important at the broadest level and small variations are of little significance. Those features, such as wall ornamentation or construction, which are not so apparent at first sight are often more reliable taxonomic markers.

This does not necessarily mean that there is no place for light microscope studies of spores. Provided such studies are carried out with due attention to technique, there is a great deal still to be learnt, particularly about those species, such as the primitive Hepaticae, which have large spores. The important thing to appreciate is how to use the data once collected; this survey of the literature clearly shows the type of investigation for which spores have proved most useful.

Nevertheless, the introduction of electron microscopy has increased the potential importance of spore morphology immensely. Observations on wall layering and on ornamentation patterns at the ultrastructural level have brought out variations of a basic kind which have proved significant systematically. So far, work of this type has hardly progressed beyond the exploratory phase. We now have a good idea of the extent of variation present but many more observations are needed to give a reasonable representation of the Bryophyta as a whole.

There have been a number of attempts to review the evidence from spore morphology for the evolutionary history of bryophytes (Jovet-Ast, 1954; Ştefureac, 1971). These were based solely on light microscope evidence from a restricted range of species and were necessarily very preliminary. Once more detailed data are available for a more comprehensive range of taxa it should be possible to deduce a great deal more about bryophyte phylogeny and about the relationships between the Bryophyta and other groups.

I mentioned at the beginning of this review that pollen morphology had played a significant role in flowering plant systematics and asked whether spore morphology might ever be equally useful to bryophyte systematics. The functions of pollen and spores are very similar but there is an important difference between them which affects their structure and hence their use in taxonomy. Pollen grains are closely connected with the system of compatibility which is an integral part of angiosperm reproduction. The wall of pollen grains is more complex in its structure than that of bryophyte spores. It houses many of the proteins which are responsible for the recognition of the pollen grains by the stigma and provides a mechanism for their release (Heslop-Harrison, 1975). The structures associated with this process are absent from bryophyte spores and cannot be used as taxonomic characters. In addition, the structure of a pollen grain is adapted for a process which, by definition, is capable of discriminating between species and this is not the case in bryophytes where the spores have a much more general function common to many species.

It is therefore unlikely that spores will ever be as significant to bryophyte systematics as pollen is to angiosperms but there is now ample evidence to suggest that spore morphology is one of the most promising lines of research into the systematics of bryophytes at higher levels.

REFERENCES

BISCHLER, H. (1978). *Plagiochasma* Lehm. et Lindenb., II. Les taxa européens et africains. *Revue bryol. lichén.* **44**, 223–300.

BOROS, Á. and JÁRAI-KOMLÓDI, M. (1975). "An Atlas of Recent European Moss Spores." Akadémiai Kiadó, Budapest.

BROOKS, J. and SHAW, G. (1978). Sporopollenin: a review of its chemistry, palaeo-chemistry and geochemistry. *Grana* **17**, 91–97.

CHAMBERLAIN, D. F. (1978). *Pottia. In* A. J. E. Smith "The Moss Flora of Britain and Ireland" pp. 234–242. Cambridge University Press, Cambridge.

DENIZOT, J. (1971). Morphologie et anatomie des parois de spores et d'élatères chez quelques Sphaerocarpales et Marchantiales. *Naturalia monspel.*, Sér. bot., **22**, 51–127.

DENIZOT, J. (1976). Remarques sur l'edification des differentes couches de la paroi sporale à exine lamellaire de quelques Marchantiales et Sphaerocarpales. *In* "The Evolutionary Significance of the Exine" (I. K. Ferguson and J. Muller, eds) pp. 185–210. Academic Press, London and New York.

DICKSON, J. H. (1969). Scanning reflexion microscopy of bryophyte spores with special reference to *Polytrichum. Trans. Br. bryol. Soc.* **5**, 902. (Abstr.)

DICKSON, J. H. (1973). "Bryophytes of the Pleistocene. The British Record and its Chorological and Ecological Implications." Cambridge University Press, Cambridge.

DOYLE, W. T. (1975). Spores of *Sphaerocarpos donnellii. Bryologist* **78**, 80–4.

ERDTMAN, G. (1957). "Pollen and Spore Morphology/Plant Taxonomy. Gymnospermae, Pteridophyta, Bryophyta (Illustrations). An Introduction to Palynology, II." Almqvist and Wiksell, Stockholm.

ERDTMAN, G. (1965). "Pollen and Spore Morphology/Plant Taxonomy. Gymno-spermae, Bryophyta (Text). An Introduction to Palynology, III." Almqvist and Wiksell, Stockholm.

ERDTMAN, G. (1960). The acetolysis method, a revised description. *Svensk bot. Tidskr.* **54**, 561–564.

ERNST-SCHWARZENBACH, M. (1939). Zur Kenntnis des sexuellen Dimorphismus der Laubmoose. *Arch. Julius Klaus-Stift. VererbForsch.* **14**, 361–474.

GROLLE, R. (1967). Monographie der Lepidolaenaceae. *J. Hattori bot. Lab.* **30**, 1–53.

GULLVÅG, B. M. (1966). The fine structure of pollen grains and spores: a selective review from the last twenty years of research. *Phytomorphology* **16**, 211–227.

GULLVÅG, B. M. (1966–1967). The fine structure of the spore of *Hylocomium loreum* (Hedw.). *J. Palynol.* **2–3**, 49–65.

HECKMAN, C. A. (1970). Spore wall structure in the Jungermanniales. *Grana* **10**, 109–119.

HESLOP-HARRISON, J. (1975). The physiology of the pollen grain surface. *Proc. R. Soc. Lond.* **B190**, 275–299.

HIROHAMA, T. (1976). Spore morphology of bryophytes observed by scanning electron microscopy, I. Dicranaceae. *Bull. natn. Sci. Mus., Tokyo*, Ser. B, Bot., **2**, 61–72.

HIROHAMA, T. (1977a). Spore morphology of bryophytes observed by scanning electron microscope, II. Bartramiaceae. *Bull. natn. Sci. Mus., Tokyo*, Ser. B, Bot., **3**, 37–44.

HIROHAMA, T. (1977b). Spore morphology of bryophytes observed by scanning electron microscope, III. Orthotrichaceae. *Bull. natn. Sci. Mus., Tokyo*, Ser. B, Bot., **3**, 113–122.

HIROHAMA, T. (1978). Spore morphology of bryophytes observed by scanning electron microscope, IV. Grimmiaceae. *Bull. natn. Sci. Mus., Tokyo*, Ser. B, Bot., **4**, 33–42.

INOUE, H. (1960). Studies in spore germination and the earlier stages of gametophyte development in the Marchantiales. *J. Hattori bot. Lab.* **23**, 148–191.

JÁRAI-KOMLÓDI, M. and ORBÁN, S. (1975). Spore morphological studies on recent European *Encalypta* species. *Acta bot. hung.* **21**, 305–345.

JONAS, F. (1952). Atlas zur Bestimmung rezenter und fossiler Pollen und Sporen. *Beih. Repert. nov. spec. Regni veg.* **133**, 5–60.

JOVET-AST, S. (1954). L'intérêt de l'étude des spores pour les théories de l'évolution chez les bryophytes. *In* "Huitième Congrès International de Botanique—Paris 1954. Rapports et Communications Parvenus avant le Congrès aux Sections 2, 4, 5 et 6". pp. 248–250. Paris.

JOVET-AST, S. (1971–1972). Distinction de *Riccia gougetina* Mont. et de *Riccia ciliifera* Link d'après les spores. *Revue bryol. lichén.* **38**, 161–175.

JOVET-AST, S. and ZIGLIARA, M. (1966). La paroi des spores de *Targionia lorbeeriana* et de *Targionia hypophylla*: sa valeur taxonomique. *Revue bryol. lichén.* **34**, 816–820.

KNOX, E. M. (1939). The spores of the Bryophyta compared with those of Carboniferous age. *Trans. Proc. bot. Soc. Edinb.* **32**, 477–487.

KOPONEN, A. (1978). The peristome and spores in Splachnaceae and their evolutionary and systematic significance. *Bryophyt. Biblthca* **13**, 535–567.

LADYZHENSKAYA, K. I. (1970). Novitates ad cognitionem generis *Fossombronia* Raddi (Hepaticae) in flora URSS. *Nov. Sist. Nizsh. Rast.* **7**, 306–312.

LADYZHENSKAYA, K. I. and ZENKOVA, E. J. (1955). Spores of hepatics as a systematic character on the example of the genus *Fossombronia* Raddi. *Bot. Zh. SSSR* **40**, 853–857.

LEITGEB, H. (1884). "Bau und Entwicklung der Sporenhäute und deren Verhalten bei der Keimung." Leuschner und Lubensky, Graz.

LARSON, D. A. (1964). Further electron microscope studies of exine structure and stratification. *Grana Palynol.* **5**, 265–276.

LEWINSKY, J. (1974a). The family Plagiotheciaceae in Denmark. *Lindbergia* **2**, 185–217.

LEWINSKY, J. (1974b). An electron microscopical study of the genus *Tortula* Hedw., with special reference to exine ornamentation. *J. Bryol.* **8**, 269–274.

LEWINSKY, J. (1977). The genus *Orthotrichum*. Morphological studies and evolutionary remarks. *J. Hattori bot. Lab.* **43**, 31–61.

LINDENBERG, J. B. W. (1836). Monographie der Riccieen. *Nova Acta physico-med.* **18**, 361–501.

MCCLYMONT, J. W. (1955). Spore studies in the Musci with special reference to the genus *Bruchia*. *Bryologist* **58**, 287–306.

MCCLYMONT, J. W. and LARSON, D. A. (1964). An electron-microscopic study of spore wall structure in the Musci. *Am. J. Bot.* **51**, 195–200.

Macvicar, S. M. (1912). "The Student's Handbook of British Hepatics." Sumfield, Eastbourne.

Mårtensson, O. (1956). Bryophytes of the Torneträsk area, northern Swedish Lappland, II. Musci. *K. svenska VetenskAkad. Avh. Naturskydd.* **14**, 1–321.

Miyoshi, N. (1966). Spore morphology of Hepaticae in Japan. *Bull. Okayama Coll. Sci.* **2**, 1–46.

Mogensen, G. S. (1978). Spore development and germination in *Cinclidium* (Mniaceae, Bryophyta), with special reference to spore mortality and false anisospory. *Can. J. Bot.* **56**, 1032–1060.

Mueller, D. M. J. (1974). Spore wall formation and chloroplast development during sporogenesis in the moss *Fissidens limbatus*. *Am. J. Bot.* **61**, 525–534.

Müller, K. (1942). *Targionia lorbeeriana. Hedwigia* **81**, 95–99.

Müller, K. (1954). Die Lebermoose Europas. *In* "Dr. L. Rabenhorst's Kryptogamen-Flora von Deutschland, Österreich und der Schweiz". Vol. 6 (4–5) Geest and Portig, Leipzig.

Ono, M. (1966). Spore morphology of Jungermanniales (Hepaticae). *Jap. J. Bot.* **41**, 17–22; 111–119; 233–236.

Reitsma, T. (1969). Size modifications of recent pollen grains under different treatments. *Rev. Palaeobot. Palynol.* **9**, 175–202.

Reitsma, T. (1970). Suggestions towards unification of descriptive terminology of angiosperm pollen grains. *Rev. Palaeobot. Palynol.* **10**, 39–60.

Rejment-Grochowska, J. (1978). Valeur taxonomique des caractères morphologiques des spores chez les Pottiacées. *Bryophyt. Biblthca* **13**, 509–534.

Roth, G. (1904–1905). "Die Europäischen Laubmoose." W. Engelmann, Leipzig.

Saito, K. and Hirohama, T. (1974). A comparative study of the spores of taxa in the Pottiaceae by use of the scanning electron microscope. *J. Hattori bot. Lab.* **38**, 475–488.

Schimper, W. P. (1876). "Synopsis Muscorum Europaeorum." Schweizerbart, Stuttgart.

Schuster, R. M. (1960). Studies on Hepaticae, II. The new family Chaetophyllopsidaceae. *J. Hattori bot. Lab.* **23**, 68–76.

Schuster, R. M. (1966). "The Hepaticae and Anthocerotae of North America East of the Hundredth Meridian." Vol. 1. Columbia University Press, New York.

Schuster, R. M. and Engel, J. J. (1977). Austral Hepaticae, V. The Schistochilaceae of South America. *J. Hattori bot. Lab.* **42**, 273–423.

Shin, T. (1963). Electron-microscopical observations of the peristome and spores of mosses, 1. *Hikobia* **3**, 202.

Smith, G. L. (1974). New developments in the taxonomy of Polytrichaceae. Epiphragm structure and spore morphology as generic characters. *J. Hattori bot. Lab.* **38**, 143–150.

Sorsa, P. (1976). Spore wall structure in the Mniaceae and some adjacent bryophytes. *In* "The Evolutionary Significance of the Exine" (I. K. Ferguson and J. Muller, eds) pp. 211–229. Academic Press, London and New York.

Sorsa, P. and Koponen, T. (1973). Spore morphology of Mniaceae Mitt. (Bryophyta) and its taxonomic significance. *Annls bot. fenn.* **10**, 187–200.

Stanley, R. G. and Linskens, H. F. (1974). "Pollen: Biology, Biochemistry, Management." Springer, Berlin.

Ştefureac, T. I. (1971). Les particularités des spores chez les bryophytes et leur importance dans quelques considérations phylogénétiques et phytohistoriques. *Anal. Univ. Bucureşti*, Biol. Veg., **20**, 21–42.

STEPHANI, F. (1900). "Species Hepaticarum." Vol. 1, Anacrogynae. George, Genève.

TALLIS, J. H. (1962). The identification of *Sphagnum* spores. *Trans. Br. bryol. Soc.* **4**, 209–213.

TAYLOR, J., HOLLINGSWORTH, P. J. and BIGELOW, W. C. (1974). Scanning electron microscopy of liverwort spores and elaters. *Bryologist* **77**, 281–327.

TAYLOR, J., KAUFMAN, P. B., ALLARD, L. and BIGELOW, W. C. (1972). Scanning electron microscope observations of surface structure of isolated spores of *Asterella tenera*. *J. Hattori bot. Lab.* **36**, 406–410.

TERASMAE, J. (1955). On the spore morphology of some *Sphagnum* species. *Bryologist* **58**, 306–311.

VALANNE, N. (1966). The germination phases of moss spores and their control by light. *Annls bot. fenn.* **3**, 1–60.

VITT, D. H. (1968). Sex determination in mosses. *Mich. Bot.* **7**, 195–203.

VITT, D. H. and HAMILTON, C. D. (1974). A scanning electron microscope study of the spores and selected peristomes of North American Encalyptaceae (Musci). *Can. J. Bot.* **52**, 1973–1981.

ZIGLIARA, M. (1971–1972). Le genre *Targionia* L., II. Etude de la spore. *Revue bryol. lichén.* **38**, 241–264.

12 | Comparative Studies of Sporogenesis in Bryophytes

H. V. NEIDHART

Botanisches Institut, Tierärztliche Hochschule Hannover, Bünteweg 17 d, D-3000 Hannover 71, West Germany

Abstract: Aspects of spore formation have been studied for many years but there has been a recent upsurge in research at the ultrastructural level. Studies with a taxonomic aim have tended to concentrate on the structure of the mature spore but in order to appreciate the significance of differences found between taxa, it is necessary to study the formation of the spore wall.

A review of the available evidence shows that within the Bryophyta there is great variation in the morphology of the spore mother cells and in the formation of the spore walls. This suggests that the Bryophyta is a heterogeneous group of plants.

The Musci is a relatively homogeneous group. With the exception of *Polytrichum*, spore development is the same for all taxa for which data are available. The Anthocerotales have some features in common with the Musci, the Marchantiales and Sphaerocarpales, and the Metzgeriales and Jungermanniales, but the shared features are different in each case. The Anthocerotales are distinct from all these groups. The Marchantiales and Sphaerocarpales have a great deal in common and the Metzgeriales and Jungermanniales seem to form another pair of closely allied orders although more data are needed here.

At our present state of knowledge it is hard to elucidate relationships between the bryophytes and the different groups of pteridophytes. One reason for this is the paucity of detailed information on spore wall ontogeny in the pteridophytes. Only the Anthocerotales seem to have much in common with spermatophytes in the structure of their spores.

INTRODUCTION

The analysis of sporogenesis in the bryophytes began with Mohl's investigations on *Anthoceros* and Gotsche's on *Haplomitrium* in 1839 and 1841 respectively,

Systematics Association Special Volume No. 14, "Bryophyte Systematics", edited by G. C. S. Clarke and J. G. Duckett, 1979, pp. 251–280, Academic Press, London and New York.

as cited in Leitgeb (1884). Although numerous investigations have since been carried out, several important groups of bryophytes still lack any detailed ultrastructural investigation, especially moss orders such as the Andreaeales, Buxbaumiales, Grimmiales, Schistostegales, Tetraphidales, Isobryales and Hookeriales. Sporogenesis has been better investigated in the Hepaticae, especially in those species with large spores (Leitgeb, 1884; Beer, 1906; Black, 1913; Siler, 1934).

Apart from sporoderm formation, the older reports mainly concentrate on the partitioning of the spore mother cells and on nuclear division or spindle formation during meiosis (Farmer, 1895; Davis, 1899, 1901; Moore, 1905; Blair, 1926; Vaarama, 1954). Recently, spindle formation and dynamics in the mosses have been re-investigated and compared with data from other organisms (Lambert, 1968, 1978).

In the older accounts, details of spore maturation were neglected largely because of the low resolution of the light microscope and insufficient structural preservation. However, several authors have recently studied spore maturation at the ultrastructural level in the Musci (Paolillo and Kass, 1973, 1977; Mueller, 1974; Nurit, 1974; Neidhart, 1975, 1978a; Chevallier et al., 1977; Nurit and Chevallier, 1978) and in the Hepaticae (Horner et al., 1966; Neidhart, 1978b). However, this work has been largely physiological rather than taxonomic.

There has been a considerable amount of recent work on the taxonomic importance of spore characters (see review by Clarke in this volume, Chapter 11) but this has concentrated largely on the structure of the mature spore rather than on its development.

In order to understand differences in structure and sculpturing, it is necessary to study the process of formation of the spore wall. The aim of this account is to shed further light on the mechanisms which are involved in specific patterning, following on from earlier attempts by Genevès (1972b), Mueller (1974), Jarvis (1974), Denizot (1976) and Neidhart (1978b), and in addition, by a combination of detailed cytological investigations and improved histo-chemical methods, to reveal the homologies of the different spore wall layers in the Musci, Hepaticae and Anthocerotales.

THE SPORE MOTHER CELL (SMC)

1. General Morphology

In the Anthocerotales (Davis, 1899), the Marchantiales (Leitgeb, 1884; Beer, 1906; Black, 1913; Blair, 1926) and the Sphaerocarpales (Leitgeb, 1884; Kelley and Doyle, 1975; Neidhart, 1978b) the SMCs are spherical. When their archesporial walls are degraded most Musci have spherical SMCs (Fig. 1). Only the Polytrichales have quadrilobed SMCs (Weier, 1931; Paolillo, 1964). After each of the two successive and independent nuclear divisions, the partitioning of the tetrad cells starts with cell plate formation (Lambert, 1978). The plastids play an important role in the orientation of the spindles (Sapehin, 1911; Lambert, 1978).

In the Metzgeriales and Jungermanniales SMC protoplasts are typically quadrilobed (Humphrey, 1906; Clapp, 1912; Horner *et al.*, 1966; Suire, 1970; Oltmann, 1974; Lambert, 1978) although in *Lepidozia*, reduced lobes have been described (Lambert, 1978). It is still unclear whether SMC partitioning is by cell plate formation alone or whether furrowing also plays a part (Oltmann, 1974).

A prominent feature of moss SMCs is that their protoplasts form cytoplasmic protrusions (Figs 1, 2). Similar protrusions are to be found in *Lophocolea* (Jungermanniales) (Oltmann, 1974) and *Riccardia* (Metzgeriales) (Horner *et al.*, 1966). Protrusions of greater dimensions are described for the SMCs of the Sphaerocarpales and Marchantiales (Leitgeb, 1884; Beer, 1906; Siler, 1934). In *Riella* (Sphaerocarpales) they have an important role in spore wall sculpturing (Neidhart, 1978b). However, their function in the SMCs of the mosses is still unclear since they play no part in spore wall sculpturing. A hypothetical functional explanation is possible: as the protrusions enlarge the surface area of the SMC, there is a certain morphological similarity with "transfer cells" (Gunning and Pate, 1974).

KEY TO ABBREVIATIONS USED IN FIGURES

In the legends the following abbreviations are used to describe the methods for specimen preparation:

Ac+GA = prefixation with a mixture (1:1) of 1% acrolein and 1% glutaraldehyde in 0·05 m cacodylate buffer

GA = prefixation with 1% glutaraldehyde on 0·05 M cacodylate buffer at pH 6·9

RR = ruthenium red was added to the aldehyde medium according to Luft (1974)

Os = postfixation with 1% buffered OsO_4 solution

UB = block staining with uranyl acetate

US = sections stained with uranyl acetate

Pb = sections stained with lead citrate

Th = sections treated with periodic acid, thiosemicarbazide and silver proteinate according to Thiéry (1967)

Th–PA = reaction as above but without periodic acid treatment

Line scales equal 1 μm unless stated otherwise.

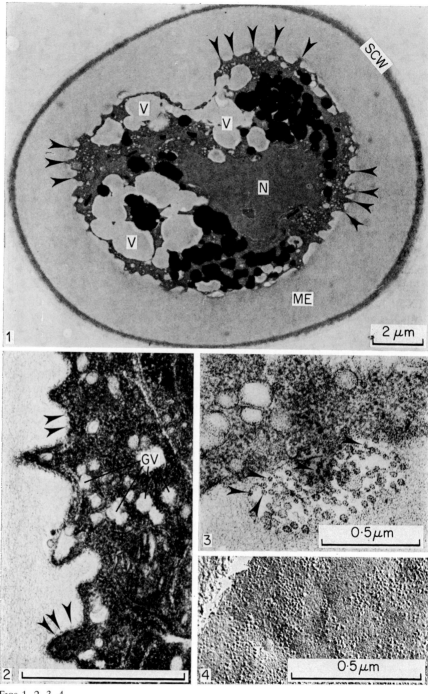

FIGS 1, 2, 3, 4

2. The SMC Walls

SMCs in the bryophytes are very active in secreting cell wall material in successive layers, as revealed by histochemical tests for different polysaccharide moieties (Genevès, 1971a, b, 1972a, 1974; Neidhart, 1975; Bienfait and Waterkeyn, 1976). Shortly before and during meiosis additional cell wall material is synthesized and this surrounds the newly formed tetrad of spores. This wall is called the "special mother cell wall" (Leitgeb, 1884) or the "special wall" (Beer, 1911).

In some members of the Metzgeriales and Jungermanniales striking differences in cell wall formation seem to occur at this stage. Suire (1970) found that young SMCs of *Pellia epiphylla* (Metzgeriales) lack any well-defined cell wall but lie in a gelatinous spore sac fluid. This is in agreement with the findings of Heckmann (1970) in *Pallavicinia lyelli* (Metzgeriales) and *Cephalozia connivens* (Jungermanniales). When the quadrilobed stage is established, the first wall of the SMC has already been laid down (Kelley and Doyle, 1975).

In the spermatophytes this "special wall" is of callosic nature (Beer, 1911; Currier, 1957; Waterkeyn and Bienfait, 1970). This is also the case in the Sphaerocarpales (Siler, 1934; Denizot, 1971c; Neidhart, 1978b) and the Marchantiales (Beer, 1906; Denizot, 1971c) as well as in the Filicales and in the Selaginellales (Bienfait and Waterkeyn, 1976). Callose is absent in the SMCs of the Lycopodiales and the eusporangiate ferns (Bienfait and Waterkeyn, 1976). Similarly the SMCs of the Metzgeriales (Horner *et al.*, 1966), the Jungermanniales (Oltmann, 1974) and mosses do not show any aniline blue reaction. These facts may indicate that the presence or absence of callose in the SMCs should be taken into account in phylogenetic speculation on possible relationships between bryophytes and ferns.

Fig. 1. SMC of *Funaria hygrometrica* in the pre-meiotic stage. The SMC wall is composed of a mucous envelope (ME) and a two layered outer wall (SCW). Shortly after liberation of the SMC from the archesporial tissue callose is only found in the outer part of the SCW. Arrows indicate cytoplasmic protrusions. Vacuoles (V) contain polysaccharides. N=nucleus. GA, Os, UB, Th. ×7500.

Fig. 2. Detail of cytoplasmic protrusions. Arrows indicate some particles on the plasmalemma surface. They are in contact with fibrils of the mucous envelope. Golgi vesicles (GV) also contain polysaccharides. GA, Os, UB, Pb. ×50 000.

Fig. 3. Detail of Fig. 1 showing an oblique view near the plasmalemma surface. Arrows indicate particles corresponding to those in Fig. 2. GA, Os, UB, Th. ×59 500.

Fig. 4. Freeze-etched SMC of *F. hygrometrica*. Outer surface of a smooth plasmalemma region with particles corresponding to those in Figs 2 and 3. GA, RR, 25% glycerol. ×80 000.

Callosic substances are also to be found in the Musci, but they are secreted much earlier, at the end of the proliferation phase of the archesporial tissue. This is why callose is found near the archesporial walls (Genevès, 1971a; Neidhart, 1975). Outside the SMC traces of callose can be found, but only immediately after liberation of the SMCs into the sporangial fluid (Neidhart, 1975). After meiosis moss SMCs are completely without callosic substances (Neidhart, 1975; Bienfait and Waterkeyn, 1976). Thus, it becomes of considerable interest to determine the chemical nature of the different layers in the SMC wall and which organelles take part in their formation.

The fibrils of the SMC wall stain with Thiéry's (1967) method (Genevès, 1971a, b, 1972a, 1974; Neidhart, 1975; Figs 1, 3) with KMnO$_4$ (Genevès, 1974), with phosphotungstic acid at pH 3·0 (Genevès, 1974), with lead salts (Fig. 3) and are affected by proteolytic solutions (Genevès, 1974). Genevès (1974) concluded that the fibrils consist of 1,4-linked polysaccharides. The fibrils are embedded in a matrix, which does not stain with the above methods but gives good contrast when ruthenium red or alcian blue are added to the fixative (Neidhart, 1975). This indicates the presence of pectic substances (Luft, 1964). When toluidine blue is used as a stain, metachromatic effects indicate the presence of acid mucopolysaccharides (Bienfait and Waterkeyn, 1976).

Similar investigations at the ultrastructural level are lacking for the Hepaticae where histochemistry has so far only been studied with the light microscope (Horner et al., 1966).

The production of the fibril precursors is connected with the dictyosomes, cytoplasmic vesicles, nucleoplasmic regions and to the plasmalemma (Genevès, 1971a, b, 1974). It is now established that fibrils are formed in connection with particles on the outer surface of the plasmalemma (Figs 2–4) in line with Preston's (1964) hypothesis. That polysaccharide fibrils are formed by enzyme particles on the outer surface of the plasmalemma has been demonstrated in various algae (Staehelin, 1966; Robinson and Preston, 1972) and higher plants (Preston and Goodman, 1968; Robards, 1969; Willison, 1976).

3. Plastids and Other Cell Constituents

During SMC formation in the Musci the number of plastids is reduced to one (Weier, 1931; Paolillo, 1964, 1969; Eymé and Suire, 1969; 1971, Neidhart, 1975). This plastid is dish-like in shape and encloses the nucleus. Although the diameter of the "plastid-dish" can be 8–12 μm, cross-sections reveal only narrow profiles, measuring 0·2–0·5 μm (Paolillo, 1969; Eymé and Suire, 1971; Neidhart, 1975). The number of thylakoids can be reduced to one and starch is rare. Phytoferritin inclusions are sometimes present (Neidhart, 1975). This plastid type is the so-called "plaste foliacé" of French authors (Eymé and Suire, 1969, 1971).

Outside the Musci, only the Anthocerotales have a similarly reduced number of plastids in the SMC (Davis, 1899). In the Musci and the Anthocerotales this single plastid divides into four parts during the premeiotic phase. The daughter plastids separate and are distributed to the four meiospores (Sapehin, 1911; Weier, 1931; Paolillo, 1969). The Hepaticae have a different plastome development (Table I).

In the Jungermanniales (Oltmann, 1974), the Metzgeriales (Horner *et al.*, 1966; Suire, 1970) and the Sphaerocarpales (Kelley and Doyle, 1975; Neidhart 1978b) the SMCs contain several plastids and the drawings of Beer (1906)

TABLE I. Morphology, callose content, plastid number and type in the SMCs of the bryophytes

	Morphology	Callose	Plastid number	Plastid type
Musci	Spherical	Absent	One	"Plaste-foliacé"
Anthocerotales	Spherical	Present	One	Amyloplast
Sphaerocarpales	Spherical	Present	Several	Amyloplasts
Marchantiales	Spherical	Present	Several	?
Metzgeriales	Quadrilobed	Absent	Several	Chloroplasts Amylochloroplasts
Jungermanniales	Quadrilobed	Absent	Several	Etioplasts Proplastids

clearly indicate the same for the Marchantiales. In the pre-meiotic SMC and the newly formed tetrad of the Sphaerocarpales, the number of plastids seems to be constant, which means that before and during meiosis no plastid divisions take place (Neidhart, 1978b). The plastid number subsequently increases during the development of the SMCs (Suire, 1970).

There is considerable morphological variation between the plastids of different groups (Table I). The SMCs of the Sphaerocarpales (Neidhart, 1978b) and of the Anthocerotales (Davis, 1899) contain typical amyloplasts. In the Jungermanniales etioplasts (Oltmann, 1974) or proplastids (e.g. in *Frullania*, Neidhart, unpublished data) are common. Typical chloroplasts (Horner *et al.*, 1966) or amylochloroplasts (Suire, 1970) are known from the Metzgeriales.

In thylakoid formation the similarity between the plastids of the Hepaticae and the "plaste foliacé" of the mosses seems to be greater. In the Musci (Paolillo, 1964, 1969; Eymé and Suire, 1969, 1971) and in the Hepaticae (Oltmann, 1974; see Fig. 24) thylakoids are formed in association with "plastid centres" or "prolamellar bodies". But the thylakoids can also be formed by the inner membrane of the plastid envelope (Gullvåg, 1966–1967; Neidhart, 1975). This means that both mechanisms of thylakoid formation known in higher plants are also present in bryophytes.

Data on other cell constituents such as mitochondria, endoplasmic reticulum, vacuoles and the Golgi system are as yet insufficient for them to be used taxonomically; the morphology of these organelles is more likely to be related to their function than to taxonomy.

SPORE WALL FORMATION

1. Musci

Although our present knowledge of wall structure of mature spores in the mosses is good there is a striking lack of information on spore wall formation. Leitgeb's (1884) short report provides little experimental data and contains much misinterpretation because the author extrapolates his results on the

---

FIG. 5. Young spore of *Funaria hygrometrica* after degradation of the SMC wall. A mucous envelope (ME) is still present. The exine (Ex) has no ornamentation. Plastids (P) are of the proplastid type. M=mitochondrion, N=nucleus, Lv= lipid vacuole. GA, RR, Os, US, Pb. ×12 000.

FIG. 6. Near the outer surface of a young spore of *F. hygrometrica* within a carbohydrate network, electron-dense substances (arrows) are precipitated which later form the ornamentation of the perine. GA, RR, Os, US, Pb. ×40 000.

FIG. 7. Same stage of development as in Fig. 5. Detail of a spore wall after acetolysis treatment. The outer side of the exine (Ex) is coated by a thin electron-dense layer, which may already be a perine element. Os, UB, Pb. ×20 000.

FIGS 8–12. Sections through a typical spore wall of *Funaria hygrometrica* after various histochemical treatments. Pe=perine, Ex=exine, In=intine. ×51 000.

FIG. 8. GA, RR, Os, UB, Th. The intine (In) stains positively for polysaccharides. Precipitations in the perine (Pe) are not specific; they are also present in the controls.

FIG. 9. Ac+GA, RR, UB, Th–PA. Intine (In) without precipitations. The perine (Pe) is stained as in Fig. 8.

FIG. 10. Ac+GA, RR, UB. An electron-dense coat (arrowed) is revealed on the surface of the ornamentation elements (Pe).

FIG. 11. Os, UB, after acetolysis treatment. Remnants of the electron-dense coating substance (arrowed) are preserved.

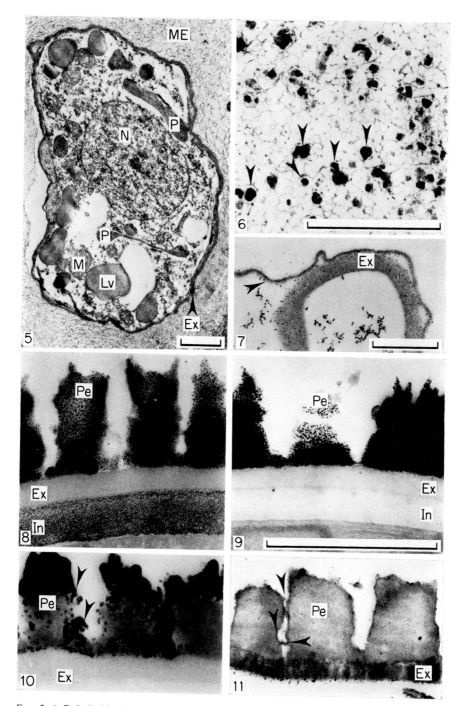

FIGS 5, 6, 7, 8, 9, 10, 11

Hepaticae to the Musci. The formation of spore walls has been investigated in detail with the electron microscope in only three species: *Hypnum rusciformis* (= *Eurhynchium rusciforme*), Hypnales (Genevès, 1972b); *Fissidens limbatus* Sull., Fissidentales (Mueller, 1974) and *Funaria hygrometrica*, Funariales (Jarvis, 1974; Neidhart, 1975).

In these species the exine of the spore wall forms no part of the wall ornamentation. This is quite the reverse of the situation in *Archidium*, *Bruchia* and *Ephemerum* (McClymont and Larson, 1964) where the ornamentation is wholly exinous. The observations that follow are only applicable to mosses with non-exinous ornamentation or to those in which the exine only forms the extreme basal part of the sculpturing elements.

Mature spore walls which have non-exinous ornamentation have also been found in *Polytrichum*, *Encalypta*, *Physcomitrium* and *Phascum* (McClymont and Larson, 1964), *Ceratodon* (Valanne, 1971), *Dicranum* (Valanne *et al.*, 1976), *Bryum* (Nurit, 1974), *Eurhynchium* (Neidhart, unpublished data), *Grimmia* (Fig. 21), *Ulota* (Fig. 23) and *Hylocomium* (Gullvåg, 1966–1967). This seems to be the most usual type of spore wall in the mosses.

Genevès (1972b) reported that directly after SMC division, the young tetraspores are surrounded by a carbohydrate envelope. The tetraspores then separate and their fibrillar component is progressively degraded. Finally they lie at the periphery of the old SMC wall and have an ellipsoid outline. In this phase the first layer of the spore wall becomes visible (Genevès, 1972b). Looking closely at his pictures one can see that this layer is identical with the inner part of the exospore and so has to be called exine. In *Fissidens* (Mueller, 1974) and *Funaria* (Jarvis, 1974; Neidhart, 1975; Figs 5, 7) the exine is the layer of the spore wall which is formed first. After exine deposition has started the wall of the SMC is fully degraded, and the spores are free within the spore sac (Genevès, 1972b; Mueller, 1974; Neidhart, 1975). Even at this stage, no evidence of wall ornamentation can be traced (Jarvis, 1974; Neidhart, 1975).

In the spore wall of young tetraspores of *Desmatodon* and *Pottia* (Ripetsky and Matasov, 1975) and of *Funaria* (Niedhart, 1975) there is yellow–green fluorescence. The colour of this fluorescence is identical with that of the exine elements of *Anthoceros* (Ridgway and Larson, 1966) and of higher plants (Shellhorn *et al.*, 1964). This is further evidence that the exine of bryophytes is comparable with that of higher plants.

Spores of *Hypnum* reach their final size within the old SMC wall (Genevès, 1972b). In contrast, after being liberated from the tetrad, the spores of *Funaria* reach their final size as a result of the formation of large vacuoles (Neidhart, 1975). The expansion of spores during this phase has also been described for

Fissidens (Mueller, 1974). The exine seems to have a certain plasticity which allows the spore to expand. The conclusion from this is that the spore walls become sculptured after the spores reach their final size. However, this may be different in *Funaria "mediterranea"* where the ornamentation is restricted to small areas (Figs 18, 19). (*F. mediterranea* (L.) Lindb. is regarded as a "nomen illegitimum". The figured spore could belong to a subspecies of *F. mühlenbergii*.)

There is no indication that it is the spore protoplast itself which produces the ornamentation or provides material for its formation (Mueller, 1974). Mueller (1974), Jarvis (1974) and Neidhart (1975) agree that this material derives solely from external sources. Thus, as in most pteridophytes (Leitgeb, 1884; Pettitt, 1971; Lugardon, 1976), the outermost part of the sporoderm of the mosses is a true "perine".

In *Funaria*, the deposition of the perine begins with an acetolysis-resistant basal layer (Fig. 7). *Hypnum* is comparable in this (Genevès, 1972b) but, when stained, its perine is less electron dense. Figure 6 shows that the perine elements of *Funaria* are set in a network of carbohydrate.

Although both perine and exine are resistant to acetolysis, histochemical studies show that in many ways they are quite different (Figs 8–11). In non-acetolysed spores the perine, unlike the exine, reacts with silver proteinate after periodic acid treatment coupled with thiosemicarbazide (Thiéry, 1967). The same happens in the control with thiosemicarbazide and silver proteinate (Figs 8, 9). The outer parts of the perine, again unlike the exine, are electron-opaque after uranyl acetate staining (Fig. 10). After acetolysis, however, both perine and nexine show elements of a unit membrane structure and the two layers become more similar in their staining properties (Fig. 11). The electron density of the non-acetolysed exine in mosses is caused only by its osmophilic properties (compare Fig. 8 with Figs 9, 10), and the same seems to be true for hepatics. These histochemical differences between nexine and perine add further support to the hypothesis of an external origin of the perine.

Spore wall sculpturing seems to be specific for some members of the genus *Funaria* (Figs 12–19). In *F. arctica* (Figs 14, 15) there are small, regular ornamentations of equal distribution (Fig. 14) or arranged in rows (Fig. 15). Both types are common in the spores of our laboratory culture of *F. hygrometrica* (Fig. 12). Here the ornamentation elements seem to be slightly thicker (compare Figs 13–15). In *F. convexa* the outline of the elements is very irregular (Fig. 16). Smaller elements seem to cluster to form the larger elements. Often large areas of the spore wall lack any ornamentation (Fig. 17); this tendency is more pronounced in *F. "mediterranea"* (Figs 19, 20) where the sporoderm is

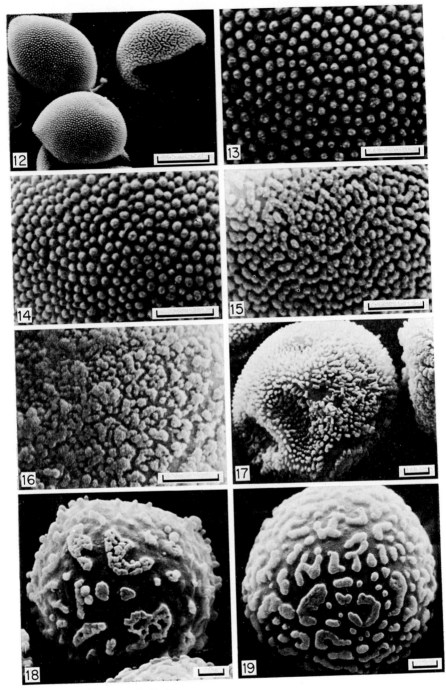

Figs 12, 13, 14, 15, 16, 17, 18, 19

sculptured with large plaques of irregular outline (Fig. 20) which are formed by the fusion of smaller elements (Fig. 19) as in *F. convexa*. It seems that in the species with a more southerly geographical distribution, the spore wall structures are larger and more irregular.

How is the development of specific patterns controlled? Is it a temperature effect? Spores from sporophytes cultivated in summer under relatively high temperatures have more irregularly ornamented walls (Neidhart, unpublished data). Could it be that the pattern is predetermined in the exine by the formation of special perine binding sites? This seems unlikely as the first layer of perine material is a thin uniform layer. Preliminary results suggest that the polysaccharide network in the spore sac fluid (Jarvis, 1974; Neidhart, 1975) might be responsible for the patterning. This has still to be confirmed.

In those genera such as *Archidium*, *Bruchia* and *Ephemerum* (McClymont and Larson, 1964), where the exine forms the ornamentation we must look for differences of spore wall formation. The question of the stage of exine development at which these spores leave the SMC wall seems to be of special interest. Only after this time is deposition of perine material possible; the mould for the sculpturing of the exine would otherwise be lost.

2. Anthocerotales

Although *Anthoceros* is the most suitable genus for the study of spore development since all stages of development are present in one sporangium, only the

---

FIGS 12–19. Scanning micrographs of the spore wall of different *Funaria* species.

FIG. 12. *F. hygrometrica* spores from greenhouse culture, Hannover. The finely granulated sculpturing (two spores on the left) is typical. In some spores the perine elements are arranged in rows (upper right spore). ×1570. Scale indicates 10 μm.

FIG. 13. Detail of the finely granulate spore wall. ×8700. Scale indicates 2 μm.

FIG. 14. *F. arctica* (Berggr.) Kindb. from West Spitzbergen (Isfjorden) with a finely granulate spore wall, as in *F. hygrometrica*. ×8700. Scale indicates 2 μm.

FIG. 15. Spore from the same capsule as in Fig. 14. The ornamentation elements tend to be arranged in rows. ×8700. Scale indicates 2 μm.

FIG. 16. *F. convexa* Spruce. from La Chiappa, South Corsica. Ornamentation elements are irregular in outline. Normal type. ×8700. Scale indicates 2 μm.

FIG. 17. A spore from the same collection as in Fig. 16. Small areas are free of ornamentation. ×3475.

FIG. 18. *F. "mediterranea"* from Sicily. Large plaques are formed by the fusion of smaller elements. Large areas are free of ornamentation. ×3475.

FIG. 19. *F. "mediterranea"* from Foggia, Italy. Subunits are not visible in the plaques. ×3745. Scale indicates 2 μm as in Figs 17 and 18.

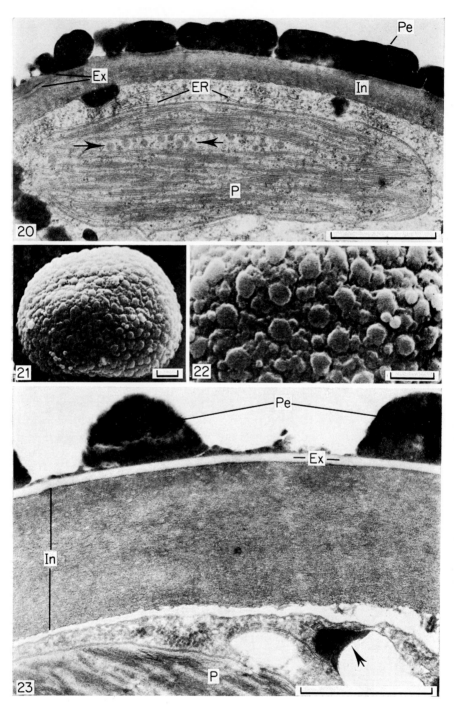

Figs 20, 21, 22, 23

works of Leitgeb (1884) and Ridgway and Larson (1966) deal with spore wall formation. Ridgway's thesis (1965) was not available for the present account, it would no doubt supplement what follows.

The sporoderm and the different types of ornamentation processes, the "gekörnte Außenhaut" in the sense of Leitgeb (1884), are formed within the SMC wall. We can therefore conclude that the exospore of *Anthoceros*, like that in some other Hepaticae, is a typical exine. The "baculum-like or murus-like processes" (Erdtman, 1965) are considered to be of sexinous nature. These are borne on a more or less undulating "nexinous" surface, as Erdtman (1965) pointed out. After pre-treatment with chromic acid, Leitgeb (1884) succeeded in staining the "gekörnte Außenhaut" blue with ClZnI. The sexinous nature of this layer, which may be preformed by carbohydrates as in the spermatophytes (Dickinson and Heslop-Harrison, 1968; Heslop-Harrison, 1968; Dickinson, 1970, 1976), some Jungermanniales (Heckmann, 1970; Oltmann, 1974) and one member of the Metzgeriales (Horner *et al.*, 1966), is supported by this. Later, a "cutininzation" (Leitgeb, 1884), i.e. impregnation with sporopollenin, occurs. Sporopollenin-coated lamellae of unit membrane dimension (the so called "ensembles élémentaires" of Denizot, 1971a, b, 1974, 1976) were not found in this layer (Heckman, 1970) although they are common in the Jungermanniales and Metzgeriales. The intine is formed after the exospore.

3. Marchantiales and Sphaerocarpales

The formation and structure of the spore wall has been better investigated in these two groups than in other bryophytes (Denizot, 1971a, b, c, 1974, 1976). In principle, spore wall formation seems to be similar in both groups although the morphological variation between, for example, *Targionia* (Zigliara, 1971–

FIG. 20. *Grimmia orbicularis*, from Chateauneuf-sur-Charente, France. The perine (Pe) forms large plaques or plates. The exine (Ex) is thin and two-layered. The plastids (P) are typical chloroplasts with stacked regions and osmiophilic globules (arrows). Endoplasmatic reticulum (ER) is present. In=intine. GA, Os, UB, Pb. ×32 000.

FIGS 21–23. *Ulota bruchii*.

FIG. 21. Typical aspect of a spore. ×2375. Scale indicates 2 μm.

FIG. 22. Detail of the spore wall. ×6600. Scale indicates 2 μm.

FIG. 23. Thin section through the spore wall with two perine elements (Pe) connected by a basal layer. Ex=exine. The intine (In) is not fully formed. The lipid reserves become degraded (arrow). The plastids (P) are of the chloroplast type. GA, Os, UB, Pb. ×39 400.

1972; Denizot, 1974), *Sphaerocarpos* (Doyle, 1975) and *Corsinia* (Erdtman, 1957) is great.

The exospore is a typical exine; it is formed within the special wall and no further sculpturing elements are added from the outside (Denizot, 1976; Neidhart, 1978b). The main aspects are summarized by Denizot (1976):

> The mode of spore wall formation in this group is different from the general mode of spore wall formation in other groups of plants. It differs from the pteridophytes as no perine, in the actual sense of this term, is formed. By its totally centripetal formation it differs from the pollen of angiosperms. The exine of the Marchantiales and Sphaerocarpales resembles the endexine (nexine) of the angiosperm pollen regarding its morphology, the appearance of elements of unit membrane dimensions (the so called "ensembles exiniques élémentaires") and their insolubility in amino-2-ethanol.

The exine can be two-layered as in *Targionia, Marchantia, Conocephalum* and *Lunularia* (Denizot, 1971a, b, c), mainly three-layered as in *Riella* (Neidhart, 1978b) or multilayered as in *Sphaerocarpos* (Siler, 1934) and *Riccia* (Denizot, 1971a, b, c). The different exine layers are produced in contact with the spore protoplast (Leitgeb, 1884; Beer, 1906; Siler, 1934; Denizot, 1976; Neidhart, 1978b). The sculpturing of the spore wall is mostly related to the outermost exine layer (the exine 1 of Denizot, 1971a, b, c, 1974, 1976) although in *Riella*, the exine 2 can also participate. The sculpturing elements are not pre-formed by a carbohydrate material but are deposited directly onto the surface of the spore protoplast (Neidhart, 1978b).

Granular or fibrillar materials which later become impregnated with an electron-dense substance lie between the exine layers (Denizot, 1971a, b, c, 1974, 1976; Neidhart, 1978b). The innermost layer of the exine prevents impregnation from the endospore (Neidhart, 1978b). The electron dense material must therefore come from outside the spore.

As in the mosses and the Anthocerotales, the intine is formed last (Leitgeb, 1884; Siler, 1934; Neidhart, 1978b). The only contradictory report is by Wiermann and Weinert (1969), who stated that spore wall formation in *Corsinia* begins with the endospore. Their results therefore need careful discussion. Leitgeb's (1884) report on *Corsinia* agrees with the description given above, although he uses the inappropriate term "perine" for the outermost sporoderm layer. On the distal pole of the spore this layer forms the channelled surface. In this part Leitgeb (1884) already differentiated three regions of the spore wall: a so-called "Außenschicht", which is very thin (perhaps identical with layer A in Fig. 1 of Wiermann and Weinert, 1969), a middle layer, which later becomes granular (comparable to layer B in Fig. 3 of Weirmann and Weinert, 1969) and a very thin inner layer of lamellae (identical with layer L in Fig. 3 of Wiermann and Weinert, 1969, the so-called heavy staining lamellae). Inside these spore wall structures is a layer which Leitgeb (1884) calls "exine" and Erdtman (1965) calls "nexinous level". In my opinion the same layer is shown in Fig. 3 of Weirmann and Weinert (1969), who call it "layer A". This is the cause of the confusion. Wiermann and Weinert (1969) used the same term "layer A" for two ontogenetically different spore wall structures: (1) for the so-called "Außenschicht" and (2) for the innermost exine layer. The true intine is not shown in their pictures and is not identical with one of the "A-layers", especially that of Fig. 1. Leitgeb's careful histochemical investigations clearly indicate that the intine is the last element of the spore wall to be formed.

If this is true, it follows that in *Corsinia*, formation of the sporoderm is by the typical method of the Marchantiales and Sphaerocarpales. There seems to be only one difference: the lack of structures comparable to the "ensembles exiniques élémentaries". We must conclude that in *Corsinia* the spore wall is formed centripetally.

4. Metzgeriales and Jungermanniales

There are excellent reports on spore wall formation on *Riccardia* (Horner et al., 1966) and on *Lophocolea* (Oltmann, 1974). These clearly indicate that there is a certain concordance between the Metzgeriales and the Jungermanniales because, as in the spermatophytes, a typical primexine (sexine) is formed (Figs 24, 27). The older reports on *Riccardia* (Clapp, 1912), *Fossombronia* (Humphrey, 1906) and *Pallavicinia* (Moore, 1905) indicate that the spore wall is formed within the SMC wall and thus is a typical exine. Normally no perine-like elements are formed outside the exine as they are in mosses and most ferns (c.f. Lugardon, 1976). However, in *Diplophyllum*, *Cephalozia* and *Nowellia* Heckman (1970) reports granules of sporopollenin nature on the outer surface of the spore wall. That they are perine is still not proved because ontogenetic information is lacking.

An exception can be noted in the multicellular spores of *Frullania*. Their outer wall layers, including the nexine, are described as frequently broken (Heckman, 1970). One reason for this is the expansion of the spores in the ripening period (Neidhart, unpublished data). Only the projections within the rosettes (Figs 26, 27) seem to be truly sexinous in nature. When the rosettes begin to develop they form a compact mass sunk into the underlying intine

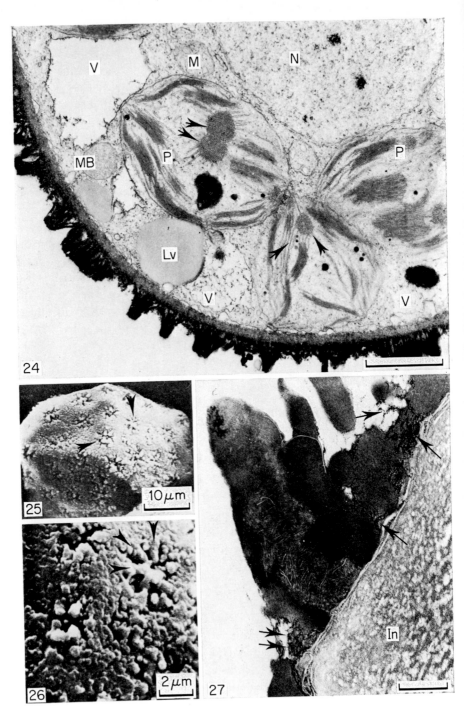

FIGS 24, 25, 26, 27

and into the spore protoplast. Later, some material seems to be dissolved away and as the spore expands, the rosettes appear. Between the typical sexinous elements (with their "ensembles exiniques élémentaires" in younger stages) there are deposits of a different chemical nature (Fig. 27). Similar material also covers some parts of the smooth regions of the spore wall (Neidhart, unpublished data). Heckman (1970) further stated that the outer surface of the spore wall is covered by non-sporonine material. Are these structures perine or are they comparable to the "Pollenkitt" of some spermatophyte pollen? This question has still to be answered.

The primexine is preformed by carbohydrates and later impregnated with substances of a sporopollenin nature. Sporopollenin is deposited on "platelets" (Horner et al., 1966) or "lamellae" (Rowley and Southworth, 1967; Oltmann, 1974) of unit membrane dimension; Heckman (1970) has called corresponding structures "slip-lamellae". Similar structures have not been described for the Anthocerotales, Marchantiales and Sphaerocarpales (Heckman, 1970) or for the mosses (Neidhart, 1975). During subsequent ripening of the spore, the spaces between the lamellae can become electron dense (Fig. 27) or may appear granular by additional deposition of sporopollenin (Heckman, 1970). This might be the result of impregnation with brown pigments.

The primexine (sexine) is normally underlain by a layer of one or several "laminae" (Horner et al., 1966; Suire, 1970; Fig. 27) which Heckman (1970) calls "sheet lamellae". They could be identical with the so-called "ensembles exiniques élémentaires" of Denizot (1971, 1974, 1976). This layer is called nexine (Horner et al., 1966; Suire, 1970). Oltmann (1974) states that a nexine is also present in *Lophocolea*, but Heckman (1970) did not find one (compare also Fig. 24).

FIG. 24. Detail of a nearly-ripe spore of *Lophocolea heterophylla*. The spore wall is two-layered. The inner, smooth layer is the intine. The outer, sculptured layer is the exine (of sexinous nature) in which slip-lamellae are visible. The plastids (P) have prolamellar bodies from which the thylakoids develop (arrowed). Mitochondria (M) and microbodies (MB) are present. The vacuoles (V) contain phenolic compounds. Lv=lipid reserves, N=nucleus. GA, Os, UB, Pb. ×23 500.

FIG. 25. A typical spore of *Frullania dilatata*. Arrows indicate the typical rosettes. ×1350.

FIG. 26. Detail of Fig. 25, arrows, indicate three sexine processes. ×4040.

FIG. 27. Median section through a process of the spore wall of *F. dilatata*. The intine (In) is bordered by some sheet lamellae. An electron-dense substance (arrowed) of unknown nature is precipitated between the sexine elements. GA, Os, UB, Pb. ×14 400.

The last sporoderm layer, again formed by the spore protoplast, is the intine as in all other mosses and liverworts.

Maturation phenomena are related to the following physiological processes and morphological changes.

1. Germination

Freshly isolated young spores of *Riella* (Tenge, 1959) and *Funaria* (Neidhart, 1978a) need a longer time to germinate than fully ripe spores which have been allowed to dry out. In *Polytrichum*, the mortality of young pale spores on mineral agar is higher than that of ripe spores. Sucrose can facilitate germination of such young spores but it increases mortality in green, wet spores (Paolillo and Kass, 1973). The ability of spores to germinate is therefore developed during the after-ripening period. In *Funaria* after-ripening is correlated with the development of plastids, microbodies and enzymes. Degradation of endogenous inhibitors may also be important (Neidhart, 1978a).

2. The Ability to Withstand Environmental Stress

This ability is of great importance for the distribution of a species, especially with regard to intercontinental distribution by air streaming (van Zanten, 1976). The development of temperature resistance in *Funaria* seems mainly to be correlated with the state of desiccation. Low temperatures between $-20°C$ and $-195°C$ can be survived in a relatively hydrated stage (Neidhart, 1975), but in other cases desiccation can itself act as a stress.

When complete sporophytes are used, young spores of *Riella* (Tenge, 1959) and of *Funaria* can overcome desiccation (Neidhart, 1975), but isolated young spores of *Funaria* lose their viability during desiccation (Neidhart, unpublished data). Young *Riella* spores germinate more rapidly after desiccation than when they are freshly isolated (Tenge, 1959).

At the moment we can only speculate on the reasons for this behaviour. Freeze-etching experiments suggest that desiccation effects the organization of the cytoplasmatic membranes, especially the plasmalemma (Neidhart, 1978c). Stress resistance also seems to be affected by the content and the size of vacuoles within a spore, especially when they contain phenolic compounds (Neidhart, 1975). As phenols are also responsible for the pigmentation of the spore wall, their possible role in spore wall formation should not be over-looked.

3. Organelle Development and Growth

Most of the ripe spores of the Musci and the Hepaticae contain green plastids with a well-developed thylakoid system (Figs 20, 23, 24, 29 and Table II). Exceptions are *Sphagnum* (Kofler and Chevallier, 1973), *Dawsonia* (Stetler and DeMaggio, 1976), *Funaria* (Schulz, 1971, 1972; Nurit, 1974; Neidhart, 1975) and *Riella* (Neidhart, 1978b) where the spores contain proplastids. In *Funaria* typical chloroplasts are developed during sporogenesis, and these are later reduced to proplastids (Nurit, 1974; Neidhart, 1975). In *Riella* the proplastids develop directly from the amyloplasts of the SMCs (Neidhart, 1978b).

In several genera such as *Polytrichum* (Weier, 1931; Paolillo and Kass, 1973, 1977), *Funaria* (Neidhart, 1975, 1978a), *Lophocolea* (Neidhart, unpublished data) and *Frullania* (Figs 28, 29) plastid numbers change during sporogenesis. In *Funaria* plastids divide, as they are reduced to proplastids (Neidhart, 1975). In *Polytrichum* plastid division is further correlated with an expansion of the spore (Paolillo and Kass, 1973). The division of chloroplasts is affected by light and dark treatment, and by different inhibitors. Paolillo and Kass (1977) concluded that within a spore there must be two types of chloroplast; phase-I chloroplasts need light to convert them into phase-II chloroplasts. They are then able to divide in the dark and to become phase-I chloroplasts again.

This tendency for plastid division, and even cellular growth, during the sporogenesis of unicellular spores can also shed light on the origin of multi-cellular spores. Multicellular spores are present in various genera of the Hepaticae and the Musci (Clarke, this volume, Chapter 11). In *Frullania* the number of plastids has already increased before new cell walls are formed (Figs 28, 29). In principle, this behaviour can be compared with the plastid divisions in unicellular spores, but there is a difference in the extent of the divisions. It may be that in multicellular spores this development is not interrupted by desiccation or other inhibitory phenomena. In *Funaria* (Neidhart, 1978a) and *Polytrichum* (Paolillo and Kass, 1973) desiccation is considered one reason for the cessation of metabolic activities.

In *Frullania*, microbodies develop before the spores become multicellular. Microbodies are a common organelle in spores of many mosses and liverworts (Table II) but have not been described for the protein-storing spores of *Dawsonia* and *Dicranum*, and are certainly absent in spores of *Riella* (Table II). During germination, microbodies in the fat-storing spores of the mosses function as

TABLE II. Presence of chloroplasts and microbodies in ripe spores of some mosses and liverworts

Genus	Chloroplasts		Microbodies	
Sphagnum	(−)	Kofler and Chevallier, 1973	(?)	
Dawsonia	(−)	Stetler and DeMaggio, 1976	(?)	
Atrichum	(+)	Neidhart, unpublished	(?)	
Polytrichum	(+)	Paolillo, 1969	(+)	Karunen, 1972
		Karunen, 1972		
Fissidens	(+)	Mueller, 1974	(−)	Mueller, 1974
Ceratodon	(+)	Valanne, 1971	(+)	Valanne, 1971
Dicranum	(+)	Valanne *et al.*, 1976	(−)	Valanne *et al.*, 1976
Tortula	(+)	Kofler *et al.*, 1970	(?)	
		Neidhart, unpublished		
Desmatodon	(+)	Ripetsky and Matasov, 1975	(?)	
Phascum	(+)	Ripetsky and Matasov, 1975	(?)	
Grimmia	(+)	Neidhart, unpublished	(+)	Neidhart, unpublished
Funaria	(−)	Kofler *et al.*, 1970	(+)	Monroe, 1968
		Schulz, 1971, 1972		Schulz, 1971, 1972
		Nurit, 1974		Neidhart, 1975, 1978a
		Neidhart, 1975, 1978a		
Bryum	(+)	Kofler *et al.*, 1970	(?)	
		Nurit, 1974		
Ulota	(+)	Neidhart, unpublished	(+)	Neidhart, unpublished
Thuidium	(+)	Neidhart, unpublished	(?)	
Eurhynchium	(+)	Neidhart, unpublished	(+)	Neidhart, unpublished
Hypnum	(+)	Neidhart, unpublished	(?)	
Hylocomium	(+)	Gulvåg, 1966–1967	(+)	Gullvåg, 1966–1967
Riella	(−)	Neidhart, 1978b	(−)	Neidhart, 1978b
Riccardia	(+)	Horner *et al.*, 1966	(?)	
Pellia	(+)	Suire, 1970	(?)	
Lophocolea	(+)	Oltmann, 1974	(+)	Neidhart, unpublished
		Neidhart, unpublished		
Radula	(+)	Suire, 1970	(?)	
Frullania	(+)	Neidhart, unpublished	(+)	Neidhart, unpublished

glyoxysomes (Schulz, 1971; Karunen, 1972; Neidhart, 1978a) and provide material for cell wall synthesis. It is possible that in unicellular spores this glyoxysomal function is blocked, possibly by desiccation. In the multicellular spores of *Frullania*, however, this function develops to a much greater extent. Because of this, and because the endoplasmic reticulum (Suire, 1970) and the Golgi apparatus stay active in multicellular spores, further cell wall synthesis is possible. In unicellular spores (e.g. *Funaria*) however, the Golgi apparatus is reduced to an inactive state (Schulz, 1971, 1972; Neidhart, 1975).

Thus, most unicellular spores do not differ from multicellular spores in their organelles. *Riella* spores are exceptional in that the ion concentration of the environment seems to inhibit metabolic activities at a very early stage so that no microbodies are developed. Spores may become inactive and stay unicellular for several reasons such as the development of endogenous inhibitors, the senescence of the sporophyte tissues or the influence of environmental conditions like drought or special ionic conditions. Eventually, multicellular spores also reach a quiescent state and do not continue to germinate within the sporangium.

CONCLUDING REMARKS

Assuming that the differences mentioned above have a genetic basis, one must conclude that the bryophytes are a very heterogeneous group of plants in their SMC morphology and in the formation of their spore walls.

The Musci seem to form a relatively homogeneous group because of their spherical SMCs (except in the genus *Polytrichum*), the absence of callose in the special wall, the so-called "plaste foliacé", the reduction of plastids to one before meiosis and the formation of a typical perine which is underlain by nexine.

The Anthocerotales have a mixture of characters which they share with other groups. Reduction of the number of plastids to one before meiosis, and the spherical SMCs are found in Musci; the spherical SMCs, presence of callose in the special wall, and amyloplasts in the Marchantiales and Sphaerocarpales, and formation of a typical sexine underlain by nexine and intine is formed in the Metzgeriales and Jungermanniales.

The Marchantiales and Sphaerocarpales are closely allied because of their spherical SMCs, the presence of callose in the special wall, the SMCs with several plastids of the amyloplast type and the unique mode of spore wall formation with a typical exine of nexinous nature.

The Metzgeriales and Jungermanniales seem to have close affinities because of their quadrilobed SMCs, the absence of callose in the special wall, the SMCs with several plastids of different types and the formation of a typical sexine and nexine (ultimately lacking in *Lophocolea*). Since experimental data available for this group are scanty, the above generalizations must be regarded as preliminary. We urgently need histochemical investigations of the formation of the SMC wall and the spore wall in other genera.

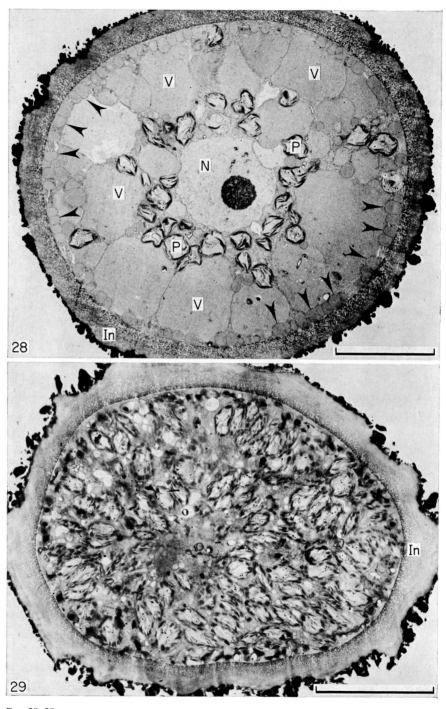

Figs 28, 29

If we look amongst the bryophytes for a possible ancestor of the spermato-phytes it is only the Anthocerotales which fulfil two necessary requirements in the structure of their spores:

(a) The special wall is formed of callose.

(b) The typical spore wall stratification is present, composed of sexine, nexine and intine.

All the other groups of bryophytes are very different in one or other of these respects.

With our present knowledge of spore development, relationships between the bryophytes and the different groups of pteridophytes are difficult to elucidate. There is little detailed information on spore wall ontogeny in the pteridophytes (Pettitt, 1971; Lugardon, 1976) with the result that the homologies of spore wall structures and their correct terminology have not been finally determined (see also Denizot, 1974). With the exception of the Lycopodiales, pteridophytes develop a typical perine; in the bryophytes this spore wall element is only present in the mosses. Moss SMCs do not have callose in their special walls. This is true of the Lycopodiales and the eusporangiate ferns (Bienfait and Waterkeyn, 1976), but not of the Selaginellales. There is a certain similarity between the morphology and the patterning of the lamellated exine of the Lycopodiales (see Lugardon, 1976) and the Marchantiales and Sphaero-carpales. But these hepatics, unlike the Lycopodiales, have callose in their special walls (Bienfait and Waterkeyn, 1976).

Representatives of different taxa are similar in their maturation, so phenomena connected with this process are of little taxonomic use. This may be because the environmental requirements of the process are so closely defined that there is little room for variation.

FIG. 28. Young unicellular spore of *Frullania dilatata*. The intine (In) has reached its final thickness. The outer, electron-dense sexine layer often has gaps. Arrows indicate lipid vacuoles which later become degraded when the plastids (P) increase in number and the cell walls of the multicellular spores are formed. Vacuoles (V) contain phenolic compounds. N=nucleus with a condensed nucleolus. GA, Os, UB, Pb. ×2900. Scale indicates 10 μm.

FIG. 29. Nearly-ripe spore of *F. dilatata*. The plastids have increased in number but new cell walls have still not formed. A nucleus is difficult to detect at this stage. In=intine. GA, Os, UB, Pb. ×3600. Scale indicates 10 μm.

ACKNOWLEDGEMENTS

This work was supported by the Deutsche Forschungsgemeinschaft and the Elektronen-mikroskopische Arbeitsgemeinschaft at the Tierärztliche Hochschule Hannover. Thanks are due to Maria Ullmann for correcting the English text and to J.-P. Frahm for providing *Funaria* species from his herbarium. Thanks are also due to Professor Reale of the Medizinische Hochschule Hannover for permission to use the scanning microscope and to Mr. Konitz of that institute for technical assistance.

REFERENCES

BEER, R. (1906). On the development of the spores of *Riccia glauca*. *Ann. Bot.* **20**, 275–291.
BEER, R. (1911). Studies in spore development. *Ann. Bot.* **25**, 199–214.
BIENFAIT, A. and WATERKEYN, L. (1976). Sur la nature des parois sporocytaires chez les Mousses et chez quelques Ptéridophytes. Etude comparative. *C. r. hebd. Séanc. Acad. Sci. Paris*, sér. D, **282**, 2079–2081.
BLACK, C. A. (1913). The morphology of *Riccia frostii*. *Ann. Bot.* **27**, 511–532.
BLAIR, M. C. (1926). Sporogenesis in *Reboulia hemisphaerica*. *Bot. Gaz.* **81**, 377–400.
CHEVALLIER, D., NURIT, F. and PESEY, H. (1977). Orthoposphate absorption by the sporophyte of *Funaria hygrometrica* during maturation. *Ann. Bot.* **41**, 527–531.
CLAPP, G. L. (1912). The life history of *Aneura pinguis*. *Bot. Gaz.* **54**, 177–193.
CURRIER, H. B. (1957). Callose substance in plant cells. *Am. J. Bot.* **44**, 478–488.
DAVIS, B. M. (1899). The spore-mother-cell of *Anthoceros*. *Bot. Gaz.* **28**, 89–109.
DAVIS, B. M. (1901). Nuclear studies on *Pellia*. *Ann. Bot.* **15**, 147–180.
DENIZOT, J. (1971a). Sur la présence d'ensembles exiniques élémentaires dans le sporoderme de quelques Marchantiales et Sphaerocarpales. *C. r. hebd. Séanc. Acad. Sci. Paris*, sér. D, **272**, 2166–2169.
DENIZOT, J. (1971b). Recherches sur l'origine des ensembles exiniques élémentaires des sporodermes de quelques Marchantiales et Sphaerocarpales. *C. r. hebd. Séanc. Acad. Sci. Paris*, sér. D, **272**, 2305–2308.
DENIZOT, J. (1971c). Recherches sur les formations callosiques au cours de la sporogénèse de quelques Marchantiales et Sphaerocarpales. *C. r. hebd. Séanc. Acad. Sci. Paris*, sér. D, **272**, 2769–2772.
DENIZOT, J. (1974). Genèse des parois sporocytaires et sporales chez *Targionia hypophylla* (Marchantiales). Justification de la terminologie utilisée. *Pollen Spores* **16**, 303–371.
DENIZOT, J. (1976). Remarques sur l'édification des différentes couches de la paroi sporale á exine lamellaire de quelques Marchantiales et Sphaerocarpales. In "The Evolutionary Significance of the Exine" (I. K. Ferguson and J. Muller, eds), pp. 185–210. Academic Press, London and New York.
DICKINSON, H. G. (1970). Ultrastructural aspects of primexine formation in the microspore tetrad of *Lilium longiflorum*. *Cytobiologie* **1**, 437–449.
DICKINSON, H. G. (1976). Common factors in exine deposition. In "The Evolutionary Significance of the Exine" (I. K. Ferguson and J. Muller, eds), pp. 67–90. Academic Press, London and New York.
DICKINSON, H. G. and HESLOP-HARRISON, J. (1968). Common mode of deposition for the sporopollenin of sexine and nexine. *Nature, Lond.* **220**, 927–928.

DOYLE, W. T. (1975). Spores of *Sphaerocarpos donnellii*. *Bryologist* **78**, 80–84.

ERDTMAN, G. (1957). "Pollen and Spore Morphology/Plant Taxonomy. Gymnospermae, Pteridophyta, Bryophyta (Illustrations). An Introduction to Palynology, II." Almqvist and Wiksell, Stockholm.

ERDTMAN, G. (1965). "Pollen and Spore Morphology/Plant Taxonomy. Gymnospermae, Bryophyta (Text). An Introduction to Palynology, III." Almqvist and Wiksell, Stockholm.

EYMÉ, J. and SUIRE, S. (1969). Ultrastructure des cellules sporogènes des Mousses: observations sur le plastidome et le chondriome. *C. r. hebd. Séanc. Acad. Sci. Paris*, sér. D, **268**, 290–293.

EYMÉ, J. and SUIRE, S. (1971). Recherches sur l'évolution ultra-structurale des constituants cellulaires durant la sporogénèse des Mousses. *Botaniste* **54**, 109–151.

FARMER, J. B. (1895). On spore-formation and nuclear division in the Hepaticae. *Ann. Bot.* **9**, 469–523.

GENEVÈS, L. (1971a). Mise en place de polysaccharides membranaires dans le tissu sporogène de l'*Hypnum rusciforme* (Hypnacées) au cours de la phase de prolifération et au début de la différenciation. *C. r. hebd. Séanc. Acad. Sci. Paris*, sér. D, **273**, 723–726.

GENEVÈS, L. (1971b). Sur la présence de polysaccharides dans diverses inclusions du cytoplasme et du noyau, pendant la prophase réductionelle, chez l'*Hypnum rusciforme* (Hypnacées). *C. r. hebd. Séanc. Acad. Sci. Paris*, sér. D, **273**, 2508–2511.

GENEVÈS, L. (1972a). Etude cytochemique de la genèse des parois intersporales pendant la méiose, chez l'*Hypnum rusciforme*. *C. r. hebd. Séanc. Acad. Sci. Paris*, sér. D, **274**, 2875–2878.

GENEVÈS, L. (1972b). Aspects ultrastructuraux de la formation de l'enveloppe des spores chez l'*Hypnum rusciforme*. (Hypnacées). *C. r. hebd. Séanc. Acad. Sci. Paris*, sér. D, **275**, 197–200.

GENEVÈS, L. (1974). Caractères ultrastructuraux et cytochimique de l'enveloppe mucilagineuse des meiocytes d'une Bryophyte: le *Platyhypnidium riparoides*. *Proc. VIII int. Congr. el. Micr. Canberra* **2**, 622–623.

GULLVÅG, B. M. (1966–1967). The fine structure of the spore of *Hylocomium loreum* (Hedw.). *J. Palynol.* **2–3**, 49–65.

GUNNING, B. E. S. and PATE, J. S. (1974). Transfer cells. *In* "Dynamic Aspects of Plant Ultrastructure" (A. W. Robards, ed.) pp. 441–480. McGraw-Hill, Maidenhead, Berkshire.

HECKMAN, C. A. (1970). Spore wall structure in the Jungermanniales. *Grana* **10**, 109–119.

HESLOP-HARRISON, J. (1968). Pollen wall development. *Science, N.Y.*, **161**, 230–237.

HORNER, H. T., Jr, LERSTEN, N. R. and BOWEN, C. C. (1966). Spore development in the liverwort *Riccardia pinguis*. *Am. J. Bot.* **53**, 1048–1064.

HUMPHREY, H. B. (1906). The development of *Fossombronia longiseta* Aust. *Ann. Bot.* **20**, 83–108.

JARVIS, L. R. (1974). Electron microscope observations of spore wall development in *Funaria hygrometrica*. *Proc. VIII int. Congr. el. Micr. Canberra* **2**, 620–621.

KARUNEN, P. (1972). Studies on moss spores, I. The triglycerides of *Polytrichum commune* spores and their mobilization and degradation in relation to the germination phases. *Annls Univ. turku.*, Ser. A, II, **51**, 1–70.

KELLEY, C. B. and DOYLE, W. T. (1975). Differentiation of intracapsular cells in the sporophyte of *Sphaerocarpos donellii. Am. J. Bot.* **62**, 547–559.

KOFLER, L. and CHEVALLIER, D. (1973). Comparison des effects de la DCMU et d'une carence en manganèse sur la germination des spores et la croissance du jeune protonéma de Mousses. *Physiol. Vég.* **11**, 121–136.

KOFLER, L., NURIT, F. and VIAL, A.-M. (1970). Germination des spores de Mousses et évolution de leurs plastes en présence de DCMU. *Mém. Soc. bot. Fr.*, **1970**, Colloq. Cytol. exp., 285–310.

LAMBERT, A. M. (1968). Differenciation des spores et elatères chez quelques Hépatiques. *Bull. Soc. bot. N. Fr.* **21**, 79–86.

LAMBERT, A. M. (1978). La méiose chez les Bryophytes: ultrastructure et dynamique du fuseau chez une Mousse, *Mnium hornum* Hedw. *Bryophyt. Biblthca* **13**, 113–145.

LEITGEB, H. (1884). "Bau und Entwicklung der Sporenhäute und deren Verhalten bei der Keimung." Leuschner und Lubensky, Graz.

LUFT, J. H. (1964). Electron microscopy of cell extraneous coats as revealed by ruthenium red staining. *J. Cell Biol.* **23**(2), 54A–55A. (Abstr.)

LUGARDON, B. (1976). Sur la structure fine de l'exospore dans les divers groupes de Ptéridophytes actuelles (microspores et isospores). *In* "The Evolutionary Significance of the Exine" (I. K. Ferguson and J. Muller, eds) pp. 231–250, Academic Press, London and New York.

McCLYMONT, J. and LARSON, D. A. (1964). An electron-microscopic study of spore wall structure in the Musci. *Am. J. Bot.* **51**, 195–200.

MONROE, J. H. (1968). Light- and electron-microscopic observations on spore germination in *Funaria hygrometrica. Bot. Gaz.* **129**, 247–258.

MOORE, A. C. (1905). Sporogenesis in *Pallavicinia. Bot. Gaz.* **40**, 81–96.

MUELLER, D. M. J. (1974). Spore wall formation and chloroplast development during sporogenesis in the moss *Fissidens limbatus. Am. J. Bot.* **61**, 525–534.

NEIDHART, H. V. (1975). "Elektronenmikroskopische und physiologische Untersuchungen zur Sporogenese von *Funaria hygrometrica* Sibth." Ph.D. Thesis, Technische Universität, Hannover.

NEIDHART, H. V. (1978). Sporogenese bei Archegoniaten: Zur Natur eines Nachreifephänomens bei Sporen von *Funaria hygrometrica* Hedw. (Musci). *Bryophyt. Biblthca* **13**, 169–194.

NEIDHART, H. V. (1978b). Ultrastructural aspects of sporogenesis in *Riella affinis* Howe and Underwood (Hepaticae). *J. Bryol.* **10**, 145–154.

NEIDHART, H. V. (1978c). Ultrastructural changes of the plasmalemma surface during desiccation of *Funaria hygrometrica* spores. *Flora* **167**, 445–450.

NURIT, F. (1974). Ultrastructure et maturation des spores de *Funaria hygrometrica* et *Bryum capillare. Bull. Soc. bot. Fr.* **121**, Colloq. Bryol., 169–177.

NURIT, F. and CHEVALLIER, D. (1978). La sporogenèse chez *Funaria hygrometrica*: étude oxygraphique en rapport avec les modifications ultrastructurales. *Bryophyt. Biblthca* **13**, 195–210.

OLTMANN, O. (1974). Licht- und elektronenmikroskopische Untersuchungen zur Sporogenese von *Lophocolea heterophylla* (Schrad.) Dum., I. Die Entwicklung von der Sporenmutterzelle bis zum Tetradenstadium. *Pollen Spores* **16**, 5–25.

PAOLILLO, D. J., Jr (1964). The plastids of *Polytrichum commune*, I. The capsule at meiosis. *Protoplasma* **58**, 667–680.

PAOLILLO, D. J., Jr (1969). The plastids of *Polytrichum commune*, II. The sporogenous cells. *Cytologia* **34**, 133–144.

PAOLILLO, D. J., Jr and KASS, L. B. (1973). The germinability of immature spores in *Polytrichum. Bryologist* **76**, 163–168.

PAOLILLO, D. J., Jr and KASS, L. B. (1977). The relationship between cell size and chloroplast number in the spores of a moss, *Polytrichum. J. exp. Bot.* **28**, 457–467.

PETTITT, J. M. (1971). Some ultrastructural aspects of sporoderm formation in Pteridophytes. In "Pollen and Spore Morphology/Plant Taxonomy. Pteridophyta. An Introduction to Palynology IV" (G. Erdtman and P. Sorsa, eds), Almqvist and Wiksell, Uppsala.

PRESTON, R. D. (1964). Structural and mechanical aspects of plant cell walls with particular reference to synthesis and growth. In "Formation of Wood in Forest Trees" (M. H. Zimmermann, ed.). Academic Press, New York and London.

PRESTON, R. D. and GOODMAN, R. N. (1968). Structural aspects of cellulose microfibril biosynthesis. *Jl R. microsc. Soc.* **88**, 513–527.

RIDGWAY, J. E. (1965). "Some Aspects of Morphogenesis, Biotic Coaction, and Ultrastructure in the Genus *Anthoceros*." Ph.D. Thesis, Univ. Texas.

RIDGWAY, J. E. and LARSON, D. A. (1966). Sporogenesis in the bryophyte *Anthoceros*: features shown by fluorescence microscopy. *Nature, Lond.* **209**, 1154.

RIPETSKY, R. T. and MATASOV, V. I. (1975). A luminescence-microscopic study of sporogenesis in true mosses. *Tsitol. Genet.* **9**, 307–309.

ROBARDS, A. Q. (1969). Particles associated with developing plant cell walls. *Planta* **88**, 376–379.

ROBINSON, D. G. and PRESTON, R. D. (1972). Plasmalemma structure in relation to microfibril biosynthesis in *Oocystis. Planta* **104**, 234–246.

ROWLEY, J. R. and SOUTHWORTH, D. (1967). Deposition of sporopollenin on lamellae of unit membrane dimensions. *Nature, Lond.* **213**, 703–704.

SAPEHIN, A. (1911). Über das Verhalten der Plastiden im sporogenen Gewebe. *Ber. dt. bot. Ges.* **29**, 491–496.

SCHULZ, D. (1971). "Elektronenmikroskopische Untersuchungen zur Entwicklung des Protonemas von *Funaria hygrometrica*." Ph.D. Thesis, Technische Universität, Hannover.

SCHULZ, D. (1972). Darstellung der submikroskopischen Strukturen lufttrockener Moossporen. *Ber. dt. bot. Ges.* **85**, 193–202.

SHELLHORN, S. J., HULL, H. M. and MARTIN, P. S. (1964). Detection of fresh and fossil pollen with fluorochromes. *Nature, Lond.* **202**, 315–316.

SILER, M. B. (1934). Development of spore walls in the *Sphaerocarpos donellii. Bot. Gaz.* **95**, 563–591.

STAEHELIN, A. (1966). Die Ultrastruktur der Zellwand und des Chloroplasten von *Chlorella. Z. Zellforsch.* **74**, 325–350.

STETLER, D. A. and DEMAGGIO, A. E. (1976). Ultrastructural characteristics of spore germination in the moss *Dawsonia superba. Am. J. Bot.* **63**, 438–442.

SUIRE, C. (1970). Recherches cytologiques sur deux Hépatiques: *Pellia epiphylla* (L.) Corda (Metzgériale) et *Radula complanata* (L.) Dum. (Jungermanniale). Ergastome, sporogénèse et spermatogénèse. *Botaniste* **53**, 125–392.

TENGE, F.-K. (1959). Zur Physiologie der Sporenkeimung von *Riella affinis*. *Z. Bot.* **47**, 287–305.

THIÉRY, J.-P. (1967). Mise en évidence des polysaccharides sur coupes fines en microscopie électronique. *J. Microscopie* **6**, 987–1018.

VAARAMA, A. (1954). Cytological observations on *Pleurozium schreberi*, with special reference to centromere evolution. *Suomal. eläin-ja kasvit. Seur. van. Kasvit. Julk.* **28**(1), 1–59.

VALANNE, N. (1971). The effects of prolonged darkness and light on the fine structure of *Ceratodon purpureus*. *Can J. Bot.* **49**, 547–554.

VALANNE, N., TOIVONEN, S. and SAARINEN, R. (1976). Ultrastructural changes in germinating *Dicranum scoparium*: a moss containing protein storage material. *Bryologist* **79**, 188–198.

WATERKEYN, L. and BIENFAIT, A. (1970). On a possible function of the callosic special wall in *Ipomoea purpurea* (L.) Roth. *Grana.* **10**, 13–20.

WEIER, T. E. (1931). A study of the moss plastid after fixation by mitochondrial osmium and silver techniques, I. The plastid during sporogenesis in *Polytrichum commune*. *Cellule* **40**, 261–289.

WIERMANN, R. and WEINERT, H. (1969). Untersuchungen zur Sporodermentwicklung bei *Corsinia coriandrum* (Sprengel) Lindb. *Ber. dt. bot. Ges.* **82**, 175–182.

WILLISON, J. H. M. (1976). Examination of relationship between freeze-fractured plasmalemma and cell-wall microfibrils. *Protoplasma* **88**, 187–200.

ZANTEN, B. O. van (1976). Preliminary report on germination experiments designed to estimate the survival chances of moss spores during aerial trans-oceanic long-range dispersal in the southern hemisphere. *J. Hattori bot. Lab.* **41**, 133–140.

ZIGLIARA, M. (1971–1972). Le genre *Targionia* L., II. Etude de la spore. *Revue bryol. lichén.* **38**, 241–264.

13 | Anisospory and Sexual Dimorphism in the Musci

HELEN P. RAMSAY

Faculty of Biological Sciences, University of New South Wales, Kensington 2033, N.S.W., Australia

Abstract: In numerous members of the tropical and subtropical genera *Macromitrium* and *Schlotheimia* (Orthotrichaceae) the size of the spores has a bimodal distribution. In contrast to heterosporous vascular plants where the mega- and microspores are borne in different sporangia, in these mosses they occur together in the same sporangium, each tetrad producing two large and two small spores. The term anisospory has been used to describe this situation because of its very different origin from the heterospory of pteridophytes.

In some dioecious species of *Macromitrium* and *Schlotheimia* the bimodal distributions of spore size can be correlated with the presence of heteromorphic sex chromosomes and sexual dimorphism where the males (derived from the small spores) are dwarf epiphytes on the leaves of the females (derived from the large spores). However, other dioecious taxa with sexual and bivalent dimorphism are isosporous. Further anisosporous species produce both normal and dwarf males whereas monoecious taxa lack dimorphic bivalents and are invariably isosporous. Two additional terms "functional" and "morphological" anisospory are necessary to describe these complexities. The only way to elucidate the evolutionary significance, origin and nature of anisospory in bryophytes is by a multidisciplinary approach involving electron microscopy of sporogenesis, chromosome cytology, spore germination and gametophyte culture.

Differences in suface ornamentation between large and small spores probably arise from differences in swelling during maturation. Striking differences between the germination patterns of *Macromitrium* and *Macrocoma* support the elevation of the latter to generic status as does karyotype analysis. Antheridial morphology is very useful in the taxonomy of dwarf male plants which may well be of two fundamentally different types in mosses: genetically and phenotypically determined. The prevalence of sexual dimorphism in

Systematics Association Special Volume No. 14, "Bryophyte Systematics", edited by G. C. S. Clarke and J. G. Duckett, 1979, pp. 281–316, Academic Press, London and New York.

tropical and sub-tropical epiphytes and the rarity of morphological anisospory in bryophytes cannot yet be explained.

INTRODUCTION

The term heterospory has been widely applied in the plant kingdom to the production of two sizes of spores by a single plant. These spores give rise to separate male and female gametophytes. Some vascular plants (e.g. the Filicales and Lycopodiales) are homosporous; all their spores are the same size and produce monoecious gametophytes. Others, such as the water ferns and Selaginellales are heterosporous (Doyle, 1970; Sussex, 1966). Incipient heterospory has been described in *Platyzoma*, a specialized Queensland fern (Tryon, 1964; Sussex, 1966), and has sometimes been claimed to occur in *Equisetum* (Duckett, 1970).

In all extant heterosporous vascular plants the two kinds of spores, "macro- or megaspores" and "microspores", are produced in separate sporangia so that each meiotic division gives rise to spores of the same kind (i.e. only macro- or microspores). The macrosporangia and microsporangia in which meiosis occurs may be produced on separate plants or in different positions on the same plant. In general, the large macrospores are produced in small numbers, frequently by abortion of three of the four products of meiosis, while the small microspores tend to be produced in large numbers. Thus increase in spore size is often associated with decrease in the number of spores. These vascular groups all produce their gametophytes endosporically.

In bryophytes Fleischer (1920) reported that in some dioecious genera, particularly *Macromitrium*, spores of unequal sizes were produced. This led Ernst-Schwarzenbach (1936, 1938, 1939, 1942, 1943, 1944) to carry out an intensive study of heterospory in 39 species of *Macromitrium*. Her study included not only statistical analysis of spore size and type (heterosporous or homosporous), but also germination experiments designed to discover whether there was any relationship between spore size and sexuality in dioecious species. She concluded that the type of heterospory found in mosses is fundamentally different from that in pteridophytes. In the latter, the expression of sexual dimorphism is phenotypically determined, while in mosses it is genotypically controlled. In mosses both large and small spores are formed in the same sporangium and derived from one spore mother cell, while in pteridophytes the two kinds of spores arise from separate spore mother cells that are already physiologically distinct.

The question of whether the same term was appropriate for both mosses and higher plants was also raised by Ernst-Schwarzenbach (1939) and the term anisospory was used in reference to *Macromitrium*. Vitt (1968) revived the argument and pointed out that since:

(1) spores in bryophytes are derived from only one type of sporangium;

(2) both kinds of spores, when size differences do occur, are derived from the same meiotic division; and

(3) gametophytes produced are exosporic in origin,

the differences between heterospory in mosses and higher plants warrant the use of a separate term. He suggested anisospory as the most appropriate.

Vitt interpreted the difference between heterospory and anisospory as probably indicating parallel evolution rather than direct evolution between bryophytes and higher plants. He also noted that the "mechanism of sex and differential spore size as well as development is not the same in higher plants as in bryophytes". This does not mean, however, that the situation in bryophytes should be considered in isolation from other groups, for similar problems in understanding spore differentiation and sex determination occur in studies of all plant systems.

Although more than half of all bryophytes are dioecious, anisospory, as defined by spore size differences, is unknown in hepatics (Schuster, 1966) and rare in mosses being mainly confined to a few genera such as *Macromitrium* and *Schlotheimia* in the family Orthotrichaceae. In some cases its occurrence is clearly linked with sexual dimorphism but the converse is not true. Thus *Eurhynchium pulchellum* has dwarf males but is isosporous (Vitt, 1968) and the same is true of *Sphaerocarpos* species in the Hepaticae.

Within the genus *Macromitrium* the situation is extremely complex (Ernst-Schwarzenbach, 1939, 1943, 1944; Miyoshi, 1963; Ramsay, 1966; Vitt, 1968). Ernst-Schwarzenbach measured the size of spores taken from herbarium specimens of 39 *Macromitrium* species, at least one from each of the regions where the genus is distributed (Java, India, Ceylon, Japan, Australia, Pacific Islands, Africa, Central and South America). The ten monoecious species were all homosporous, but seven of the dioecious species also gave unimodal distributions of spore size. Surprisingly, these seven also exhibited strong sexual dimorphism (reduced males epiphytic on the larger females) very similar to the remaining dioecious species which were anisosporous. *M. blumei* Schwaegr. was of particular interest in having two close but separate peaks for spore sizes. This species produces both normal and dwarf males (Dening, 1935) and the size distribution was interpreted as reflecting this polyoecism.

Ramsay (1966) found in Australia that while all the monoecious species of *Macromitrium* had unimodal distributions of spore size, the same was also true of the dioecious taxa *M. archeri* Mitt. (gathering 29/65) and *M. ligulare* Mitt. whereas *M. archeri* (40/63) and *M. daemelii* C. Muell. (41/64) gave weak bimodal curves similar to *M. blumei*. Anisospory appeared to be related to sex chromosome differentiation.

Miyoshi (1963) and Vitt (1968) showed that of five Japanese species, two which are said to be dioecious (*M. comatum* Mitt., *M. prolongatum* Mitt.) gave weakly bimodal distributions of spore size while the other three (*M. ferrieri* Card. and Thér., *M. incurvum* (Lindb.) Thér. and *M. holomitriodes* Nog.) which are said to be monoecious, were isosporous. On the other hand, Noguchi (1967) states that *M. holomitriodes* is the only species in Japan with dwarf males. Thus the only Japanese species with sexual dimorphism is isosporous while *M. comatum* and *M. prolongatum* are interpreted by Noguchi (1967) as autoecious with axillary antheridial branches. However, his drawings of the male shoots closely resemble dwarf males and might perhaps have been misinterpreted.

It will be apparent from these examples that the nature of what at first sight appears a well defined phenomenon, is far from clearly understood. The only way to shed further light on the possible determining mechanisms and to find the biological significance of anisospory is by studies involving not only spore size measurements but also spore structure, chromosome cytology, sexuality and culture experiments to discover which sized spores give rise to male and female plants. This paper presents an attempted synthesis of results from these multiple approaches on several Australian Orthotrichaceae (Table I).

MATERIALS

The specimens used in these studies were collected by the author in the following localities in New South Wales, Australia, unless otherwise indicated.

Macrocoma tenue (Hook. and Grev.) Vitt, 3/76, trees, Mt Wilson 13/1/76; S.477, Fitzroy Falls 17/9/77; S.777, Belmore Falls, Robertson 17/9/77; S.977 and S.1077, apple trees, Hampton 29/11/77; S.1177, roof, Jenolan Caves 29/10/77; S.1277, trees near school, Blackheath 30/10/77; *Macromitrium wattsii* Broth., 1/76, rocks, Garie Beach, Royal National Park (leg. P. Wallin); *M. pugionifolium* C. Muell., 14/75, trees, Bribie Island, Queensland 11/10/75; *M. weymouthii* Broth., 5/76, trees, Katoomba Falls, 9/2/76; *M. involutifolium* (Hook. and Grev.) Schwaegr., 34/77 and 35/77, trees, Echo Point, 28/10/77; *Macromitrium* sp. 6/76, Echo Point, 9/2/76; S.177, near Forestry Camp, Wauchope (leg. C. J. Quinn); S.377 Royal National Park, 17/9/77; *Schlotheimia brownii* Schwaegr. 2/77, rocks, Apple Tree Bay, Bobbin Head, 15/7/76; S.877, rocks Apple Tree Bay, 17/9/77.

Taxon	Voucher	Chromosome number (n)	Spore-size	Spore germination	Sexuality	Morphology of dwarf males
Macrocoma tenue (*Macromitrium eucalyptorum*)	3/76	11 (10+m)	Isosporous	Multiporose	Autoecious	Autoicous, no dwarf males. Antheridium wall cells elongated in neat rows, paraphyses present. Males normal.
,, ,,	S 477	—				
,, ,,	S 777	—				
,, ,,	S 977	—				
,, ,,	S1077	—				
,, ,,	S1177	—				
,, ,,	S1277	—				
Schlotheimia brownii	2/77	11 (10+m)	Anisosporous (weak)	1–2 pores	Dioecious (dwarf males)	0·5–1 mm. Leaf 2 mm. Wall cells of antheridium regularly arranged in rows, stalk single tier of 3 cells, no paraphyses.
,, ,,	S 877	—		—		
Macromitrium wattsii	1/76	10	Anisosporous (weak)	—	Dioecious (dwarf males)	1–1·5 mm. Leaf 2·5 mm. Antheridial wall cells elongate in neat rows, stalk 2 cells wide, paraphyses present.
Macromitrium pugionifolium	14/75	9 (8+X/y)	Anisosporous (weak)	—	Dioecious (dwarf males)	1 cm. Leaf 1 cm. Wall cells of antheridium regularly arranged in rows, stalk single tier of cells, several antheridial heads per plant, no paraphyses.
Macromitrium weymouthii	5/76	9 (8+X/y)	Anisosporous (weak)	—	Dioecious (dwarf males)	0·5 mm. Leaf 2·0 mm. Antheridial wall cells elongate in neat rows, stalk more than one tier cells, no paraphyses.
Macromitrium involutifolium	35/77	8 (7+X/y)	Anisosporous	—[1]	Dioecious (dwarf males)	0·5–1 mm. Leaf 1·5–2 mm. Antheridial wall cells elongate in neat rows, stalk single tier cells, no paraphyses.
Macromitrium involutifolium	34/77	8 (7+X/y)	Anisosporous	—[1]	Dioecious (dwarf males)	0·5 mm. Leaf 1·5 mm. Antheridial wall cells irregular, normal males size of female branches, antheridia very much larger. No paraphyses in either.
Macromitrium sp.	6/76	9 (8+X/y)	Anisosporous (strong)	—[1]	Dioecious (normal and dwarf males)	
Macromitrium sp.	S 177	—	Anisosporous	1–2 pores	Dioecious (normal and dwarf males)	1·0 mm. Leaf 3 mm. No paraphyses.
Macromitrium sp.	S 377	—	Anisosporous	1–2 pores	Dioecious (normal and dwarf males)	0·25 mm. Leaf 2 mm. No paraphyses.

[1] High spore mortality

Voucher specimens have been placed in the Herbarium, School of Botany, University of New South Wales. All the above names should be regarded as tentative since the identification of Australian *Macromitrium* spp. is notoriously difficult (Ramsay, 1966; Sainsbury, 1955; Scott and Stone, 1976) and many specimens bearing specific names in herbaria do not always agree with type specimens and descriptions.

OBSERVATIONS AND DISCUSSION

1. Spore Studies

(*a*) *Spore size.* Particular care is needed when using the size of spores to diagnose possible anisospory. Sporogenesis is influenced not only by genetic, but also by environmental factors. The resources of the sporogonium can easily be depleted by adverse climatic conditions in the vital period from meiosis to spore maturity with profound effects on spore size (Mogensen, 1978; Neidhart, 1975, 1978, this volume, Chapter 12). The reports of sporogonia of different sizes producing different sized spores (Fleischer, 1904–1923, 1920; Greguss, 1964) are probably due to failures in the maturation process rather than heterospory (Ernst-Schwarzenbach, 1939). A further problem is how to decide when variation in spore size within a species can be defined as anisospory. This is impossible from the data given by Erdtman (1965) and Boros and Jàrai-Komlódi (1975). It is critical that all measurements should be made on mature spores (Mogensen, 1978) and highly desirable that other information be obtained from the same specimens (Table I). All the measurements given here are of mature spores from capsules from which the operculum had just fallen, or was about to be released, mounted in Hoyers solution (Anderson, 1954). The size distributions (Figs 1–9) are of a representative sample of 200 spores from each population. Viable spores are readily distinguished from abortive ones by their bright green contents, even in the smaller spores where the wall becomes yellow-brown in colour. In herbarium specimens spores can still be readily classified as living or dead at the time of collection.

The monoecious species *Macrocoma tenue* (Fig. 1) gave a unimodal distribution curve similar to that recorded previously (Ramsay, 1966) while the dioecious taxa presented various patterns from weakly (Figs 2, 3, 6) to strongly bimodal (Fig. 4). The remaining taxa (Figs 5, 7–9) exhibited distributions with more than two modes and a wider variation in the size of the spores including a small percentage which were very large (over 30 μm). In *M. involutifolium* (Fig. 8) this variability was almost certainly the result of some

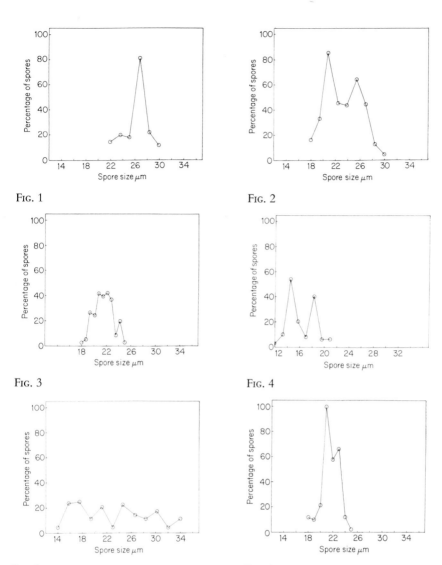

FIGS 1–9. Spore sizes in *Macromitrium* and *Schlotheimia*.

FIG. 1. *Macrocoma tenue* (3/76). Unimodal distribution; size range 22–30 μm.

FIG. 2. *Schlotheimia brownii* (2/77). Bimodal distribution; size range 18–30 μm. (18–25 and 25–30 μm.) Spores in the range 20–26 μm are probably a mixture of males and females.

FIG. 3. *Macromitrium pugionifolium* (14/75). Slightly bimodal distribution with subpeaks in each half; size range 18–25 μm. (18–20 and 20–25 μm.)

FIG. 4. *Macromitrium* sp. (6/76). Strongly bimodal distribution; size range 12–21 μm. (12–17 and 17–21 μm.)

FIG. 5. *Macromitrium weymouthii* (5/76). Bimodal distribution with subpeaks in each half; size range 14–34 μm. (14–20 and 20–34 μm.)

FIG. 6. *Macromitrium wattsii* (1/76). Bimodal distribution but considerable overlap in the range 19–22 μm; size range 18–25 μm. (18–20 and 20–25 μm.)

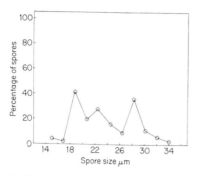

Fig. 7

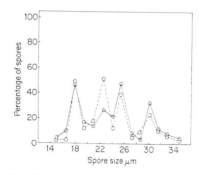

Fig. 8

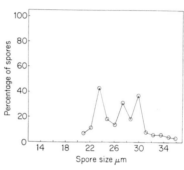

Fig. 9

Fig. 7. *Macromitrium* sp. (S377). Possibly bimodal with subdivision of the lower range; size range 15–34 μm. (15–25 and 25–34 μm.)

Fig. 8. *Macromitrium involutifolium* (34/77 – – – – and 35/77 ——). Bimodal with subdivision of each major class; size range 15–35 μm. (15–23 and 23–35 μm.) High spore mortality may account for the quadrimodal curves.

Fig. 9. *Macromitrium* sp. (S177). Strongly bimodal with two peaks in the large spore range; size range 21–36 μm. (21–26 and 26–36 μm.)

genetic abnormality since a high proportion of the spores (33% in 34/77 and 41% in 35/77) were aborted.

These results show that spore measurements alone provide insufficient information to define anisospory clearly.

(*b*) *Spore structure.* A somewhat different approach is to investigate spore structure using both scanning (SEM) and transmission electron microscopy (TEM) to complement observations under the light microscope. Ernst-Schwarzenbach (1939, 1943, 1944) and Ramsay (1966) both noted some differences in appearance between the large and small spores of *Macromitrium* but as yet no comparative studies have been published. Under both the light microscope and the TEM possible differences which warrant investigation include the numbers of plastids (about twice as many in large spores (Mogensen and Olesen, personal communication)), thickness and structure of the wall and the type (using histochemical techniques) and distribution of storage substances. Accurate recognition and separation of male and female spores in all examples of anisosporous species that do not show two clear size classes of spores is impossible unless other diagnostic structural features can be identified.

In the absence of comparative TEM studies on spore wall structure and organelles in *Macromitrium*, SEM provides the means by which surface ornamentation can be observed in more detail. Although Erdtman (1965) noted the presence of large and small spores in some of the 20 species of *Macromitrium* he studied under the light microscope (e.g. *M. crenulatum* Hamp. and *M. eriomitrium* C. Muell.) he made no mention of possible differences in surface structure. Hirohama (1977) in a survey of the spores of the Orthotrichaceae stated that ornamentation is characteristic for each genus, with the most complicated and advanced type occurring in *Macromitrium*. Unfortunately SEMs of only one type of spore were included for each anisosporous species so that no comparison of large and small spores is possible.

I have now gathered SEM data on the spores of 18 of the 50 or so *Macromitrium* species in Australia, and have examined several populations of some species (e.g. six of *M. wattsii*, five of *M. daemelii* and seven of *Macrocoma tenue*) as well as several capsules from each population, to determine the reliability of surface ornamentation as a taxonomic character (Figs 10–21). Although some variation can be seen between the shape, size and density of papillae within and between populations, this is much less than the differences between species. A major problem is the production of taxonomically meaningful descriptions of the ornamentations. It is essential however, to compare equivalent spore surfaces as some spore types have clearly distinct concave and convex faces (Fig. 11) or thinner regions of the wall which may represent germ pore areas. The general pattern emerging is that although the papillae on the perine are more densely distributed on the small spores and more widely spaced on the large, there appears to be no absolute distinction between the two types.

Understanding the sequence of events in wall formation now becomes essential for any interpretation of the differences in patterning between small and large spores. The spores of mosses have thin walls when released from the spore mother cell wall following the completion of tetrad development (Neidhart, 1975; Mogensen, 1978). Surface ornamentation (perine formation) is deposited during maturation (Jarvis, 1974; Neidhart, 1975, 1978, this volume,

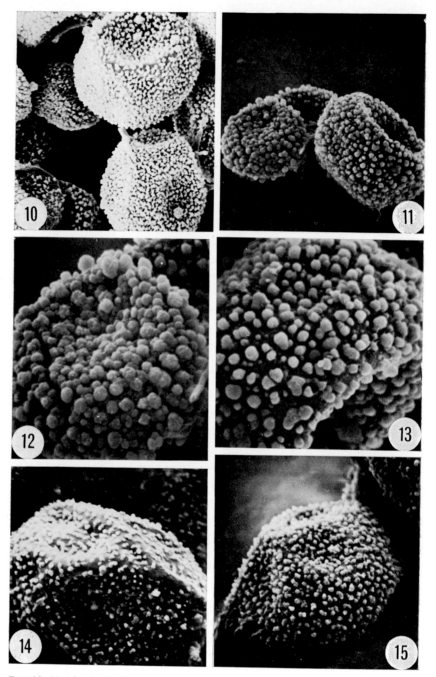

FIGS 10, 11, 12, 13, 14, 15

Chapter 12; Hirohama, 1977; Mogensen, 1978) in association with globular bodies which appear in the spore sac fluid at the time of release of the spores from the tetrad. In *Funaria* and *Cinclidium* the globular bodies adhere to the exine and spore size increases continuously. Perine formation terminates when the spore mass changes from hyaline to yellowish white and the surface of the spores is completely covered with granules (Mogensen, 1978). However, the final stages of maturation are marked by further expansion and colour changes to brown and eventually green. On the other hand, in *Fissidens limbatus* Sull. the perine is apparently deposited *after* the spores have reached full size (Mueller, 1974).

Thus it is necessary to determine for *Macromitrium* precisely when, in spore ontogeny, the papillae are deposited. If this takes place after the expansion phase then ornamentation is clearly being deposited in different ways on the spores of different sizes. Alternatively the greater spacing of the papillae on the large spores may result from the greater expansion of their walls during maturational swelling after perine formation earlier in development. The information available suggests that the latter hypothesis is more likely.

Globular bodies are present in young capsules of all *Macromitrium* species while the spore mass changes from hyaline to yellow and they provide a useful guide in selecting capsules for chromosome studies. Thus, there is some evidence that patterning in *Macromitrium*, as in *Cinclidium* and *Funaria*, may be completed before the spores swell. Hirohama (1977) also recognized that wall patterning immediately following the tetrad stage is general in the Orthotrichaceae. Moreover, Ernst-Schwarzenbach (1939) remarked that the size differences in anisosporous species were not manifest until after the release of spores from the tetrad. Thus, differential swelling of large and small spores, as can be seen in Figs 22–25, would produce the observed differences in papillar density. Further studies are now required to elucidate the cytoplasmic changes underlying the

---

FIGS 10–15. Scanning electron micrographs of spore surfaces.
FIG. 10. *Macromitrium archeri*, large and small spores. Surface patterning distinct and more widely spaced on the large spores. ×1000.
FIGS 11–13. *Schlotheimia brownii*
FIG. 11. Group of two large and one small spore. ×1000.
FIG. 12. Small spore. ×2500.
FIG. 13. Large spore. Note the wider spacing of the ornamentations compared with Fig. 12 ×2500.
FIGS 14–15. *Macromitrium* sp. (S.177) ×2500.
FIG. 14. Large spore, surface pattern clearly defined and well spaced.
FIG. 15. Small spore, spacing not so great.

differential swelling of the spores. Preliminary TEM observations by Mogensen and Olesen (personal communication) show that the chloroplasts replicate during the maturational swelling phase, which suggests that the male spores may reach maturity at an earlier stage of sporogenesis than the females.

(c) *False anisospory*. This term was proposed by Mogensen (1978) to describe a situation in three species of *Cinclidium* where about 50% of the spore mass died during an intermediate stage of sporogenesis. The aborted spores were smaller than the viable ones. Various explanations proposed by Mogensen include irregularities in tetrad formation (meiotic abnormalities) problems of synchronization of the development of the spore mass before the formation of the perine and environmental factors. However, he concluded that although some mortality may result from any of these factors, the 50% mortality which regularly occurs in *Cinclidium* could be better explained as due to two genes, one on a sex chromosome and the other on an autosome.

Spore abortion has often been noted in mosses but its basis has largely remained a mystery. For example, Newton (1972) noted mortality of 45% of the spores of *Plagiomnium undulatum* but did not find any meiotic irregularities.

In *Macromitrium* spore mortality as high as 41% was recorded in *M. pugionifolium* and 60% in *Macromitrium* sp. (6/76), but in neither case were meiotic irregularities seen. Thus, both false and true anisospory may occur in the same population or species and this could well explain some of the multimodal distributions of spore size (Figs 3, 5 and 8) as has been demonstrated in *Equisetum* (Duckett, 1970). In *M. pugionifolium* (Fig. 3) the size of the spores has a bimodal distribution in which each half is subdivided into two groups. Peaks 1 and 3 might represent aborted male and female spores and peaks 2 and 4 viable males and females respectively. Whether this situation is ubiquitous in *M. pugionifolium* or due purely to environmental factors requires further study. However, in other cases (*Macromitrium* sp. 6/76; Fig. 4) the bimodal distribution

FIGS 16–21. Photomicrographs of spores and tetrads.
FIG. 16. *Schlotheimia brownii* (S877), mature spores. Note the papillose walls and the size range. There is no clear cut separation of two size classes. ×400.
FIG. 17. *Macromitrium weymouthii* (5/76), mature spores. The walls are very finely papillose. Large and small spores are clearly distinct. ×400.
FIGS 18–21. *Macromitrium pugionifolium* (14/75).
FIG. 18. Normal and aborted spores. ×400.
FIGS 19–21. Various examples of normal and abnormal tetrads. ×1600. The giant cells may have resulted from cytomixis or may be enlarged normal cells. Some enlarged cells produce 8 (Fig. 20) or more (Fig. 21) spores.

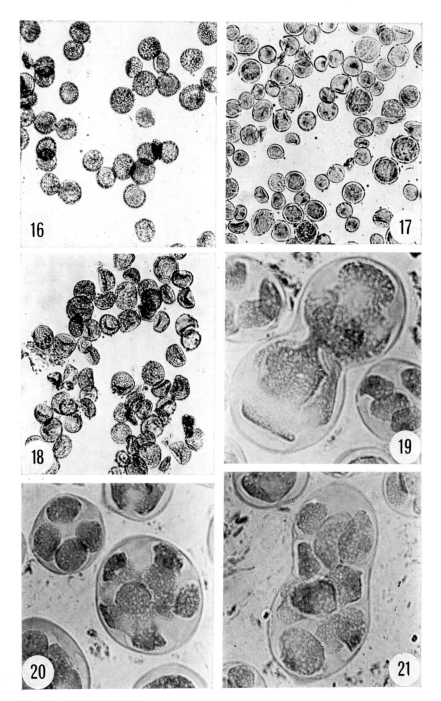

Figs 16, 17, 18, 19, 20, 21

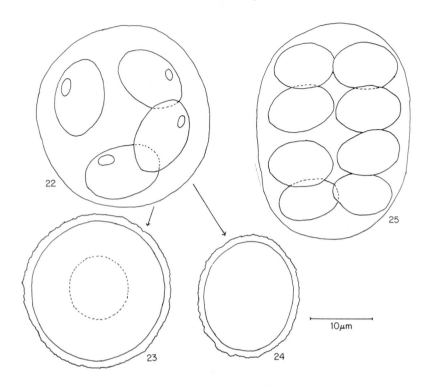

Figs 22–25. *Macromitrium pugionifolium* (14/75). Tetrads and spores.
Fig. 22. Normal tetrad with spores at the point of release from the spore mother cell
         wall. Compare their size at this stage with the mature spores in Figs 23 and 24.
Figs 23 and 24. Mature spores with thick walls and an ornamented perine.
Fig. 23. Large spore; the dotted region represents an area on the surface where the
         perine is less well developed and may correspond to the region through which
         the germ tube emerges.
Fig. 24. Small spore.
Fig. 25. Abnormal tetrad containing 8 spores.

of spore size gives no indication that a large number of the spores are aborted.

Since the spores of *Macromitrium* tend to remain associated in tetrads it is
possible to investigate the frequency of abortion in individual tetrads. No
clear picture emerges: sometimes all four spores are aborted, sometimes only
one or two. Moreover, aborted spores, in particular tetrads, are not always
the small ones although these abort more frequently.

Regardless of the amount of abortion, in some species I have studied (Figs
2, 4, 6), and in *M. salakanum* C. Muell. studied by Ernst-Schwarzenbach (1939)

the larger spores are in the minority. It would be interesting to discover whether this is due purely to sampling error or indicates preferential abortion of one type of spore, as is the case in some truly heterosporous pteridophytes (Goebel, 1928). The similar occurrence of small abortive spores has also been noted in other Orthotrichaceae (Lewinsky, 1977) and a detailed investigation to determine whether there is any parallel evolutionary trend towards a reduction in numbers of large female spores and increase in numbers of small male spores should be carried out. Further work on spore viability in mosses perhaps using methods such as enzymatically induced fluorescence (Heslop-Harrison and Heslop-Harrison, 1970), may help elucidate the extent of false anisospory, while studies in sporogenesis may be able to analyse the events which lead to the death of spores during development.

2. Germination Studies

The study of germination and of the types of sporelings in the Musci has in recent years been shown to give results which are of significance to moss taxonomy (Nishida, 1978).

The germination and subsequent culture of spores of *M. salakanum* over a period of $1\frac{1}{2}$–$3\frac{1}{2}$ years by Ernst-Schwarzenbach (1939, 1943, 1944) established that small spores gave rise to dwarf males, and large ones to female plants. Sexual dimorphism was not apparent in the early protonemal stage, but developed only when secondary protonema and gametophore stages were produced. The difference in size of the leaves in males and females was a function of mean cell number and not of cell size, the male dwarfs being 33·5 cells long and 23·2 cells wide, while the females were 204 cells long 74·7 cells wide.

I have cultivated specimens with known chromosome numbers to see if different patterns of development could be distinguished. Spores were grown in multispore cultures for several months on Bold's medium (Bold, 1967) under a 16h light regime of 16RJm^{-2} at 20°C with parallel cultures of *Funaria hygrometrica* for which protonemal morphogenesis is well documented (Valanne, 1966; Monroe, 1968).

In *Macromitrium* sp. S.177 (Figs 28, 32) and S.377 (Fig. 33) the larger spores germinated through one or two germ pores 2–3 days earlier than the smaller spores which normally had only a single germ pore. The filamentous sporelings produced numerous side branches close to the spore while the ends of some of the filaments cut off cells with oblique cross walls. These subsequently became coloured and differentiated into rhizoids. As a result of this close branching

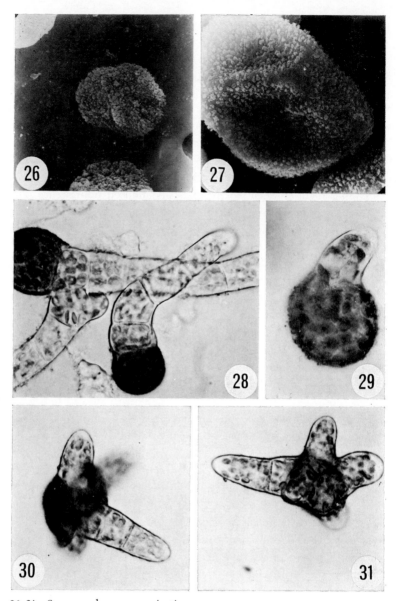

Figs 26–31. Spores and spore germination.

Figs 26 and 27. *Macromitrium involutifolium* (35/77). Scanning electron micrographs of spore surfaces. ×3020.   Fig. 26. Small spore.   Fig. 27. Large spore; note the well spaced surface ornamentations.   Figs 28–31. Spore germination. Fig. 28. *Macromitrium* sp. (S177); Large and small spores. Note the protonema from the large spore is well developed and branches close to the spore, while that from the small spore is a single filament. ×650.   Figs 29–31. *Macrocoma tenue*.   Fig. 29. Initially a single germ tube emerges (S1277). ×625. Figs 30, 31. (S977). Later stages showing a multiporose type of germination with several germ tubes emerging in different planes. Both ×625.

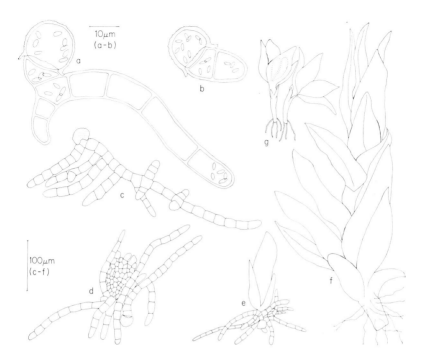

Fig. 32. Spore germination in *Macromitrium* sp. (S177). a, germinating large spore. b, germinating small spore. c, older sporeling showing branching of the protonema near the spore. d, bud development near the spore wall. e, older bud. f, young female gametophore. g, dwarf male from wild collection (compare its size with "f").

pattern the protonemata were compact and did not extend over the surface of the agar like those of *Funaria*.

Gametophores began to differentiate within three (S.377) to six (S.177) weeks and after 2–4 months well developed plantlets with their own rhizoid systems had formed. Comparison with dwarf males on adult plants reveals these to be far too large to be male (Fig. 32).

In all seven populations the growth pattern of *Macrocoma tenue* sporelings (Table I) was quite different from that in the *Macromitrium* species studied here and by previous workers (Ernst-Schwarzenbach, 1939; Nehira, 1976). Although germination began with the emergence of one or two germ tubes, the spore tissue remaining within the spore wall subsequently divided to produce several more tubes sequentially (Fig. 31). Thus the spore becomes multicellular with one basal cell inside the wall for each germ tube (up to eight) or protonemal

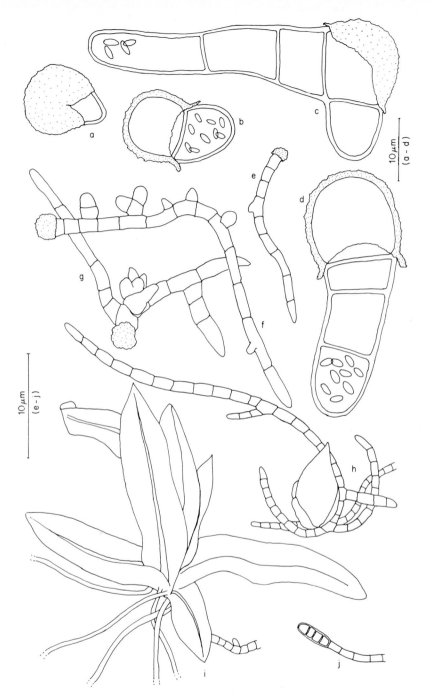

FIG. 33. Spore germination in *Macromitrium* sp. (S377). a, b, germinating small spores. c, d, germinating large spores. e, germinating small spore with an unbranched protonema. f, germinating large spore of the same age as "e". g, h, bud formation adjacent to the spore. i, gametophore with leaves and rhizoids. j, gemma at the tip of a rhizoid.

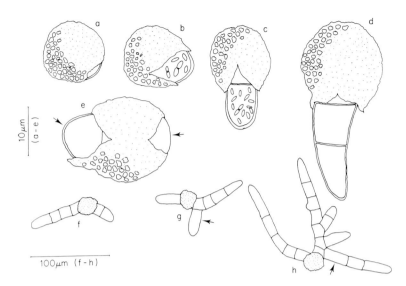

FIG. 34. Spore germination in *Schlotheimia brownii* (S877). a–e, early stages; note the warty nature of the perine. e, large spore producing two germ tubes (arrowed). f–h, young filamentous protonemata. No more than two germ tubes are produced but branching occurs close to the spore (arrowed).

filament. Buds appeared within 30 days and well developed gametophores within seven weeks. The protonemal filaments were all chloronemal and no rhizoids were developed until the gametophores produced their own (Fig. 35).

As in *Macromitrium*, large spores of *Schlotheimia brownii* germinated by one or two germ tubes 1–2 days earlier than the smaller spores and secondary protonemal branches arose close to the spore (Fig. 34). However, much longer filaments were formed which spread extensively over the agar surface and produced oblique-walled rhizoids at intervals. No gametophores developed within the 7–8 week period of culture.

These studies confirm the findings of Ernst-Schwarzenbach (1939) that the larger spores of *Macromitrium* produce germ tubes earlier than the smaller ones. Protonemal morphogenesis in *Macromitrium* sp. (6/76) and *Schlotheimia* was very similar to that previously recorded in *M. cuspidatum* Hamp. The very different germination pattern seen in *Macrocoma tenue* adds weight to its generic separation from *Macromitrium* (Vitt, 1973), a separation which has sometimes been contested (Scott and Stone, 1976).

Although the present cultures were not maintained long enough for sex organs to be formed, there was evidence from the size of the gametophores

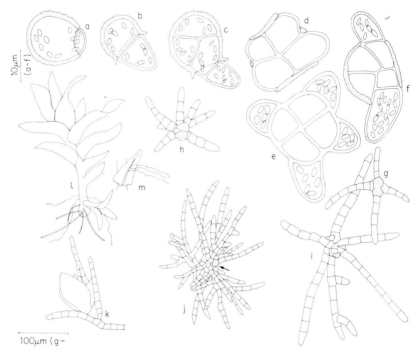

FIG. 35. Spore germination in *Macrocome tenue*. a, germ tube emergence. b, first cell division. c, initiation of cell division within the spore wall. d, emergence of four germ tubes. e, first cell division of four germ tubes. f, two germ tubes and divisions within the spore wall. g–i, production of several uniseriate filaments from the spore. j, later stage showing numerous filaments radiating from the central spore area (arrowed). All the filaments are chloronemal. Adjacent to the spore wall is a central plate of cells formed as a result of division within the spore during early germination. k, bud formation. l, young gametophore. m, older leaf from "l" showing rhizoids produced from its lower surface.

that the female gametophores are formed first. However, the longer term cultures of Ernst-Schwarzenbach (1939, 1943) readily produced antheridia but rarely developed archegonia. These results suggest that the factors responsible for gametophore induction and sex organ formation might well be different in male and female plants.

3. Chromosome Cytology

Details of the chromosomes of *Macrocoma*, *Macromitrium* and *Schlotheimia* in Australia are summarized in Tables I and II. Dioecious species of *Macromitrium*

TABLE II. Some chromosome numbers recorded for the Orthotrichaceae

	n	Spores	Sex	Author
Orthotrichum spp.	6, 11, 18	Isosporous	Autoicous Seldom Dioicous	Fritsch (1972)
Ulota spp.	9, 11, 12 20, 22, 10, 16	Isosporous	Autoicous Seldom Dioicous	Fritsch (1972)
Zygodon spp.	11, 12	Isosporous	Dioicous or Autoicous	Fritsch (1972)
Macrocoma tenue	11	Isosporous	Autiocous	Ramsay (1966)
M. hymenostomum	12	Isosporous	Autiocous	
Macromitrium makinoi	11	Isosporous	Autoicous	Inoue (1964)
M. scottiae	11	Isosporous	Autoicous	Ramsay (1966)
M. daemelii	8 (7+X/Y)	Anisosporous	Dioicous	Ramsay (1966)
M. archeri	8 (7+X/Y)	Anisosporous	Dioicous	Ramsay (1966)
M. archeri	9 (8+X/Y)	Anisosporous	Dioicous	Ramsay (1966)
M. ligulare	9 (8+X/Y)	Anisosporous	Dioicous	Ramsay (1966)
Schlotheimia rugifolia	9 (8+X/Y)	Anisosporous	Dioicous	Anderson (1979)
S. lancifolia	9 (8+X/Y)	Anisosporous	Dioicous	Anderson (1979)

have chromosome numbers of $n = 8$ or 9 and in monoecious ones $n = 11$.

A heteromorphic bivalent thought to represent morphologically distinct X and Y chromosomes can be seen in species with $n = 8$ or 9. Associated with six of the chromosomes are unevenly sized heteropycnotic bodies at interphase, and precocious staining of a localized region of the nucleus in early prophase. Tetrads in these species contain two spores with a large heteropycnotic body and two with a small one indicating cytological anisospory. Pecularities of particular taxa are summarized below.

Macromitrium wattsii is the only species of the genus so far examined cytologically with $n = 10$ (Fig. 36). A dimorphic bivalent is absent, but one precociously staining heteropycnotic body is present at early prophase I, although this is paler than in species with heteromorphic bivalents. The bivalents are larger and more uniform in size than in other species.

The heteromorphic bivalent in *M. weymouthii* (Figs 37 and 38) resembles that in *M. daemelii* (Ramsay, 1966): the smaller member is less than half the size of its homologue. *M. pugionifolium* (Figs 41 and 42) possesses a hetero-morphic bivalent similar to those in *M. ligulare* and *M. archeri*. The complement also contains a small bivalent which sometimes exhibits precocious disjunction. In a number of the sporogonia I encountered giant spore mother cells (Figs 19–21, 25). These underwent several divisions to produce variable numbers of spores. In some cases the mother cells were round and about twice the size of the normal tetrad. They contained eight spores, presumably arising from one regular mitosis followed by meiosis. In other instances the mother cells were elongate with larger number of spores which probably derived from a combination of mitosis, meiosis and cytomixis as reported in *Dicnemo-loma pallidum* (Hook.) Wijk and Marg. (Ramsay, 1973). Despite these develop-mental peculiarities less than 10% of the spores of *M. pugionifolium* were abnormal at maturity.

The dimorphic bivalent in *Macromitrium* sp. (6/76) is rod-shaped with chiasmata in only one arm and it frequently disjoins precociously (Figs 39 and 40). The smaller chromosome of this pair is more than half the size of its homologue and is thus larger than in all other species with $n=9$. By contrast,

---

Figs 36–48. Chromosome analysis.

Fig. 36. *Macromitrium wattsii* (1/76) $n=10$.

Figs 37, 38. *M. weymouthii* (5/76) $n=9$ $(8+X/y)$.

Fig. 37. Prometaphase, the heteromorphic bivalent is arrowed.

Fig. 38. "Parachute" formation of the paired homologues, one large the other small, is clearly visible.

Figs 39, 40. *Macromitrium* sp. (6/76) $n=9$ $(8+X/y)$.

Fig. 39. Metaphase I, nine bivalents are present with possible heteromorphy in the one arrowed.

Fig. 40. Precocious separation of one slightly dimorphic bivalent (both components arrowed).

Figs 41, 42. *M. pugionifolium* (14/75) $n=9$ $(8+X/y)$, both metaphase I.

Fig. 42. The dimorphic bivalent has separated (arrowed) while in Fig. 41 it remains attached to one arm of its homologue (arrowed).

Figs 43–45. *M. involutifolium* (34/77, 35/77) $n=8$ $(7+X/y)$.

Figs 43, 44. Metaphase I; the two components of the heteromorphic bivalent are disjoined (arrowed).

Fig. 45. Metaphase I showing the heteromorphic bivalents still attached (arrowed).

Fig. 46. *Macrocoma tenue* (3/76) $n=11$ $(10+m)$, metaphase I; note the early disjunction of the smallest bivalent (arrowed).

Figs 47, 48. *Schlotheimia brownii* $n=11$ $(10+m)$, the small bivalent may be fully con-tracted (Fig. 48) or not contracted (Fig. 47). In both cases it is much paler than the remainder.

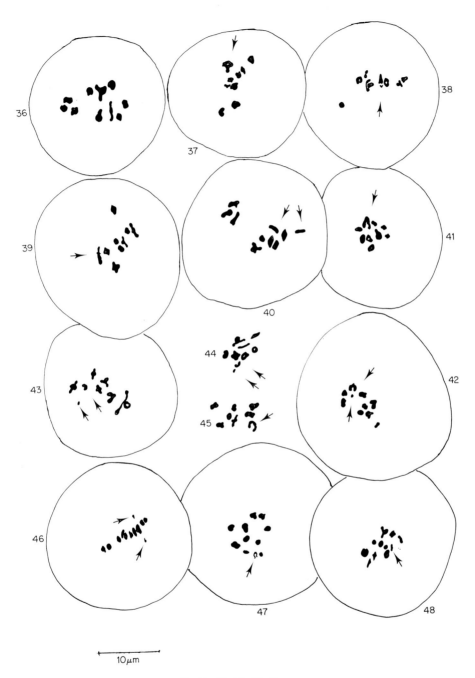

10μm

Figs 36, 37, 38, 39, 40, 41, 42, 43, 44, 45, 46, 47, 48

the smaller member of the dimorphic bivalent in *M. involutifolium* is only
about one-third the size of the larger. (Figs 43–45). Early disjunction of this
bivalent is particularly well marked.

The chromosomes of *Macrocoma tenue* (Fig. 46) are smaller than those in
all species of *Macromitrium* and the smallest bivalents (*m*) disjoin early in all
populations. These observations together with a count of $n = 12$ for *Macrocoma
hymenostomum* (Mont.) Grout (Anderson and Bryan, 1958) which is also
autoecious and isosporous (Table II), support Vitt's (1973) raising of *Macrocoma*
to generic status. It would now be of interest to find out if *M. tenue* from
different parts of its extremely wide distribution conforms in cytology and
in spore germination pattern to the Australian populations.

As in *Macrocoma tenue*, $n = 11$ in *Schlotheimia brownii* and the complement
includes a small, pale and precociously disjoining bivalent (Figs 47 and 48).
Heteropycnotic bodies are not clearly discernible at prophase I. In contrast,
the chromosome number of $n = 9$ ($8 + X/y$) have been recorded for *S. rugifolia*
(Hook.) Schwaegr. and *S. lancifolia* Bartr. (Anderson, 1979). Both these
species from southeastern North America are phyllodioecious and anisosporous
(Table II). Their heteromorphic bivalent consists of a large chromosome
paired with a dot-like small chromosome similar to that in some *Macromitrium*
species.

The present cytological studies confuse rather than clarify our understanding
of the relationship between chromosome numbers, karyotypes, sex chromosomes, anisospory, dioecism and sexual dimorphism. Although correlations
between these features are apparent in Tables I and II these are far from absolute.
Sexual dimorphism, indicated by the presence of dwarf males, occurs in species
with $n = 8$, 9, 10 (*Macromitrium*) and $n = 9$ and 11 (*Schlotheimia*) whether a
dimorphic bivalent is present or not, thus suggesting that sex determination
is controlled by heteromorphic X/y sex chromosomes in some species but
not all. However, in the dioecious and anisosporous *Macromitrium* sp. (6/76)
although a dimorphic sex bivalent is present, both similar and dissimilar male
and female plants are produced. Here the chromosomal determination of sexual
dimorphism may perhaps be over-ridden by environmental factors. The results
also show that anisospory may be detected by chromosomal differences, with
the presence of distinct X and Y chromosomes or alternatively may exist
even when dissimilar bivalents are not detectable. We are far from understanding
precisely what induces anisospory in *Macromitrium* and its fundamental significance. Cytological studies of many more species are now required, including
the comparison of karyotypes in males and females. Such investigations are
particularly difficult to carry out because of the tiny cells and heavily thickened

walls in the females, and the small size, early maturation and short life span of the dwarf males. Techniques must be developed for cytological analysis of both male and female gametophores produced in culture where sex and spore type can be identified in advance. Alternatively, regenerants from leaves of adult plants may provide suitable materials as demonstrated by Ernst-Schwarzenbach (1939, 1942).

4. Sexual Dimorphism

Not only do cytological and spore size studies show that anisospory is more complex than appears at first sight, but a critical examination of sex expression in mosses indicates that the same is also true for sexual dimorphism. Phenotypic expression of sex in mosses may take several forms (Vitt, 1968; Khanna, 1971): although field collections can usually be classified as monoecious or dioecious, there are a number of ways by which a particular condition may be achieved. Monoecious taxa may produce antheridia and archegonia in close proximity (synoecious and paroecious inflorescences) or on separate branches (autoecism). Alternatively, either antheridia or archegonia, but not both, may be present on an individual plant at different times (pseudodioecism). Without recourse to cytological or culture studies it is almost impossible to discriminate between this last pattern of sexual behaviour and true dioecism which is determined by the segregation of sex chromosomes. Thus a correct decision on whether a species is monoecious or dioecious will ultimately depend on a demonstration that antheridia and archegonia develop from the same or separate spores. Both true and pseudodioecism may or may not be accompanied by sexual dimorphism.

Since the original discovery of dwarf males in *Fissidens* and *Camptothecium* (Philibert, 1883), numerous other examples of sexual dimorphism in mosses representing a wide range of genera from diverse families have been described (Table III, Fleischer, 1904–1923, 1920; Brotherus, 1924–1925; Woesler, 1935a, b; Dening 1935; Allen, 1935, 1945; During, 1977). This reaches its most extreme expression where male plants, reduced to a few leaves and antheridia, occur on the leaves of, or associated with the larger females as in *Macromitrium*. Other species such as *Orthotrichum lyellii* are only slightly dimorphic.

Despite the diversity of the known examples (Table III) little attempt has been made to elucidate the basis of the dimorphism. However, what little information there is suggests that the dwarf males may be of different kinds in different taxa: those which are genetically determined, and those where environmental influences are responsible for the dwarfness so that the dwarf

TABLE III. The occurrence of dwarf males in mosses
General References—Schellenberg (1920); Brotherus (1924–1925); Fleischer (1904–1923)

Family	Genus and Species	References
Fissidentaceae	*Fissidens* 5–6 spp.	Philibert 1883
		Anderson 1979
Ditrichaceae	*Cheilothela* 1 sp.	
Dicranaceae	*Trematodon* 2 spp.	
	Dicranella 1 sp.	
	Holomitrium all spp.	Loveland 1956
	Braunfelsia all spp.	Crum 1976
	Schliephackea 1 sp.	Vitt 1968
	Dicranum 22 spp.	
	Dicranoloma all spp.	
	Brotherobryum all spp.	
	Eucamptodontopsis all spp.	
	Campylopodium 1 sp.	
Leucobryaceae	*Leucobryum* most spp.	Woesler 1935b
	Schistomitrium 1 sp.	Vitt 1968
	Octoblepharum 1 sp.	Scott and Stone 1976
Calymperaceae	*Syrrhopodon* 1 sp.	
Bryaceae	*Orthodontium* 1 sp.	Fleischer 1904–1923
Mniaceae	*Rhizomnium* 1 sp.	Koponen 1971
Rhizogoniaceae	*Hymenodon* sp.	
Pottiaceae	*Leptodontium* 1 sp.	Schellenberg 1920
Orthotrichaceae	*Macromitrium* 75% spp.	Noguchi 1967
	Schlotheimia all spp.	Anderson 1979
	Drummondia	
Rhacopilaceae	*Rhacopilum* spp.	
Hedwigiaceae	*Cleistostoma* 1 sp.	
Ptychomniaceae	Some genera?	
Pterobryaceae	*Garovaglia* all spp.	
	Euptychium 1 sp.	During 1977
	Endotrichella 1 sp.	
	Endotrichellopsis 1 sp.	
Myuriaceae	*Piloecium* 1 sp.	
Meteoriaceae	*Barbella* spp.	
	Aerobryum sp.	
Lembophyllaceae	*Lembophyllum* spp.	
Nemataceae	*Ephemeropsis* sp.	
Hookeriaceae	*Eriopus* spp.	
	Chaetomitrium all spp.	
Thuidiaceae	*Thuidium* 1 sp.	
	Claopodium some spp.	
Amblystegiaceae	*Hygroamblystegium* 1 sp.	

	Campylium 1 sp.	
Hypnaceae	*Hypnum* 7 spp.	Ando 1972
	Ctenidium some spp.	
	Microthamnium 1 sp.	
Stereodontaceae	*Stereodon* some spp.	
Brachytheciaceae	*Pleuropus* 1 sp.	Crum 1976
	Camptothecium 9 spp.	Wallace 1970
	Homalothecium 9 spp.	
	Eurhynchium 9 spp.	Anderson 1979
Sematophyllaceae	*Trismegistia* some spp.	
	Sematophyllum 1 sp.	
	Acanthocladium some spp.	
	Brotherella some spp.	
	Acroporium some spp.	
Hylocomiaceae	*Microthamnium* some spp.	
Buxbaumiaceae	*Buxbaumia* all spp.	

males are merely juvenile plants which have produced antheridia precociously but may subsequently grow into normal sized individuals.

Loveland (cited in Crum, 1976) showed that dwarfness of males in some *Dicranum* species is produced by the chemical influence of female plants on the germinating spores and not by strictly chromosomal or genetic means. Similarly, in wild collections of *Camptothecium megaptilum* Sull. the majority of the males were reduced and epiphytic on the older leaves of the females (Wallace, 1970). About 0·5% were large and independent. Spores grown in culture all gave large plants while dwarf males removed from females also developed into large plants. It was concluded that, in this species, there is hormonal control of sexual dimorphism.

In contrast, Ernst-Schwarzenbach (1939) found that the sexual dimorphism in the species of *Macromitrium* she cultured was genetically determined. Whatever the culture conditions the dwarf males never attained the same size as the females thus ruling out the possibility that they were merely juvenile or stunted individuals.

However, without similar culture studies it cannot be assumed that the same is true of all other species of *Macromitrium* and *Schlotheimia*. Experiments are also necessary to determine whether the dwarf males belong to the same species as the females upon which they are epiphytic; it could well be that the dwarf and normal males found on *Macromitrium* sp. (6/76) belonged to different species. Alternatively, this might be an example of dwarfism merely representing a stunted condition.

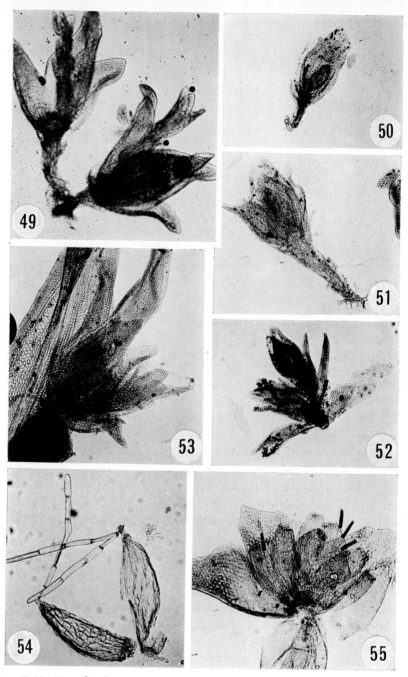

Figs 49–55. Dwarf males.

Fig. 49. *Macromitrium wattsii* ×25.

Fig. 50. *Macromitrium weymouthii* ×25.

Fig. 51. *Macromitrium involutifolium* (35/77) ×25.

Fig. 52. *Macromitrium pugionifolium* ×25.

Fig. 53. *Schlotheimia brownii* ×25.

Fig. 54. *Macromitrium wattsii* antheridia and paraphyses. ×50.

Fig. 55. *Macromitrium* sp. (6/76) antheridia from normal male. ×25.

A further problem, when examining herbarium specimens, is that one cannot always be sure that structures which at first sight appear to be dwarf males are actually separate plants. Microscopic antheridial branches occurring in the axils of leaves have been interpreted as dwarf males by some authors who therefore describe the species as dioecious. Others see them as antheridial branches of a monoecious taxon (Fleischer, 1904–1923, 1920; Brotherus, 1924–1925; Noguchi, 1967). Equally difficult to interpret are small male plants found growing attached to the rhizoids of females. Are the males developing adventitiously from the female rhizoids, or have they arisen independently from spores which fell and germinated there?

Thus, although all the available evidence suggests that sexual dimorphism is the rule in *Macromitrium* and *Schlotheimia*, in view of the possible complications outlined above, more experiments are necessary.

The size and morphology of the dwarf males in *Macromitrium* varies considerably between taxa (Table I and Figs 49–64). They may be less than half the length of the leaves of the females on which they are borne or equal in size to the leaves. Some have very few leaves and only a single antheridial head, while others have many of both. The variations in antheridial structure between taxa are particularly interesting; but they have largely been overlooked by taxonomists.

Regardless of the morphogenetic basis for sexual dimorphism in mosses a most intriguing feature of the phenomenon is that it occurs most commonly in subtropical epiphytes. Indeed a number of other species are or may be epiphytic (e.g. species of *Garovaglia*, *Leucobryum* and *Dicranum*). Although it can be readily argued that dwarfism provides an advantage as the male plant is brought close to the female so that fertilization is facilitated and the probability of outbreeding increased, it is difficult to see why it should be especially beneficial in epiphytic habitats (During, 1977).

A further ecological problem is how dwarf male plants are apparently most readily established on females of the same species. It seems very likely that there is some mechanism which inhibits the germination of foreign spores and/or stimulates those of the same species. Alternatively, the position of the males may depend on the ability of the form of the female leaf to determine the settling place for the spores along the midrib or in an axillary position (Ernst-Schwarzenbach, 1939). In *Macromitrium* spp. dwarf males often occur on leaves of young female branches or on shoots bearing sporogonia.

Perhaps one of the more surprising aspects of the present and earlier chromosome studies on *Macromitrium* (Ramsay, 1966) is that sexual dimorphism is not always associated with sex chromosomes. The two phenomena are correl-

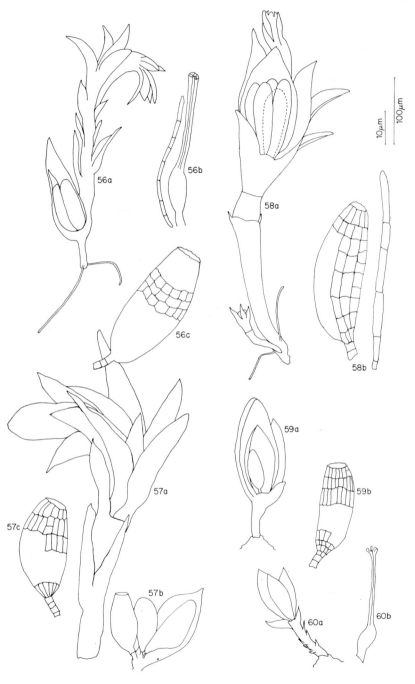

10μm

100μm

56a

56b

56c

57a

57b

57c

58a

58b

59a

59b

60a

60b

Figs 56, 57, 58, 59, 60

ated in *Macromitrium* species in which *n* = 8 or 9 (Table I) but not in *Schlotheimia brownii* or *M. wattsii.* Anderson (1978) linked heteromorphic sex chromosomes with the presence of dwarf males in *Dicranum spurium* and a species of *Schlotheimia* with *n* = 9. However, heteromorphic bivalents have also been detected in *Pleurozium schreberi, Fissidens osmundoides* and *F. cristatus* but in these species they are not associated with sexual dimorphism.

Similarly, extreme sexual dimorphism is not always associated with anisospory (Table I): isospory, weak anisospory and strong anisospory all occur in species with dwarf males. Thus it is impossible to predict from spore studies or chromosome cytology whether a species is sexually dimorphic. Conversely, sexual dimorphism does not necessarily indicate sex chromosomes or anisospory.

5. Terminology and Future Studies

These studies are in complete accord with earlier findings that the type of heterospory encountered in *Macromitrium* and *Schlotheimia* is fundamentally different from that in vascular plants. However, the two terms isospory and anisospory are completely inadequate to describe all the various permutations between sexual dimorphism, spore size and chromosome cytology now known in these mosses (Table I). Whether sex is determined at meiosis either by morphologically distinct dimorphic sex chromosomes as reported in at least 20 genera of hepatics (Allen, 1945; Lewis, 1961; Berrie, 1974) or by structural sex chromosomes with different heterochromatic properties (Tatuno, 1941; Tatuno and Segawa, 1955; Ramsay, 1966; Tatuno and Kise, 1970; Ono, 1970a, b, c; Anderson, 1979) or both (Khanna, 1971), the fact that two separate

---

Figs 56–60. Dwarf male plants.
Fig. 56. *Schlotheimia brownii* (2/77). a, dwarf male plant with a group of antheridia. Note the new growth beyond the antheridia. b, archegonium and paraphysis. s, antheridium enlarged.
Fig. 57. *Macromitrium pugionifolium* (14/75). a, dwarf male, antheridial head removed. b, group of antheridia. c, antheridium enlarged.
Fig. 58. *Macromitrium wattsii* (1/76). a, dwarf male plant, note the innovation developing beyond the antheridial group and the unsuccessful young plant attached at the base. b, antheridium.
Fig. 59. *Macromitrium weymouthii* (5/76). a, dwarf male. b, antheridium enlarged. b, archegonium drawn to the same scale as "a"—note the dwarf male plant is about the same size as the archegonium.
Fig. 60. *Macromitrium sp.* (S377). a, dwarf male plant. b, archegonium drawn to the same scale.
The larger line scale refers to Figs 56a, 57a and 58a the smaller to the remainder.

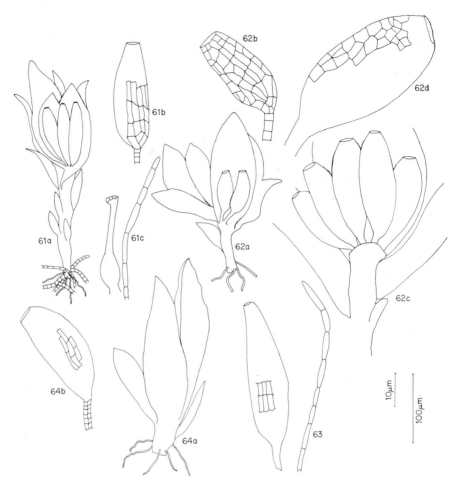

Figs 61–64. Dwarf male plants.

Fig. 61. *Macromitrium involutifolium* (34/77, 35/77). a, dwarf male. b, antheridium enlarged. c, archegonium and paraphysis.

Fig. 62. *Macromitrium* sp. (6/76) both normal and dwarf male plants present. a, dwarf male. b, antheridium from dwarf male enlarged. c, antheridial apex from normal male. d, antheridium from normal male drawn to the same scale as "b".

Fig. 63. *Macrocoma tenue* (3/76) antheridium and paraphysis from male inflorescence.

Fig. 64. *Macromitrium* sp. (S177) a, dwarf male plant. b, antheridium enlarged.

The larger line scale refers to Figs 61a, 62a, c, and 64a and the smaller to the remainder.

kinds of spore must be formed in all dioecious bryophytes led Doyle (1970) to suggest that there are two kinds of anisospory, morphological (distribution of spore sizes bimodal) and functional (unimodal). With these considerations in mind, perhaps a relatively simple way to describe species of *Macromitrium* and *Schlotheimia* would be as follows:

(a) morphologically isosporous but functionally anisosporous;
(b) both morphologically and functionally anisosporous.

There is no problem in retaining the term isosporous without qualification for monoecious species. The degree of anisospory (e.g. weak or strong) would then only be used in discussions on the variability of spores between taxa. Perhaps the chief difficulty with these new criteria is that they are based on experimental studies and therefore cannot be readily incorporated into standard taxonomic descriptions of species. Far more significant is the fundamental question: with dioecism so frequent in bryophytes, why is anisospory so rare? To find an answer almost certainly requires a multidisciplinary approach. Any investigator must be prepared to explore the fine structure of sporogenesis together with chromosome cytology and sexuality in relation to gametophore morphogenesis. It would be particularly fascinating to discover whether it is possible to regenerate females from dwarf males and vice versa by extensive subculturing of gametophore fragments, similar to that recently carried out on *Equisetum* gametophytes (Duckett, 1977). *Macromitrium* and *Schlotheimia* undoubtedly provide a most favourable experimental system for elucidating the evolutionary significance and the morphogenetic basis of the origins of anisospory.

ACKNOWLEDGEMENTS

Much of this work received financial support from the Australian Research Grants Committee. Dr C. J. Quinn and Mr P. Wallin each provided specimens of one species. Mr P. Wallin, Mrs S. Lowry and staff in the electron microscope units at the University of New South Wales and the University of Reading assisted with SEM studies and photography. Dr A. Touw at the Rijksherbarium, Leiden, Mr A. Eddy and Dr A. Harrington at the British Museum (Natural History) and the Director and staff at the New South Wales National Herbarium allowed me free access to specimens. I am grateful to all these people and particularly to Dr E. V. Watson and other staff at the University of Reading who assisted in many ways and provided the facilities necessary to enable completion of the work.

REFERENCES

ALLEN, C. E. (1935). The genetics of bryophytes, I. *Bot. Rev.* **1**, 269–291.
ALLEN, C. E. (1945). The genetics of bryophytes II. *Bot. Rev.* **11**, 260–287.

ANDERSON, L. E. (1954). Hoyer's solution as a rapid permanent mounting medium for bryophytes. *Bryologist* **57**, 242–244.

ANDERSON, L. E. (1979). Cytology and reproductive biology of mosses. (In press.)

ANDERSON, L. E. and BRYAN, S. (1958). Chromosome numbers in mosses of North America. *J. Elisha Mitchell scient. Soc.* **74**, 173–199.

ANDO, H. (1972). Studies on the genus *Hypnum* Hedw., I. *J. Sci. Hiroshima Univ.*, Ser. B, Div. 2, **14**, 53–73.

BERRIE, G. K. (1974). Sex chromosomes of *Plagiochila praemorsa* Stephani and the status of structural sex chromosomes in hepatics. *Bull. Soc. bot. Fr.* **121**, Colloq. Bryol., 129–135.

BOLD, H. (1967). "Laboratory Manual for Plant Morphology." Harper and Row, New York.

BOROS, Á. and JÁRAI-KOMLÓDI, M. (1975). "Atlas of Recent European Moss Spores." Akad. Kiadō, Budapest.

BROTHERUS, V. F. (1924–1925). Musci (Laubmoose). *In* "Die natürlichen Pflanzenfamilien" (A. Engler, ed.). Second edition, Vol. 10, 129–478; Vol. 11, 1–542. W. Engelmann, Leipzig.

CRUM, H. A. (1976). "Mosses of the Great Lakes Forest." (Revised edition) University of Michigan, Ann Arbor.

DENING, K. (1935). Untersuchungen über sexuellen Dimorphismus der Gametophyten bei heterothallischen Laubmosen. *Flora, Jena* **30**, 57–86.

DOYLE, W. T. (1970). "The Biology of Higher Cryptogams." Macmillan, New York and London.

DUCKETT, J. G. (1970). Spore size in the genus *Equisetum*. *New Phytol.* **69**, 333–346.

DUCKETT, J. G. (1977). Towards an understanding of sex determination in *Equisetum*: an analysis of regeneration in gametophytes of the subgenus *Equisetum*. *Bot. J. Linn. Soc.* **74**, 215–242.

DURING, H. J. (1977). A taxonomical revision of the Garovaglioideae (Pterobryaceae, Musci). *Bryophyt. Biblthca* **12**, 1–244.

ERDTMAN, G. (1965). "Pollen and Spore Morphology/Plant Taxonomy, Gymnospermae, Bryophyta (Text). An Introduction to Palynology, III." Almqvist and Wiksell, Stockholm.

ERNST-SCHWARZENBACH, M. (1936). Zur Heterosporie in der Laubmoos Gattung *Macromitrium*. *Verh. schweiz. naturf. Ges.* **117**, 310–311.

ERNST-SCHWARZENBACH, M. (1938). Dimorphismus der Sporen und Zwergmännchen-Problem in der Laubmoos Gattung *Macromitrium*. *Annls bryol.* **11**, 46–55.

ERNST-SCHWARZENBACH, M. (1939). Zur Kenntnis der sexuellen Dimorphismus der Laubmoos. *Arch. Julius Klaus-Stift. VererbForsch.* **14**, 362–474.

ERNST-SCHWARZENBACH, M. (1942). Weitere Mittelungen über den sexuellen Dimorphismus der tropischen Laubmoos Gattung *Macromitrium*. *Arch. Julius Klaus-Stift. VererbForsch.* **17**, 458–461.

ERNST-SCHWARZENBACH, M. (1943). The sexual dimorphism of the tropical mosses of the genus *Macromitrium*. *Farlowia* **1**, 195–198.

ERNST-SCHWARZENBACH, M. (1944). La sexualité et le dimorphisme des spores des Mousses. *Revue bryol. lichén* **14**, 105–113.

FLEISCHER, M. (1904–1923). "Die Musci der Flora von Buitenzorg." E. J. Brill, Leiden.

FLEISCHER, M. (1920). Über die Entwicklung der Zwergmännchen aus sexuell differenzierten Sporen bei den Laubmoosen. *Ber. dt. bot. Ges.* **38**, 84–92.

FRITSCH, R. (1972). Chromosomenzahlen der Bryophyten. *Wiss. Z. Friedrich-Schiller-Univ., Jena*, Math.-nat. Reihe, **21**, 839–944.

GOEBEL, K. (1928). "Organographie der Pflanzen." Third edition. G. Fischer. Jena.

GREGUSS, P. (1964). The phylogeny of sexuality and triphyletic evolution of the land plants. *Acta biol., Szeged*, N.S. **10**, 1–50.

HESLOP-HARRISON, J. and HESLOP-HARRISON, Y. (1970). Evaluation of pollen viability by enzymatically induced fluorescence—intracellular hydrolysis of fluroescein diacetate. *Stain Technol.* **45**, 115–120.

HIROHAMA, T. (1977). Spore morphology of bryophytes observed by scanning electron microscope, III. Orhotrichaceae. *Bull. natn. Sci. Mus., Tokyo*, Ser. B, Bot. **3**, 113–121.

INOUE, S. (1964). Karyological studies on mosses, I. *Bot. Mag. Tokyo* **77**, 412–417.

JARVIS, L. R. (1974). Electron microscope observations of spore wall development in *Funaria hygrometrica*. *Proc. VIII int. Congr. el. Micr. Canberra* **2**, 620–621.

KHANNA, K. R. (1971). Sex chromosomes in bryophytes. *Nucleus, Calcutta* **14**, 14–23.

KOPONEN, T. (1971). Male plants of *Rhizomnium*. *Hikobia* **6**, 47–49.

LEWINSKY, J. (1977). The genus *Orthotrichum*. Morphological studies and evolutionary remarks. *J. Hattori bot. Lab.* **43**, 31–61.

LEWIS, K. R. (1961). The genetics of bryophytes. *Trans. Br. bryol. Soc.* **4**, 111–130.

LOVELAND, H. F. (1956). "Sexual Dimorphism in the Moss Genus *Dicranum* Hedw." Ph.D. Dissertation, University of Michigan, Ann Arbor.

MIYOSHI, N. (1963). Heterospores observed in some Japanese *Macromitriums*. *Hikobia* **3**, 294.

MOGENSEN, G. S. (1978). Spore development and germination in *Cinclidium* (Mniaceae, Bryophyta) with special reference to spore mortality and false anisospory. *Can. J. Bot.* **56**, 1032–1060.

MONROE, J. H. (1968). Light- and electron-microscopic observations on spore germination in *Funaria hygrometrica*. *Bot. Gaz.* **129**, 247–258.

MUELLER, D. M. J. (1974). Spore wall formation and chloroplast development during sporogenesis in the moss *Fissidens limbatus*. *Am. J. Bot.* **61**, 525–534.

NEHIRA, K. (1976). Protonema development in mosses. *J. Hattori bot. Lab.* **41**, 157–165.

NEIDHART, H. V. (1975). "Electron–mikroskopische und physiologische Untersuchungen zur Sporogenese von *Funaria hygrometrica*. Hedw." Ph.D. dissertation, University of Hannover.

NEIDHART, H. V. (1978). Sporogenese bei Archegoniaten: zur Natur eines Nachreifephänomens bei Sporen von *Funaria hygrometrica* Hedw. (Musci). *Bryophyt. Biblthca* **13**, 169–194.

NEWTON, M. E. (1972). Sex-ratio differences in *Mnium hornum* Hedw. and *M. undulatum* Sw. in relation to spore germination and vegetative regeneration. *Ann. Bot.* **36**, 163–178.

NISHIDA, Y. (1978). Studies on the sporeling types in mosses. *J. Hattori bot. Lab.* **44**, 371–454.

NOGUCHI, A. (1967). Musci Japonici VII. The genus *Macromitrium*. *J. Hattori bot. Lab.* **30**, 205–230.

ONO, K. (1970a, b, c). Karyological studies on Mniaceae and Polytrichaceae with special reference to structural sex-chromosomes. I, II, III. *J. Sci. Hiroshima Univ.* Ser. B, Div. 2, **13**, 91–105; 107–166; 167–221.

PHILIBERT, H. (1883). Les fleurs mâles du *Fissidens decipiens*. *Revue bryol.* **10**, 65–67.

RAMSAY, H. P. (1966). Sex chromosomes in *Macromitrium*. *Bryologist* **69**, 293–311.

RAMSAY, H. P. (1973). Unusual sporocytes in *Dicnemoloma pallidum* (Hook) Wijk and Marg. *Bryologist* **76**, 178–182.

SAINSBURY, G. O. K. (1955). A handbook of the New Zealand mosses. *Bull. R. Soc. N.Z.* **5**, 1–490.

SCHELLENBERG, G. (1920). Verteilung der Geschlechtsorgane bei den Bryophyten. *Beih. bot. Zbl.* **37**, 115–153.

SCHUSTER, R. M. (1966). "The Hepaticae and Anthocerotae of North America." Vol. 1, Columbia University Press, New York.

SCOTT, G. A. M. and STONE, I. G. (1976). "The Mosses of Southern Australia." Academic Press, London and New York.

SUSSEX, I. M. (1966). The origin and development of heterospory in vascular plants. *In* "Trends in Plant Morphogenesis" (E. G. Cutter, ed.) pp. 140–152. Longmans, London.

TATUNO, S. (1941). Zytologische Untersuchungen über die Lebermoose von Japan. *J. Sci. Hiroshima Univ.* Ser. B, Div. 2, **48**, 73–187.

TATUNO, S. and KISE, Y. (1970). Cytological studies on *Atrichum* with special reference to structural sex chromosomes and sexuality. *Bot. Mag.* Tokyo **83**, 163–172.

TATUNO, S. and SEGAWA, M. (1955). Über strukturelle Geschlechtschromosomen bei *Mnium maximowiczii* und Nukleolinuschromosomen bei einigen Bryophyten. *J. Sci. Hiroshima Univ.*, Ser. B, Div. 2, **7**, 1–9.

TRYON, A. F. (1964). *Platyzoma*—a Queensland fern with incipient heterospory. *Am. J. Bot.* **51**, 939–942.

VALANNE, N. (1966). The germination phases of moss spores and their control by light. *Annls bot. fenn.* **3**, 1–60.

VITT, D. H. (1968). Sex determination in mosses. *Mich. Bot.* **7**, 195–203.

VITT, D. H. (1973). A revisionary study of the genus *Macrocoma*. *Revue bryol. lichén.* **39**, 205–220.

WALLACE, M. H. (1970). "Developmental morphology and sexual dimorphism of *Homalothecium megaptilum*. (Sull.) Robins." Ph.D. dissertation, Washington State University.

WOESLER, A. (1935a). Zur Frage der Sexualdimorphismus bei Laubmoosen. *In* "Proceedings VI Internationaal Botanisch Congress" Vol. 2, pp. 143–145. E. J. Brill, Leiden.

WOESLER, A. (1935b). Zur Zwergmännchenfrage bei *Leucobryum glaucum* Schpr, 1. *Planta* **24**, 1–13.

14 | Taxonomic Implications of Cell Patterns in Haplolepidous Moss Peristomes

S. R. EDWARDS

Manchester Museum, The University, Oxford Road, Manchester M13 9PL, Great Britain

Abstract: Since Philibert divided the Bryopsida into the Haplolepidae and Diplolepidae in 1884, bryologists have questioned whether these groups are natural. The problem has been confused by the false assumption that the terms haplolepidous and diplolepidous refer to single and double peristomes respectively. This has led to some families being assigned to the wrong group.

An easily observable arrangement of the cells that form the peristome has been noted as characteristic of the Haplolepidae. It is present in all five haplolepidous orders, but has not with certainty been found elsewhere. It is therefore suggested that the Haplolepidae is a natural, monophyletic group which can be characterized by this cell pattern, except in species where the peristome is very degenerate or otherwise secondarily aberrant. The possible occurrence of the pattern in two diplolepidous orders might suggest a link between the two groups but this seems unlikely.

In addition, a distinctive type of double haplolepidous peristome has been recognized in *Seligeria* and *Glyphomitrium*, which genera may well be related. The peristomes of *Dicranoweisia* and *Venturiella* have features in common with those of the *Seligeria* type and the implications of this are considered.

INTRODUCTION

1. Brief Taxonomic History

In the Musci, the higher taxonomic ranks such as classes and orders have for a long time been largely defined by peristome characters; in recent years,

Systematics Association Special Volume No. 14, "Bryophyte Systematics", edited by G. C. S. Clarke and J. G. Duckett, 1979, pp. 317–346, Academic Press, London and New York.

however, there have been very few studies on peristome structure as a taxonomic character. Now that many bryologists are turning their attention to taxa of higher rank than the family (Miller, this volume, Chapter 2), it is important that the characters which have traditionally been used to define such taxa should be reassessed and, where appropriate, that new observations should be made.

This paper is concerned with peristome structure in the Bryopsida, the class of Musci which contains the great majority of moss species. Philibert, at the turn of the century, produced a monumental series of 18 papers in the *Revue Bryologique* which dealt at length with the so-called arthrodontous peristome of the Bryopsida. Fortunately, this remarkable work has recently been abridged and translated into English by Taylor (1962) and, for convenience, it is Taylor's version of Philibert's papers which will be referred to here. Serious students of peristome structure should of course refer to Philibert's original works.

Philibert coined the terms Haplolepidae and Diplolepidae for the two major groups within the Bryopsida, which are characterized by haplolepidous and diplolepidous peristomes; these terms are explained below. This basic division was adopted by Fleischer (1904–1923) and subsequently by Brotherus (1924–1925) who, however, did not use the terms Haplolepidae and Diplolepidae. Dixon (1932) used the terms as the names for subclans (see p. 34), but since then they seem to have fallen into disrepute with taxonomists, as have the growth forms acrocarpous and pleurocarpous.

Reimers (1954) referred to "Haplolepidae" and "Diplolepidae" in his introduction and key to the orders, but always in inverted commas and never as taxa. He did, however, describe as haplolepidous those orders which were traditionally placed in the Haplolepidae. Lewinsky (1976) compared the two groups but using characters other than those of the peristome.

Crosby and Magill (1977) have produced the most recent review of moss classification above the level of genus, but they mention no category above the level of family and give an undivided list of the 87 currently accepted families from the Sphagnaceae to the Polytrichaceae (see pp. 38–39). Thus we now have a sequence of families which has been handed down with progressively less understanding, and which includes some misfits that might have stood out if the families had been segregated into orders. Most haplolepidous mosses are indeed grouped together, but some odd placements appear amongst the diplolepidous families.

2. Misconceptions about the Haplolepidae and Diplolepidae

The neglect of the Haplolepidae and Diplolepidae as taxonomic groups is not so much due to any error on the part of Philibert, but to the fact that his work has been misunderstood by later workers. This is, perhaps, not surprising since none of his 18 papers was illustrated. Thus most bryologists have assumed that "haplolepidous" describes a peristome with a single rank of teeth around the capsule mouth and that "diplolepidous" indicates a peristome with two such ranks; this supposition seems to be confirmed by the wrong assumption that the prefixes "haplo-" and "diplo-" refer to the number of these ranks.

Philibert's original concept was not primarily concerned with the number of ranks of teeth, but with the way the teeth are formed from columns (or vertical "rows") of cells arranged around the circumference of the mouth:

> Essentially, then, the outer peristome of the Arthrodontae is of two types, the teeth of one having double external plates and single internal ones and the teeth of the other having single external plates and usually double internal ones. I denote the outer plates by the Greek word *lepos*, scale; and will call HAPLOLEPIDAE those mosses with an articulated peristome having a single row of external plates and DIPLOLEPIDAE those where this outer row is double. The DIPLOLEPIDAE are the only one which have a double peristome, one of the indications that these divisions are natural ones. (Taylor, 1962, pp. 178–179.)

Thus, according to Philibert, in order to decide whether a species is haplolepidous or diplolepidous, it is necessary to examine the rank of teeth lying on the outside of the peristome. But in the Haplolepidae, this peristome rank is generally the sole or main structure, and this is homologous with the inner of the two ranks usually found in the Diplolepidae; the outer diplolepidous rank generally has no (or only a rudimentary) homologue in the Haplolepidae. So, as Philibert was certainly aware (Taylor, 1962, pp. 199–200), his classification of species into Haplolepidae and Diplolepidae depended on comparison of non-homologous structures. Although his observations were essentially correct (Blomquist and Robertson, 1941), the only effective conclusion for which he is generally remembered is that the Haplolepidae has one rank of teeth whereas the Diplolepidae has two. It so happens that this is usually the case, but it was the many exceptions in both groups, together with the apparently intermediate group Heterolepidae, that have particularly helped to discredit his work.

It is ironic that Philibert's division is none the less essentially sound, and that despite his interest in cell patterns he never recognized a far sounder and more basic distinction between the two groups. It is the aim of this paper to clarify the distinction by demonstrating the difference between the patterns of cells in homologous parts of the peristome, and to show the fundamental significance to taxonomy of these cell patterns.

1. Summary of Previous Work

The peristome is a repetitive structure found round the mouth of the ripe moss capsule; it is developed internally and revealed when the lid has fallen off. The peristome generally has 2^n elements, i.e. 4, 8, 16, 32 or 64 teeth, and in the Bryopsida is formed of one, two or occasionally more concentric ranks. The teeth are formed from thickenings of the periclinal wall-pairs of the inner three amphithecial layers which, starting from the outside, have been designated the OPL (Outer Peristomial Layer), **PPL** (Primary Peristomial Layer) and IPL (Inner Peristomial Layer), by Blomquist and Robertson (1941). Additional layers to the outside may be designated OPL2, OPL3 and so on.

The equivalence of these layers between the haplolepidous and diplolepidous type of peristome has been well summarized by Blomquist and Robertson (1941), and by Kreulen (1972) who has shown how in both cases the IPL may be traced down the capsule to the outer spore sac. Such equivalence appears to be true in those mosses examined for this paper.

The peristome formed between the OPL and the **PPL** is referred to as the exostome (or outer peristome) and the peristome formed between the **PPL** and the IPL is referred to as the endostome, whether it is the inner rank of a double peristome, or the sole peristome rank. The term prostome is sometimes used to indicate a rudimentary structure outside, and generally adhering to, the base of the main peristome teeth, irrespective of whether these are exostome or endostome.

Since Philibert's time little significance has been placed on the patterns of the cells that constitute the peristome, i.e. their number and arrangement. As an illustration of this attitude, it is worth quoting Taylor (1962) in his own introduction to the translation of Philibert's articles: "Between various types of tooth and segment the differences are mainly not in the number and arrangement of the cells, but in the amount and location of the thickening deposited on the inner surfaces of the hollow cells." The significance of these thickenings is illustrated by Kreulen (1972), and their taxonomic value is not to be underestimated.

2. The Peristomial Formula

This paper is not primarily concerned with the thickenings of the peristomial cell walls, but rather with the cell patterns. In order to codify these patterns, a peristomial formula has been devised which can be used in an analagous fashion to the floral formula of flowering plants. Taking transverse sections and counting the number of cells all the way round each peristomial layer results in large, cumbersome numbers which, moreover, are unreliable because of irregularities in the development of the peristome and the limited vertical extent of any particular cell pattern. It has therefore been found most convenient always to work from a segment equivalent to an arc of 45°, i.e. one eighth of the circumference of the mouth; this would, for example, comprise two teeth of a peristome with 16 teeth. The cell counts are given starting from the outside and taking each layer in turn as if the peristome were seen in transverse section, although it will be seen below that in practice the pattern may best be deciphered from mature teeth viewed from the inside or outside. For clarity, the **PPL** has been indicated here in bold type.

Various refinements can be incorporated into this basic system as the following examples show. *Bryum capillare* and *Rhizogonium spiniforme* (Hedw.) Bruch have the formula 4:**2**:6c, indicating the numbers of cells per 45° arc in the OPL, **PPL** and IPL respectively. The "c" indicates that in these species the IPL is displaced laterally by half a cell and thus occurs as a vertical column of cells between the outer teeth. *Funaria hygrometrica* would be designated 4:**2**:4z. In this case the "z" indicates that the IPL is not displaced laterally and so appears between the outer teeth as a vertical zig-zag line of wall-pairs.

It is also possible to indicate the degree of thickening of the walls by means of single or double inverted commas. *Funaria hygrometrica* could thus be indicated 4':"**2**:4z, and *Brachydontium trichodes* by 4:**2**":3.

Finally, prostomial development and other reduced or occasional variations generally restricted to the base of the teeth, can be shown by means of parentheses. Thus the peristomial formula of *Schistidium apocarpum* could be given as (4":)**2**":'3 indicating that the OPL development is prostomial, and *Octoblepharum albidum* Hedw. could be given as **2**(−3)":"2z indicating that the variation to the **PPL** is occasional and restricted to the base of the teeth.

The devising of a peristomial formula does not necessarily place more significance on cell patterns than on gross peristome morphology, which can easily be described in the conventional fashion. But certainly cell patterns can have a deep underlying taxonomic significance across a wide range of superficially disparate peristomes, particularly in the case of the Haplolepidae.

3. The Examination of the Teeth

It is an advantage in this kind of study to work with mature teeth. The reasons for this are twofold; firstly it is particularly convenient when working with herbarium material, and secondly it is of great value for seeing cell patterns in perspective since they frequently vary from the base of the teeth to the tip.

The pattern is generally taken from the lower portion of the teeth; the variations that occur above may be disregarded for the purpose of the peristomial formula. The basal membrane, when present, generally exhibits the **PPL**:IPL pattern more or less uniformly. It is best to assess, after examining several peristomes, what basic patterns are present per 45° arc, and to use these rather than a single random count that does not take into consideration any possible variations from the characteristic pattern of the species.

The disadvantage of working with mature teeth lies in the possibility of misinterpreting the equivalence of the peristomial layers, though in practice there is rarely much risk of this. As will be seen, the parts of most peristomes are readily identifiable, but a few comments may help in difficult cases. A 4:2 count almost invariably indicates OPL:**PPL**, and, as I hope to show below, a 2:3 count indicates **PPL**:IPL of the Haplolepidae. The reverse assumptions do not always hold good.

As a last resort, it is possible to take longitudinal sections, after thorough soaking, of well-dried but slightly immature capsules from the herbarium, and to trace the outer spore sac (usually two layers) up the capsule until it forms the IPL (one layer). The illustration of *Hypodontium dregei* (Fig. 9a) was taken from such material and the cell layers are clearly recognizable.

THE HAPLOLEPIDAE

1. The Haplolepidous Pattern

As mentioned above, Philibert's main criterion for distinguishing the Haplolepidae was that the outer face (**PPL**) of each endostome tooth is composed of a single column of cells. Figure 1 is a diagrammatic representation of part of the haplolepidous peristome of *Dicranella heteromalla* and it shows clearly that there is one peristome tooth present for each column of **PPL** cells. There is, however, another very characteristic feature of dicranoid peristomes which can be seen here. For each pair of **PPL** cell columns, and hence for every two teeth, there are three columns of IPL cells, and so the inner surface of each tooth is composed of 1½ columns of cells. The teeth are thus asymmetric and form pairs; one member of each pair is left-handed and the other right-handed, and this asymmetry is often most distinctive. The horizontal wall-pairs of the columns of IPL cells (trabeculae) can be traced across the gap between the teeth of each pair since they were once joined, but there is no such continuity

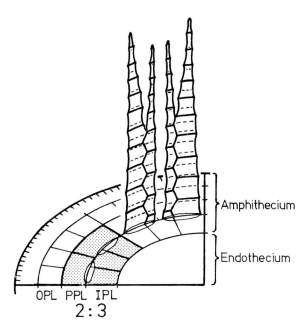

Fig. 1. Stereogram of a haplolepidous peristome, loosely based on *Dicranella hetero-malla*; 2 out of 16 teeth are shown on a 90° segment of the capsule mouth. OPL=Outer Peristomial Layer, **PPL**=Primary Peristomial Layer, IPL=Inner Peristomial Layer.

between the pairs. Expressed in terms of the peristomial formula, such a pattern would be **2**:3.

Philibert, however, believed that the inner face of each tooth in *Dicranum*, which has the same structure as in *Dicranella*, was composed of 2 IPL cells rather than 1½; according to him the formula would thus be **2**:4z, differing essentially from that of *Funaria hygrometrica* (Diplolepidae) only in the lack of outer teeth (Taylor, 1962, p. 199). Apparently he never appreciated that the inner two columns of IPL cells in each pair of teeth were in fact a single column split between the teeth, although he had evidently once seen the phenomenon in *Grimmia* without recognizing it as anything more than a peculiarity (Philibert, 1888). There, he observed, the two columns of internal plates of each tooth were very unequal, and "quelquefois alors" a plate that starts on one tooth continues on the neighbouring tooth. (This paper was one of many published by Philibert concurrently with, but not included in, his series of peristome articles, and thus was not translated by Taylor (1962).)

But the 2:3 pattern has been noted before on several occasions, by Kienitz-Gerloff (1878) and Evans and Hooker (1913) in *Ceratodon purpureus*, by Kreulen (1972) in *Dicranella heteromalla* and *Leucobryum glaucum*, and most recently by Mueller (1973) in *Fissidens limbatus* Sull. In almost all these examples the pattern was noted only in section, rather than in mature teeth where it would have been more easily seen. The ratio was always given for the whole circumference (i.e. 16:24) but the figures often varied somewhat from these because of irregularities in the development of the peristome and probably also because of variations in the precise level of the section.

All of the above authors except Mueller reported the pattern in species of the Dicranales. Most of them considered but a single species and even Kreulen in his more wide-ranging paper investigated only two further haplolepidous species. Mueller's (1973) observations were on a species from the only other order that shows a typical dicranoid peristome, the Fissidentales.

Thus none of these authors was in a position to propose the 2:3 pattern as a haplolepidous character. The pattern certainly seems characteristic of the dicranoid peristome, but is it present in the other haplolepidous orders, and is it diagnostic for the Haplolepidae as a whole? It is clearly time to conduct a wide-ranging survey of peristomes, selecting examples not only from all five haplolepidous orders (Dicranales, Fissidentales, Pottiales, Grimmiales and Syrrhopdontales), but also from all eight of the other orders of the Bryopsida as well.

2. Variation in the Haplolepidous Orders

(a) *Dicranales*. The Dicranales possess the classic haplolepidous peristome, as found in *Campylopus, Ceratodon, Dicranella* (Figs 2a, 7a), *Dicranum, Leucobryum* and many other genera. Each of the 16 teeth is typically forked into two prongs (hence the name *Dicranum* and the term dicranoid), and the basal membrane is low or absent. The ventral surface of the teeth clearly shows the IPL pattern as thickened trabeculae, and although the dorsal surface is usually almost equally thickened, the **PPL** pattern is only visible as distinct lines rather than trabeculae; the dorsal plates are frequently ornamented with striae.

Several genera such as *Rhabdoweisia* have a reduced or somewhat irregular peristome, in which both dorsal and ventral plates may be equally thin (total thickness about 4 μm), both the dicranoid shape and the 2:3 pattern may be irregular and almost lost (Fig. 3b), and the vertical striae may become weak and oblique or horizontal. However, the structure remains essentially similar to

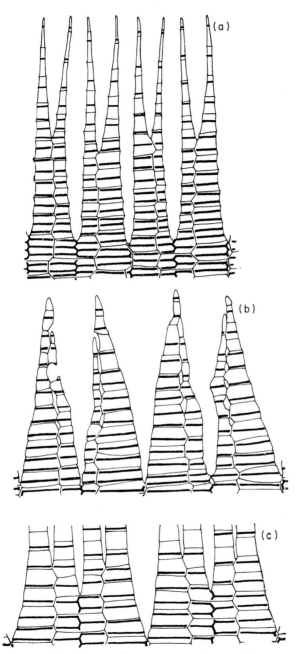

Fig. 2. Semidiagrams of peristomes showing 4 out of 16 teeth from the inside. (a) *Dicranella heteromalla*; (b) *Schistidium apocarpum*; (c) *Ptychomitrium polyphyllum*. Heavy lines=**PPL**, thin lines=IPL.

that of the normal dicranoid peristome, even to the presence of reduced ventral trabeculae, and there is no reason to regard it as distinct.

Octoblepharum, Leucophanes and related genera have recently been separated from Leucobryum (Leucobryaceae) and placed in the Calymperaceae (Crosby and Magill, 1977). Leucobryum has a typically dicranoid peristome and remains in the Dicranales, but the transferred genera are considered under the Syrrhopo-dontales in this paper. Some genera, such as Seligeria, have a peristome of a significantly different structure, and these are dealt with later.

(b) Fissidentales. The Fissidentales possess a peristome that is essentially the same as that described for the Dicranales. A full dicranoid peristome is found in Fissidens, and may perhaps best be distinguished from that of the Dicranales by the ornamentation which is more typically papillose than striate dorsally, and which may appear to form spiral thickenings round the distal filaments. Reduced forms have been described for Octodiceras and Moenkemeyera.

(c) Pottiales. The peristome in the Pottiales generally has a distinct membrane, which may be very short and hidden within the capsule mouth, or a structure 1 mm or more high. The teeth are frequently forked or divided into 32 fila-ments which arise more or less equally from the basal membrane. Both faces may be thickened and ornamented, but the **PPL** pattern on the outer face may be particularly prominent in the form of ridges or trabeculae (Fig. 7b, c).

The **2:3** pattern is usually well shown, even in those peristomes with a small basal membrane such as Barbula (Fig. 4b). Those species with a tall membrane, such as in Tortula subgenus Tortula, show the pattern in a particularly attrac-tive fashion (Fig. 4a); the tube can easily be slit lengthwise and viewed from the inside.

(d) Grimmiales. Teeth in the Grimmiales may be simple, perforate (Fig. 2b), or divided into 32 somewhat paired filiform segments (Fig. 3c); the basal membrane may be low or absent. Both faces may be thickened and ornamented but, as in the Pottiales, the outer face is typically more strongly developed and trabeculate. The **2:3** pattern is usually well shown, except in genera such as Coscinodon where it is obscured.

The Ptychomitriaceae are sometimes placed here, for example by Smith (1978). Ptychomitrium and Campylostelium are clearly haplolepidous (Fig. 2c), and they both show a marked similarity to the teeth of Racomitrium (Fig. 3c). In section, the peristome of Campylostelium (Fig. 9c) agrees well with that of Racomitrium (Fig. 9d), but Ptychomitrium (Fig. 7e) lacks the dorsal trabeculae

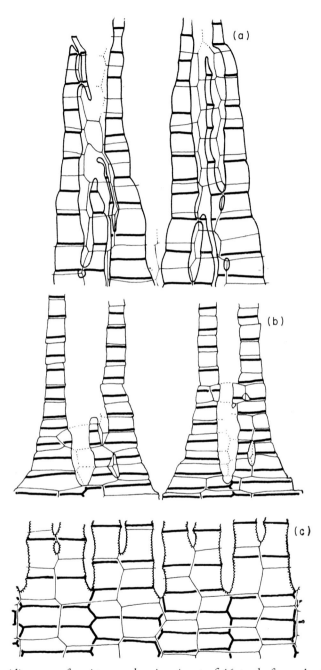

Fig. 3. Semidiagrams of peristomes showing 4 out of 16 teeth from the inside. (a) *Eustichia longirostris* (Brid.) Brid.; (b) *Rhabdoweisia crispata*; (c) *Racomitrium canescens*. Heavy lines = **PPL**, thin lines = IPL.

and also possesses a characteristic air-gap at the base. Altogether, this family seems quite at home in the Grimmiales, and is certainly wrongly placed in its traditional position amongst the diplolepidous mosses.

(e) *Syrrhopodontales.* The peristome of the largely tropical order Syrrhopodontales has usually been ignored by bryologists based in temperate regions; it has been reported as being double in *Mitthyridium* (Mitten, 1868), and is single in *Syrrhopodon*. Teeth may be simple, perforated or divided, as well as papillose, striate or smooth, and in the large genus *Calymperes* the peristome is absent.

Hypodontium, a genus split from *Syrrhopodon*, has a more complex peristome than might appear at first examination (Fig. 5b). The 16 endostome teeth are typically haplolepidous, with strong dorsal thickenings and somewhat thinner ventral plates, but outside the endostome are 32 narrow exostome teeth, thickened exclusively on the dorsal surface (Fig. 9a); these however are frequently shed with the operculum. Other peristome developments also remain within the operculum: further up the OPL, above the exostome, there occur scattered thickenings on the opposite (outer) wall, and moreover similar thickenings may be repeated on the OPL2, although these are variable. If these last mentioned structures had remained as part of the functional peristome, they would be represented as 64 very narrow teeth. The peristomial formula (including the OPL2 when present) is $(''8:'4'':)2'':'2(-3)$; the IPL figure of 3 is often irregular and weak, but is occasionally well marked enough (Fig. 5b) to be regarded as present.

In species of *Syrrhopodon* the IPL pattern may be lost completely and the ventral plates may be relatively thicker and even trabeculate, as in *S. gaudichaudii* Mont. The report by Mitten (1868) that the peristome of *Mitthyridium* is double ("Peristomium duplex, externum dentibus 16, internum ciliis totidem, membranae parum exsertae impostis, conniventibus, nullisve.") seems odd; the species upon which Mitten based his description, *M. fasciculatum* (Hook. and Grev.) Robins, as well as others, have been examined, and the peristome is similar to that of *Syrrhopodon*.

Of the genera recently transferred from the Dicranales to the Syrrhopodontales (Crosby and Magill, 1977), two have been examined. *Octoblepharum albidum* has only eight simple teeth, so the formula $2(-3)'':'2z$ refers to a single tooth; there are no trabeculae since both dorsal and ventral plates are slightly convex. Such reduction leaves little clue as to relationships, but *Leucophanes candidum* (Schwaegr.) Lindb. has a peristome that is typical of the Syrrhopodontales. It has 16 simple teeth, a prostome of 32 teeth (Brotherus, 1924, Fig.

179), and a peristomial formula of $(4'':)2'':'2(-3)$ with a very poorly represented **2**:3 pattern. Despite certain similarities in section (Fig. 9e) to the dicranoid peristome (Fig. 7a), and strong papillosity (which is common in *Syrrhopodon*), it shows the same fundamental structure as that of *Hypodontium* (Figs 5b, 9a).

(f) *Peristomes of the* Seligeria *type*. In addition to the mosses that fit into these five orders (Dicranales, Fissidentales, Pottiales, Grimmiales, Syrrhopodontales), a number of haplolepidous species have been found to possess a distinctive type of double peristome. The endostome has little or no thickening ventrally, and the IPL pattern is only decipherable as fine lines. The **PPL** pattern is represented by strong dorsal trabeculae, and an exostome adheres to the margins of these as a thin membrane; the membrane is quite without any thickening for most of the tooth, but at the base there are often OPL depositions in the form of a prostome. None the less, the OPL pattern is visible as a median vertical zig-zag line, and is quite clear if the teeth are viewed from the outside with transmitted light (or particularly if a scanning electron microscope is used, Fig. 11). The teeth are usually simple and reduced, or irregularly divided, and they may be smooth or variously papillose.

The **2**:3 pattern is usually very reduced, and detectable only as an irregular third column (often represented by only one or two cells) between alternate teeth (Fig. 12). But even when the third column is quite absent, the zig-zag line separating the remaining two columns is very irregular and displaced from side to side, whereas the zig-zag line between the pairs is relatively regular. Such teeth may thus be designated $4('')$:**2**$'':2(-3)$, and this formula applies equally to *Seligeria acutifolia* (Fig. 8c) and *Glyphomitrium daviesii* (Figs 6a, 8d). This type of tooth is characteristic of the Seligeriaceae, and similar reduced forms may be found in *Brachydontium* and *Blindia*. Stone (1973) observed the double nature of the peristome in *Brachydontium intermedium* Stone, but although she illustrates a longitudinal section, she clearly interprets both the exostome and the trabeculae as being ventral, which they are not. Confusion may have arisen for two reasons:

(i) there is also a quite distinct ventral membrane formed by the relatively unthickened IPL walls, which may laterally exceed the **PPL** thickenings (as in *Blindia*, Fig. 6c), and which may become detached in *Brachydontium* (Fig. 8b); and

(ii) it is traditionally assumed that thin membranes are inner, and that teeth in the Dicranales have ventral trabeculae. This problem does emphasize the need to illustrate peristome sections in the context of the capsule mouth.

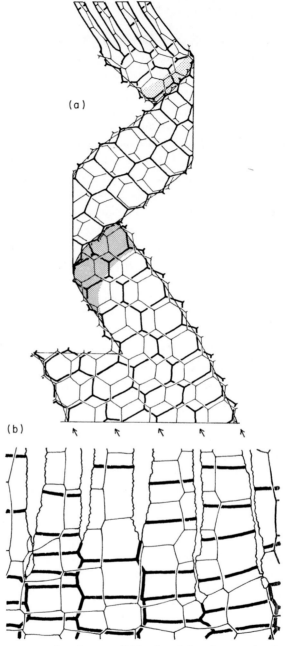

FIG. 4. (a) Stereogram of peristome of *Tortula subulata* showing basal membrane for one quarter (below) and one eighth (above) peristome; (b) semidiagram of peristome of *Barbula spadicea* showing 4 out of 16 teeth from the inside. Heavy lines=**PPL**, thin lines=IPL.

The teeth of *Glyphomitrium* (Figs 6a, 8d, 11, 12) are remarkably similar to those of the Seligeriaceae. The main reason why this similarity has not been noticed may be the fact that the tooth pairs (as generally understood here with regard to the **2**:3 pattern) are splayed to such an extent that each half is fused with the adjacent tooth of a neighbouring pair; in other words, the naturally visible tooth pairs are misleading. In *Blindia* teeth are normally or hardly paired, but in *Seligeria acutifolia*, reverse pairing does occur distinctly and regularly, and this unusual character confirms the relationship with *Glyphomitrium*.

The peristome of *Dicranoweisia* (Figs 5a, 8a) has also been found to have a similar structure, despite three notable differences: (i) the teeth have a particularly distinct and regular third IPL column, (ii) the ventral face is somewhat thickened, and (iii) the overall shape of the teeth is somewhat dicranoid. None the less, *Dicranoweisia* has teeth that appear to be essentially of the *Seligeria* type, rather than of the type found in true dicranoid peristomes.

The peristome of *Venturiella* needs to be considered here, if only because of its historical association with *Glyphomitrium*. The taxonomic position of the two genera has recently been discussed, first by Noguchi (1952) who placed both genera in the Erpodiaceae (Orthotrichales), and then by Crum (1972) who argues that *Glyphomitrium* is best left in the Ptychomitriaceae. The teeth of *Venturiella* are strongly dorsally trabeculate, and also have a rudimentary unthickened basal exostome (Fig. 7g); these are characters of a haplolepidous peristome, although not of the dicranoid type. The ventral surface of the teeth is somewhat thickened as in *Dicranoweisia*, and both surfaces are densely papillose, but the peristomial formula is (4:')**2**":'2z with no sign of a **2**:3 pattern (Fig. 6b). This is effectively the pattern indicated in sections taken by Noguchi, except that he interpreted the teeth of both *Venturiella* and *Glyphomitrium* as being exostomal. As has been shown above, Noguchi is almost certainly wrong about *Glyphomitrium* and, although the teeth of *Venturiella* are lacking in much definitive information, it seems at least as reasonable to conclude that they are probably endostomal and possibly haplolepidous. The peristomial formula and the location of secondary thickening in *Venturiella* does fit well with that of the Orthotrichales which, as will be seen, may have more in common with the Haplolepidae than with the other diplolepidous orders. But at least the problem as to whether the teeth are endostomal or exostomal might be resolved by taking longitudinal sections of nearly mature capsules. This was done by Noguchi, but he gives no indication of the relationship between the peristomial layers and the outer spore sac. Material seen for the present paper was too old, and the relevant cell walls had broken down.

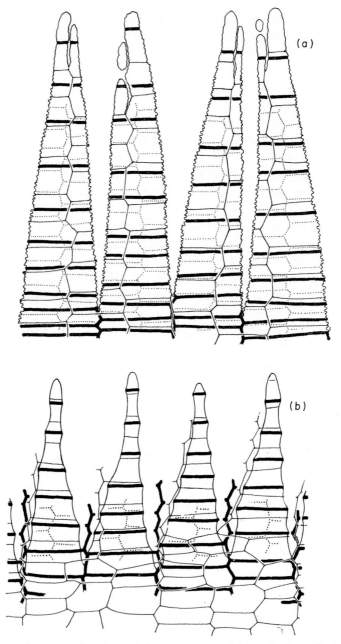

FIG. 5. Semidiagrams of peristomes showing 4 out of 16 teeth from the inside. (a) *Dicranoweisia cirrata*; (b) *Hypodontium dregei* (Hornsch.) C. Muell. Dotted lines = OPL, heavy lines = **PPL**, thin lines = IPL.

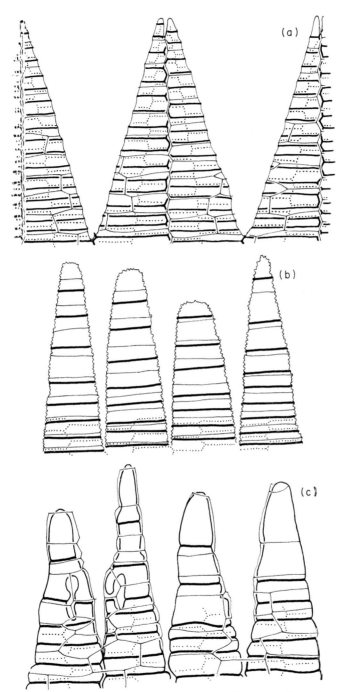

F<small>IG</small>. 6. Semidiagrams of peristomes showing 4 out of 16 teeth from the inside. (a) *Glyphomitrium daviesii*; (b) *Venturiella sinensis* (Vent.) C. Muell. var. *angusti-annulata* Griffin and Sharp; (c) *Blindia acuta*. Dotted lines = OPL, heavy lines = **PPL**, thin lines = IPL.

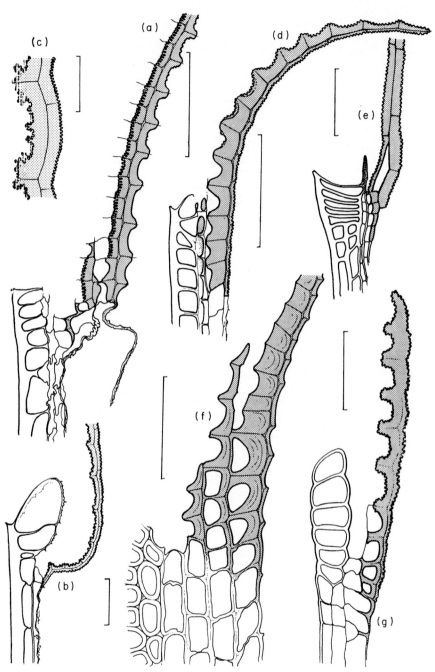

Fig. 7. Longitudinal sections of peristomes. (a) *Dicranella heteromalla*; (b), (c) *Tortula subulata*; (d) *Schistidium apocarpum*; (e) *Ptychomitrium polyphyllum*; (f) *Scouleria aquatica* Hook.; (g) *Venturiella sinensis* var. *angusti-annulata*. Scale lines = 50 μm.

Scouleria (traditionally placed in the Grimmiales) has an unusual peristome of 32 irregular teeth, and an unusual formula of (8″:'4–8″:')4″:'8c. Such proliferation of the cell pattern occurs elsewhere in mosses (e.g. *Splachnum* and some *Bryum* species) and is likely to be secondary, but what remains is a marked similarity to the *Seligeria* type of peristome, especially when seen in section (Fig. 7f). Differences lie in the particularly strong development of one or two prostomes, and in the failure of the unthickened exostomal membrane to remain above the limit of secondary thickening.

3. Genera Transferred to the Haplolepidae

Apart from *Ptychomitrium* and *Campylostelium*, already discussed under the Grimmiales, and *Glyphomitrium*, already discussed under the *Seligeria* type, there remains a small group of three families that has consistently been placed amongst diplolepidous families. In fact the Phyllodrepaniaceae, Eustichiaceae and Sorapillaceae are traditionally placed between the Bryaceae and Rhizogoniaceae, which are two families of the Bryales that develop a perfect diplolepidous bryoid peristome.

Eustichia shows a peristome that is typically haplolepidous with a clearly marked **2:3** pattern (Fig. 3a). The inner face is slightly thickened (1–1½ μm) and although the IPL pattern is easy to see, the cross-walls only form very slight trabeculae (Fig. 9b). The dorsal surface of the teeth is similarly thickened, and additionally ornamented with strong vertical striae. The teeth themselves are somewhat irregular, varying from being simple, through perforate, to bifid, but altogether they agree closely to a slightly reduced dicranoid type.

All three families have leaves vertically oriented in a regularly distichous arrangement that appears superficially similar to that of *Fissidens*, and the leaves of *Sorapilla* prove to be structurally similar as well. Unfortunately the peristome of *Sorapilla* is reduced to 16 filiform teeth borne on a narrow basal membrane, and little information is available other than the fact that a **2:3** pattern is probably present in a reduced form. Peristome material seen of *Phyllodrepanium* was so poor as to be useless. None the less, assuming that there is no reason to separate the three families for the first time, they must clearly all be moved to either the Dicranales or Fissidentales, the final position possibly hanging on an investigation of the gametophore apical cell.

THE HETEROLEPIDAE

The so-called heterolepidous order Encalyptales falls outside the scope of this

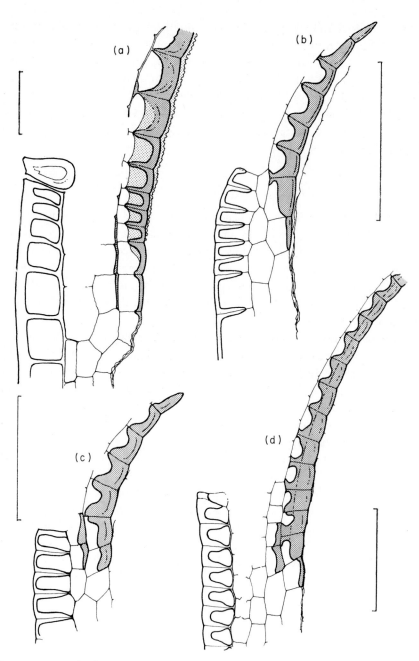

FIG. 8. Longitudinal sections of peristomes. (a) *Dicranoweisia cirrata*; (b) *Brachydontium trichodes*; (c) *Seligeria acutifolia*; (d) *Glyphomitrium daviesii*. Scale lines = 50 μm.

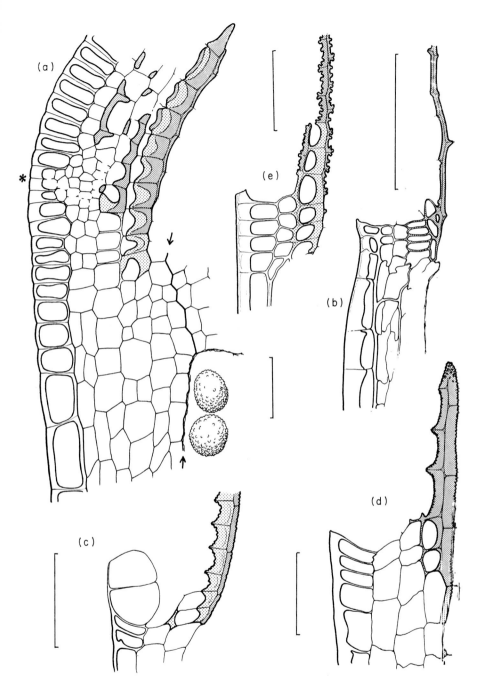

FIG. 9. Longitudinal sections of peristomes. (a) *Hypodontium dregei*; (b) *Eustichia longiro-stris* (Brid.) Brid.; (c) *Campylostelium saxicola*; (d) *Racomitrium canescens*; (e) *Leucophanes candidum* (Schwaegr.) Lindb. Arrowed heavy line = amphithecium/endothecium interface, * = point of operculum separation. Scale lines = 50 μm.

paper, but it is relevant enough to merit comment. As in *Hypodontium* (and various other genera), additional peristome layers may be developed, and this has given rise to the statement by Philibert (Taylor, 1962, p. 180) that the Encalyptaceae should be considered as the central point from which the Haplolepidae, Diplolepidae, and indeed the Nematodontae (in which he included the Polytrichales, Buxbaumiales and Tetraphidales) have diverged.

The endostome only may be developed as in *Encalypta rhaptocarpa*, or a more typical double peristome may be found as in *E. streptocarpa*. Some species like *E. longicollis* develop additional peristome layers and these may adhere to each other radially to form complex solid teeth. All *Encalypta* species so far examined have a peristomial pattern that is generally 4:2:4, and although occasional doublings and irregularities have been seen, no sign of a 2:3 pattern has yet been found. The absence of a 2:3 pattern certainly does not conclusively remove the order from the Haplolepidae, and the discovery of that pattern in just one member would dramatically alter the situation; but for the moment it seems best to exclude the Encalyptales, at least until more definite links are demonstrated.

THE DIPLOLEPIDAE

As with the Heterolepidae, the Diplolepidae falls outside the scope of this paper, but before the 2:3 pattern can be proposed as essentially haplolepidous, it is necessary to demonstrate that it is not found in any of the seven diplolepidous orders. To prove this conclusively would require endless research, but a selective survey of over 50 mosses from these orders has revealed only two traces of the haplolepidous pattern, in the Orthotrichales and Isobryales, and these were equivocal.

The diplolepidous peristome generally has 16 sturdy exostome teeth that are mostly thickened ventrally, and a fragile endostome which is usually relatively unthickened. The range of variation is enormous, and this is reflected in the variety of IPL cell counts per 45° arc, which may be almost anything except for three since the teeth are symmetrical and the number will always be even. The OPL:**PPL** ratio generally remains constant (as in the Haplolepidae) at 4:2. The more complex peristomes such as the bryoid or hypnoid structure present a regular pattern with a high number of IPL columns per 45° arc; the more reduced peristomes found, for example, in the Orthotrichales or some Isobryales and Hookeriales, present smaller, less regular numbers of IPL columns. At this stage it is not certain whether peristomial cell patterns will prove to be of much taxonomic value in the diplolepidous mosses, but

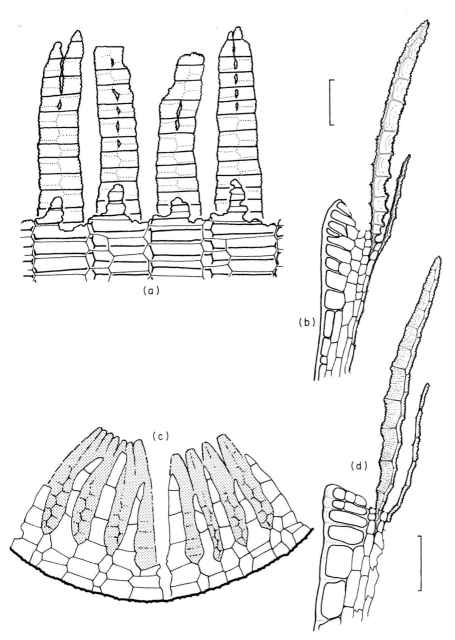

Fig. 10. Diplolepidous peristomes. (a), (b) *Leucodon maritimus* (Hook.) Wijk. and Marg. (a) semidiagram of one quarter peristome viewed from the inside (dotted lines = OPL, heavy lines = **PPL**, thin lines = IPL); (b) longitudinal section; (c) *Zygodon obtusifolius* Hook., approximately half a peristome viewed from the inside (redrawn from Malta, 1926); (d) *Orthotrichum striatum* longitudinal section (peristomial formula 4″:′**2**:′4c–z). Scale lines = 50 μm.

it is hoped that workers will record the peristomial formulae as a matter of course in the future. A very brief review of the orders is given below.

In the Bryales and Hypnales, a pattern of 4:2:6c is common, but the IPL number may vary from 4–8, and lateral displacement need not occur. The endostome typically has a well-developed basal membrane that often develops groups of two or three cilia opposite the outer teeth and usually has perforated teeth (or processes) between them. The exostome teeth are sturdy and trabeculate on the ventral surface. This peristome is generally known as the bryoid or hypnoid peristome, but many variations and reductions occur, especially in aquatic mosses.

In the Hookeriales the peristome is essentially bryoid; a pattern of 4:2:4c is common, but the IPL number is often 6, and lateral displacement need not occur. Endostome cilia are typically reduced or absent in the order, and the thickenings of the two OPL columns that face the dorsal surface of each exostome tooth are frequently separated by a median furrow.

The Isobryales is characterized by a considerable variation in peristome structure, and this is reflected by a similar variation in cell pattern; 4:2:6–8c is common in a typically bryoid peristome, but the IPL number varies from 2–10. Displacement need not occur, and either endostome or exostome may be severely reduced or absent. *Leucodon maritimus* (Hook.) Wijk and Marg. may be regarded as having a basic peristomial formula of 4:2:2z, in which the IPL columns may each irregularly cut off a smaller additional truncated column to one side. This would normally be designated as 4:2:2–4z, but in some capsules the additional column is produced fairly regularly only in alternate teeth, giving the appearance of a 4′:″2:′3 pattern (Fig. 10a, b).

The peristome of the Funariales is often very reduced or absent, but the most complex development may be found in plants such as *Funaria hygrometrica*, which has a pattern of 4:2:4z. The endostome teeth are opposite the exostome teeth and they generally adhere to them in the proximal region. The outer teeth are strongly trabeculate on the inside, as in the bryoid peristome, and are joined at their tips by a small perforated epiphragm. In other genera such as *Discelium* the endostome is lacking and the exostome reduced.

In the Splachnales, the genus *Splachnum* is noted for its aberrant peristome in which the **PPL** proliferates beyond the normal number of **2**. The three species so far examined share a pattern of 4″:′4–8:′8z, with the endostome both opposite to, and adhering to the exostome. *Tetraplodon* lacks an endostome and has the normal **PPL** number of **2**, so that altogether the peristome of the Splachnales has much in common with that of the Funariales.

Members of the Orthotrichales so far seen have a reduced peristome with

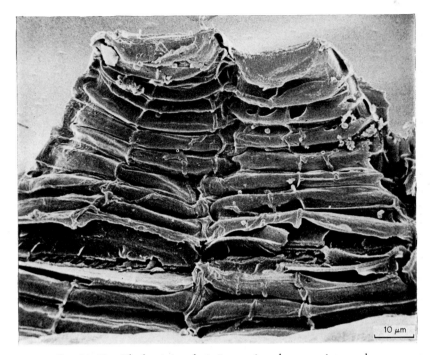

FIGS 11, 12. *Glyphomitrium daviesii*, scanning electron micrographs.

FIG. 11. Naturally occurring pair of teeth from the outside showing trabeculae and also a median zig-zag line up the middle of each tooth, ×1070.

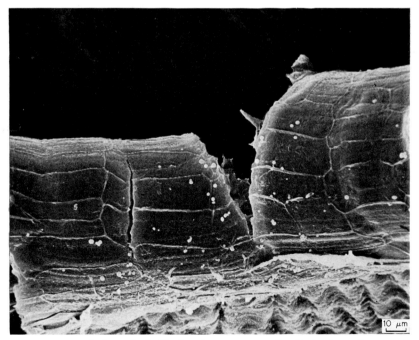

FIG. 12. Four teeth from the inside showing remnants of a third column between two naturally occurring pairs (i.e. within one **2**:3 pair, see text p. 331). n.b. there are also fungal hyphae and spores present. ×530.

a formula that varies from 4:**2**:2 to 4:**2**:6, with variable displacement. The endo-stome is typically reduced or absent, or may exist only as a basal membrane. Although the exostome is thickened and trabeculate dorsally rather than ventrally, the dorsal thickening when seen in longitudinal section (Fig. 10d) still maintains the striate markings typical of the other diplolepidous orders. The teeth may be paired as in *Zygodon* Hook. and Tayl., or lacking altogether as often in *Macromitrium* Malta (1926) illustrates an interesting peristome from *Zygodon obtusifolius*, which appears to show a distinct **2**:3 pattern (Fig. 10c). The figure is convincing, but I have so far failed to find such a pattern in this or any other species of the order.

These seven orders of diplolepidous mosses form a heterogeneous group. The Splachnales and Funariales are possibly closely related, and the four orders that possess a bryoid peristome, the Bryales, Hypnales, Hookeriales and at least some of the Isobryales, are likely to form another group. The Orthotrichales may represent a third, more isolated group, but links may be sought on one side with the Isobryales, and on the other side with the three pottioid orders (Pottiales, Syrrhopodontales and Encalyptales).

DISCUSSION

Although the present survey of about 100 arthrodontous species is hardly comprehensive, it is wide-ranging and is, I hope, sufficient to establish the **2**:3 pattern as a diagnostic haplolepidous character that is not found elsewhere. The five orders of the Haplolepidae therefore appear to form a natural, mono-phyletic group, characterized by a **PPL**:IPL ratio of **2**:3, except in species where the peristome is very degenerate or otherwise secondarily aberrant.

For separating orders within the Haplolepidae, peristome types have been found to be of limited value. Those observations that are of use may be sum-marized as follows. The greater thickening and presence of trabeculae on the ventral face rather than the dorsal face is typical only of the dicranoid peristome of the Dicranales and Fissidentales; the Dicranales typically have vertical striae on the dorsal face which, in the Fissidentales is more often papillose. Division of the peristome into 32 filiform segments rather than into 16 divided teeth occurs particularly in the Pottiales and Grimmiales, and both these orders have teeth that are often papillose on both sides; an extensive basal membrane is characteristic of the Pottiales, but many of the Grimmiales have large, solid, triangular teeth. Strong dorsal thickening with trabeculae, exostomal development, and a **2**:3 pattern which is very reduced or absent is typical of the Syrrhopodontales and peristomes of the *Seligeria* type; peristomes of the *Seligeria* type have an adhering, relatively unthickened exostomal membrane, whereas in the Syrrhopodontales the development of the exostome is usually restricted to the regions of marked secondary thickening at the base of the teeth.

Peristomes of the *Seligeria* type are of particular interest since they do not seem to have been described before. Such peristomes have been found in genera placed in the Dicranales, Grimmiales, and also amongst diplolepidous mosses. Gametophyte and other sporophyte characters have not been fully investigated, but it does seem probable that *Glyphomitrium* and *Seligeria* are related. Although this type of peristome could have been developed on a number of separate occasions, it is at least as probable that genera with *Seligeria* type peristomes are monophyletic. If the latter should be the case, then *Scouleria* could be seen as a derived form which is, at least in part, an adaptation to an aquatic habitat; the method of dehiscence is somewhat similar to that in the genus *Trochobryum* of the Seligeriaceae. *Dicranoweisia* and *Venturiella* both have teeth which, although they are papillose, also have similarities with the *Seligeria* type of peristome. The possibility of a relationship between all these genera should be considered; if they were brought together in a single order, it would be no more disparate than many others.

The similarity between the peristomes of *Leucophanes* and *Hypodontium* is of particular significance, since it confirms the moving of *Leucophanes* and related genera (including *Octoblepharum* and *Exodictyon*) from the Dicranales to the Syrrhopodontales (Crosby and Magill, 1977). Certain arguable similarities seen in section between the peristome of *Leucophanes* and dicranoid teeth can hardly be held as evidence of a close connection between two such basically different structures. Vegetatively, although the leaf base of *Exodictyon* is remarkably similar to that of *Syrrhopodon*, the sandwiching of chlorocysts by hyalocysts in the leaf of both *Leucobryum* (which has an exemplary dicranoid peristome and remains in the Dicranales) and of the transferred genera, is unlike anything found in the Calymperaceae. The question thus remains as to whether the transferred genera do form a link between the Dicranales and the Syrrhopodontales, or whether the unusual leaf structure has arisen twice in disparate orders or even three times if *Theriotia* (Buxbaumiales) is considered.

Since all the haplolepidous orders have been shown to possess a **PPL**:IPL ratio of **2**:3 despite the remarkable range in form and function, this pattern is likely to indicate ancient relationships. If this is so, then the occurrence of a similar pattern in the Orthotrichales and Leucodontaceae (Isobryales) could have one of four implications:

(1) the pattern has evolved separately in these cases;

(2)   the pattern in these taxa represents a primitive condition from which both the Haplolepidae and the remaining diplolepidous mosses have evolved;

(3)   the pattern in these taxa is intermediate between an ancestral haplolepidous peristome and a derived diplolepidous one; or

(4)   it is intermediate between an ancestral diplolepidous peristome and a derived haplolepidous peristome.

Until the **2**:3 pattern is more convincingly demonstrated in the Orthotrichales or Leucodontaceae, it must be assumed that the cell arrangements in these taxa have a purely superficial similarity to that of the Haplolepidae; this is effectively the first possibility listed above. The second suggestion seems unlikely since there are few associated characters to indicate that the structure of the peristome in the Orthotrichales, etc., might be ancestral to all arthrodontous peristomes. The remaining two alternatives may be considered irrespective of whether the peristome of the Orthotrichales, etc., represents a link condition. The first, suggesting that the haplolepidous condition is primitive, is in accord with the traditional linear sequence of taxa. But the second, suggesting that the diplolepidous condition is primitive, fits best with Crosby's proposal of a hypothetical ancestor for the Bryopsida that is not unlike *Bryum* (Crosby, 1980). In this case the traditional sequence of taxa must be reversed, probably starting with the Splachnales and Funariales, moving through the bryoid orders to the Isobryales, the Orthotrichales, the pottioid orders, and ending with the Dicranales and Fissidentales. It must be stressed that such a linear sequence is not intended to be a representation of the course of evolution, but rather an indication of the relationships of living groups.

Other characters which bear on this problem can briefly be discussed here. In searching for the ancestor of the Haplolepidae, Philibert unequivocally decided that it lay within the Encalyptales; he also noted that the similarity of the peristome of *Encalypta* to that of *Syrrhopodon* (Taylor, 1962, p. 212) but did not suggest that the Syrrhopodontales forms a link between the Encalyptales and the rest of the Haplolepidae. Links between the three pottioid orders, the Encalyptales, Syrrhopodontales and Pottiales, are strong. They all have hob-nail papillae on the chlorocysts, similar pores in the hyalocysts (not generally known to be present in *Tortula* subgenus *Tortula*), a strong, often excurrent costa, and a large calyptra. Interestingly, all these characters have been reported from the diplolepidous Orthotrichales, although I could not find the hyalocyst pores reported by Lorch (1931) in *Ulota*. In addition, the structure and often position of the foliar gemmae in *Ulota* are very similar to those in the Syrrhopodontales.

Against such arguments for the reversed linear sequence suggested above is Dixon's statement (1932): "Philibert's suggestion that *Encalypta* . . . marks the point of divergence of the Haplolepidae from the Diplolepidae, cannot be upheld". Also, it might be expected that the ancestral haplolepidous moss would be likely to possess a clear and regular **2**:3 pattern; most haplolepidous orders exhibit reduction series from such a peristome. The Pottiales, which arguably possesses the most primitive characters of the Haplolepidae, shows the pattern well, but the putative modern representatives of the ancestors of this group within the Diplolepidae do not. For some reason, the **2**:3 pattern is distinctly shown only in the absence of a marked exostome.

Despite Dixon's (1932) widely accepted view that no classification based on gametophytic characters must cut across one based on peristome types, any systematic arrangement must take all available evidence into account. This paper is restricted to one aspect of peristome structure, and does not propose any taxonomic revisions on the strength of this character alone. But new information and some new lines of thought are provided here and it is hoped that these may be taken into account when a broader view is taken. Peristomes have largely been ignored since Philibert, and those studies which have been made have generally been concerned with precise investigations of one species. Such studies may hide valuable information that can be revealed by cruder, wider surveys.

ACKNOWLEDGEMENTS

I am grateful to the Keeper of Botany at the British Museum (Natural History) and to Professor P. W. Richards for the loan of material, and to Dr A. Harrington for his help.

REFERENCES

BLOMQUIST, H. L. and ROBERTSON, L. L. (1941). The development of the peristome in *Aulacomnium heterostichum. Bull. Torrey bot. Club* **68**, 569–584.

BROTHERUS, V. F. (1924–1925). Musci (Laubmoose). *In* "Die natürlichen Pflanzenfamilien" (A. Engler, ed.). Second edition, Vol. 10, 129–478; Vol. 11, 1–542. W. Engelmann, Leipzig.

CROSBY, M. R. (1980). The diversity and relationships of mosses. *In* "Mosses of North America: a Symposium." American Association of Science. (In press.)

CROSBY, M. R. and MAGILL, R. E. (1977). "A Dictionary of Mosses." Missouri Botanical Garden, St Louis.

CRUM, H. (1972). The dubious origin of *Glyphomitrium canadense* Mitt. *J. Bryol.* **7**, 165–168.

DIXON, H. N. (1932). Classification of Mosses. *In* "Manual of Bryology" (F. Verdoorn, ed.) pp. 397–412. Martius Nijhoff, The Hague.

EVANS, A. W. and HOOKER, H. D., Jr (1913). Development of the peristome in *Ceratodon purpureus. Bull. Torrey bot. Club* **40**, 97–109.

FLEISCHER, M. (1904–1923). "Die Musci der Flora von Buitenzorg." E. J. Brill, Leiden.

KIENITZ-GERLOFF, F. (1878). Untersuchungen über die Entwickelungsgeschichte der Laubmoos-Kapsel und die Embryo-Entwickelung einiger Polypodiaceen. *Bot. Ztg* **36**, 33–48, 49–64.

KREULEN, D. J. W. (1972). Features of single- and double-peristomate capsules. Homology of layers and ontogeny of outer spore sac. *Lindbergia* **1**, 153–160.

LEWINSKY, J. (1976). On the systematic position of *Amphidium* Schimp. *Lindbergia* **3**, 227–231.

LORCH, W. (1931). Anatomie der Laubmoose. *In* "Handbüch der Pflanzenanatomie" (K. Linsbauer, ed.) Vol. 7. Borntraeger, Berlin.

MALTA, N. (1926). Die Gattung *Zygodon* Hook. et Tayl. Ein monographische Studie. *Latv. Univ. bot. Darza Darbi* **1**, 1–185.

MITTEN, W. (1868). A list of the Musci collected by the Rev. Thomas Powell in the Samoa or Navigator's Islands. *J. Linn. Soc. Bot.* **10**, 188.

MUELLER, D. M. J. (1973). "The Peristome of *Fissidens limbatus* Sullivant." University of California Press, California.

NOGUCHI, A. (1952). Musci Japonici II. Erpodiaceae. *J. Hattori bot. Lab.* **8**, 5–17.

PHILIBERT, H. (1888). La fructification du *Grimmia Hartmanni*. *Revue bryol.* **14**, 49–52.

REIMERS, H. (1954). "Bryophyta. Moose." *In* "Syllabus der Pflanzenfamilien". Twelfth edition (H. Melchior and E. Werdermann. eds) Vol. 1, 242–268. Borntraeger, Berlin.

SMITH, A. J. E. (1978). "The Moss Flora of Britain and Ireland." Cambridge University Press, Cambridge.

STONE, I. G. (1973). A new species of *Brachydontium* from Australia. *J. Bryol.* **7**, 343–351.

TAYLOR, E. C. (1962). The Philibert peristome articles. An abridged translation. *Bryologist* **65**, 175–212.

15 | Rhizoids and Moss Taxonomy

A. C. CRUNDWELL

Department of Botany, The University,
Glasgow G12 8QQ, Great Britain

Abstract: Although it has long been realized that rhizoid characters can have taxonomic value they have been largely neglected. Most rhizoids are stem-borne and arise from initials surrounding the buds, which are usually axillary; but as well as these macronemata there may also be micronemata, unrelated to the buds. Micronemata are to macronemata as paraphyllia to pseudoparaphyllia. Leaf-borne rhizoids are commoner than is generally realized, especially from the lower part of the abaxial surface of the midrib. Papillose rhizoids are significant mainly at the species level, though those of the Bartramiaceae are distinctive. Rhizoid pigments fall into two groups differing in colour and pH reactions, but little is known of their chemistry. The influence of environmental factors on rhizoid development needs further study. The use of rhizoid characters in taxonomy is illustrated in the Campylopodioideae, Splachnaceae and Plagiotheciaceae.

INTRODUCTION

It is a truism of taxonomy that all the characters of the organism should be studied, that all have something to contribute and that none can safely be ignored. Most specimens of mosses bear rhizoids, and the characters of the rhizoids are not usually difficult to observe; but rhizoid characters, unless they are very striking, are rarely included in the descriptions of species, and to most taxonomists rhizoids are like bubbles in the preparation, noticed only when they get in the way. Yet from the earliest days of moss taxonomy it has been realized that rhizoid characters can be useful. Hedwig in 1794 described a *Hypnum tomentosum* (now *Racopilum tomentosum*). *Bryum radiculosum* was described by Bridel in 1817 and *Brachythecium erythrorhizon* by the authors of

Systematics Association Special Volume No. 14, "Bryophyte Systematics", edited by G. C. S. Clarke and J. G. Duckett, 1979, pp. 347–363, Academic Press, London and New York.

the *Bryologia Europaea* in 1853. It is typical of the superficial treatment that rhizoid characters have received the *Bryum radiculosum* has rhizoids no larger or more numerous than in most other species of *Bryum* and that in *Brachythecium erythrorhizon* they are exactly the same colour as in most other species of *Brachythecium*. The flora of Limpricht (1886–1904) which contains much useful information on rhizoid morphology and distribution, is a noteworthy exception.

The treatment adopted here is firstly to discuss the various characters of rhizoids, what is known of them, how they vary and what use they are in taxonomy. This survey is followed by studies of a few particular taxonomic groups to illustrate the uses and limitations of rhizoid characters. It must be emphasized that rhizoid characters are of no special taxonomic significance and that many reproductive characters are undoubtedly much better guides to relationships; but rhizoids may have a practical use in identification, for they are nearly always present, while reproductive structures are often absent.

MOSSES WITHOUT RHIZOIDS

Nearly all moss species are known to produce rhizoids, but there are many in which they are rare or produced only in small numbers. In some pleurocarpous moss species the majority of specimens are without rhizoids and these species fall into two ecological categories. Firstly there are a number of mosses of wet peaty places with a permanently high water table: species of *Sphagnum*, *Drepanocladus*, *Hygrohypnum*, *Calliergon* and *Scorpidium*. Secondly there are some terrestrial species that grow in turf, or at least not normally closely attached to soil or rock: *Pseudoscleropodium purum*, *Pleurozium schreberi*, *Rhytidium rugosum* and *Hylocomium* spp. Probably nearly all of these produce rhizoids occasionally, either when juvenile or when, exceptionally, they are closely attached to wood, rock or soil. There are however a few species, such as *Calliergon wickesii* Grout, in which rhizoids have never been observed and in which the ability to produce them may have been lost.

THE DISTRIBUTION OF RHIZOIDS

It is not within the scope of this study to consider rhizoids that have been reported as growing from the sporophyte, the calyptra or the vaginula. References to these are given by Lorch (1931). Neither will rhizoids borne directly upon the protonema be considered, though comparative study of these, based upon the cultivation under standard conditions of a wide range of species, would probably be very rewarding. Attention is here confined to rhizoids occurring on the stems and leaves of the gametophyte.

In many pleurocarpous mosses there is differentiation into primary and secondary stems and into stems of indefinite growth and branches of definite growth. In some species there are stoloniferous stems or branches on which the rhizoids are especially numerous; and in very many species there are branches from which rhizoids are absent or in which they are confined to the base. These features are of great interest to the taxonomist but the distribution of rhizoids is here an incidental part of the wider subject of differentiation of the plant body. It would be inappropriate to deal with it in isolation, and these phenomena are therefore not considered here.

1. Stem-borne Rhizoids

In most mosses, whether the rhizoids are restricted to the lower parts of the shoot or whether they are also present at higher levels in the form of a tomentum, they are borne on the stem, not on the leaves, and arise from initial cells that surround the buds. These buds are usually situated in the axils of the leaves as for instance in *Bryum* and in *Isopterygium pulchellum* (Fig. 8) but sometimes they occupy other positions, as for instance in *Tortula* subgenus *Syntrichia*, where they are at the upper ends of the internodes, just below the junction of stem and midrib. A systematic survey of the positions of buds in mosses would be most interesting. In some species and genera the rhizoid initials do not merely form a ring round the bud but extend up the internode in the form of a double line, as in *Aulacomnium palustre* and sometimes in *Splachnum vasculosum* (Fig. 1). In this and other species the rhizoid initials are sometimes differentiated only on one side of the bud, when they are arranged in the form of an arc or of an oblique line. Sometimes the rhizoid-bearing area stretches out over a much larger proportion of the stem surface as in *Cinclidium* (Koponen, 1974) and in *Aulacomnium heterostichum* (Hedw.) Bry. eur. At the extreme bases of the stems of acrocarpous mosses, and often too at the points of origin of branches, rhizoids also may be borne over larger areas of stem surface, and the relationship of the rhizoids to the buds and leaves may be obscured.

In prostrate pleurocarpous mosses, and on the plagiotropic shoots of the Mniaceae (Koponen, 1968a), rhizoids are not developed from the cells surrounding the buds on all sides of the stem, but only from the neighbourhood of those buds on the side of the stem toward the substrate. Frequently, as at the bases of acrocarpous shoots, the rhizoid-bearing areas may be extended so that

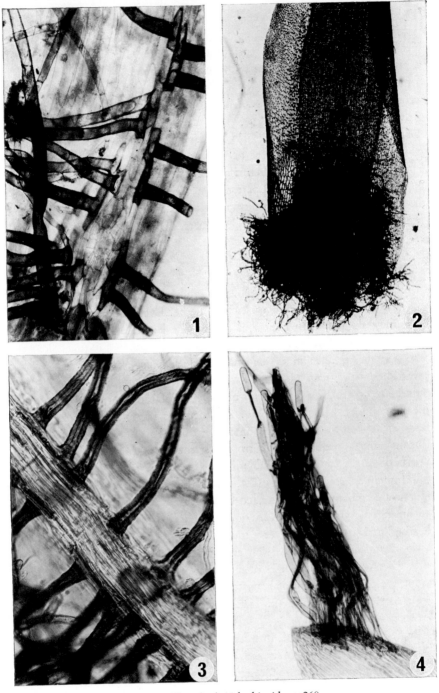

FIG. 1. *Splachnum vasculosum*, axillary bud with rhizoids. ×260.

FIG. 2. *Campylopus shawii*, leaf with rhizoids from base of midrib. ×30.

FIG. 3. *Homalothecium nitens*, rhizoids from back of midrib. ×260.

FIG. 4. *Calliergon stramineum*, rhizoids from leaf apex. ×160.

neighbouring ones become contiguous. This may happen all along the length of the stem, so that the rhizoids are evenly distributed, or there may be an alternation of rhizoid-bearing and non-rhizoid-bearing lengths of stem.

In a study of the family Mniaceae, Koponen (1968a) reported that in some species there were two types of stem-borne rhizoid, which he termed macronemata and micronemata. The macronemata, which are large and often profusely branched and arise from large initial cells surrounding the buds, are similar to the stem-borne rhizoids of most other mosses. The micronemata are thinner, shorter and less branched, and arise from initials with no special relationship to leaves or buds. Macronemata are present in all species. Micronemata are absent or confined to the extreme base of the stem in the genera *Mnium*, *Trachycystis*, *Leucolepis*, *Cinclidium* and *Cyrtomnium*. Micronemata develop on the upper parts of the stem in all the species of *Plagiomnium*, *Pseudobryum*, *Orthomnion* and *Orthomniopsis*, though in the last two genera they are morphologically indistinguishable from the macronemata. Only in *Rhizomnium* are there both species with micronemata and species without them. In this genus the rhizoids provide a most useful and important character, distinguishing *R. magnifolium* (*R. perssonii* Kop.) and *R. pseudopunctatum*, species with micronemata, from *R. punctatum*, which has none (Koponen, 1968b). The presence of micronemata also separates *R. gracile* Kop. from the related *R. andrewsianum* (Steere) Kop.

This differentiation of rhizoids in the Mniaceae may well not be unique, but I have been quite unable to trace any record of similar differentiation in other families. The nearest approach that I have seen is in the genus *Tayloria*. In all species examined except the South American and West African *T. jamesonii* (Tayl.) C. Müll. the rhizoids are exclusively macronematal in nature, and arise only from initials immediately surrounding the axillary buds. In *T. jamesonii* however they are borne all over the surface of the internode, but with no difference in size and form between macronemata and micronemata.

There is an interesting and possibly significant comparison between the micronemata and macronemata of the Mniaceae and the paraphyllia and pseudoparaphyllia of many pleurocarpous mosses, these terms being used in the sense of Ireland (1971), whose study of pseudoparaphyllia in North American mosses has thrown so much light on these hitherto rather obscure structures. Ireland says of pseudoparaphyllia: "They are present only around the branch primordia and on mature branches, where they occur in small numbers. Quite often they have been confused with paraphyllia, which differ by not being restricted to a specific site and by usually occurring in abundance everywhere on the stems and branches." Pseudoparaphyllia thus have to paraphyllia approximately the same relationship as macronemata have to micronemata though this is much obscured by the terminology. "Macronemata" and "Micronemata" are not themselves the best of terms, for the essential and only constant difference between them is in position of origin, not size. "Pseudoparaphyllia" and "paraphyllia" likewise differ only in position, not in morphology; and Ireland himself suggests that filamentous and foliose paraphyllia were derived separately and independently from pseudoparaphyllia. There is need for a single inclusive term to cover both types of organ—perhaps it would be best to revert to the traditional more comprehensive use of the term "paraphyllia"—with appropriate subordinate categories for pseudoparaphyllia and paraphyllia as understood by Ireland.

2. *Leaf-borne Rhizoids*

Although moss rhizoids are normally borne on the stem, there are many more genera and species with leaf-borne rhizoids than is generally realized. Classification of these into micronemata and macronemata is only occasionally possible. Thus in the Mniaceae there are a limited number of species that bear rhizoids on the leaves as well as on the stems. These rhizoids usually arise from the decurrent wings of the leaves or from the back of the midrib near the base of the leaf. Leaf-borne rhizoids are found only in the genera *Orthomnion*, *Orthomniopsis* and *Plagiomnium*, all of which have micronemata, and to some of the micronema-bearing species of *Rhizomnium*. These rhizoids are similar morphologically to micronemata in those species in which the micronemata and macronemata are morphologically differentiated.

On the other hand, the leaf-borne rhizoids in those species of *Tortula* subgenus *Syntrichia* which have them are certainly macronemata. In *Tortula* the buds are not in the leaf axils but are at the upper ends of the internodes just below the leaf insertions. These buds are surrounded by rhizoid initials and some may be borne on the leaf, at the base of the midrib, rather than on the stem. Sometimes they are developed only at the upper sides of the buds, so that all the rhizoids are leaf-borne. From a rather limited number of observations it appears that in *Syntrichia* these minor differences in the arrangement of the rhizoids are not very constant and are of no great taxonomic significance. Limpricht (1896–1904) noted the occurrence of leaf-borne rhizoids in *T. papillosa* and *T. princeps* and I have seen them also in *T. laevipila* and *T. ruralis* subsp. *ruraliformis*.

In *Tortula* and in the Mniaceae the rhizoids on the leaves appear to have spread to them from the stem; but there are many other mosses in which there are rhizoids more rigidly limited to sites on the leaves. Thus in the genus *Campylopus* and its closest relatives the rhizoids of the tomentum always arise from the back of the midrib, usually all from the basal region (Fig. 2), but sometimes also from higher levels on the leaf. At the base of the stem and at the points of origin of branches there are stem-borne rhizoids in some species of *Campylopus* and in the related genus *Atractylocarpus*, so that here there is a morphological distinction between the tomentum and the basal rhizoid system. In *Homalothecium nitens* (Fig. 3) all the rhizoids arise from the back of the midrib, sometimes covering it from the base to well above half way. No other species of *Homalothecium* has leaf-borne rhizoids and this character supports the segregation of this species in a separate genus, *Toment-hypnum*.

Rhizoids arise from the back of the leaf apex in members of the three families Amblystegiaceae, Plagiotheciaceae and Leucobryaceae. In the Amblystegiaceae the phenomenon is confined to the two genera *Drepanocladus* and *Calliergon*. Warnstorf (1904–1906) made a not very successful use of this feature, characteristic of *D. fluitans* and *D. exannulatus*, in his key to the species of *Drepanocladus* in Brandenburg. In *Calliergon* apical groups of rhizoid initials are found in several species, e.g. *C. cordifolium*, *C. giganteum*, *C. richardsonii* (Mitt.) Kindb. ex Warnst., *C. megalophyllum* Mik. and *C. stramineum*. These initials can give rise to rhizoids in all these species (Fig. 4), but in several they do so only very rarely. Other species are quite without apical rhizoid initials, e.g. *C. sarmentosum*, *C. trifarium*, *C. wickesii* Grout, *C. cuspidatum*. Karczmarz (1971) pointed out that apical rhizoid initials were present only in species with long midribs.

Plagiothecium itself is the only genus of the Plagiotheciaceae in which apical rhizoids occur, and this is discussed in greater detail below. Finally in the Leucobryaceae apical rhizoids have been found in all the species examined of *Leucobryum*, *Leucophanes*, *Exodictyon* and *Arthrocormus*; and they also occur in *Octoblepharum* (A. J. Harrington, personal communication).

In almost all those species with apical rhizoids, rhizoids can also emerge from initials on or on either side of the back of the midrib, mainly in the upper part of the leaf. Development of rhizoids from the adaxial surface of the leaf is rare, but I have observed it in all four British species of *Dicranodontium*.

RHIZOID MORPHOLOGY

Little taxonomic use has so far been made of the gross morphology of the rhizoid system, and it will therefore only be touched upon briefly here. It is well known that the normal structure is one of uniseriate laterally branched filaments with oblique cross-walls. With the exception of *Andreaea* there are few departures from this pattern, though the cross-walls are not invariably oblique. Sometimes, as in *Aulacomnium palustre* (Fig. 5) they are transverse in some of the finer branches of the rhizoid system. Paul (1903) figures the tips of rhizoids of *Hypnum cupressiforme* penetrating birch bark and having cells with very thick walls, the cross-walls all being transverse. Paul's study of the specialized rhizoids attaching mosses to stone and bark needs extending. The rhizoid "wicks" of *Polytrichum commune* and other species (Vaupel, 1903; Wigglesworth, 1947) should also be noted. Beside these rather striking departures from the typical there is within the normal rhizoid system considerable range in rhizoid diameter and in the extent and manner of the branching, whether it is markedly unequal, producing a trunk-and-branches pattern of growth, as in *Aulacomnium* and *Breutelia*, or whether it is pseudodichotomous as in *Campylopus* and *Dicranum* spp. These characters certainly vary from plant to plant, but there is only a limited range of variation within each species and they could no doubt be put to some taxonomic use.

PAPILLOSE RHIZOIDS

Little attention has been paid to the structure of the rhizoid wall, but there are scattered observations in the literature on the occurrence of papillae on the surfaces of rhizoids. Although in the majority of mosses the rhizoids are smooth, papillose ones have been observed or have previously been reported in 15 European moss families, and further study would undoubtedly increase this number. The papillae are most striking in some members of the Bartramiaceae, such as *Plagiopus oederi* and *Anacolia webbii* (Mont.) Schimp., where the rhizoids may be described as spiny-papillose. In other families the papillae, though often conspicuous, project less sharply.

The degree of development of the papillae may vary greatly, even on the same plant. In the Bartramiaceae the variation is often discontinuous, the spiny-papillose rhizoid surfaces being interrupted here and there by lengths that are quite smooth. In most mosses, however, the variation is more or less continuous, the coarsest being the most papillose, the finest the least. Sometimes, as in *Bryum ruderale*, all the rhizoids are distinctly papillose except the very finest. In other species, such as *Isopterygium elegans*, all the rhizoids in many specimens are quite smooth, but in a few the largest are slightly though distinctly papillose.

There have not been many instances when the papillosity of rhizoids has been of practical use in taxonomy. In the *Bryum erythrocarpum* complex *B. radiculosum* has the rhizoids markedly more papillose than any related species, and the distinctly papillose rhizoids of *B. ruderale* provide a useful distinction from *B. violaceum* in which they are almost smooth (Crundwell and Nyholm, 1964). *B. elegans* can at times be troublesome to separate from *B. capillare* and *B. stirtonii*, but its highly papillose rhizoids are very distinctive (Syed, 1973).

THE COLOURS OF RHIZOIDS

We have no chemical analyses of the compounds of the rhizoid cell wall, but we do have some optical observations on those few chemical compounds that are coloured. We must not forget that in reporting on the colours of rhizoids we are using only one tool of the many available to study only a very small proportion of the different chemical compounds contained in the cell walls. The chemistry of rhizoid walls, and of their pigments in particular, is likely to remain inadequately studied for a long time. The cell-wall pigments of bryophytes are difficult to extract, and the quantities obtainable from rhizoids are minute. There are few workers in the field of the chemistry of bryophyte pigments and none of them is so masochistic as to work on rhizoids when so many larger structures await investigation.

To record the colours of rhizoids by means of the Munsell system or some other internationally accepted system of colour nomenclature is a complex process. The observed colour of a given rhizoid depends not only upon whether it is seen by reflected or by transmitted light and whether it is examined wet or dry, but also upon the intensity of the light with which it is examined. The colour also changes with the pH of the medium in which it is mounted. The density of the pigment, the diameter of the rhizoid and the thickness of the wall all affect the colour observed. Thus even when only a single pigment is involved, a large number of different colours can be recorded.

The main pigments of the rhizoids of mosses appear to fall into two groups. Firstly there is a group of red, crimson, purple and violet pigments that turn red in acid but become bluer in alkali. To this group belong the pigments of the tubers of *Bryum rubens* and *B. ruderale* (Crundwell and Nyholm, 1964). These pigments are evidently the "Membran–anthocyan" of Herzfelder (1921). They are discussed by Mårtensson and Nilsson (1974) under the heading of "Anthocyanins" and it is clear that next to nothing is known about their chemistry. The pigments of the second group, present in a slightly larger number of species than those of the first, are orange, orange-red, orange-brown

or red-brown. These become very much paler and yellower in acid, redder
and slightly darker in alkali. To this group belong the pigments of the tubers
of, for example, *Bryum alpinum* and *B. radiculosum*. The group was recognized
by Herzfelder and its chemistry is even less known than that of the first.

As well as these two main groups of pigments rhizoids also frequently
contain brown pigments, presumably belonging to the group of "Lignin type
pigments" discussed by Mårtensson and Nilsson.

Sometimes these pigments appear to be more or less pure, with seemingly
only one group represented in a particular species, but often pigments of both
groups are present in mixtures that vary in composition from plant to plant,
or even on the same plant. In some specimens of *Tetraplodon mnioides* the
pigments appear to be exclusively of the second group, but in others there
is clearly a mixture of both. In *Voitia nivalis* Hornsch. the finer branches of
the rhizoid system are orange and have pigments of the second group only,
while the main "trunks" are reddish and contain pigments of the first group
as well. In alkali the smaller rhizoids are reddish and darker, the thicker ones
violet. In acid the smaller ones are pale yellow, the thicker ones red to orange.

In spite of this variation, rhizoid colour is at times a very useful character.
Two mosses growing together rarely have rhizoids of exactly the same colour.
Tubers in the soil are often attached to rhizoids that become detached from
the parent plant in dissection but which can be identified from their colour
as belonging to a particular one of the constituents of the mixture. In all species
of *Anoectangium* examined the rhizoids are of a deep brownish crimson, with
pigments of the first group, while in *Gymnostomum* they are yellowish brown,
with pigments of the second group. This character was convenient to use
when sorting out material of *Anoectangium warburgii* from the superficially
similar *Gymnostomum calcareum* (Crundwell and Hill, 1977).

Rhizoid colour has long been used as a character to separate *Fissidens curnowii*
from *F. bryoides*. *Funaria attenuata* can be distinguished from *F. obtusa* by the
violet colour of its rhizoids, and this is a great help with stunted or juvenile
material. In the *Bryum erythrocarpum* complex *B. violaceum* and *B. ruderale*
differ from the remaining species in the colour of the rhizoids, and the latter
would be a very difficult species to recognize but for this character. In a recent
study of the *Bryum bicolor* complex in Belgium Wilczek and Demaret (1976)
made use of rhizoid colour, but Smith and Whitehouse (1978) found this
unsatisfactory.

RHIZOIDS AND THE ENVIRONMENT

To make the best use of rhizoid characters in taxonomy it is obviously desirable to know in what ways they are influenced by the environment. Unfortunately this is something we know very little about. In those pleurocarpous mosses in which the rhizoids appear at intervals along the stem we know nothing about the causative factors involved. We do not know what factors, such as light, gravity and moisture, acting singly or in combination, are responsible for the restriction of the rhizoids in plagiotropic moss shoots to the side of the stem toward the substrate (or, in *Homalothecium nitens*, to the leaves on this side of the stem). We do not even know whether such shoots have, in respect of their rhizoid production, an irreversible dorsiventrality, as in the thalli of *Marchantia*, or whether this dorsiventrality is dependent upon the continuing operation of environmental factors and is reversible.

Rhizoid production is undoubtedly influenced quantitatively by the environment. Birse *et al.* (1957) noted the production of abundant rhizoids in shoots produced by *Bryum pendulum*, *Ceratodon purpureus* and *Climacium dendroides* after burial by sand. Birse (1957) also reported that in a number of species the production of prostrate shoots, producing many rhizoids and adhering to the substrate by them, was increased by low humidity and high light intensity. There is no reason to doubt the accuracy of these observations, though these conditions are the opposite of those produced by burial in sand.

Species that normally produce few or no rhizoids often develop them when growing over stone or wood. Paul (1903) referred to a plant of *Calliergon trifarium* attached to an alder branch by rhizoids, which are otherwise quite unknown in this species. Karczmarz (1971) noted the occurrence of abundant rhizoids in the normally rhizoidless *C. cuspidatum* when it is growing over wood.

There is no unanimity about the relationship to water of the development of rhizoid tomentum in those species that have it. Paul (1903) was definite that very wet conditions lead to a reduction in tomentum. "Auch wenn acrocarpische Sumpfmoose ins Wasser geraten, bilden sie keine Rhizoiden aus; die *Meesea*-Arten, *Paludella squarrosa* Brid. und *Aulacomnium palustre* Schwägr. verlieren sogar ihren Stengelfilz." Corley (1976) in his study of *Campylopus pyriformis* and related species reached similar conclusions after prolonged cultivation of several species and reported that in the field the tomentum: "does not develop in conditions where the plant is normally immersed in water almost to the apex, nor where a continuous supply of water can be obtained from the air, as in the spray zones of waterfalls, nor where water drips off cliffs and ledges on to the plants beneath." The non-tomentose bog varieties *uliginosus* and *paludosus* of *C. flexuosus* (*C. paradoxus*) produced tomentum in cultivation. On the other hand, Koponen (1968a) reported the opposite state of affairs among those members of the Mniaceae that have micronematous stem tomentum. "In wet localities or when the stem is submerged the micronemata appear at an early stage and often are numerous. In drier conditions the micronemata sprout later and may be few in number." Confirmation of these observations seems desirable.

RHIZOIDS IN SELECTED FAMILIES AND SUBFAMILIES

1. Campylopodioideae

In the genus *Campylopus* the rhizoids arise from the back of the midrib near its base (Fig. 2). This is true both of very conspicuously tomentose species such as *C. schimperi* and of sparsely tomentose ones such as *C. schwarzii*. In most species some rhizoids also come from higher levels on the back of the midrib, though never from the leaf apex or near it. The tomentum is never stem-borne, though there may be a few stem-borne rhizoids at the bases of stems and branches in non-tomentose species, such as *C. setifolius*.

Of the other genera placed by Brotherus (1924–1925) in this subfamily, *Microcampylopus*, *Pilopogon*, *Atractylocarpus* (*Metzlerella*) and *Dicranodontium* all resemble *Campylopus* in having the rhizoids midrib-borne except for some stem-borne ones at the extreme bases of the stems. In *Atractylocarpus costaricensis* (C. Müll.) Williams and in all examined species of *Dicranodontium* the rhizoids are borne on both surfaces of the midrib, those on the adaxial surface originating later than those on the abaxial one.

Thysanomitrium is not now accepted as a good genus (Frahm, 1974) and I have not seen either of the two species of *Campylopodiella*. The species of Brotherus's three remaining genera, *Microdus*, *Dicranella* and *Campylopodium*, have all their rhizoids stem-borne and have relatively narrow midribs. The separation of *Anisothecium* and *Dicranella* is artificial, and these three genera are probably better transferred to the Anisothecioideae. If this is done the occurrence of leaf-borne rhizoids becomes a character diagnostic of the Campylopodioideae, for they apparently occur nowhere else in the Dicranaceae. As mentioned above, they do occur in most species of the related Leucobryaceae, but here they are mainly from the tip of the leaf, not from near its base.

2. Splachnaceae

This family was selected for special study because Limpricht's descriptions of the European species suggested that the colours might be interesting. In all species the rhizoids are stem-borne and, with one exception, they arise only from initials surrounding the axillary buds, these initials sometimes extending up the internode as a double line (Fig. 1). The exception is *Tayloria jamesonii* (Tayl.) C. Muell., anomalous among the 11 species of *Tayloria* examined, with rhizoids borne all over the internodes as in those Mniaceae with micronemata. In *Tetraplodon*, *Voitia* and *Haplodon* the rhizoids are always smooth, but in *Tayloria* and *Splachnum* they vary from smooth to coarsely papillose, and the character may be of some diagnostic value at the species level. In *Tayloria* the rhizoids are normally reddish purple, red in acid and violet in alkali, but there is one anomalous species, *T. magellanica* (Brid.) Mitt., which has rhizoid pigmentation of the second and commoner type, orange-brown, turning yellowish in acid, darker but not violet in alkali. In *Tetraplodon* the rhizoids are usually as in most *Tayloria* species, but *T. caulescens* (Brid.) Lindb. has pigmentation of the second type, as does *T. mnioides*—though here, as in *Voitia* both types of pigment may be present. The rhizoids of *Splachnum* and *Haplodon* are invariably brownish, never turning violet in alkali.

3. Plagiotheciaceae

In *Plagiothecium* the rhizoids are never axillary. In *P. latebricola* and *P. piliferum*, both sometimes put in a separate genus *Plagiotheciella*, they are always stem-borne, arising from just below the leaf insertions. In the other 14 species studied most or all of the rhizoids arise from the leaves, from cells on or beside the midrib (Figs 6, 7). In about half of these species they may also be borne on the stem just below the point of origin of the leaf—sometimes it may be very difficult, and indeed arbitrary, deciding whether the rhizoids are leaf- or stem-borne. There may in addition be stem-borne rhizoids at the bases of the branches, especially the sexual branches. Rhizoids also arise from initial cells at the leaf apices in *P. platyphyllum* and *P. neckeroideum* Bry. eur. Their presence is useful in distinguishing the former species from *P. succulentum* and *P. nemorale* (though I have once seen, from Jersey, *P. nemorale* with apical rhizoids) and the latter from the east Asiatic *P. euryphyllum* (Card. and Thér.) Iwats. (Iwatsuki, 1970). Jedlicka (1948) gave taxonomic recognition, under the name of forma *phyllorhizans* to plants of *P. denticulatum*, *P. ruthei*, *P. curvifolium* and *P. platyphyllum* in which he had observed leaf-borne rhizoids; but these were normal

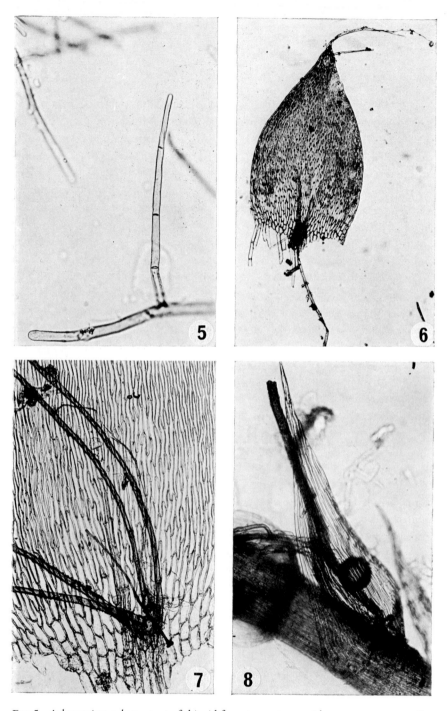

Fig. 5. *Aulacomnium palustre*, part of rhizoid from tomentum, with transverse cross-walls. ×350.

Fig. 6. *Plagiothecium neckeroideum*, with rhizoids from base and tip of leaf. ×50.

Fig. 7. *Plagiothecium neckeroideum*, rhizoids from base of leaf. ×350.

Fig. 8. *Isopterygium pulchellum*, axillary rhizoids. ×160.

states of these species and should therefore not be given special names.

The association by Brotherus of the genus *Catagonium* with *Plagiothecium* in a subfamily Plagiothecioideae is not supported by rhizoid characters, for in *Catagonium* the rhizoids are all stem-borne and are papillose, whereas in *Plagiothecium* they are smooth.

The nearest genus to *Plagiothecium* in rhizoid characters is *Taxiphyllum*. In this the rhizoids are borne at the stem-leaf junction and vary from smooth to papillose.

The species of *Herzogiella* either have all the rhizoids papillose or the coarser ones papillose and the finer ones smooth. They are always stem-borne and never arise at the stem-leaf junction but come from nearer the middle of the internode, except in *H. striatella*, in which they are axillary. They are also axillary and papillose in *Isopterygiopsis muellerana* and *Isopterygium pulchellum* (Fig. 8), though in the other seven species of *Isopterygium* examined the rhizoids, which vary from smooth to coarsely papillose, are borne higher up the internode, from the middle or at or just below the stem-leaf junction above. Rhizoid characters clearly support the removal of *Herzogiella striatella* from *Plagiothecium* but not the segregation of *Isopterygiopsis* and *Herzogiella* from *Isopterygium*.

CONCLUSIONS

The taxonomic literature of mosses contains little information about rhizoids—much less than the corresponding liverwort literature. Much of what it does contain is inaccurate as a result of superficial observation or observation based upon too few specimens. Nevertheless it is clear that rhizoid characters can be very useful at the species level, and sometimes even up to the levels of subfamily and family. It is to be hoped that a little more attention will be paid to rhizoids in the future by those writing descriptions in floras, monographs and revisions of critical groups; not only in families with conspicuous tomentum but also in others such as the Plagiotheciaceae, where the rhizoids are for the most part few and inconspicuous. Our knowledge of the influence of environmental factors on rhizoid development and of rhizoid pigmentation is at present rudimentary. While it would be optimistic to suppose that any critical study of the chemistry of rhizoid pigments is likely in the near future, the increased attention now being paid to the biochemistry of bryophytes may be expected to lead to a greater understanding of moss pigments in general, and this should be applicable to the pigments of the rhizoids.

ACKNOWLEDGEMENT

I am grateful to Mr T. Norman Tait for taking the photographs.

REFERENCES

BIRSE, E. M. (1957). Ecological studies on growth-form in bryophytes. II. Experimental studies on growth-form in mosses. *J. Ecol.* **45**, 721–733.

BIRSE, E. M., LANDSBERG, S. Y. and GIMINGHAM, C. H. (1957). The effects of burial by sand on dune mosses. *Trans. Br. bryol. Soc.* **3**, 285–301.

BROTHERUS, V. F. (1924–1925). Musci (Laubmoose). *In* "Die natürlichen Pflanzenfamilien" (A. Engler, ed.). Ed. 2, Vol. 10, 129–478; Vol. 11, 1–542. W. Engelmann, Leipzig.

CORLEY, M. F. V. (1976). The taxonomy of *Campylopus pyriformis* (Schultz.) Brid. and related species. *J. Bryol.* **9**, 193–212.

CRUNDWELL, A. C. and HILL, M. O. (1977). *Anoectangium warburgii*, a new species of moss from the British Isles. *J. Bryol.* **9**, 435–440.

CRUNDWELL, A. C. and NYHOLM, E. (1964). The European species of the *Bryum erythrocarpum* complex. *Trans. Br. bryol. Soc.* **4**, 597–637.

FRAHM, J.-P. (1974). Revision der Gattung *Thysanomitrium* Schwaegr. (*Dicranaceae*) in Südamerika. *J. Bryol.* **8**, 255–264.

HERZFELDER, H. (1921). Beiträge zur Frage der Moosfärbungen. *Beih. bot. Centralbl.* **38**(1), 355–400.

IRELAND, R. R. (1971). Moss pseudoparaphyllia. *Bryologist* **74**, 312–330.

IWATSUKI, Z. (1970). A revision of *Plagiothecium* and its related genera from Japan and her adjacent areas, I. *J. Hattori bot. Lab.* **33**, 331–380.

JEDLICKA, J. (1948). Monographia speciarum europaearum gen. *Plagiothecium* s.s. (Partis specialis I. Summarium). *Spisy Vydáv. přír. Fak. Masaryk. Univ.*, Rad. L2, **308**, 1–45.

KARCZMARZ, K. (1971). A monograph of the genus *Calliergon* (Sull.) Kindb. *Monographiae bot.* **34**, 1–209.

KOPONEN, T. (1968a). Generic revision of Mniaceae Mitt. (Bryophyta). *Annls bot. fenn.* **5**, 117–151.

KOPONEN, T. (1968b). The moss genus *Rhizomnium* (Broth.) Kop., with description of *R. perssonii*, species nova. *Memo. Soc. Fauna Flora fenn.* **44**, 33–50.

KOPONEN, T. (1974). A guide to the Mniaceae in Canada. *Lindbergia* **2**, 160–184.

LIMPRICHT, K. G. (1896–1904). Die Laubmoose. *In* "L. Rabenhorst's Kryptogamen-Flora von Deutschland, Oestrreich und der Schweiz". Second edition, Vol. 4. Kummer, Leipzig.

LORCH, W. (1931). Anatomie der Laubmoose. *In* "Handbuch der Pflanzenanatomie" (K. Linsbauer, ed.) Vol. 7(1). Gebrüder Borntraeger, Berlin.

MÅRTENSSON, O. and NILSSON, E. (1974). On the morphological colour of bryophytes. *Lindbergia* **2**, 145–159.

PAUL, H. (1903). Beiträge zur Biologie der Laubmoosrhizoiden. *Bot. Jb.* **32**, 231–274.

SMITH, A. J. E. (1978). "The Moss Flora of Britain and Ireland." Cambridge University Press, Cambridge.

SMITH, A. J. E. and WHITEHOUSE, H. L. K. (1978). An account of the British species of the *Bryum bicolor* complex including *B. dunense* sp. nov. *J. Bryol.* **10**, 29–47.

SYED, H. (1973). A taxonomic study of *Bryum capillare* Hedw. and related species. *J. Bryol.* **7**, 265–326.

VAUPEL, F. (1903). Beiträge zur Kenntnis einiger Bryophyten. *Flora, Jena* **92**, 346–370.

WARNSTORF, C. (1904–1906). "Kryptogamenflora der Mark Brandenburg, Vol. 2, Laub-moos." Gebrüder Borntraeger, Leipzig.

WIGGLESWORTH, G. (1947). Reproduction in *Polytrichum commune* L. and the significance of the rhizoid system. *Trans. Br. bryol. Soc.* **1**, 4–13.

WILCZEK, R. and DEMARET, F. (1976). Les espèces belges du "complex *Bryum bicolor*" (Musci). *Bull. Jard. bot. natn. Belg.* **46**, 511–541.

16 | Conducting Tissues in Bryophyte Systematics

C. HEBANT

Biologie végétale, Université des Sciences et Techniques du Languedoc, Place E. Bataillon, 34060 Montpellier cedex, France

Abstract: Conducting tissues can yield useful criteria for classifying bryophyte taxa of various ranks. Some difficulties are raised, however, by structural variability which is frequently encountered even within a single taxon, and by the need for either fresh or well preserved (fixed) specimens for precise histological studies. Accurate terminology is vital. On a broad level, water-conducting cells of liverworts are structurally distinct from those of mosses (hydroids). Those of liverworts (with the exception of *Moerckia*) possess small plasmodesmata-derived "pores" in their walls; such pores are absent from the hydroids of mosses. Water-conducting strands are common in mosses and provide additional criteria for the distinction of genera and species. Sieve element-like leptoids are apparently restricted to the gametophytic phase of various Polytrichaceae and Dawsoniaceae. Some further examples of the application of data on conducting tissues to systematic and evolutionary problems are given, and potential fields for future research are outlined.

INTRODUCTION

This paper reviews the use of the conducting tissues in bryophyte systematics, as seen from the point of view of a plant histologist. I shall try to outline both the scope of such an approach and the limitations inherent in it. Although I shall only consider conducting tissues here, other histological or anatomical features could be used in the same way.

The first question is what types of cells should be recognized as conducting elements, and on what basis can they be identified? The problem is that in

Systematics Association Special Volume No. 14, "Bryophyte Systematics", edited by G. C. S. Clarke and J. G. Duckett, 1979, pp. 365–383, Academic Press, London and New York.

bryophytes internal conduction very frequently occurs through scarcely specialized cells of conducting parenchyma (Haberlandt, 1886), which can hardly be compared to the xylem and phloem elements of the tracheophytes. In addition, imprecise or duplicate terminologies have frequently been used for bryophytes, with the result that a number of discrepancies are found in various published works, both old and recent. Such discrepancies are not always easy to clarify.

The first structures I shall consider are the water-conducting strands. By analogy with the tracheary elements of the vascular plants, I shall use this expression only for those internal conducting structures in bryophytes in which the cells are dead and empty when mature. Such specialized structures as the hyaline cells of *Sphagnum*, or the empty cells found in leaves and/or outer portions of stems of various mosses (e.g. the Leucobryaceae, some Syrrhopodontaceae) will not be reviewed here.

The recognition of specialized food-conducting cells, the so-called bryophytic leptoids, raises semantic problems, as testified by the many contradictory reports that have appeared recently. For this reason, caution is needed when dealing with leptoids in current taxonomic work.

WATER-CONDUCTING CELLS IN BRYOPHYTE SYSTEMATICS

1. Occurrence and Identification

The elongated cells which constitute the water-conducting strands are dead and empty at maturity. This is the result of a programmed lysosomal degeneration of the protoplast occurring at the last stage of development. Thus a sharp distinction exists between the living cortical tissues and the dead cells of the strand. This is true not only of leafy stems and setae of mosses, but also of those rare liverwort gametophytes which possess a strand (Fig. 1). Such expressions as "ill-defined" or "feebly distinct" central strand are therefore incorrect descriptions of a very clear anatomical feature and should be avoided. This applies equally to the water-conducting cells of the leaf nerves and leaf traces found in a number of mosses (Fig. 13). I have observed only a few apparent exceptions to this rule. The leaf-traces of some Polytrichales and the quite small, possibly reduced, strands of some other mosses, contain certain elongated cells which, by their position, would be expected to differentiate as hydroids. They in fact retain a living protoplast and do not complete "normal" development.

From another point of view, when free-hand sections are made of fresh or fixed specimens, contamination of the empty water-conducting elements by cellular debris from the neighbouring cortical cells may occur. This also applies to the preparation of material for electron microscopy, with the result that, for example, intact chloroplasts are sometimes encountered within otherwise empty, mature hydroids. The lack of a normal plasmalemma, in these cells, among other things, is of great help in identifying them under the electron microscope. "Collapsed" strands are also sometimes reported in the literature. I suspect them to be artefacts and seen only in aged or herbarium specimens.

Although true water-conducting strands are very clearly seen in cross-sections of well fixed and stained specimens (and *a fortiori* under the electron microscope), they are not always easy to identify when only herbarium specimens are at hand. This is the first difficulty in the utilization of anatomical features in bryophyte systematics. In practice, however, central strands are identifiable in many cases even on free-hand sections of rehydrated herbarium material, because the walls of the conducting cells are different in constitution from those of the cortical cells. A second feature facilitates the identification of the strands: normally, the living cells of the inner cortex protrude slightly into the central strand because of differences in turgor pressure between the living cortical cells and the dead conducting elements (Fig. 8). This pattern is usually retained even in dried herbarium specimens, and greatly facilitates the recognition of the limits of the strands.

2. Difficulties in the Use of Conducting Tissues in Bryophyte Systematics

It is a limitation for systematics that there is comparatively little diversity in the internal organization of the bryophytes. A wide range of patterns such as those observed in pteridophyte steles does not exist here. Strands, when present, are of a simple protostele type in most cases.

Another potential source of difficulty lies in the structural variability which appears to be considerable in bryophytes. A number of factors, both internal and external, appear to control this variability in anatomical patterns (Hébant, 1977). Although no comprehensive study of these phenomena has yet been completed, a few well-documented cases are known. One such example is that of *Moerckia flotoviana* which loses its conducting strands when natural populations develop under high humidity (De Sloover, 1959). It has also been well known since the work of Lorentz (1867, 1867–1868) that, in certain mosses which normally have one, the central strand can be absent from stems of smaller diameter. In some species, fertile stems possess a strand and sterile ones do not (Kawai, 1965, pp. 114–115). Other examples of structural variability within a single taxon have also been recognized (Hébant, 1977). Thus, if conducting strands are to be used as distinguishing characters in keys, sections taken from several specimens originating from different populations should be investigated for each taxon. Such criteria as the presence or absence, or relative development, of a strand are not good taxonomic characters unless meticulous controls of potential variations have been made.

3. Examples of the Use of Water-conducting Tissues in Taxonomic and Evolutionary Problems

(a) *Position of the bryophytes within the plant kingdom.* An increasing amount of data from widely different sources such as comparative biochemistry and comparative spermatology further support the view that there are real evolutionary affinities between bryophytes and primitive tracheophytes. In this context, the following similarities between the water-conducting cells of both groups are significant:

(1) occurrence of the water-conducting cells in the median region of stems or thalli, usually in the form of a single solid protostele (Figs 1, 8);

(2) elongated form of the conducting cells (Fig. 3);

(3) a lysosomal degeneration of the protoplast with a peak of strong acid phosphatase activity constituting the last step in development.

Whether or not these shared characteristics are inherited from common Silurian or pre-Silurian archegoniate progenitors remains open to speculation (Schuster, 1966; Miller, 1974; Hébant, 1977). Bryophytes might well exhibit relict types of conducting systems from early land invaders. It is a reasonable hypothesis that the relationship of bryophytes to tracheophytes in the plant kingdom is comparable to that of the prochordates to the vertebrates in the animal kingdom.

The demonstration of authentic lignins in the conducting cells of bryophytes would be of great phylogenetic interest. Unfortunately, convincing evidence for this has not yet been produced (Miksche and Yasuda, 1978).

(b) *Relationships between liverworts and mosses.* Although the conducting cells of only a limited number of bryophytes have been studied under the electron microscope, they include representatives of all major groups (Table I). From these studies, in conjunction with those previously performed with the light microscope, one important fact emerges: the water-conducting cells of liverworts are structurally different from the hydroids of mosses. In all but one of the liverworts investigated, the walls of the conducting cells exhibit many small, plasmodesmata-derived, perforations (Figs 2, 4, and 5) whereas such pores are normally absent from the hydroids of mosses (Figs 9, 11). The "pitted walls" depicted by Caputo and Castaldo (1968) in *Plagiomnium undulatum* most probably belong to parenchyma cells of the cortex and not to hydroids, as claimed in the captions. Contact walls between moss hydroids may show partial hydrolysis (Fig. 12) which makes them very similar in appearance to the non-lignified, hydrolysed primary walls of tracheary elements in vascular plants. These facts emphasize the distinctness of liverworts from mosses. One notable exception is presented by the relatively isolated liverwort *Moerckia* whose conducting cells do not possess pores, but show hydrolysed walls (Figs 6, 7).

(c) *Water-conducting cells and hepatic classification.* Water-conducting strands are rare in liverworts. Only a few leafy species of the genera *Haplomitrium* and *Takakia*, and a few thalloid species of the Metzgeriales, are known to possess such a strand. Significantly, the structure of this strand is very similar in *Takakia* and *Haplomitrium* (Figs 1, 2). The conducting cells in both genera retain the general appearance of elongated parenchyma cells from which they differ, however, in lacking a protoplast when mature.

On the other hand, the water-conducting cells of *Pallavicinia*, *Symphyogyna* and *Hymenophyton* are decidedly more specialized, being somewhat tracheid-like in appearance (Fig. 3). The structure of their walls (Figs 4, 5) is nevertheless quite distinct from that of the tracheids in vascular plants. Indeed, the organization of these cells is unique in the plant kingdom and suggests the affinity of all those Metzgeriales which possess them. It would be interesting to investigate the closely related genera *Makednothallus*, *Xenothallus* and *Podomitrium* with the electron microscope. *Makednothallus* is akin to *Pallavacinia*, *Xenothallus* to *Symphyogyna*, and *Podomitrium* is a segregate from *Hymenophyton* (Schuster, 1963; Smith, 1966).

At a more specialized level, the existence of a multistrand condition in various *Symphyogyna* species from the tropical and southern parts of the New World (Smith, 1966) might also be of taxonomic significance.

As already pointed out, *Moerckia* stands apart from all other Hepaticae in the curious structure of its conducting system which has two parallel strands in the thallus, and the hydrolysed walls of the conducting cells (Figs 6, 7).

Discovery of true water-conducting strands consisting of dead, empty cells in additional liverworts would be of considerable systematic and evolutionary interest.

(d) *Water-conducting cells and moss classification.* In their water-conducting cells, mosses are the most interesting group of bryophytes. Conducting tissues are

Table I. Transmission electron microscope studies on the water-conducting cells of Bryophytes

Order	Species	Location of tissues studied		References
		gametophyte	sporophyte	
Calobryales	*Haplomitrium gibbsiae* (Steph.) Schust.	+		Burr et al., 1974
	Haplomitrium rotundifolium (Mitt.) Schiffn.	+		Chapman and Grubb (unpublished; cited in Grubb, 1970); Hébant, 1973
	(*H. mnioides* (Lindb.) Schust.)			
	Takakia ceratophylla (Mitt.) Grolle	+		Hébant, 1973
	Takakia lepidozioides Hatt. and Inoue	+		Hébant, 1972
Metzgeriales	*Hymenophyton flabellatum* (Labill.) Dum. ex Trev.	+		Burr et al., 1974; Campbell et al., 1975
	Hymenophyton leptopodum (Hook and Tayl.) Steph.	+		Campbell et al., 1975
	Pallavicinia lyellii	+		This report
	Symphyogyna circinata Nees and Mont.	+		Smith, 1966
	Moerckia flotoviana	+		Hébant, 1973; this report
Polytrichales	*Atrichum* spp.	+		Hébant, 1974
	Dendroligotrichum dendroides (Hedw.) Broth.	+	+	Hébant, 1973, 1975; Scheirer, 1973
	Oligotrichum tenuirostre (Hook.) Jaeg.	+		Hébant, 1974
	Polytrichadelphus magellanicus (Hedw.) Mitt.	+		Hébant, 1974
	Pogonatum spp.	+	+	Hébant, 1974; Favali and Gianni, 1975
	Polytrichum spp.	+	+	Hébant, 1974; Favali and Bassi, 1974; Zamski and Trachtenberg, 1976
	Psilopilum australe (Hook. fil. and Wils.) Jaeg.	+		Hébant, 1974
Dawsoniales	*Dawsonia* spp.	+	+	Hébant, 1975, 1976

Order	Species			Reference
Tetraphidales	*Tetrapus pellucida*	+		[ref]
Dicranales	*Dicranum scoparium*	+	+	Hébant, 1970, 1973
	Leucophanes candidum (Hornsch.) Lindb.		+	Favali and Bassi, 1978
Pottiales	*Tortula muralis*	+	+	Favali and Gianni, 1973
Funariales	*Funaria hygrometrica*	+	+	Hébant 1969; Schulz and Wiencke, 1976
Eubryales	*Mnium orthorhynchum*		+	Bassi and Favali, 1973
	Plagiomnium undulatum	+		This report, Fig. 9
Hookeriales	*Cyathophorum bulbosum* (Hedw.) C. Muell.	+	+	This report
	Hookeria lucens	+	+	This report
	Hypopterygium setigerum (P. Beauv.) Hook. fil. and Wils.	+		This report
Hypnobryales	*Thuidium tamariscinum*	+	+	This report, Fig. 11

of common occurrence in either one or both their generations and early investigators had already considered the taxonomic implications of these tissues (Lorentz, 1867–1868; Morin, 1893). Brotherus (1924–1925) also paid great attention to internal structures.

No comprehensive survey of conducting tissues in mosses has yet been completed. This would constitute a formidable task, especially if intraspecific structural variability is to be considered. In addition, distribution patterns of conducting strands within taxa of different rank are subject to such variation that most attempts at generalization would rapidly prove unsatisfactory. I will only give here a few miscellaneous examples to emphasize the potential interest of water-conducting tissues for moss taxonomy.

On a broad level, Frey (1971) has recently stressed the fact that many pleuro-carpous mosses have reduced strands (Fig. 10) or no strand at all, whereas a

––––––––––––––––

Figs 1–7. Water-conducting cells in liverworts.

Fig. 1. *Haplomitrium gibbsiae.* Portion of thick free-hand longitudinal section of upright gametophyte. A sharp contrast exists between the central strand (CS), which consists of dead, empty water-conducting elements, and the cortex, whose cells retain a living protoplast (here darkly stained by iodine–potassium iodide reagent). ×67.

Fig. 2. *Takakia lepidozioides.* Scanning electron microscope view of contact end-wall between two water-conducting cells of the central strand. Numerous plasmodesmata-derived pores can be seen. ×12 000.

Fig. 3. *Pallavicinia lyellii.* Water-conducting cells from gametophytic strand isolated by maceration. Phase contrast. ×95.

Fig. 4. *Pallavicinia lyellii.* Detail of "pits" in wall of conducting cell (this scanning electron micrograph shows the inner surface of the wall, i.e. that directed towards lumen of cell). Note the presence of oblique, slit-shaped depressions. Small plasmodesmata-derived perforations connect adjacent conducting-cells from within the depressions. ×7800.

Fig. 5. *Hymenophyton flabellatum.* Transmission electron micrograph of longitudinal section through contact-wall between two water-conducting cells. Method of Thiéry (1967) for the demonstration of polysaccharides. The pores are indicated by arrows. ×13 000.

Fig. 6. *Moerckia flotoviana.* Portion of longitudinal section through gametophytic water-conducting strand. Note the peculiar structure of the conducting cell walls. CC, conducting cell; EW, end wall; LW, lateral wall; PAR, parenchyma cell. ×2700.

Fig. 7. *Moerckia flotoviana.* Detail of contact zone between a conducting cell, (CC) of the central strand and a parenchyma cell (PAR) of the ground tissue of the thallus. Hydrolysis of the wall is evident on the side of the conducting cell, whereas the wall of the adjacent parenchyma cell has apparently remained intact (asterisk). HW, hydrolysed wall; PAR, parenchyma cell. ×8000.

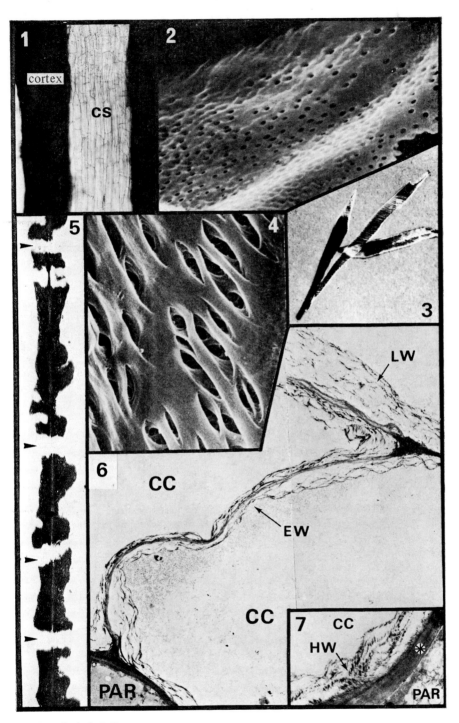

FIGS 1, 2, 3, 4, 5, 6, 7

number of acrocarpous species possess a well-developed conducting-tissue system (Fig. 8). However, no sharp distinction can be established on this basis, since a number of acrocarps without a strand are known, e.g. *Orthotrichum* spp. and *Leptodontium* spp. (Saito, 1975), whereas pleurocarps with a conspicuous strand also exist, e.g. *Spiridens* spp. and *Cyathophorum bulbosum* (Hedw.) C. Muell.

It has also long been recognized, for example by Brotherus (1924–1925), that well-developed strands predominate among representatives of certain orders or families; Eubryales and Polytrichales are typical examples. Conversely, a majority of species in other high ranking taxa such as the Hypnobryales possess a small strand or no strand at all in their gametophytes. In addition, other groups such as the Pottiales show a wide range of internal patterns in their leafy stems, which have proved to be very useful to systematists (Saito, 1975). No definitive explanation for these variations can be suggested at the moment. In particular, no simple and regular correlation with ecology or growth patterns can be established (Morin, 1893). For example, several species of *Mnium*, all of which exhibit a well-developed strand, may grow in association with *Climacium dendroides*, which possesses a highly reduced strand. Several features of bryophyte evolution, e.g. peristome structure, are more understandable if the concept of non-Darwinian evolution due to neutral mutations and genetic drift (King and Jukes, 1969) is accepted. This could also apply to certain features in conducting tissues.

At a more specialized level, the existence of striking similarities between the

---

Figs 8–13. Water-conducting cells (=hydroids) of mosses.

Fig. 8. *Mnium* sp. Cross-section of erect leafy stem showing well-developed water-conducting strand of hydroids (CS). LT, false leaf trace. ×77.

Fig. 9. *Plagiomnium undulatum.* Ultrastructure of hydroids (HYD) in central strand of leafy stem. Cross-section. EW, end wall; LW, lateral wall. ×13 500.

Fig. 10. *Thuidium tamariscinum.* Portion of thick free-hand cross-section in main stem of gametophyte. Only a small relictual strand of hydroids (CS) can be seen. ×2500.

Fig. 11. *Thuidium tamariscinum.* Detail of two hydroids (HYD) of the gametophytic central strand, as seen under the electron microscope. Cross-section. HW, hydrolysed wall; RB, residual body. ×6700.

Fig. 12. *Dawsonia longiseta.* Gametophyte. Detail of hydrolysed wall (HW) of contact between two mature hydroids. ×13 500.

Fig. 13. *Polytrichum commune.* Portion of cross-section in still immature region of leafy stem showing young leaf trace. The hydroids (H) of the leaf trace are already devoid of a protoplast at this stage and thus contrast sharply with the neighbouring living cells. ×2500.

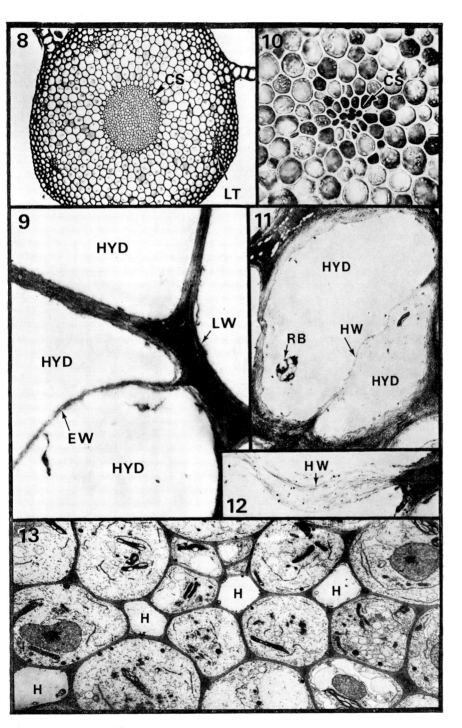

FIGS 8, 9, 10, 11, 12, 13

conducting systems of the Polytrichaceae and Dawsoniaceae is a good indication of their close relationship (Smith, 1971; Van Zanten, 1973). Among other characteristics, the root-like organization of the rhizomes in both groups is a unique and most distinctive feature.

Within the genus *Dawsonia*, the section *Dawsonia* is readily separated because in the aerial leafy portions of its gametophytes, the hydroids of the central strand are not intermixed with stereids. In the section *Superba* of the same genus, which appears to be more specialized, stereids are intermixed with the hydroids of the central strand in all parts of the gametophytes (Van Zanten, 1973). Various other examples of correlation between anatomical features and systematic position could be given. The presence, relative development, or absence, of water-conducting strands in the gametophytes constitute useful criteria for the distinction of taxa at the generic, specific or even infraspecific levels. The recent works of Saito (1975) and Kanda (1975, 1976) on Japanese Pottiaceae and Amblystegiaceae respectively, or of Seppelt and Stone (1977) on *Ditrichum* could be taken as examples of this, amongst many others. Kawai (1965, Plates 43, 44) shows a cross-section of the stem of *Grimmia gracilis* Schleich. ex Schwaegr. f. *subepilosa* without a strand, and one of *G. gracilis* f. *tenuis* with such a strand.

Other structures implicated in water-conduction, namely the leaf traces present in the gametophytes of many mosses, and the sporophytic strands, have hardly been investigated, and might yield useful criteria for taxonomic work, as might the branching of the gametophytic strand.

True leaf traces are known with certainty only from the Polytrichaceae and Dawsoniaceae. Leaf traces entering the cortical tissues of the stems to various degrees have also been reported from a number of Funariales and Eubryales, although they appear to be false leaf traces in most cases, since they do not join the central strand (Fig. 8). I was unable to replicate the observation of Lorentz (1867) of true leaf traces which join the central strand in *Splachnum* spp. and *Voitia nivalis* Hornsch.

A broad and synthetic survey of conducting-tissues in leaf nerves, as pioneered by Morin (1893), would certainly bring new insights into moss taxonomy. Here again, partly because of the existence of considerable structural variability even within a single specimen, this will be a difficult task.

Curiously enough, a strand occurs in the seta of many moss species where the gametophyte lacks such a structure (Ruhland, 1924). I have even observed a residual strand in the much reduced sporophyte of *Ephemerum serratum*.

The evolutionary significance of the distribution patterns of water-conducting tissues in mosses has been interpreted in a variety of ways. Although a number

TABLE II. Some bryophyte species in which cells, probably of conducting parenchyma, have been described as "leptoids"

Order	Species	Location of tissues studied	References
Metzgeriales	*Pallavicinia wallissii* (Hook.) Gray	gametophyte	Winkler, 1969
Dicranales	*Leucophanes candidum* (Hornsch.) Lindb.	seta	Favali and Bassi, 1978
Pottiales	*Tortula muralis*	seta	Favali and Gianni, 1973
Funariales	*Funaria hygrometrica*	seta	Schulz and Wiencke, 1976[1]
Eubryales	*Mnium orthorhynchum*	seta	Bassi and Favali, 1973

[1] Schulz and Wiencke follow Haberlandt (1886) in describing as a "sheath" surrounding the hydroids those cells which were interpreted as leptoids by Vaizey (1888) and which I tentatively interpret as residual leptoids. They use the term "leptoids" to designate the cells of conducting parenchyma of the inner cortex. For a discussion of these problems, see Hébant (1977, pp. 48–49).

of authors have tentatively interpreted the lack of a central strand as an advanced condition (Haberlandt, 1886; Lorch, 1931; Hébant, 1966, 1977; Miller, 1971), others see no ground for such an assertion (Schauer, 1967).

FOOD–CONDUCTING CELLS IN BRYOPHYTE SYSTEMATICS:
THE NEED FOR A CLEAR APPRAISAL OF THE CONCEPT OF LEPTOID

All the living cells which constitute tissues and organs of plants are capable of translocating organic substances. Elements of conducting parenchyma are present in leafy stems, thalli and sporophytes of all bryophytes, whether they also possess more specialized conducting tissues or not. Following the pioneer investigations of Marchal and Marchal (1906) on *Mnium*, there is now ample experimental evidence to demonstrate the slow translocation of organic substances through elements of conducting parenchyma in bryophytes (Guyomarc'h, 1969, on *Dicranum*; Rota and Maravolo, 1975, on *Marchantia*; Proctor, 1977, on diverse mosses).

Some authors have termed such cells of conducting parenchyma "leptoids". In several instances, this term was even applied, using only slender morphological evidence, to elongated parenchyma cells with somewhat inclined end walls (Table II). I think there is a potential danger to taxonomists and physiologists in not distinguishing clearly between true leptoids and ordinary cells of conducting parenchyma. Strictly speaking, leptoids are highly specialized cell types. Not recognizing this fact would rapidly lead to the identification of "leptoids" in all, or nearly all bryophytes. I have suggested (Hébant, 1977) that the expression "leptoids" *sensu stricto* should be restricted to those bryophyte conducting cells which show clear ultrastructural similarities to the sieve

---

FIGS 14–16. The leptoids *s. str.* of the gametophytes of the Polytrichales (for further details, see Hébant, 1977, Plates 69 to 73).

FIG. 14. *Polytrichum commune.* Portion of cross-section at periphery of central strand showing leptoids (LEP) and associated parenchyma cells (PAR). Enlarged plasmodesmata (PD) are seen on the contact end wall between two leptoids. Note the contrast in organization of the protoplast of the leptoids (e.g. degenerate plastids, stacked endoplasmic reticulum), and of the parenchyma cells. ER, endoplasmic reticulum; N, nucleus; P, plastid. ×6500.

FIG. 15. *Polytrichadelphus magellanicus.* Plastid (P) of mature leptoid. ER, endoplasmic reticulum; L, lipid droplet; W, wall. ×41 500.

FIG. 16. *Polytrichum commune.* Portion of cross-section of mature leptoid. Stacked endoplasmic reticulum (ER) is seen. Note also the occurrence of an inclusion (G) similar in appearance to the refractive spherules of pteridophyte phloem. L, lipid droplet. ×37 500.

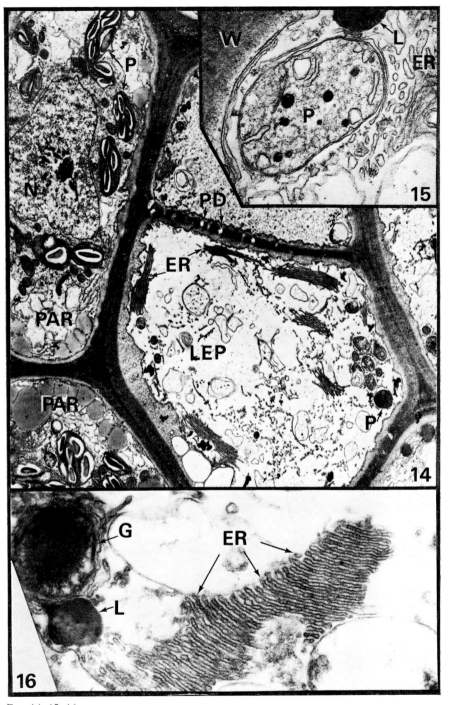

FIGS 14, 15, 16

elements of vascular plants. Most of these similarities can only be recognized through careful cytological studies, for which the transmission electron microscope is essential (Figs 14–16). This refers in particular to specialized features of their protoplast such as the occurrence of stacked endoplasmic reticulum (sieve element reticulum) (Figs 14, 16) and of regressed plastids (Fig. 15), and to the frequent, but developmentally late, degeneration of their nucleus (for details see Hébant, 1977, pp. 59 *et seq.* and Plates 69–73). Such leptoids are always found in association with parenchyma cells. Translocation in them is fairly rapid (Eschrich and Steiner, 1967; Collins, personal communication).

Up to now, leptoids *s. str.* have only been found in the gametophytes of a number of species of Polytrichaceae and Dawsoniaceae. This strengthens both the affinity of these two families and the fact that they stand apart from other contemporary mosses. Cells with characteristics intermediate between those of leptoids *s. str.* and of parenchyma cells are also found in Polytrichaceae and Dawsoniaceae. But this does not alter the taxonomic significance of the leptoids *s. str.* The occasional occurrence of nacreous walls in the leptome of certain Polytrichales (*Atrichum; Polytrichum* section *Juniperina*) might also be of taxonomic significance.

A much less typical leptome is found in the setae of the same mosses. Elements suggestive of leptoids also occur in the setae, but not in the gametophytes, of certain Funariales (*Funaria, Splachnum, Meesia*) (Vaizey, 1888, 1890; Hébant, 1977). In no way can these cells be readily compared to the sieve elements of the tracheophytes. I have tentatively interpreted them as "residual" leptoids. Such leptoids might also exist in the setae of other mosses, especially certain Eubryales. This deserves further investigation, and, if confirmed, might prove to be of great taxonomic and evolutionary interest. But, as far as we know at present, all bryophytes, apart from those cited above, only possess elements of conducting parenchyma in their gametophytes and sporophytes and not leptoids *s. str.*

CONCLUSION

Since the pioneering work of Lorentz (1867–1868), conducting tissues, especially the water-conducting strands, have yielded very valuable criteria which can be used at various levels in bryophyte taxonomy. Obstacles are created, however, by the existence of significant structural variability within taxa, and by technical difficulties such as the need of fresh specimens and special equipment to identify the conducting tissues with certainty in some cases. The use of precise terminology is most important.

Much is to be expected from the consideration of conducting tissues as additional characters in bryophyte systematics, provided careful studies are made on specimens from different populations and at different stages of development. Unfortunately, such precise studies cannot always be performed when only scarce herbarium specimens are available.

ACKNOWLEDGEMENTS

I wish to thank John Holmes and David W. Lee for reading the text, and Mme Jeannine Cochard for typing the manuscript. Daniel Rievière's assistance with the scanning electron microscope is also acknowledged.

REFERENCES

BASSI, M. and FAVALI, M. A. (1973). Seta ultrastructure in *Mnium orthorhynchum*. *Nova Hedwigia* **24**, 337–346.

BROTHERUS, V. F. (1924–1925). Musci (Laubmoose). *In* "Die natürlichen Pflanzen-familien" (A. Engler, ed.) Second edition Vol. 10, 129–478; Vol. 11, 1–542. W. Engelmann, Leipzig.

BURR, R. J., BUTTERFIELD, G. B. and HÉBANT, C. (1974). A correlated scanning and transmission electron microscope study of the water-conducting elements in the gametophytes of *Haplomitrium gibbsiae* and *Hymenophyton flabellatum*. *Bryologist* **77**, 612–617.

CAMPBELL, E. O., MARKHAM, K. R. and PORTER, L. J. (1975). Dendroid liverworts of the order Metzgeriales in New Zealand. *N.Z. Jl Bot.* **13**, 593–600.

CAPUTO, G. and CASTALDO, R. (1968). Prime osservazioni ultrastrutturali sul "sistema conduttore" di *Mnium undulatum* Weiss. *Delpinoa*, n.s. **8/9**, 85–91.

ESCHRICH, W. and STEINER, M. (1967). Autoradiographische Untersuchungen zum Stofftransport bei *Polytrichum commune*. *Planta* **74**, 330–349.

FAVALI, M. A. and BASSI, M. (1974). Seta ultrastructure in *Polytrichum commune* L. *Nova Hedwigia* **25**, 451–463.

FAVALI, M. A. and BASSI, M. (1978). Ultrastructure of the gametophyte and sporophyte of *Leucophanes candidum*. *Nova Hedwigia* **29**, 147–165.

FAVALI, M. A. and GIANNI, F. (1973). Sporophyte ultrastructure in *Tortula muralis* Hedw. *Öst. bot. Z.* **122**, 323–331.

FAVALI, M. A., and GIANNI, F. (1975). Seta ultrastructure in *Pogonatum aloides* Hedw. *G. Bot. ital.* **109**, 375–385.

FREY, W. (1971). Blattentwicklung bei Laubmoosen. *Nova Hedwigia* **20**, 463–556.

GRUBB, P. J. (1970). Observations on the structure and biology of *Haplomitrium* and *Takakia*, hepatics with roots. *New Phytol.* **69**, 303–326.

GUYOMARC'H, C. (1969). Sur les variations de l'azote soluble au cours du développement du sporophyte chez deux Mousses Bryales. *C.r. hebd. Séanc. Acad. Sci. Paris*, Sér. D, **268**, 2339–2342.

HABERLANDT, G. (1886). Beiträge zur Anatomie und Physiologie der Laubmoose. *Jb. wiss. Bot.* **17**, 359–498.

HÉBANT, C. (1966). Précisions nouvelles sur la signification et la répartition des tissus conducteurs dans la tige feuillée des mousses. *C.r. hebd. Séanc. Acad. Sci. Paris*, Sér. D, **263**, 1065–1068.

HÉBANT, C. (1969). Observations sur les traces foliaires des mousses *s. str.* (Bryopsida). I. Les hydröides et leurs relations avec le cylindre central. *Revue bryol. lichén.* **36**, 721–728.

HÉBANT, C. (1970). A new look at the conducting tissues of mosses (Bryopsida): their structure, distribution and significance. *Phytomorphology* **20**, 390–410.

HÉBANT, C. (1972). Précisions nouvelles sur la structure et la signification du faisceau conducteur irriguant le gamétophyte de *Takakia lepidozioides* Hattori et Inoue. *C.r. hebd. Séanc. Acad. Sci. Paris*, Sér. D, **275**, 189–192.

HÉBANT, C. (1973). Diversity of structure of the water-conducting elements in liverworts and mosses. *J. Hattori bot. Lab.* **37**, 229–234.

HÉBANT, C. (1974). Studies on the development of the conducting tissue-system in the gametophytes of some Polytrichales, II. Development and structure at maturity of the hydroids of the central strand. *J. Hattori bot. Lab.* **38**, 565–607.

HÉBANT, C. (1975). Organization of the conducting tissue-system in the sporophytes of *Dawsonia* and *Dendroligotrichum* (Polytrichales, Musci). *J. Hattori bot. Lab.* **39**, 235–254.

HÉBANT, C. (1976). Comparative anatomy of the gametophytes in *Dawsonia* (Polytrichales, Musci). *J. Hattori bot. Lab.* **40**, 221–246.

HÉBANT, C. (1977). The conducting tissues of bryophytes. *Bryophyt. Biblthca* **10**, 1–157.

KANDA, H. (1975). A revision of the family Amblystegiaceae of Japan, I. *J. Sci. Hiroshima Univ.*, Ser. B, Div. 2, **15**, 201–276.

KANDA, H. (1976). A revision of the family Amblystegiaceae of Japan, II. *J. Sci. Hiroshima Univ.*, Ser. B, Div. 2, **16**, 47–119.

KAWAI, I. (1965). Studies on the genus *Grimmia*, with reference to the affinity of gametophyte. *Sci. Rep. Kanazawa Univ.* **10**, 79–132.

KING, J. L. and JUKES, T. H. (1969). Non-Darwinian evolution. *Science, N.Y.* **164**, 788–798.

LORCH, W. (1931). Anatomie der Laubmoose. *In* "Handbuch der Pflanzenanatomie" (A. Linsbauer, ed.) Vol. 7. Gebrüder Borntraeger, Berlin.

LORENTZ, P. G. (1867). Studien zur vergleichenden Anatomie der Laubmoose. *Flora, Jena* **50**, 241–248; 257–264; 289–297; 305–313; 526–528; 529–540; 544–558.

LORENTZ, P. G. (1867–1868). Grundlinien zu einer vergleichenden Anatomie der Laubmoose. *Jb. wiss. Bot.* **6**, 363–466.

MARCHAL, E. L. and MARCHAL, E. M. (1906). Recherches physiologiques sur l'amidon chez les bryophytes. *Bull. Soc. r. Bot. Belg.* **43**, 115–214.

MIKSCHE, G. E. and YASUDA, S. (1978). Lignin of "giant" mosses and some related species. *Phytochemistry* **17**, 503–504.

MILLER, H. A. (1971). An overview of the Hookeriales. *Phytologia* **21**, 243–252.

MILLER, H. A. (1974). Rhyniophytina, alternation of generations and the evolution of bryophytes. *J. Hattori bot. Lab.* **38**, 161–168.

MORIN, F. (1893). "Anatomie comparée et expérimentale de la feuille des Muscinées. Anatomie de la nervure appliquée à la classification." Ph.D. Thesis, University of Rennes.

PROCTOR, M. C. F. (1977). Evidence on the carbon nutrition of moss sporophytes from

$^{14}CO_2$ uptake and the subsequent movement of labelled assimilates. *J. Bryol.* **9**, 375–386.

ROTA, J. A. and MARAVOLO, N. C. (1975). Transport and mobilization of ^{14}C-sucrose during regeneration in the hepatic *Marchantia polymorpha*. *Bot. Gaz.* **136**, 184–188.

RUHLAND, W. (1924). Musci, Allgemeiner Teil. *In* "Die natürlichen Pflanzenfamilien" (A. Engler, ed.) Second edition, Vol. 10, 1–100. W. Engelmann, Leipzig.

SAITO, K. (1975). A monograph of Japanese Pottiaceae (Musci). *J. Hattori bot. Lab.* **39**, 373–537.

SCHAUER, T. (1967). Anatomische und systematische Studien über die mitteleuropäischen Arten der Gattung *Seligeria* (Musci). *Nova Hedwigia* **14**, 313–325.

SCHEIRER, D. C. (1973). Hydrolysed walls in the water-conducting cells of *Dendroligotrichum* (Bryophyta): histochemistry and ultrastructure. *Planta* **115**, 37–46.

SCHULZ, D. and WIENCKE, C. (1976). Sporophytenentwicklung von *Funaria hygrometrica* Sibth. II. Differenzierung des Wasser-und Stoffleitungssystems in der Seta. *Flora, Jena* **165**, 47–60.

SCHUSTER, R. M. (1963). Studies on Antipodal Hepaticae, I. Annotated keys to the genera of Antipodal Hepaticae with special reference to New Zealand and Tasmania. *J. Hattori bot. Lab.* **26**, 185–309.

SCHUSTER, R. M. (1966). "The Hepaticae and Anthocerotae of North America," Vol. 1. Columbia University Press, New York and London.

SEPPELT, R. D. and STONE, I. G. (1977). A comparison of vegetative features of *Ditrichum cylindricarpum* and *Ditrichum punctulatum*. *J. Bryol.* **9**, 321–325.

SLOOVER, J. de (1959). Considération sur la valeur spécifique de *Moerckia flotoviana* (Nees) Schiffn., Dilaenacée nouvelle pour la flore belge. *Bull. Jard. bot. Etat. Brux.* **29**, 157–181.

SMITH, G. L. (1971). A conspectus of the genera of Polytrichaceae. *Mem. N.Y. bot. Gdn* **21(3)**, 1–83.

SMITH, J. L. (1966). The liverworts *Pallavicinia* and *Symphyogyna* and their conducting system. *Univ. Calif. Publs Bot.* **39**, 1–83.

THIÈRY, J. P. (1967). Mise en évidence des polysaccharides sur coupes fines en micro-scopie électronique. *J. Microsc.* **6**, 987–1018.

VAIZEY, J. R. (1888). On the anatomy and development of the sporogonium of the mosses. *J. Linn. Soc. Bot.* **24**, 262–285.

VAIZEY, J. R. (1890). On the morphology of the sporophyte of *Splachnum luteum*. *Ann. Bot.* **5**, 1–10.

WINKLER, S. (1969). Anatomische Untersuchungen über Leitstränge der Gametophyten bei den Lebermoosen. *Öst. bot. Z.* **117**, 348–364.

ZAMSKI, E. and TRACHTENBERG, S. (1976). Water movement through hydroids of a moss gametophyte. *Israel J. Bot.* **25**, 163–173.

ZANTEN, B. O. van (1973). A taxonomic revision of the genus *Dawsonia* R. Brown. *Lindbergia* **2**, 1–48.

17 | Spermatogenesis in the Systematics and Phylogeny of the Musci

J. G. DUCKETT*

*School of Plant Biology, University College of North Wales
Bangor, Wales*

and

Z. B. CAROTHERS

*Department of Botany, University of Illinois, Urbana,
Illinois 61801, USA*

Abstract: This comparative ultrastructural study reveals that the spermatids and spermatozoids of mosses possess features useful in moss systematics at all levels. A detailed reconstruction of the blepharoplast in *Sphagnum* spermatids together with data on over a dozen other genera including *Andreaea, Tetraphis, Polytrichum, Mnium, Hookeria* and *Hypnum* indicate that the male gametes of mosses have more characters in common with those of hepatics than they do with those of the Anthocerotales. In all mosses the bases of the two flagella are unequal in length and lie in a staggered configuration above a multilayered structure which is typically four-layered. The spline varies in width from 38 tubules in *Tetraphis*, 25 in *Hookeria* and *Physcomitrium*, and 23 in *Andreaea* down to only 10 in *Polytrichum juniperinum*. However, most taxa have tubule numbers in the mid-teens. Spline apertures one or two tubules wide are ubiquitous beneath the anterior basal body. On the basis of blepharoplast ultrastructure we consider that *Sphagnum* should remain a subdivision of the Musci rather than be elevated to a rank equal to that

* Present address: Department of Plant Biology and Microbiology, Queen Mary College, University of London, Mile End Road, London E1 4NS, Great Britain.

Systematics Association Special Volume No. 14, "Bryophyte Systematics", edited by G. C. S. Clarke and J. G. Duckett, 1979, pp. 385–423, Academic Press, London and New York.

of the Hepaticae. *Tetraphis* appears to have little affinity with the Polytrichales and could well merit subclass status.

The uniplastidic condition of the male gametes and the fusion of the total mitochondrial complement in the young spermatids are phenomena common to mosses, hepatics and the Anthocerotales. Although the plastid is invariably filled with starch at maturity, the number and disposition of the starch grains and their mode of formation in relation to the thylakoid system varies considerably between taxa. The distribution of wall thickenings in antheridial jacket cells, and whether these contain chromoplasts or chloroplasts requires systematic investigation.

Two distinct patterns of chromatin condensation are identified during nuclear metamorphosis in mosses. Both of these are different from anything found in the Hepaticae or Anthocerotales. The size of the spermatozoid nuclei and their degree of coiling varies considerably between genera. As in the male gametes of pteridophytes, spline width is closely correlated with nuclear volume in bryophytes. X-ray microanalysis suggests that the size of spermatozoid nuclei is related to the amount of DNA they contain and may provide a means by which polyploid races can be identified independently of chromosome counts.

INTRODUCTION

Historically, ultrastructural studies of moss spermatozoids predate those of hepatics by two years. In 1952 Manton and Clarke demonstrated the compound nature of the flagella in shadowed *Sphagnum* spermatozoids and suggested that they had the same basic structure as those of the male gametes of ferns, *Fucus* and man. Manton (1957) went on to publish the first pictures of sectioned moss spermatozoids and described the spline in *Sphagnum* as a fibrous band shortly after its discovery in several hepatics by Satô (1954, 1956) who referred to it as the filamentous appendage. The multilayered structure (MLS) in mosses was first described for *Splachnum* and *Physcomitrium* as a "Dreiergruppe" (Heitz, 1960). Paolillo (1965) similarly described the MLS of *Polytrichum* as three layered and attempted the first reconstruction of this organelle. However, the second or lamellar stratum was misinterpreted as consisting of rods, and not, as was subsequently established in other archegoniate plants, as plates orientated at approximately 45° to the axes of the overlying spline microtubules (Carothers and Kreitner, 1967; Duckett, 1975b; Duckett and Bell, 1977; Paolillo, 1974). The MLS of *Physcomitrium* also appears to comprise three strata in the micrographs of Lal and Bell (1975) in striking contrast to its four-layered nature (the "Vierergruppe") in hepatics (Carothers, 1973, 1975), pteridophytes (Duckett, 1975b) and cycads (Norstog, 1974).

To date, the two most important papers on the ultrastructure of spermatogenesis in mosses are probably those by Paolillo *et al.* (1968a, b) which outline the major events in gamete metamorphosis in *Polytrichum*. Although their description of the development of the locomotory apparatus, morphogenesis and condensation of the nucleus and loss of extraneous cytoplasm can be seen to correspond in general terms to the same phenomena in hepatics (Carothers, 1973, 1975; Carothers and Kreitner, 1967, 1968; Kreitner, 1977a, b; Kreitner and Carothers, 1976) and Anthocerotales (Carothers *et al.*, 1977; Moser *et al.*, 1977) the data are inadequate to shed light on relationships within the bryophytes. Similarly, papers on the flagellum in *Polytrichum* (Genevès, 1968; Paolillo, 1967), the plastid-mitochondrial association (limosphere or nebenkern) in spermatids of *Polytrichum* (Genevès, 1967a; Paolillo, 1965) *Bryum* and *Funaria* (Sun, 1963), chromatin condensation in *Bryum* (Bonnot, 1967) and *Physcomitrium* (Lal and Bell, 1975) provide few clues as to the relevance of male gametes in moss systematics.

However, a preliminary investigation of the comparative microanatomy of moss spermatids (Carothers and Duckett, 1978) revealed that spline widths vary considerably among different taxa (Table I). The same study also suggested that the MLS is four-layered throughout mosses, thus bringing it into line with other archegoniates. In the present paper we shall explore further the uses of moss spermatids and spermatozoids in systematics and phylogeny. A detailed reconstruction of the blepharoplast of *Sphagnum* provides new insight into the interrelationships of the Musci, Hepaticae and Anthocerotales. Comparative data on spline widths, spermatid plastids and chromatin behaviour focuses attention on differences between classes and lower taxonomic rankings within the Musci. Finally, we describe how size measurements of sperm nuclei in conjunction with x-ray microanalysis, may facilitate the accurate determination of DNA amounts in bryophyte spermatozoids.

Key to abbreviations used in figures

ABB   anterior basal body
AM   anterior mitochondrion
MLS   multilayered structure
M   mitrochondrion associated with the plastid
N   nucleus
P   plastid
PBB   posterior basal body
S_1 spline
S_1–S_4 specific layers of the MLS, dorsal to ventral
SA   spline aperture
ST   starch
TZ   transision zone between basal body and axoneme
Bracket ([) indicates multilayered structure

MATERIALS AND METHODS

Data presented in this article are from freshly collected wild material of the following taxa: *Sphagnum capillifolium* (blanket bog, Ty'n-y-maes, near Bethesda, Caernarfonshire, Wales; August 1977; Grid reference 23/629644); *Bryum bicolor, Ceratodon purpureus* and *Ditrichum heteromallum* (friable earth on stream bank, Wern Bach, Tal-y-bont, Caernarfonshire, Wales; September 1975; Grid reference 23/617696); *Racomitrium aciculare* and *R. fasciculare* (wet rock by Afon Ogwen, Ty'n-y-maes, near Bethesda, Caernarfonshire, Wales; November 1976; Grid reference 23/633643); *Mnium hornum* and *Rhizomnium punctatum* (damp woodland banks, Crymlyn near Aber, Caernarfonshire, Wales; April 1977; Grid reference 23/639712) *Hedwigia ciliata* (boulder, Wern Uchaf, Tal-y-bont, Caernarfonshire, Wales; February 1978; Grid reference 23/614688); *Hookeria lucens* (woodland bank, Aber, Caernarfonshire, Wales; January 1976; Grid reference 23/663717); *Rhytidiadelphus triquetrus* (calcareous grassland, Newborough Warren, Anglesey, Wales; February 1976; Grid reference 23/427646); *Hypnum jutlandicum* (conifer plantation, Bwlch-yr-haiarn, Betws-y-Coed, Caernarfonshire, Wales; March 1976; Grid reference

---

FIGS 1–8. *Sphagnum palustre*

FIG. 1. Diagonal section of the MLS in a very young spermatid showing the structure of its four strata. ×128 000. Bar=0·1 μm.

FIG. 2. Section of the MLS about 15° off from the long axis of the lamellar strip showing an apparently unstratified lamellar region. ×108 000. Bar=0·1 μm.

FIG. 3. Transverse section of two MLSs in an aberrant spermatid. Keels (arrowed) are present on the spline microtubules adjacent to the lamellar strip within which three strata can be recognized. ×120 000. Bar=0·1 μm.

FIG. 4. A grazing tangential section of the MLS showing the lamellar plates of the S_2 layer orientated at about 45° from the S_1 axis. The ABB is sectioned longitudinally revealing the absence of a stellate profile in its transition region. ×54 000. Bar=0·2 μm.

FIG. 5. Tangential section showing the disposition of the basal bodies. The transition region lacking a stellate profile is arrowed. ×54 000. Bar=0·2 μm.

FIG. 6. Near-median, longitudinal section showing the extreme stagger of the basal bodies. The arrow indicates the level at which the axoneme of the ABB emerges from the cell surface. ×30 000. Bar=0·2 μm.

FIG. 7. A grazing tangential section showing the subapical disposition of the basal bodies over the MLS. Arrows indicate the convergence of the S_1 tubules behind the spline aperture. Note the curved leading edge of the MLS. ×60 000. Bar= 0·2 μm.

FIG. 8. Median longitudinal section showing the PBB running parallel to the S_1 tubules over the external surface of the nucleus. ×26 000. Bar=0·2 μm.

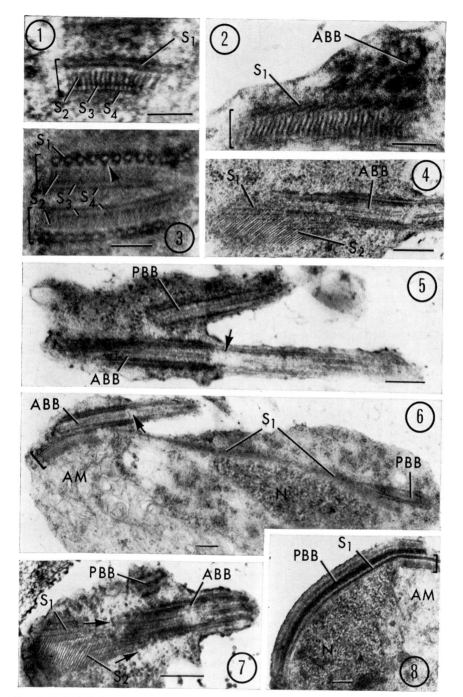

Figs 1, 2, 3, 4, 5, 6, 7, 8

23/778595). Localities for the remainder are listed in Carothers and Duckett (1978).

Excised antheridial shoots were fixed in 3% glutaraldehyde in 0·05 M cacodylate buffer, pH 6·9, for 3 h at room temperature, postfixed for 2 h at 4°C in 2% osmium tetroxide, embedded in Taab resin via a graded acetone series and propylene oxide. Sections were stained for 20 min each with saturated uranyl acetate and basic lead citrate (Reynolds, 1963). Specimen preparation for the X-ray microanalysis is described in Duckett and Chescoe (1976).

OBSERVATIONS AND DISCUSSION

1. The Microanatomy of the Sphagnum Multilayered Structure

The MLS of *Sphagnum* comprises four strata (Fig. 1). Outermost is the band of parallel spline (S_1) microtubules which have outside and inside diameters of 26 and 13 nm respectively and a centre-to-centre-spacing of 31 nm. As in other archegoniates, the tubules bear ventral keels (Fig. 3) where they are underlain by the lamellar strip. The stratification of the lamellar region is not obvious in sections which are not perpendicular to its long axis (Fig. 2) but in profiles where the S_1 tubules are displayed longitudinally, obliquely (Fig. 1) or transversely (Fig. 3) three strata are clearly visible. The alternating light and dark plates of the upper (S_2) and middle (S_3) layers, 40 and 20 nm thick respectively, are separated by a horizontal layer of electron-dense material. In sections perpendicular to the S_1 tubules (Fig. 3) S_3 is conspicuously lighter than both S_2 and the innermost S_3 layer. The latter, about 15 nm thick, consists of rods or downward extensions of S_1 lamellae with very regular periodicity (see Fig. 6 in Carothers and Duckett, 1978). Thus a substructure akin to the tubules and fins seen in hepatics and pteridophytes also occurs in *Sphagnum*. Orientation of the lamellar plates at approximately 45° to the axes of the S_1 tubules can be seen in grazing tangential sections (Figs 4 and 7).

In about 1% of the spermatids of *Sphagnum* two MLSs are present (Fig. 2) and other aberrations such as supernumerary basal bodies (e.g. in *Atrichum undulatum*, Fig. 13 in Carothers and Duckett, 1978) have been seen at lower frequencies in the majority of moss antheridia we have examined.

Various explanations come to mind for these aberrations, but these are all highly speculative and require more detailed examination.

(1)   Extra centrosomes and centrioles might be produced sporadically in spermatid mother cells, and similarly extra MLSs in spermatids.

(2)   In the antheridia of most mosses, but particularly conspicuous in *Sphagnum*, are cytomictic channels which link sister spermatids and

which often persist long after the final cytokinesis. The chance migration along these connections of the MLS, the basal bodies and more rarely the plastid would simultaneously result in additional components in one spermatid and corresponding deficiencies in its sister. This notion fails to explain the virtual absence of supernumerary basal bodies in hepatics despite the presence of cytomictic channels.

(3) Teratologies have sometimes been claimed to represent atavistic traits and they have thus figured prominently in evolutionary speculations. Such ideas are unfashionable today but on this basis supernumerary spermatid components could suggest the origin of mosses from a green algal ancestor (or a primitive vascular plant) with two MLSs and more than two flagella. *Trentepohlia*, the only green alga so far found to possess two MLSs (Graham and McBride, 1975; Stewart and Mattox, 1978) could be an extant example of such an algal line. Taking the atavism idea one stage further, the absence of aberrations in hepatics perhaps suggests that they had a completely different origin from the mosses.

2. *The Morphology of the* Sphagnum *Blepharoplast*

As in other bryophytes (Carothers and Duckett, this volume, Chapter 18), the *Sphagnum* blepharoplast, comprising the two basal bodies and subjacent MLS, and lying just beneath the cell surface, reaches its maximum degree of structural complexity in the young spermatids. To facilitate comparison with hepatics and Anthocerotales its three dimensional morphology has been reconstructed from over 500 sections of this stage.

Grazing tangential sections (Fig. 5) show the staggered position and divergent orientation of the basal bodies. The extreme nature of the stagger, first observed in shadowed spermatozoids (Manton, 1970), is highlighted by near-median longitudinal sections which show the presence of the posterior basal body (PBB) on the opposite side of the spermatid to the lamellar strip (Figs 6 and 9). Figures 11–27 are a series of transverse sections progressing from the anterior tip of the blepharoplast to behind the point of emergence of the PBB. They show details of the substructure of the basal bodies and the exact location of these bodies in relation to the spline, together with information on the precise geometry of the underlying lamellar strata.

The spline reaches its maximum width (13–14 tubules) very near the anterior tip of the blepharoplast (Fig. 12). Termination of the outermost tubules on either side of the spline forward of this point (only the seven central tubules

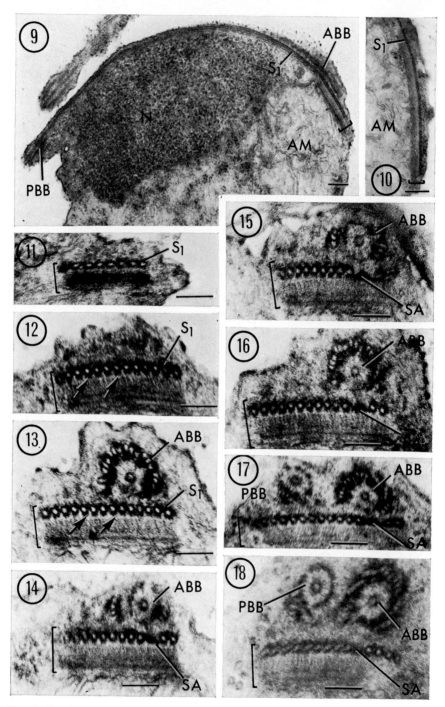

Figs 9, 10, 11, 12, 13, 14, 15, 16, 17, 18

are present in Fig. 11) produces a symmetrically curved leading edge to the MLS as can also be seen in suitable tangential sections (Fig. 7). The absence of basal bodies from these anterior transverse sections (Figs 11 and 12) confirm their subapical location as suggested from longitudinal profiles (Figs 6 and 9).

At a slightly more posterior level, the anterior basal body (ABB) first appears as a variable number of imbricating triplet tubules around a central core, often described as the cartwheel or pinwheel (Figs 13–16). Most frequently there are two (Fig. 14) or three triplets (Fig. 15) separately located to the right and left of the core but occasionally a complete dorsal arc of five or six triplets is present (Figs 13 and 16). However, at the level where the PBB first appears as a central core devoid of triplets (Fig. 17), the ABB invariably consists of a six-triplet arc. The ventral gap in the ABB is further reduced by addition of a seventh triplet at the level where the PBB possesses one triplet (Fig. 18). The latter is always on the lower left of the ABB. Slightly more posteriorly the ABB has a full complement of nine triplets (Fig. 19) which in turn give way to doublets in the transition zone (Figs 20–22). Here the central core is lost and

---

Figs 9–18. *Sphagnum palustre*

Fig. 9. Near-median, longitudinal section showing the posterior basal body emerging behind the nucleus and a voluminous anterior mitochondrion subjacent to the lamellar strip. $\times 32\ 000$. Bar $= 0.2\ \mu$m.

Fig. 10. Longitudinal section showing the maximum longitudinal extent of the lamellar strip. $\times 32\ 000$. Bar $= 0.4\ \mu$m.

Figs 11–23. Representative transverse sections of the blepharoplast from the extreme anterior tip to behind the point of emergence of the posterior axoneme. All $\times 100\ 000$. Bar $= 0.1\ \mu$m.

Fig. 11. Extreme anterior tip. The spline is 7 microtubules wide and is exactly subtended by the lamellar strip.

Fig. 12. Section slightly posterior to that of Fig. 11 showing a spline comprising 13 tubules. Keels on the spline tubules are arrowed.

Fig. 13. Section posterior to Fig. 11 showing the ABB consisting of 5 imbricated triplets surrounding the central core, and a spline of 14 tubules. Keels on the spline tubules are arrowed.

Fig. 14. Section showing the ABB comprising 2 triplets and a central core located above the spline aperture to the right of which are 2 spline tubules.

Fig. 15. Section just posterior to Fig. 14 showing the ABB comprising 3 triplets.

Fig. 16. The ABB now comprises 6 triplets dorsally.

Fig. 17. Section through the extreme anterior tip of the PBB which consists of a central core and a single incipient triplet.

Fig. 18. Section just posterior to Fig. 17 showing the PBB with a single triplet extension. The circle of ABB triplets remains incomplete adjacent to the spline aperture. The lamellar strip is narrower than in the preceding sections, having disappeared beneath the S_1 tubules to the right of the aperture.

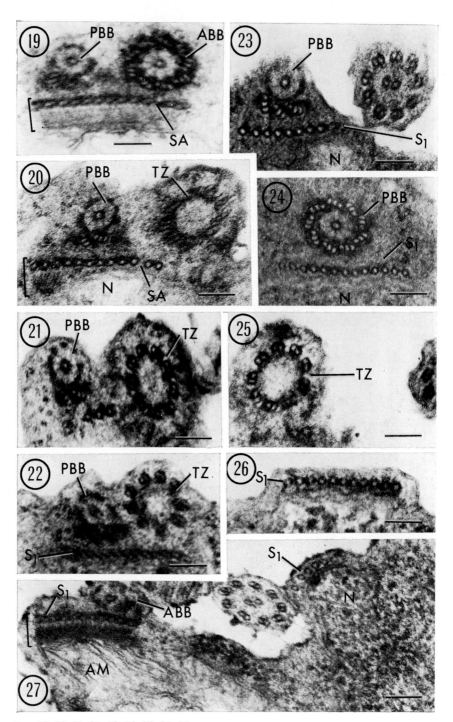

Figs 19, 20, 21, 22, 23, 24, 25, 26, 27

the basal body lumen is devoid of apparent substructure. Fibres radiate from each of the doublets and extend dorsally to the plasma membrane. At the level of the ABB transition zone, the PBB consists of but two ventral triplets. A third is added just prior to the emergence of the anterior axoneme (Fig. 23) and the PBB then retains this morphology for a distance of about 2 μm over the entire external surface of the nucleus. The full triplet complement (Fig. 24) is found about $\frac{1}{2}$ gyres from the anterior tip of the spline (Fig. 11). As in the ABB, this complement changes to doublets around a structureless central core prior to the start of the two central tubules of the axoneme. Throughout its length the PBB lies parallel to the underlying spline tubules (Fig. 8).

At the level of the anterior tip of the ABB, a gap of 60–70 nm containing osmiophilic material appears in the spline. Its centre lies immediately beneath the central core of the ABB (Figs 14–17). To the left of the gap the mean number S_1 tubules is ten (occasionally nine or eleven) and to the right two (occasionally four as in Fig. 18). Since the spline contains 14 tubules anterior to this discontinuity we interpret its appearance as resulting from the early termination of two S_1 tubules. It is thus homologous with the spline apertures seen in hepatics (Carothers and Duckett, this volume, Chapter 18).

Figs 19–27. *Sphagnum palustre*. All ×100 000

Fig. 19. Section showing the ABB with its full complement of 9 imbricating triplets. The SA is no longer occluded with dense material.

Fig. 20. The transition zone of the ABB showing the absence of a stellate pattern. Fibres radiate from the triplets to the plasma membrane. This basal body is displaced to the right of the S_1 where a spline aperture is still visible. The lamellar strip is now restricted to beneath the 3 tubules on the extreme left of the S_1. The PBB comprises a central core and the 2 lowermost triplets.

Fig. 21. Transition zone of the anterior flagellum showing doublets around an apparently structureless core.

Fig. 22. Section immediately proximal to the point of emergence of the anterior flagellum. The SA has closed and the lamellar strip disappeared.

Fig. 23. Section midway along the external surface of the nucleus. The S_1 retains 13 tubules but the PBB still only comprises 3 triplets.

Fig. 24. Section through the opposite side of a spermatid from Fig. 11, showing the PBB with its complete triplet complement overlying an S_1 of 13 tubules.

Fig. 25. Transition region of the posterior flagellum showing 9 doublets but no stellate profile.

Fig. 26. S_1 of 11 tubules overlying the nucleus approximately $\frac{3}{4}$ gyre from the anterior tip of the blepharoplast.

Fig. 27. Section showing two profiles of the S_1; one at the level of the anterior tip of the ABB (between Figs 12 and 13), the other a whole gyre removed showing 6 tubules overlying the nucleus.

TABLE I. Spline widths in *Musci*. Nomenclature follows Smith (1978)

Taxon	Maximum spline width (Number of microtubules)	References
Class Sphagnopsida		
Sphagnum palustre	14(15)	Carothers and Duckett (1978)
S. capillifolium	14	Duckett and Carothers
Class Andreaeopsida		
Andreaea rothii	23	Carothers and Duckett (1978)
Class Bryopsida		
Subclass Polytrichideae		
Order Tetraphidales		
Tetraphis pellucida	38	Carothers and Duckett (1978)
Order Polytrichales		
Polytrichum commune	11	Paolillo (1965)
P. formosum	12	Genevès (1968)
P. juniperinum	10	Paolillo *et al.* (1968)
Atrichum undulatum	14	Carothers and Duckett (1978)
Subclass Eubryidae		
Order Dicranales		
Ditrichum heteromallum		Duckett and Carothers
Order Grimmiales		
Racomitrium aciculare	14	Duckett and Carothers
R. fasciculare	14	Duckett and Carothers
Order Funariales		
Family Funariaceae		
Physcomitrium coorgense	25	Lal and Bell (1975)
Family Splachnaceae		
Splachnum rubrum	12	Heitz (1960)
Order Bryales		
Family Bryaceae		
Bryum capillare	14	Bonnot (1967)
B. bicolor	14	Duckett and Carothers
Family Mniaceae		
Mnium hornum	14	Duckett and Carothers
Rhizomnium punctatum	14	Duckett and Carothers
Order Orthotrichales		
Hedwigia ciliata	13	Duckett and Carothers
Order Hookeriales		
Hookeria lucens	25	Duckett and Carothers
Order Hypnobryales		
Family Plagiotheciaceae		
Plagiothecium nemorale	15	Duckett and Carothers
Family Hypnaceae		
Rhytidiadelphus triquetrus	15	Duckett and Carothers
Hypnum jutlandicum	15	Duckett and Carothers

TABLE II. Summary of blepharoplast characters in mosses

	Sphagnum	*Polytrichum* (from Paolillo *et al.*, 1968a)
Spline shape (anterior end)	Slightly spathulate	Spathulate
Spline width (microtubules)	14 (*c.* 0·35 μm)	10 (*c.* 0·3 μm)
Spline aperture	Present	Present
Number of tubules to the left of the aperture	10	5–6
Number of tubules to the right of the aperture	2(4)	3–4
Lamellar strip shape	Longitudinally elongated	Longitudinally elongated
Lamellar strip length (approx.)	1·25 μm	At least 1 μm
Lamellar strip max. width (approx.)	0·35 μm	0·3 μm
Basal body position	Subapical staggered	Subapical staggered
ABB approx. length (inc. extended triplets)	1·25 μm	1 μm
PBB approx. length (inc. extended triplets)	3·5–4 μm	8·0 μm
Basal body orientation	Unequally divergent	Unequally divergent
Stellate pattern	Absent	Present

The lamellar strip is exactly the same width as the spline back to the level where the triplet first appears in the PBB (Fig. 17). Then it narrows, extending on the right only as far as the spline aperture (Figs 18 and 19). From this point the strip narrows progressively until at its posterior tip it subtends only the three tubules on the far left of the spline. Throughout its length the left margin of the lamellar strip is aligned with the left edge of the spline. Thus, as a result of the unilateral tapering, the strip reaches its maximum longitudinal dimension (1·25 μm) on the left of the spline (Fig. 10). The regression from the right begins about 0·9 μm from the anterior tip of the blepharoplast. The outline of the lamellar strip including its asymmetrical posterior taper accurately reflects the position of the voluminous anterior mitochondrion (Figs 6 and 9) and the anterior edge of the nucleus. As soon as the lamellar strip regresses in width from the right-hand side of the spline it is closely underlain by the nucleus (Fig. 20) rather than by the mitochondrion (Figs 8 and 9). Behind the strip the entire spline is closely subtended by the nuclear envelope (Figs 24, 26, 27).

These details are shown as a diagram in Fig. 67 and the salient characteristics of the blepharoplast are listed in Table II. Apart from the convergence of the two right-hand tubules behind the aperture, the anterior portion of the spine in *Sphagnum* is symmetrical and of uniform width. The lamellar strip is the same width as the spline and about $3\frac{1}{2}$ times longer than wide. The PBB is 3–4 times longer than the ABB, mainly due to the ventral triplet extensions which are about 2–3 μm in length. The distance between the levels of flagellar emergence is 3 μm.

3. Comparative Morphology of the Bryophyte Blepharoplast

The MLS of *Sphagnum*, as described earlier, has the same basic organization as in other bryophytes. The thickness of each of its four strata and their detailed morphology is very similar to those in hepatics and anthocerotales. As in *Sphagnum*, the S_3 layer in both *Hookeria* (Fig. 51) and *Hypnum* (Fig. 57) is conspicuously lighter than the S_2. The S_4 in *Hookeria* is composed of either downward extensions of the S_{2-3} lamellae or small tubules, but in *Hypnum* comprises fins which alternate with the S_2 and S_3 lamellae. The horizontal layer of electron-dense material separating the S_2 and S_3 of both *Hookeria* and *Sphagnum* is absent in the illustration of *Hypnum*. In the light of the known maturational changes in the lamellar strata of *Marchantia* (Kreitner and Carothers, 1976) and various pteridophytes (Carothers *et al.*, 1975; Duckett, 1975b; Duckett and Bell, 1977; Robbins and Carothers, 1978) and in the absence of a complete ontogenetic sequence for any of these mosses, we feel that, at present, differences in S_{2-4} are more likely to reflect different developmental stages rather than constant structural features of diagnostic significance.

If the 13 component features of the blepharoplast are considered (Table II) many features of considerable systematic interest emerge. It is apparent that the blepharoplast of *Sphagnum* resembles that of *Marchantia* more closely than that of either *Haplomitrium* or *Phaeoceros*. Indeed the only common attribute with *Phaeoceros* is the absence of a stellate pattern in the transition zone. Since the stellate pattern is ubiquitous in the flagella of pteridophytes, cycads, all chlorophyll a and b containing algae except euglenoids (Moestrup, 1978), and all the other moss genera which we have so far examined (Table I), its absence in *Sphagnum* would appear to have considerable systematic significance. On the other hand, its presence in all hepatics except *Blasia* (Carothers, 1973) and *Pellia* (Suire, 1970) underlines the necessity of examining more genera before making generalizations.

The general shape of the lamellar strip in *Sphagnum* is closer to that in *Haplomitrium* than in *Marchantia* or *Phaeoceros*. The similarity extends to the PBB. In both genera, it is extremely long as a result of the ventral triplet extensions, and parallel to the straight left hand margin of the spline. The three triplet extensions in the PBB of *Marchantia* are also ventral.

The spline aperture is another feature which links *Sphagnum* with *Marchantia* and other hepatics including *Haplomitrium*, if our tentative interpretation of a one tubule gap beneath the ABB is correct. On the basis of blepharoplast morphology, the relationship between *Sphagnum* and hepatics appears to be closer than between either group and the *Anthocerotales*. Indeed, moss and hepatic spermatozoids more closely resemble those of *Selaginella* (Robert, 1974) than those of the Anthocerotales.

It is now clear that blepharoplast structure should be considered in any assessment of the highest taxonomic hierarchy in the Bryophyta. Does it suggest whether *Sphagnum* merits elevation to a rank equal to that of the Hepaticae and the remainder of the Musci, or should *Sphagnum* remain a subdivision of the Musci? If our interpretation of the micrographic data in Paolillo *et al.* (1968a) is correct, then as far as *Polytrichum* is concerned, the second alternative is the more acceptable (Table II). The most striking difference between the two genera is the presence or absence of a stellate pattern. Although apparently not identified as such by Paolillo *et al.* (1968a), there is evidence in all their relevant micrographs of a one tubule aperture beneath the ABB. The resemblance to *Sphagnum* is heightened by the presence of osmiophilic material in this discontinuity in the spline. The lamellar strip in *Polytrichum* has the same shape as that in *Sphagnum*: it is aligned parallel with the left-hand side of the spine throughout its length but tapers posteriorly from the right below the transition zone of the ABB. An extremely long PBB is characteristic of both genera.

Although precise three-dimensional relationships still have to be elucidated, these same features are characteristic of the blepharoplasts of all the mosses we have so far examined (Table I). The spline aperture beneath the ABB appears to be two tubules wide in *Andreaea* (Figs 36–38) and *Tetraphis* (Fig. 31) but only one tubule in extent in *Bryum*, *Mnium*, *Rhizomnium* and *Hypnum* (Fig. 56). A uniform arc of six or seven dorsal triplets is situated above the anterior portion of the aperture in all three genera (Fig. 37) whereas the three extended triplets in the PBB are invariably ventral. As pictured here (flagellar imbrication clockwise), the ABB is always on the right-hand side of the spline, thus the stagger is in the same direction as in hepatics. The extreme attenuation of the PBB in *Rhizomnium* and *Hypnum* can be seen in Figs 42 and 59 respectively. In each case the presence of a PBB profile on the opposite side of the cell from the lamellar strip indicates a length of not less than 6 μm.

Although the disposition of the extended basal body triplets appears to be

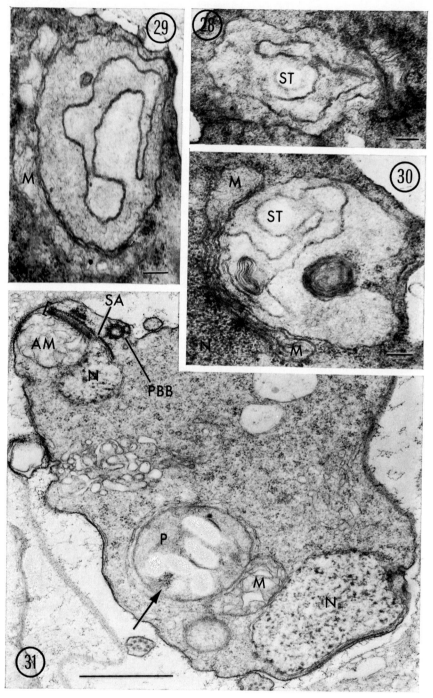

FIGS 28, 29, 30, 31

uniform throughout the mosses, the precise orientation of the PBB is variable It is parallel to the spline tubules in *Sphagnum* and *Tetraphis* but diverges from the tubules at about 5° in *Polytrichum* and 10° in *Rhizomnium* (Fig. 42) and *Hypnum* (Figs 58 and 59). Similarly, there are intergeneric differences in the position of the spline aperture. In *Sphagnum* it is normally two tubules from the in the mid-teens occur in the majority of mosses. The wider splines of *Andreaea* (Figs 37 and 38) and *Polytrichum* (Paolillo *et al.*, 1978a, Fig. 2), five in *Physcomitrium* (Lal and Bell, 1975, Plate 4A) and 12 in *Tetraphis* (Fig. 31).

The pattern emerging from the data in Table I is that spline tubule numbers in the mid-teens occur in the majority of mosses. The wider splines of *Andreaea* (Figs 36–38), *Physcomitrium* and *Hookeria* (Fig. 52) fit no obvious taxonomic framework; by far the widest spline occurs in *Tetraphis*. On the basis of a single ultrastructural character it would be unrealistic to propose the reordering of the hierarchical ranking of the Tetraphidales, but this new information adds fuel to the argument that they merit subclass or even class status and have little affinity with the Polytrichales (Cavers, 1911).

4. Mitochondria during Spermatogenesis

The characteristic behaviour of the mitochondria in the spermatids strongly links all three groups of bryophytes (Bonnot, 1967; Duckett, 1975a; Moser, 1970; Paolillo, 1965, 1974; Suire, 1970; Sun, 1964). The total mitochondrial complement undergoes fusion to form a single mitochondrion. Before nuclear metamorphosis this extends from beneath the lamellar strip (Figs 39 and 60) along the inner surface of the nucleus and separates the latter from the plastid near the centre of the spermatid. In mosses and hepatics the mitochondrium subsequently divides into two unequal parts. The larger forms the elongate anterior mitochondrion in the mature gametes and the smaller accompanies the plastid to the posteriorly located cytoplasmic remnant. *Phaeoceros* is different

---

Figs 28–30. *Sphagnum palustre*. Spermatid plastids, each showing an irregular ring of membrane internally. The single starch grain always appears adjacent to this membrane. Membranous vesicles are visible around the periphery of the plastid in Fig. 29. A cup-shaped mitochondrion separates the plastid from the nucleus (Fig. 30). Fig. 28, ×31 000. Figs 29, 30, ×34 000. Bars= 0·1 μm.

Fig. 31. *Tetraphis pellucida*. Mid-spermatid showing a spline of about 30 tubules at the anterior end associated with the anterior mitochondrion and nucleus. A spline aperture is present at the edge of the lamellar strip adjacent to the nucleus. An aggregation of phytoferritin particles is present in the plastid (arrowed). ×25 300. Bar=1·0 μm.

in that the entire mitochondrial complement becomes the anterior mitochondrion.

Despite this similar overall pattern distinct differences exist between taxa in the precise disposition of the mitochondrium. In the young spermatids of *Polytrichum* (Paolillo, 1965), *Bryum* and *Funaria* (Bonnot, 1967; Sun, 1964) at least half the mitochondrial volume coalesces around the plastid to form a nebenkern or limosphere. We have found a similar situation in all the other moss genera we have so far examined with the exception of *Sphagnum* where over 75% of the mitochondrial material condenses beneath the blepharoplast (Fig. 9) leaving only an attentuated layer with swollen margins extending approximately halfway around the periphery of the plastid and separating it from the nucleus (Figs 29 and 30). The mitochondrium is similarly organized in the young spermatids of *Phaeoceros* (Duckett, 1975a) and *Marchantia* (Kreitner, 1970), but the same condition is not attained until the mid-spermatid stage in other mosses (Fig. 49, for *Rhizomnium*, Fig. 54 for *Hookeria*). In the young spermatids of all mosses except *Sphagnum*, up to three-quarters of the periphery of the plastid may be ensheathed by mitochondrion. This has numerous separate lobes in *Polytrichum* (Paolillo *et al.*, 1968a) but is cup-shaped with an entire margin in *Bryum*, *Mnium* (Figs 39 and 40) and *Rhizomnium* (Fig. 41) and forms an open bowl in *Hypnum* (Figs 58–61). In both members of the *Mniaceae* (Figs 40 and 41) the nebenkern occupies a deep depression in the nucleus.

Figs 32–35. *Tetraphis pellucida*

Fig. 32. Transverse section of a mid-spermatid showing the nucleus containing scattered masses of condensed chromatin and a superposed spline of 38 tubules. ×47 100. Bar=0·5 μm.

Fig. 33. Transverse section of a later spermatid showing two nuclear profiles about ¾ filled with masses of condensed chromatin. ×28 300. Bar=1·0 μm.

Fig. 34. Transverse section of a mature spermatozoid showing the nucleus with a single mass of condensed chromatin. The spline contains 34 tubules. ×50 800. Bar=0·25 μm.

Fig. 35. Transverse section of a mature spermatozoid showing a spline of 33 tubules and the starch filled plastid. ×44 700. Bar=0·5 μm.

Figs 36–38. *Andreaea rothii*. Transverse sections through the MLS in young spermatids. Bars=0·25 μm.

Fig. 36. Section showing the absence of the spline aperture anterior to the triplet region of the ABB. ×50 000.

Fig. 37. Spline comprising 23 tubules and an aperture below the ABB. ×50 600.

Fig. 38. Spline comprising 22 tubules and an aperture. The ABB contains the complete complement of 9 triplets but the PBB only 2. ×72 200.

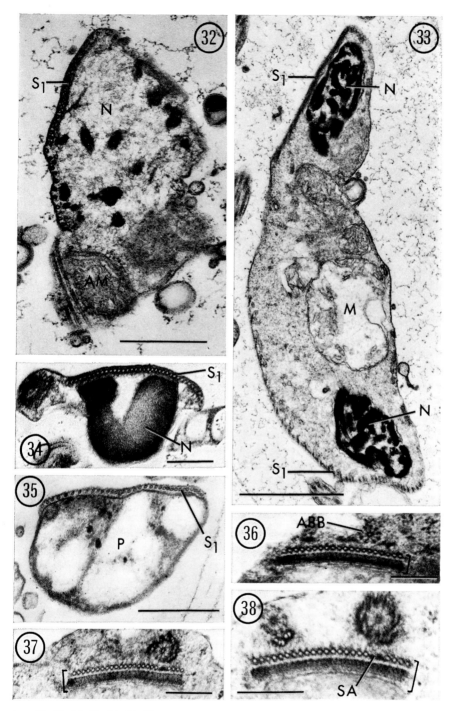

Figs 32, 33, 34, 35, 36, 37, 38

5. Plastids during Spermatogenesis

The presence of a single plastid in the male gametes of bryophytes is a feature of a great diagnostic significance. Elsewhere in archegoniate plants this mono-plastidic condition occurs only in *Selaginella* (Robert, 1974) and *Isoetes* (Thomas, 1976; Thomas and Duckett, 1976). Immediately after the last antheridial mitosis the plastid is highly pleomorphic and positioned at the periphery of the young spermatid. However, by the time the MLS is associated with the nucleus and anterior mitochondrion it is centrally located and more regular in outline (Figs 58–61). Internally, the thylakoid system is rudimentary and starch begins to accumulate. At maturity the plastid is filled with starch and hence is probably best described as an amyloplast (Paolillo, 1974; Duckett, 1975a). It is located in the cytoplasmic remnant near the posterior end of the nucleus.

Despite their numerical uniformity and similar overall ontogeny spermatid plastids appear to provide a number of taxonomically useful characters (Carothers, 1975; Duckett, 1975a). Indeed, when an exhaustive survey has eventually been carried out these could well prove as diagnostic of particular families as has been recently demonstrated for the sieve tube amyloplasts of angiosperms (Behnke, 1972). At present we can only point out the differences we have detected between a mere handful of taxa.

At the inception of starch formation in the spermatid plastids in *Bryum* (Bonnot, 1967), *Polytrichum* (Paolillo, 1965), *Physcomitrium* (Lal and Bell, 1975),

---

Figs 39, 40. *Mnium hornum*; young spermatids.

Fig. 39. Transverse section through the anterior end of the MLS. A voluminous anterior mitochondrion below the MLS is continuous with the cup-shaped mitochondrial profile lying between the nucleus and the plastid. ×18 000. Bar = 1·0 μm.

Fig. 40. Transverse section at right angles to Fig. 39 showing the posterior basal body, comprising 3 triplets lying alongside the spline. The plastid (containing several starch grains) together with the cup-shaped mitochondrion lie within a deep depression in the surface of the nucleus which contains a central area of granular material (arrowed). ×23 000. Bar = 1·0 μm.

Figs 41, 42. *Rhizomnium punctatum*; young spermatids

Fig. 41. Transverse section through MLS and anterior mitochondrion at the top of the cell, and the plastid and mitochondrion within a cup-shaped nuclear depression at the bottom. ×18 000. Bar = 1·0 μm.

Fig. 42. Transverse section showing the posterior basal body overlying both the MLS, at the anterior end of the cell and the nucleus $\frac{1}{2}$ gyre away where the PBB is laterally displaced from the spline. In both cases this basal body comprises 3 triplets only. ×14 000. Bar = 1·0 μm.

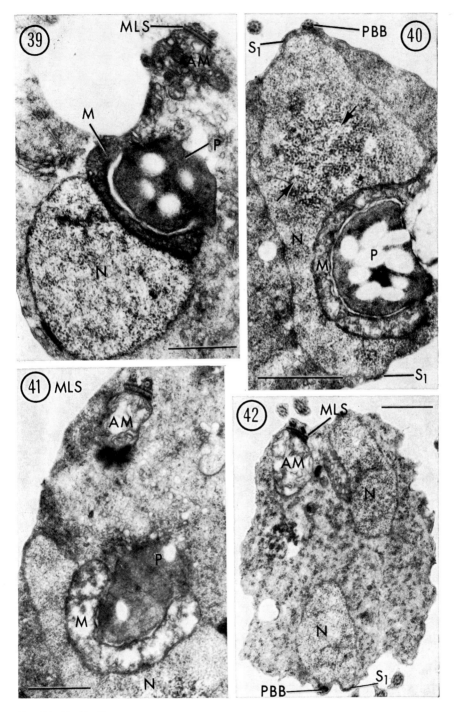

FIGS 39, 40, 41, 42

Mnium, *Rhizomnium*, *Hypnum* and *Plagiothecium* several elongate lamellae occupy a central position while the peripheral stroma contains scattered membranous vesicles. By contrast, a convoluted single thylakoid delimits the central region of the stroma in *Sphagnum* (Figs 28 and 30) and peripheral vesicles are rarely visible (Fig. 29). Frequently conspicuous in *Tetraphis* (Fig. 31) are aggregations of fine particles, morphologically very similar to the phytoferritin deposits recently described in dedifferentiated plastids in the eggs, spore mother cells and young spores of *Pteridium* (Sheffield and Bell, 1978). Initially, the starch grains are closely invested by the lamellae in *Polytrichum*, *Bryum* and *Physcomitrium* but this association has not been observed in other genera (Figs 39, 40, 54, 58–61). Peculiarities of *Hypnum* at this stage are a series of parallel bars of fibrillar material (Fig. 61) and the L-shape of the plastid (Fig. 59). Osmiophilic globuli are particularly conspicuous in *Hookeria* (Fig. 54). In all genera so far examined except *Sphagnum* the mature gametes contain upwards of 20 starch grains, each less than 1 μm in diameter. The grains are clumped in the almost spherical plastids (Figs 35 and 66). The plastid in *Sphagnum* contains a single starch grain (Manton, 1957), a highly unusual attribute shared only with *Phaeoceros* (Duckett, 1975a). Multiple starch grains appear to be the rule in hepatics. These may be linearly disposed (e.g. *Pellia* (Suire, 1970)) or

FIGS 43–47. *Mnium hornum*; condensation of the chromatin

FIG. 43. Nucleus of young spermatid showing a central aggregation of osmiophilic globuli. $\times 18\,500$. Bar $= 1 \cdot 0 \, \mu$m.

FIG. 44. Slightly older spermatid showing fine strands of chromatin attached to sheets of granular material (arrowed). $\times 22\,000$. Bar $= 1 \cdot 0 \, \mu$m.

FIG. 45. Older spermatid with thicker rods of chromatin associated with a sheet of fibrillar material (arrowed). $\times 41\,000$. Bar $= 0 \cdot 2 \, \mu$m.

FIG. 46. Transverse section of nucleus showing rods of chromatin largely sectioned transversely and adhering to a sheet of less dense material (arrowed). $\times 32\,000$. Bar $= 0 \cdot 2 \, \mu$m.

FIG. 47. Longitudinal section of nucleus showing rods of chromatin associated with two helical sheets of less dense material (arrowed). $\times 24\,000$. Bar $= 1 \cdot 0 \, \mu$m.

FIGS 48–50. *Rhizomnium punctatum*

FIG. 48. Longitudinal section of the nucleus of a late spermatid showing a central core of helically arranged condensed chromatin blocks. $\times 26\,000$. Bar $= 1 \cdot 0 \, \mu$m.

FIG. 49. Transverse section of a late spermatid showing three nuclear profiles (N1–N3 from the anterior) each containing a central core of condensed chromatin. The central cytoplasm contains the plastid still associated with a cup-shaped mitochondrion. $\times 12\,000$. Bar $= 1 \cdot 0 \, \mu$m.

FIG. 50. Transverse section of a mature spermatozoid showing four nuclear profiles (N1–N4). The nucleus now comprises a single mass of completely condensed chromatin. $\times 18\,000$. Bar $= 1 \cdot 0 \, \mu$m.

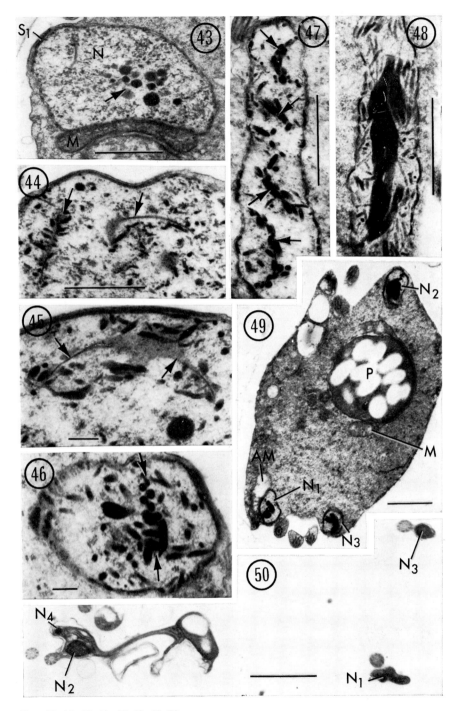

F<small>IGS</small> 43, 44, 45, 46, 47, 48, 49, 50

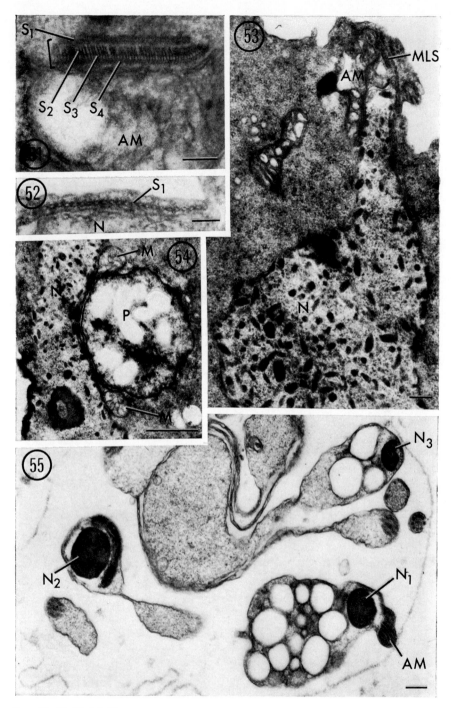

Figs 51, 52, 53, 54, 55

clumped (e.g. *Sphaerocarpos* (Zimmerman, 1973)). A further plastid feature whose distribution may well have systematic significance is the "Fibrillen-scheide", or fibrillar sheath (Diers, 1967). This tubule-containing, flap-like extension of the plastid in *Sphaerocarpos* (Diers, 1967; Zimmermann, 1973), *Marchantia* (Kreitner, 1970; Carothers, 1975) and *Blasia* (Carothers, 1975) has not yet been found in mosses or the Anthocerotales.

6. Antheridial Jacket Cells

Although the differentiating spermatids provide many more characters of systematic value, the possible uses of the jacket cells of the antheridia should not be overlooked. The vivid orange or yellow colouration of the mature antheridia of diverse bryophytes results from a transformation of the jacket cell plastids into chromoplasts by the accumulation of accessory photosynthetic pigment as droplets in the stroma (Duckett, 1975a). Since these changes are particularly conspicuous in taxa where the antheridia are exposed at maturity (e.g. *Fossombronia*, *Haplomitrium*, *Discelium*, *Anthoceros*) the function of the chromoplasts may be to shield the spermatozoids from ionizing solar radiation (Duckett, 1975a). Indeed, protection from radiation damage might also be a major function of the chromoplasts in the paraphyses which overtop the antheridia in many mosses. There are considerable differences between the chromoplasts of different taxa. In *Haplomitrium* globuli fill the plastids and there is extensive breakdown of the thylakoid systems. Globuli are also numerous in *Phaeoceros* and most mosses but the structural integrity of the thylakoids in grana is retained. In *Sphagnum*, however, the mature antheridia are bright green and the jacket cells contain chloroplasts with numerous starch grains just like those in the photosynthetic cells of mature leaves. Similar jacket cell plastids also occur in *Polytrichum* (Hausmann and Paolillo, 1978a) and *Atrichum*.

FIGS 51–55. *Hookeria lucens*
FIG. 51. Diagonal section of the MLS in a mid-spermatid showing the structure of its four strata. ×96 000. Bar=0·1 μm.
FIG. 52. Transverse section of the spline comprising 23 tubules overlying the nucleus. ×78 000. Bar=0·1 μm.
FIG. 53. Transverse section of a mid-spermatid showing scattered rods of condensed chromatin in the nucleus. ×29 000. Bar=0·2 μm.
FIG. 54. The association between the plastid, cup-shaped mitochondrion and nucleus in a mid-spermatid. ×14 000. Bar=1·0 μm.
FIG. 55. Transverse section of a mature spermatozoid showing three nuclear profiles (N1–3). The nucleus is broadest in the centre (N2). ×29 000. Bar=0·2 μm.

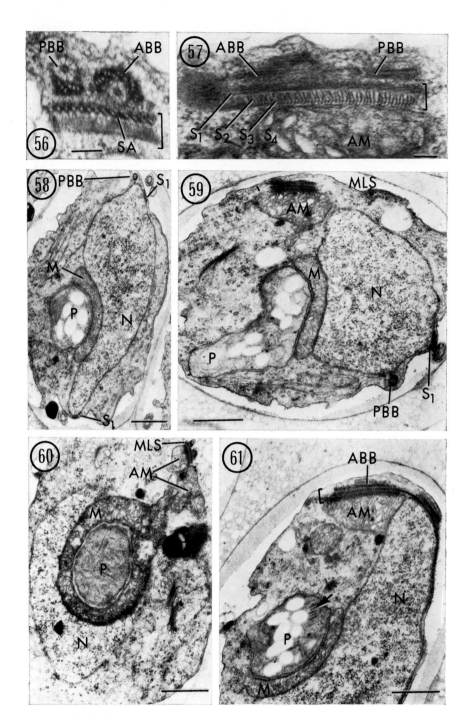

Figs 56, 57, 58, 59, 60, 61

In the light of the profound differences demonstrated by Niedhart (this volume, Chapter 12) in the chemistry, layering and origins of the spore walls in different groups of bryophytes it is pertinent to enquire whether similar differences exist in the antheridia.

As in sporogenesis, the apparently capricious appearance and disappearance of callose has been recorded in the antheridia of two hepatics, *Marchantia* and *Riccia* (Górska-Brylass, 1969). Callose is initially detectable in the walls of the sperm mother cells when the cell plate of the diagonal division producing the spermatids is beginning to form. Subsequently, this last wall also becomes impregnated with a thick layer of callose. However, it begins to dissolve when the spermatids become rounded until at maturity none remains. Nothing is yet known about callose during spermatogenesis in mosses and the Anthocerotales.

Irrespective of whether the distribution of callose during spermatogenesis ultimately turns out to vary from taxon to taxon, it is rapidly becoming apparent that the jacket walls possess several characters of systematic interest. Apart from the differentiation of thickened walls in the tip cells, recently described in detail for *Polytrichum* (Hausmann and Paolillo, 1978b), in the vast majority of moss antheridia the jacket cells are uniformly thin walled. The one exception is *Andreaea* where the antheridia possess three longitudinal grooves, extending from their tips to the base of the gamete cavity. The jacket cells subtending the grooves are smaller and thicker than those elsewhere. This feature, which appears to be peculiar to *Andreaea*, is not mentioned in the

---

FIGS 56–61. *Hypnum jutlandicum*

FIG. 56. Transverse section of the MLS in a young spermatid showing the two basal bodies and the spline aperture. $\times 85\,000$. Bar$=0\cdot1\,\mu$m.

FIG. 57. Diagonal section of the MLS showing its four strata. $\times 65\,000$. Bar$=0\cdot1\,\mu$m.

FIG. 58. Transverse section of a young spermatid approximately $\frac{1}{4}$ gyre from the anterior tip showing the PBB displaced from the S_1 overlying the nucleus. The plastid and mitochondrion occupy a shallow depression in the nucleus which contains a central region of granular material (arrowed). $\times 9500$. Bar$= 1\cdot0\,\mu$m.

FIG. 59. Longitudinal section of a young spermatid showing the L-shaped plastid characteristic of *Hypnum*. The PBB is displaced from the S_1 on the opposite side of the cell from the MLS. $\times 14\,000$. Bar$=1\cdot0\,\mu$m.

FIG. 60. Transverse section showing the continuity of the anterior mitochondrion with that surrounding the plastid adjacent to the nucleus. $\times 13\,000$. Bar$=1\cdot0\,\mu$m.

FIG. 61. Longitudinal section showing parallel bars of fibrillar material (arrowed). in the plastid. The nucleus contains a central region of granular matrial (also visible in Figs 58 and 59). $\times 13\,000$. Bar$=1\cdot0\,\mu$m.

previous account of antheridial development (Kühn, 1870). Similarly, in the majority of hepatics the jacket cells are thin walled. The extensive development of wall ingrowths by these cells in *Phaeoceros* (Duckett, 1973) may be cited as yet one more feature separating the hornworts from other bryophytes.

Hausmann and Paolillo (1977, 1978a, 1978b) have recently drawn attention to the presence of a cuticle-like layer, containing materials resembling the suberized lamellae of root endodermis and leaf bundle sheaths in Gramineae, across the stalk of the antheridium of *Polytrichum*. They conclude their account (1978a) by stating that this boundary is widespread in mosses and ferns and is worthy of study on a comparative basis. An investigation of its possible occurrence in hepatics and the Anthocerotales would thus be of considerable interest.

7. *The Nucleus*

Within the antheridia of bryophytes the initially spherical nucleus of a young spermatid undergoes metamorphosis into a helically coiled rod. Concomitant with the change in shape, the chromatin condenses and nucleoplasm is eliminated. Both the pattern of chromatin condensation and the shape and size of the nuclei at maturity vary considerably between taxa. Initially, however, the nuclei of the young spermatids of mosses, hepatics and *Phaeoceros* are very similar. Nucleoli are either absent or ill-defined and the central region of the nucleus is occupied by a reticulum of granular material (e.g. Fig. 40 for *Mnium* and Fig. 58 for *Hypnum*). This latter feature is then temporarily replaced by prominent osmiophilic globuli (Fig. 43) and from this time different types of chromatin behaviour are discernible. In the mosses we have so far detected two patterns. The commoner, illustrated in Figs 44–49 for *Mnium* and *Rhizomnium*, also occur in *Atrichum*, *Ditrichum*, *Ceratodon*, *Racomitrium*, *Bryum*, *Hedwigia*, *Plagiothecium*, *Hypnum* and probably *Polytrichum* as far as can be judged from micrographic data of Genevès (1967b, 1968, 1970) and Paolillo and Cukierski (1976). Condensed chromatin first appears as short fibres up to *c*. 50 nm in diameter (Fig. 44). These are both scattered throughout the nucleoplasm and aggregated against the edges of a sheet of granular material which traces a helical course through the nucleus roughly parallel to its long axis (Fig. 47). By the time the nucleus is coiled into about one gyre and has its maximum diameter reduced to about 1·5 μm, the chromatin comprises fewer but larger rods, up to 0·1 μm in diameter, the majority of which are attached to the sheet of granular material (Figs 46 and 47). Other condensed chromatin forms a discontinuous layer lining of the nuclear envelope. In nuclei of $1\frac{1}{2}$–2 gyres are central masses of two to four blocks of spirally twisted, condensed chromatin (Figs 48 and 49) lying parallel with the nuclear axes. At maturity the nuclei are filled with a single homogeneous mass of condensed chromatin (Fig. 50).

The condensing chromatin has a different disposition in *Hookeria* (Figs 53 and 54), *Physcomitrium* (Lal and Bell, 1975) and *Tetraphis* (Fig. 32). At the one gyre stage, rodlets with no preferred orientation are scattered throughout the nucleoplasm. These gradually fuse together (Fig. 33) into a solid mass as the nuclear volume diminishes but at no stage is a distinct central mass of chromatin present.

Nothing identical to either of these two chromatin patterns has yet been seen in hepatics. The nuclei of mid-spermatids of *Blasia* (Carothers, 1973) *Marchantia* (Kreitner, 1977b), *Pellia* (Suire, 1970), *Southbya* (Carothers and Duckett, 1978), *Dumortiera* (Carothers and Duckett, unpublished data) and *Fossombronia* (Figs 63 and 64) contain an abundance of flexuous rods of condensed chromatin of extremely uniform diameter and predominantly orientated parallel to the nuclear axes. Further compaction into a solid rod of chromatin takes place in an anterior-posterior direction. In sharp contrast, the nuclei of mid-spermatids of *Haplomitrium* are packed with condensed chromatin rods of diverse sizes with no obvious preferred orientation (Figs 62 and 65). The condensation pattern found in *Phaeoceros* (Duckett and Chescoe, 1976; Duckett, Carothers and Moser, unpublished data) has more in common with *Haplomitrium*, *Hookeria*, *Physcomitrium* and *Tetraphis* than with other hepatics or mosses.

In the spermatozoids of all mosses and hepatics, whatever their chromatin condensation pattern, there is a considerable overlap between the anterior mitochondrion and the nucleus. Whereas the nucleus of the mature spermatozoids in *Blasia* and *Marchantia* is widest at its extreme posterior end (Carothers, 1973, 1975), in all the mosses we have so far examined the nucleus tapers at both ends. Once again, *Phaeoceros* is different (Carothers, 1975): its nucleus has blunt extremities, is constricted over the middle third of its length and terminates anteriorly directly behind the anterior mitochondrion. Thus, although we still need much more information to make definitive generalizations, one major conclusion from gamete ultrastructure is inescapable: hepatics and mosses share more characters in common with each other than either group does with the Anthocerotales.

A convenient way to describe the overall shape of the nuclei of archegoniate spermatozoids is by the number of gyres they circumscribe. In *Mnium* (Fig. 50) the nucleus extends through more than $1\frac{1}{2}$ gyres (i.e. 4 profiles in sections), in

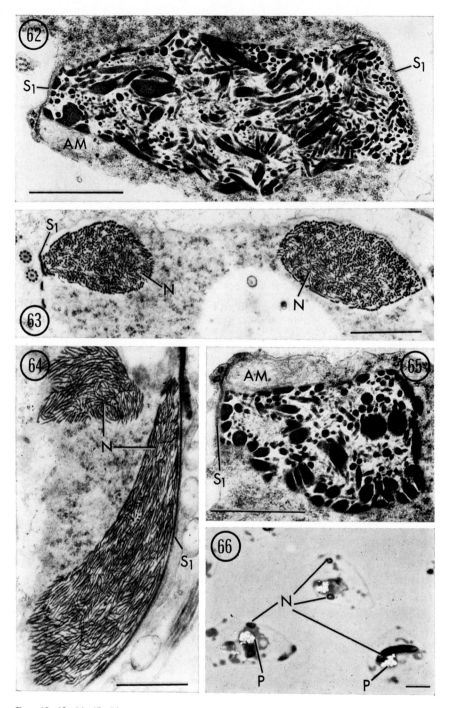

Figs 62, 63, 64, 65, 66

Hookeria (Fig. 55), *Ceratodon* and *Polytrichum* (Paolillo *et al.*, 1968b) about 1½ (i.e. 3 profiles in sections) and in *Tetraphis* about one. Similar coiling is found in *Phaeoceros* (Carothers, 1975) and the two hepatics *Marchantia* (Kreitner, 1977b) and *Blasia* (Carothers, 1973) but just over three gyres have been recorded in *Pellia* (Suire, 1970). Thus the degree of coiling appears to be more variable in hepatics than in mosses.

With a knowledge of the degree of coiling, it can be readily deduced from transverse sections of mature spermatozoids that the total volume of the nucleus varies considerably between different mosses. The smallest nuclei recorded to date are in *Polytrichum* (Paolillo *et al.*, 1968b) and *Ceratodon* where the maximum diameter is scarcely more than 0·2 μm. In *Mnium* (Fig. 50) the nuclei attain a diameter of about 0·3 μm one gyre from the anterior tip of the nucleus but in *Hookeria* (Fig. 55) and *Tetraphis* (Fig. 34) the corresponding dimensions are 0·5 μm and 0·55 μm respectively. It is interesting to note the close correlation between these nuclear diameters and spline widths in mosses (Table I). This relationship is also found between the spermatozoids of different phyla of pteridophytes (Duckett and Bell, 1977).

8. Spermatozoids and DNA in archegoniate plants

So far in this account and its companion on hepatics and Anthocerotales (Carothers and Duckett, this volume, Chapter 18) we have attempted to demonstrate how the application of the standard techniques of transmission electron microscopy to spermatogenesis in bryophytes should add a new dimension to many phylogenetic and systematic problems. However, because of the complexities of the data, most of which have never been reviewed previously, male gamete ultrastructure has largely been discussed in isolation

---

Fig. 62. *Haplomitrium hookeri*; transverse section of a mid-spermatid showing randomly arranged rods of condensed chromatin of different sizes. ×26 000. Bar = 1·0 μm.

Figs 63, 64. *Fossombronia incurva*; mid-spermatids

Fig. 63. Transverse section showing two nuclear profiles. The nucleus is filled with highly uniform rods of condensed chromatin which here are principally sectioned transversely. ×19 500. Bar = 1·0 μm.

Fig. 64. Longitudinal section showing the rods of condensed chromatin sectioned longitudinally. ×19 500. Bar = 1·0 μm.

Fig. 65. *Haplomitrium hookeri*; transverse section of mid-spermatid showing rods of condensed chromatin of different diameters. ×26 000. Bar = 1·0 μm.

Fig. 66. Non-osmicated and unstained 100 nm thick, carbon coated, section of a mature antheridium of *Hookeria lucens* used for X-ray microanalysis. The nuclei and plastids are clearly visible. ×3500. Bar = 2·0 μm.

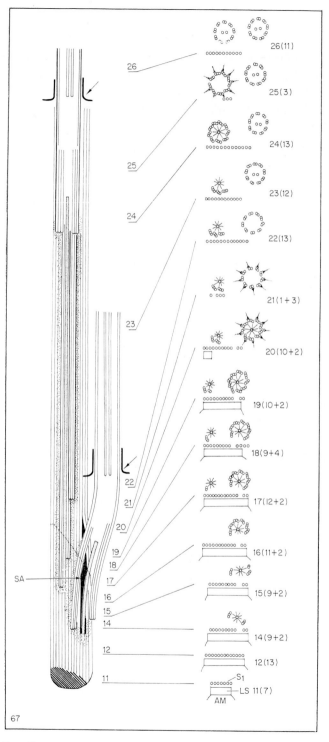

Fig. 67

from other major sources of information. Thus in conclusion it is appropriate to explore how, in the future, the relatively new cytological technique of X-ray microanalysis may provide a much needed accurate picture of the DNA content of bryophytes and other archegoniates. Not only is this likely to prove fundamental in any future assessment of phylogenetic relationships but, somewhat surprisingly, it also indicates that spermatozoids may become useful in the everyday taxonomy of bryophytes.

Sparrow *et al.* (1972) and Price (1976) have collated the available data on the DNA content per cell in eukaryotic organisms. The DNA contents per cell of the four bryophytes included in that survey (*Marchantia polymorpha*, *Riccia* sp., *Mnium* sp. and *Sphagnum* sp.) are, on the one hand, near the upper limit found in green algae and, on the other, relatively low compared with other land plants. What is particularly interesting from the point of view of the present study is the correlation which emerges when the DNA values are compared with the sizes of the spermatozoid nuclei in different archegoniate phyla. The sequence: bryophytes—*Selaginella* (Robert, 1974)—*Lycopodium* (Robbins and Carothers, 1978)—Filicales (Duckett, 1975b)—*Equisetum* (Duckett and Bell, 1977) represents both an increase in nuclear dimensions of the spermatozoids and the DNA content in meristematic cells. Thus it seems likely that the size of the nuclei of male gametes is closely related to DNA amounts. If, at present, the blepharoplast holds pride of place in speculations on the phylogeny of archegoniate plants which involve details of gamete ultrastructure (Moestrup, 1978; Pickett-Heaps, 1975; Stewart and Mattox, 1978), the nucleus could well become equally important in the future.

The routine technique for obtaining DNA amounts per cell is based on measurements of nuclear volumes in meristematic tissues coupled with cytophotometry and autoradiography. However, in bryophytes nuclear volume

FIG. 67. Diagrammatic interpretation of blepharoplast structure in *Sphagnum*. The cross-sectional profiles on the right correspond to Figs 11, 12, 14–26, with the numbers of spline tubules in parentheses. Longitudinal dimensions were derived from Figs 4–10 and numerous other micrographs. The drawing on the left is an idealized reconstruction based on all our micrographic data. A portion of the spline anterior has been omitted to show the orientation of the subtending S₂ lamellae. Arrows indicate the levels at which the two flagella emerge from the cell surface. Certain of the cross-sectional profiles (Figs 18, 21 and 25) do not correspond exactly in spline dimensions to the drawing on the left. They illustrate the range of variability encountered in the morphology of the *Sphagnum* blepharoplast. As a result, several hundred transverse sections had to be examined before the reconstruction could be attempted with any confidence.

TABLE III. Elemental analysis of *Hookeria lucens* spermatids and spermatozoids for phosphorus performed in an AEI Cora, operated at 80 keV with a 40 nm probe diameter (see Duckett and Chescoe, 1976, for technical procedures). Note the increase in nuclear phosphorus readings during spermatogenesis

Material	Phosphorus Readings			
	Peak	Background	Peak–Background	Integral
Nucleus, young spermatid	71	25	46	364
Nucleus, mid-spermatid	222	65	157	983
Nucleus, mature spermatozoid	413	105	308	2232
	409	85	324	2006
Nucleus, mature spermatozoid after RNAase treatment	389	94	295	1993
Nucleus, mixture spermatozoid after pronase treatment	398	108	290	2103
Nucleus, mature spermatozoid after DNAase treatment	37	27	10	—
Control, resin	—	—	—	—

appears to be particularly dependent on the metabolic state of the cell. Moreover, endopolyploidy is common, not only in highly specialized cells (Hallet, 1972, 1974) but also in normally growing tissues (Knoop, 1978). With its completely condensed inactive chromatin the nucleus of the male gamete is neither subject to metabolic fluctuations nor endoreplication of the DNA. Thus, providing that it can be shown that the degree of chromatin compaction is the same across a broad spectrum of taxa, then comparisons of the nuclear volumes of spermatozoids should more accurately reflect relative DNA amounts than those from meristematic cells.

As recently described for spermatogenesis in *Phaeoceros* (Duckett and Chescoe, 1976), X-ray elemental microanalysis of chromatin is a direct means of quantifying the degree of chromatin condensation. Phosphorus is a fundamental component of DNA which is not removed by aldehyde fixation and resin embedding. The progressively higher phosphorus readings obtained from spermatids at different degrees of maturity indicate the progressive compaction of the DNA (Table III). This notion is reinforced by the fact that the phosphorus is virtually undetectable following DNAase digestion but not after RNAase and pronase treatment. High phosphorus signals have also been obtained from the nuclei of human spermatozoids (Lacy and Pettitt, 1972) and mitotic chromosomes of mice and *Chironomus* (Cameron *et al.*, 1977; Trötsch, 1977). One drawback to the elemental analysis of nuclei is that the X-ray spectrum for osmium, the standard post-fixative, overlaps with that of phosphorus (P Kα and Os Mα at 1·9 keV). Osmium must therefore be omitted from specimens prepared for the probe with a consequent lack of resolution and contrast (Fig. 66). Nevertheless, the probe beam may still be accurately located over the nucleus. A second more serious deficiency of the technique is that no standard spectra have yet been worked out for pure DNA samples (or for any other natural substances) so that it is impossible to convert a phosphorus signal for a given volume of nuclear material to absolute DNA content. Consequently, all the present data are relative. Nevertheless, the similar phosphorus signals already obtained from mature spermatozoid nuclei of *Hookeria* (Table III), *Mnium*, *Atrichum*, *Fossombronia* and *Phaeoceros* indicate that the degree of chromatin condensation is much the same in all groups of bryophytes with the clear implication that taxa with larger nuclei contain more DNA. On this basis *Tetraphis* is but distantly related to *Polytrichum*, reinforcing the conclusion reached from the very different spline widths (Table I) and patterns of chromatin condensation.

Current studies on the fern *Ceratopteris* (Duckett *et al.*, 1979 and unpublished data on aposporous gametophytes) have shown that doubling the chromosome complement results in spermatozoids with larger nuclei but unchanged levels of chromatin condensation. Should the same be true in bryophytes, then critical measurements of sperm nuclei with the light microscope could well provide the means by which polyploid races, previously indistinguishable without chromosome counts (Smith, 1978; Smith and Newton, 1966, 1967, 1978), could be more readily recognized.

ACKNOWLEDGEMENTS

We are indebted to D. Andrew Davies, Garry Stuart and Mrs Sheila Walker for skilled technical assistance and David G. Long for providing the fertile material of *Tetraphis pellucida*.

We gratefully acknowledge the financial support of National Science Foundation grant PCM 76–04495; a grant from the University of Illinois Research Board to Z.B.C. is also acknowledged with gratitude.

REFERENCES

BEHNKE, H.-D. (1972). Sieve-tube plastids in relation to angiosperm systematics—an attempt towards a classification by ultrastructural analysis. *Bot. Rev.* **38**, 155–197.

BELL, P. R. and DUCKETT, J. G. (1976). Gametogenesis and fertilization in *Pteridium*. *Bot. J. Linn. Soc.* **73**, 47–78.

BONNOT, E.-J. (1967). Le plan d'organisation fondamental de la spermatide de *Bryum capillare* (L.) Hedw. *C.r. hebd. Séanc. Acad. Sci. Paris*, Sér. D, **265**, 958–961.

CAMERON, I. L., SPARKS, R. L., HORN, K. L. and SMITH, N. R. (1977). Concentration of elements in mitotic chromatin as measured by X-ray microanalysis. *J. Cell Biol.* **73**, 193–199.

CAROTHERS, Z. B. (1973). Studies of spermatogenesis in the Hepaticae. IV. On the blepharoplast of *Blasia*. *Am. J. Bot.* **60**, 819–828.

CAROTHERS, Z. B. (1975). Comparative studies on spermatogenesis in bryophytes. *In* "The Biology of the Male Gamete" (J. G. Duckett and P. A. Racey, eds), pp. 71–84. Academic Press, London and New York.

CAROTHERS, Z. B. and DUCKETT, J. G. (1978). A comparative study of the multilayered structure in developing bryophyte spermatozoids. *Bryophyt. Biblthca* **13**, 95–112.

CAROTHERS, Z. B. and KREITNER, G. L. (1967). Studies of spermatogenesis in the Hepaticae. I. Ultrastructure of the *Vierergruppe* in *Marchantia*. *J. Cell Biol.* **33**, 43–51.

CAROTHERS, Z. B. and KREITNER, G. L. (1968). Studies of spermatogenesis in the Hepaticae. II. Blepharoplast structure in the spermatid of *Marchantia*. *J. Cell Biol.* **36**, 603–616.

CAROTHERS, Z. B., MOSER, J. W. and DUCKETT, J. G. (1977). Ultrastructural studies of spermatogenesis in the Anthocerotales. II. The blepharoplast and anterior mitochondrion in *Phaeoceros laevis*: later development. *Am. J. Bot.* **64**, 1107–1116.

CAROTHERS, Z. B., ROBBINS, R. R. and HAAS, D. L. (1975). Some ultrastructural aspects of spermatogenesis in *Lycopodium complanatum*. *Protoplasma* **86**, 339–350.

CAVERS, F. (1911). "The Inter-relationships of the Bryophyta." New Phytologist Reprint No. 4. Cambridge.

DIERS, L. (1967). Der Feinbau des Spermatozoids von *Sphaerocarpos donnellii* Aust. (Hepaticae.) *Planta* **72**, 119–145.

DUCKETT, J. G. (1973). Wall ingrowths in the jacket cells of the antheridia of *Anthoceros laevis* L. *J. Bryol.* **7**, 405–412.

DUCKETT, J. G. (1975a). An ultrastructural study of the differentiation of antheridial plastids in *Anthoceros laevis* L. *Cytobiologie* **10**, 432–448.

DUCKETT, J. G. (1975b). Spermatogenesis in pteridophytes. *In* "The Biology of the Male Gamete" (J. G. Duckett and P. A. Racey, eds), pp. 97–127. Academic Press, London and New York.

DUCKETT, J. G. and BELL, P. R. (1977). An ultrastructural study of the mature spermatozoid of *Equisetum*. *Phil. Trans. R. Soc. Lond.* B, **277**, 131–158.

DUCKETT, J. G. and CHESCOE, D. (1976). A combined ultrastructural and X-ray microanalytical study of spermatogenesis in the bryophyte, *Phaeoceros laevis* (L.) Prosk. *Cytobiologie* **13**, 322–340.

DUCKETT, J. G., KLEWOWSKI, E. J., Jr and HICKOK, L. (1979). Ultrastructural studies of mutant spermatozoids in ferns, I. The mature non-motile spermatozoid of mutation 230X in *Ceratoperis thalictroides* (L.) Brogn. *Gamete Res.* (In press.)

GENEVÈS, L. (1967a). Evolution comparée des ultrastructures nucléaires et des ribosomes cytoplasmiques au cours de la maturation des spermatozoïdes de *Polytrichum formosum* (Bryacées). *C. r. hebd. Séanc. Acad. Sci. Paris*, Sér. D, **265**, 602–605.

GENEVÈS, L. (1967b). Sur le groupement des plastes et des mitochondries pendant la différenciation du spermatozoïde de *Polytrichum formosum* (Bryacées). *C. r. hebd. Séanc. Acad. Sci. Paris*, Sér. D, **265**, 1679–1682.

GENEVÈS, L. (1968). Modalités de l'edification de l'appareil flagellaire des spermatozoïdes de *Polytrichum formosum* (Bryacées). *C. r. hebd. Séanc. Acad. Sci. Paris*, Sér. D, **267**, 849–852.

GENEVÈS, L. (1970). Différenciation de l'appareil nucléaire a l'échelle infrastructurale, pendant la spermatogenèse, chez une bryophyte (*Polytrichum formosum*). *In* "Comparative Spermatology" (B. Baccetti, ed.) pp. 159–168. Academic Press, New York and London.

GÓRSKA-BRYLASS, A. (1969). Callose in gametogenesis in liverworts. *Bull. Acad. pol. Sci.*, Sér. biol., **17**, 549–554.

GRAHAM, L. E. and McBRIDE, G. E. (1975). The ultrastructure of multilayered structures associated with flagellar bases in motile cells of *Trentepohlia aurea*. *J. Phycol.* **11**, 86–96.

HALLET, J.-N. (1972). Morphogenèse du gamétophyte feuillé du *Polytrichum formosum* Hedw. I. Etude histochemique, histoautoradiographique et cytophotométrique du point végétatif. *Annls Sci. nat.*, Bot. sér. 12, **13**, 19–118.

HALLET, J.-N. (1974). Etude cytophotométrique du DNA nucléaire dans les différents territoires apicaux du gamétophyte mâle du *Polytrichum formosum* Hedw. pendant la phase reproductrice. *Bull. Soc. bot. Fr.* **121**, Coll. Bryologie, 101–110.

HAUSMANN, M. K. and PAOLILLO, D. J., Jr (1977). On the development and maturation of antheridia in *Polytrichum*. *Bryologist* **80**, 143–148.

HAUSMANN, M. K. and PAOLILLO, D. J., Jr (1978a). The ultrastructure of the stalk and base of the antheridium of *Polytrichum*. *Am. J. Bot.* **65**, 646–653.

HAUSMANN, M. K. and PAOLILLO, D. J., Jr (1978b). The tip cells of antheridia of *Polytrichum juniperinum*. *Can. J. Bot.* **56**, 1394–1399.

HEITZ, E. (1960). Über die Geisselstruktur sowie die Dreiergruppe in der Spermatiden der Leber- und Laubmoose. *In* "Proc. Eur. reg. Conf. Electron Microsc. Delft" (A. L. Houwink and B. J. Spit, eds). Vol. 2, 934–937. Ned. Veren. Electronemikroscopie, Delft.

KNOOP, B. (1978). Multiple DNA contents in the haploid protonema of the moss *Funaria hygrometrica* Sibth. *Protoplasma* **94**, 307–314.

KREITNER, G. L. (1970). "The Ultrastructure of Spermatogenesis in the Liverwort, *Marchantia polymorpha*." Ph.D. Diss. Univ. Illinois, Urbana.

KREITNER, G. L. (1977a). Influence of the multilayered structure on the morphogenesis of *Marchantia* spermatids. *Am. J. Bot.* **64**, 57–64.

KREITNER, G. L. (1977b). Transformation of the nucleus in *Marchantia* spermatids: morphogenesis. *Am. J. Bot.* **64**, 464–475.

KREITNER, G. L. and CAROTHERS, Z. B. (1976). Studies of spermatogenesis in the Hepaticae. V. Blepharoplast development in *Marchantia polymorpha*. *Am. J. Bot.* **63**, 545–557.

KÜHN, E. (1870). Zur Entwicklungsgeschichte der Andreaeaceen. Inaugural-dissertation. Univ. Leipzig.

LACY, D. and PETTITT, A. J. (1972). Biological applications of combined transmission electron microscopy and X-ray microanalysis with special reference to studies on the mammalian testis. *Micron* **3**, 113–129.

LAL, M. and BELL, P. R. (1975). Spermatogenesis in mosses. *In* "The Biology of the Male Gamete", (J. G. Duckett and P. A. Racey, eds) pp. 85–95. Academic Press, London and New York.

MANTON, I. (1957). Observations with the electron microscope on the cell structure of the antheridium and spermatozoid of *Sphagnum*. *J. exp. Bot.* **8**, 382–400.

MANTON, I. (1970). Plant spermatozoids. In "Comparative Spermatology" (B. Baccetti, ed.), pp. 143–158. Academic Press, New York and London.

MANTON, I. and CLARKE, B. (1952). An electron microscope study of the spermatozoid of *Sphagnum*. *J. exp. Bot.* **3**, 265–275.

MOESTRUP, Ø. (1978). On the phylogenetic validity of the flagellar apparatus in green algae and other chlorophyll A and B containing plants. *BiosyStems* **10**, 117–144.

MOSER, J. W. (1970). "An Ultrastructural Study of Spermatogenesis in *Phaeoceros laevis* subsp. *carolinianus*." Ph.D. Diss., Univ. Illinois, Urbana.

MOSER, J. W., DUCKETT, J. G. and CAROTHERS, Z. B. (1977). Ultrastructural studies of spermatogenesis in the Anthocerotales. I. The blepharoplast and anterior mitochondrion in *Phaeoceros laevis*: early development. *Am. J. Bot.* **64**, 1097–1106.

NORSTOG, K. (1974). Fine structure of the spermatozoid of *Zamia*: the *Vierergruppe*. *Am. J. Bot.* **61**, 449–456.

PAOLILLO, D. J., Jr (1965). On the androcyte of *Polytrichum*, with special reference to the *Dreiergruppe* and the limosphere (Nebenkern). *Can. J. Bot.* **43**, 669–676.

PAOLILLO, D. J., Jr (1967). On the structure of the axoneme in flagella of *Polytrichum juniperinum*. *Trans. Am. microsc. Soc.* **86**, 428–433.

PAOLILLO, D. J., Jr (1974). Motile male gametes of plants. In "Dynamic Aspects of Plant Ultrastructure" (A. W. Robards, ed.), pp. 504–531. McGraw-Hill, London.

PAOLILLO, D. J., Jr and CUKIERSKI, M. (1976). Wall developments and coordinated cytoplasmic changes in spermatogenous cells of *Polytrichum* (Musci). *Bryologist* **79**, 466–479.

PAOLILLO, D. J., Jr, KREITNER, G. L. and REIGHARD, J. A. (1968a). Spermatogenesis in *Polytrichum juniperinum*. I. The origin of the apical body and the elongation of the nucleus. *Planta* **78**, 226–247.

PAOLILLO, D. J., Jr, KREITNER, G. L. and REIGHARD, J. A. (1968b). Spermatogenesis in *Polytrichum juniperinum*. II. The mature sperm. *Planta* **78**, 248–261.

PICKETT-HEAPS, J. D. (1975). "Green Algae: Structure, Reproduction and Evolution in Selected Genera." Sinauer Associates Inc. Sunderland, Massachusetts.

PRICE, H. J. (1976). Evolution of DNA content in higher plants. *Bot. Rev.* **42**, 27–52.

REYNOLDS, E. S. (1963). The use of lead citrate at high pH as an electron-opaque stain in electron microscopy. *J. Cell Biol.* **17**, 208–212.

ROBBINS, R. R. and CAROTHERS, Z. B. (1978). Spermatogenesis in *Lycopodium*: the mature spermatozoid. *Am. J. Bot.* **65**, 433–440.

ROBERT, D. (1974). Etude ultrastructurale de la spermiogénèse, notamment de la différenciation de l'appareil nucléaire, chez le *Selaginella kraussiana* (Kunze) A. Br. *Annls Sci. nat.*, Bot. Sér. 12, **15**, 65–118.

SATÔ, S. (1954). On the filamentous appendage, a new fine structure of the spermatozoid of *Conocephalum conicum* disclosed by means of the electron microscope. *J. Hattori bot. Lab.* **12**, 113–115.

SATÔ, S. (1956). The filamentous appendage in the spermatozoids of Hepaticae as revealed by the electron microscope. *Bot. Mag. Tokyo* **69**, 435–438.

SHEFFIELD, E. and BELL, P. R. (1978). Phytoferritin in the reproductive cells of a fern, *Pteridium aquilinum* (L.) Kuhn. *Proc. R. Soc. Lond.* B, **202**, 297–306.

SMITH, A. J. E. (1978). "The Moss Flora of Britain and Ireland." Cambridge University Press, Cambridge.

SMITH, A. J. E. and NEWTON, M. E. (1966). Chromosome studies on some British and Irish mosses. I. *Trans. Br. bryol. Soc.* **5**, 117–130.

SMITH, A. J. E. and NEWTON, M. E. (1967). Chromosome studies on some British and Irish mosses. II. *Trans. Br. bryol. Soc.* **5**, 245–270.

SMITH, A. J. E. and NEWTON, M. E. (1968). Chromosome studies on some British and Irish mosses. III. *Trans. Br. bryol. Soc.* **5**, 463–522.

SPARROW, A. H., PRICE, H. J. and UNDERBRINK, A. G. (1972). A survey of DNA content per cell and per chromosome of prokaryotic and eukaryotic organisms: some evolutionary considerations. *In* "Evolution of Genetic Systems" (H. H. Smith, ed.), pp. 451–494. Gordon and Breach, New York, London and Paris.

STEWART, K. D. and MATTOX, K. R. (1978). Structural evolution in the flagellated cells of green algae and land plants. *BioSystems* **10**, 145–152.

SUIRE, C. (1970). Recherches cytologiques sur deux Hépatiques: *Pellia epiphylla* (L.) Corda (Metzgeriale) et *Radula complanata* (L.) Dum. (Jungermanniale). Ergastome, sporogénèse et spermatogénèse. *Botaniste* **53**, 125–392.

SUN, C. N. (1964). Fine structure of the spermatozoid of two mosses with special reference to the so-called "Nebenkern". *Protoplasma* **58**, 663–666.

THOMAS, D. W. (1976). "Studies on the Reproductive Biology of Selected Lycopsida." Ph.D. Thesis, Univ. Coll. N. Wales.

THOMAS, D. W. and DUCKETT, J. G. (1976). Ultrastructural studies of spermatogenesis in *Isoetes lacustris*. *Am. J. Bot.* **63**, (Suppl.), 44.

TRÖSCH, W. (1977). Condensation states of giant chromosomes studied by electronbeam X-ray microanalysis. *Cytobiologie* **15**, 335–356.

ZIMMERMANN, H.-P. (1973). Elektronenmikroskopische Untersuchungen zur Spermiogenese von *Sphaerocarpos donnellii* Aust. (Hepaticae). I. Mitochondrion und Plastide. *Cytobiologie* **7**, 42–54.

ZIMMERMANN, H.-P. (1974). Elektronenmikroskopische Untersuchungen zur Spermiogenese von *Sphaerocarpos donnellii* Aust. (Hepaticae). II. Der Kern. *Cytobiologie* **9**, 144–161.

18 | Spermatogenesis in the Systematics and Phylogeny of the Hepaticae and Anthocerotae

Z. B. CAROTHERS

Department of Botany, University of Illinois, Urbana, Illinois 61801, USA

and

J. G. DUCKETT

School of Plant Biology, University College of North Wales, Bangor, Wales*

Abstract: This comparative study of the incipient locomotory apparatus of male gametes elicits unconventional data representing a new line of inquiry into bryophyte evolution. It describes the microanatomy and morphology of the *Haplomitrium* blepharoplast, an aggregation of two flagellar basal bodies and the subtending organelle called a multi-layered structure. The blepharoplast reaches its greatest degree of structural complexity in young spermatids, those cells undergoing cytological transformation to become motile spermatozoids. Tabular and diagrammatic comparisons of some 12 blepharoplast characters are given for *Haplomitrium*, *Marchantia* and *Phaeoceros*. These characters include the shape and width of the spline (a microtubular cytoskeleton); absence or presence of a spline aperture; shape, length and width of the lamellar strip which subtends the spline; lengths of anterior and posterior basal bodies; position and orientation of basal bodies relative to the spline; and the absence or presence and length of the stellate pattern in the flagellar transition zone. It is postulated that from the putatively primitive *Haplomitrium*-type blepharoplast there has been a trend toward reduction of structural complexity in the more advanced taxa.

Systematics Association Special Volume No. 14, "Bryophyte Systematics", edited by G. C. S. Clarke and J. G. Duckett, 1979, pp. 425–445, Academic Press, London and New York.

* Present address: Department of Plant Biology and Microbiology, Queen Mary College, University of London, Mile End Road, London E1 4NS, Great Britain.

INTRODUCTION

The general course of male gamete development in liverworts and hornworts has long been known. Early studies of Hepaticae (Strasburger, 1869; Campbell, 1895; Ikeno, 1903) and Anthocerotae (Campbell, 1907, 1908; Bagchee, 1924) have depicted series of anatomical and cytological changes that culminate in the production of biflagellate spermatozoids. The light microscope, however, has been able to reveal few qualitative differences among spermatozoids (or spermatogenesis) of the various taxa. A modern example of this is the comparative, light microscope study by Mehra and Sokhi (1976) whose findings are largely tabulations of lengths and widths of spermatozoid cell bodies and flagella, differences in these dimensions, size ratios, etc., for some 29 species of liverworts and hornworts. Considering the restricted resolving power and magnification inherent in optical microscopy it is understandable that male gametes are commonly believed to exhibit little variation in structure. However, by utilizing the advantages afforded by electron microscopy, re-examination of the structure and development of male gametes has deepened our understanding of these vitally important links in the reproductive cycle. It is especially noteworthy that the spermatid, a product of the final division of a spermatozoid precursor cell, is actually the most structurally complex cell produced in any bryophyte. The spermatid, which at the time of its initiation has a simplified structure, undergoes a remarkable cytological transformation associated with development of the locomotory apparatus, morphogenesis and condensation of the nucleus, and loss of extraneous cytoplasm.

The locomotory apparatus of a mature spermatozoid comprises two flagella, their proximal basal bodies, a subtending spline (a curved band of parallel microtubules that serves as a cytoskeleton), possible remnants of the complex lamellar strip and the anterior mitochondrion. This transitory lamellar strip reaches its maximum degree of development in young spermatids; together with the spline it constitutes the so-called multilayered structure (MLS). It is the pair of basal bodies and the subtending MLS that constitute the blepharoplast described by early investigators of bryophyte spermatogenesis (Woodburn, 1911). At present detailed information regarding blepharoplast ultrastructure and development in liverworts and hornworts is available for only a few taxa: *Marchantia* (Carothers and Kreitner, 1967, 1968; Kreitner and Carothers, 1976), *Pellia* (Suire, 1970), *Blasia* (Carothers, 1973), *Anthoceros* (Duckett, 1974 and unpublished data) and *Phaeoceros* (Moser *et al.*, 1977; Carothers *et al.*, 1977); nevertheless, these studies demonstrate that significant structural differences do exist.

TABLE I. Spline maximum widths, expressed as numbers of microtubules, in spermatids of liverworts and hornworts. (Modified from Carothers and Duckett, 1978)

Class Hepaticae		
Calobryales	*Haplomitrium hookeri*	57
Jungermanniales	*Southbya tophacea*	16
Metzgeriales	*Blasia pusilla*	18
	Pellia epiphylla	15
Sphaerocarpales	*Riella americana* Howe and Underwood	20
	Sphaerocarpos donnellii Aust.	20
Marchantiales	*Marchantia polymorpha*	18
	Reboulia hemisphaerica	19
Class Anthocerotae	*Anthoceros punctatus*	12
	Phaeoceros laevis subsp. *laevis*	12
	P. laevis subsp. *carolinianus* (Michx.) Prosk.	12

In order to determine whether certain of these differences are also manifested in other taxa we sought comparative data by examining spermatids at that stage when the MLS was fully developed and found that the primitive *Haplomitrium* had a very broad spline (56 microtubules wide), while other species representing the Jungermanniales, Metzgeriales and Marchantiales (Table I) had much narrower splines ranging from 15–20 microtubules in width (Carothers and Duckett, 1978). As part of our continuing comparative investigation of bryophyte spermatogenesis and spermatozoid ultrastructure we have chosen to determine the morphology of the *Haplomitrium* blepharoplast—the principal subject of this paper—in order to see if it might provide clues reflecting dossible phylogenetic relationships.

MATERIALS AND METHODS

Specimens of *Haplomitrium hookeri* were collected on 4 February 1976 from flushed gravel by an inflow stream to the reservoir near Fachell, near Dolgarrog in Caernarfonshire, Wales; grid reference 23/754665. Excised antheridia were fixed in 3% glutaraldehyde in 0·05 M cacodylate buffer, pH 6·9 for 3 h at room temperature. After repeated washing in buffer the antheridia were postfixed in 2% osmium tetroxide for 2 h at 4°C, dehydrated in a graded acetone series and embedded in Durcupan ACM resin. Thin sections were stained for 30 min each with saturated uranyl acetate and basic lead citrate.

All micrographs showing cross-sectional profiles of the flagellar bases are presented as they would appear when viewed from the cell's anterior end, and all references to left or right pertain to positions occupied in the micrographs. The drawing of a *Marchantia* blepharoplast shown in Fig. 22 is modified from Carothers and Kreitner (1968); that of *Phaeoceros* in Fig. 22 is adapted from Carothers *et al.* (1977).

OBSERVATIONS

1. *Microanatomy of the* Haplomitrium *Multilayered Structure*

The multilayered structure comprises four strata, the dorsalmost of which consists of a band of parallel microtubules (Fig. 1). Measurements taken from transverse sections show that these S_1 tubules have outside and inside diameters of about 26 and 13 nm, respectively. Adjacent tubules have centre-to-centre spacing of about 31 nm. The ventral side of each S_1 microtubule contacts an accumulation of osmiophilic material which thus forms a dark, longitudinal ridge or keel beneath it (Fig. 1 inset). Evidence of MLS stratification below the S_1 layer tends to be indistinct in both transverse and longitudinal (Fig. 2) sections. However, a section cut diagonally with respect to the long axes of the S_1 microtubules reveals three subtending strata, each with vertically lamellate structure (Fig. 3). These alternating light and dark striae are common to the three layers which are collectively termed the lamellar strip. The upper or S_2 layer of the strip is about 40 nm thick, the middle and lower layers each half that amount.

Key to abbreviations used in figures
ABB   anterior basal body
AF   anterior flagellum
AM   anterior mitochondrion
LS   lamellar strip
N   nucleus
PBB   posterior basal body
S_1 spline
S_1–S_4 specific layers of the MLS, dorsal to ventral
SP   stellate pattern
TE   triplet extension from a basal body's anterior end
TZ   transition zone between the basal body and flagellar shaft
Bracket ([) indicates multilayered structure (MLS)
Asterisk (*) indicates spline aperture (SA)
Note: All magnification bars equal $0\cdot2\,\mu$m except that of Fig. 1 inset which equals $0\cdot1\,\mu$m.

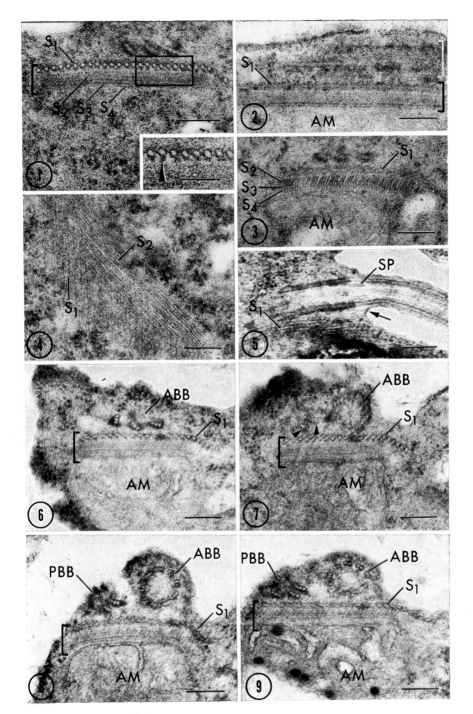

FIGS 1, 2, 3, 4, 5, 6, 7, 8, 9

FIGS 1–9. *Haplomitrium hookeri*

FIG. 1. Transverse section of the MLS. The spline comprises the S_1 microtubules. The area outlined is shown enlarged in the inset. An accumulation of osmiophilic material (arrowhead) forms a keel-like ridge below each microtubule where the tubules are subtended by the lamellar strip portion of the MLS. $\times 50\,000$. Inset $\times 100\,000$.

FIG. 2. Longitudinal section of the MLS beneath the anterior basal body (white bracket) which is shown in nonmedian section. $\times 50\,000$.

FIG. 3. Diagonal section (about 45° from the spline's long axis) of the MLS showing the structure of its four strata. $\times 50\,000$.

FIG. 4. A grazing, tangential section of the MLS. The lamellar plates of the S_2 layer are oriented at about 44° from the S_1 axis. $\times 50\,000$.

FIG. 5. Near-median, longitudinal section through a basal body showing the stellate profile region; cf. Fig. 9. The arrow indicates the level at which the flagellar shaft emerges from the cell surface. $\times 50\,000$.

FIG. 6. Transverse section near the blepharoplast's anterior end. The spline is 21 microtubules wide and extends laterally to the right beyond the subtending lamellar strip. Three forward-projecting triplet extensions of the ABB are evident. $\times 50\,000$.

FIG. 7. Transverse section slightly posterior to that of Fig. 6 showing the ABB to consist of a circle of imbricated triplets surrounding the central core; to their left lie incipient triplet extensions (arrowheads) of the PBB. $\times 50\,000$.

FIG. 8. Transverse section through the ABB at a level just posterior to the central core. The triplets are now largely tangential to the circle. The spline is here about 25 microtubules in width. $\times 50\,000$.

FIG. 9. Transverse section through the transition zone of the anterior flagellum showing the internal stellate pattern. The spline is about 26 microtubules wide at this level. $\times 50\,000$.

FIGS 10–18 show transverse sections through the MLS of *Haplomitrium* at progressively posterior levels. Long triplet extensions from the PBB are evident in Figs 10–15.

FIG. 10. The transition zone of the anterior flagellum is sectioned here through its distal (posterior) portion where it lacks the two central tubules shown in Fig. 11. The spline is about 27 tubules wide and the lamellar strip is at its maximum width of about 19 microtubules. $\times 50\,000$.

FIG. 11. At this level the spline has widened to 32 microtubules and the lamellar strip width is still at its maximum. Note the close proximity of the blepharoplast-anterior mitochondrion complex to the nucleus. $\times 50\,000$.

FIG. 12. At this level the spline is about 38 microtubules wide while the lamellar strip has narrowed to about 14. The right side of the spline is subtended by the nucleus. $\times 50\,000$.

FIGS 13, 14, 15. The spline at widths of 42, 50 and 52 microtubules, respectively. In Figs 14 and 15 about half of the spline is underlain by the nucleus. $\times 50\,000$.

FIG. 16. Transverse section through the PBB. At this level the spline is about 53 tubules wide and the lamellar strip is no longer subtended by the anterior mitochondrion. $\times 50\,000$.

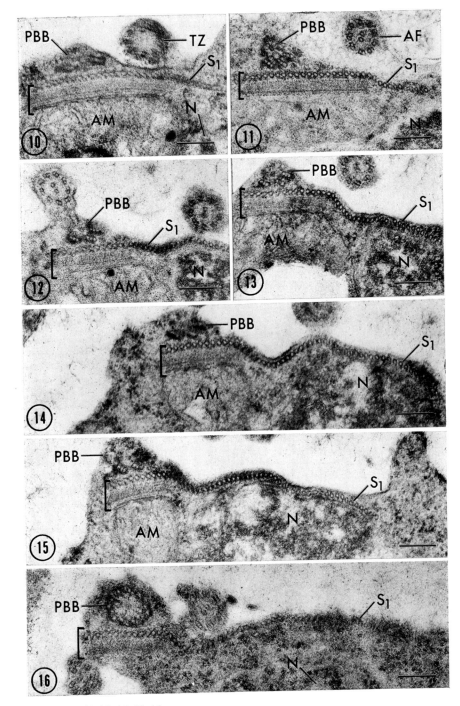

FIGS 10, 11, 12, 13, 14, 15, 16

Horizontal stratification of the lamellar strip is set off by faint lines that delimit the middle layer from those adjacent (Fig. 1). Further, the S_3 stratum is usually less osmiophilic than either S_2 or S_4 and, therefore, it appears more lightly stained (Fig. 3). Confirmation of the diagonal orientation of the dark and light lamellar plates is provided by a grazing, tangential section which passes through the S_1 and lower layers (Fig. 4). This micrograph shows that the lamellar plates lie at an angle of about 44° to the S_1 axis.

2. *Morphology of the* Haplomitrium *Blepharoplast*

The blepharoplast comprising the two basal bodies and the subtending MLS lies just beneath the surface of the spermatid where it follows the curvature of the cell surface. This asymmetric complex reaches its greatest degree of structural elaboration early in spermatid development, and its three-dimensional morphology is best reconstructed from selected transverse and longitudinal sections. The micrographs shown in Figs 6–18 represent a series of such transverse sections from several spermatids at the same developmental age. The series progresses from anterior to posterior levels of the blepharoplast, and shows clearly that the lamellar strip is subtended for nearly all its length by an elongate, anterior mitochondrion that lies closely appressed to the S_4 layer.

Near the anterior end of the blepharoplast (Fig. 6) the spline is about 21 microtubules wide and is overlain by elements of the anterior basal body (ABB) including three ventrally located, tubular triplets and a portion of the central core. The left margin of the lamellar strip is aligned with the left edge of the spline, but at this level the strip is only about three quarters as wide; thus, the spline overlaps the strip along the right margin. At a slightly more posterior level (Fig. 7) the ABB is seen as a circle—it is actually a cylinder—of imbricated, tubular triplets enclosing the central core. Immediately to the left of the ABB are osmiophilic depositions representing incipient core and triplets of the posterior basal body (PBB). More distally (Fig. 8) the ABB triplets appear oriented tangentially in a circle, with only slight evidence of imbrication. A pair of PBB ventral triplets and its central cores are clearly differentiated, and

FIG. 17. Transverse section of the PBB revealing its circle of nine imbricated triplets. With about 55 microtubules the spline has reached its full width. ×50 000.

FIG. 18. A transverse section slightly posterior to Fig. 17 showing the stellate profile in the transition zone of the posterior flagellum. The spline is about 55 microtubules wide. ×51 500.

FIG. 19. A cell in transverse section which, because of the blepharoplast's curvature, shows portions of the MLS and locomotory structures in two regions. The posterior tapering of both the lamellar strip and the anterior mitochondrion are evident. ×38 600.

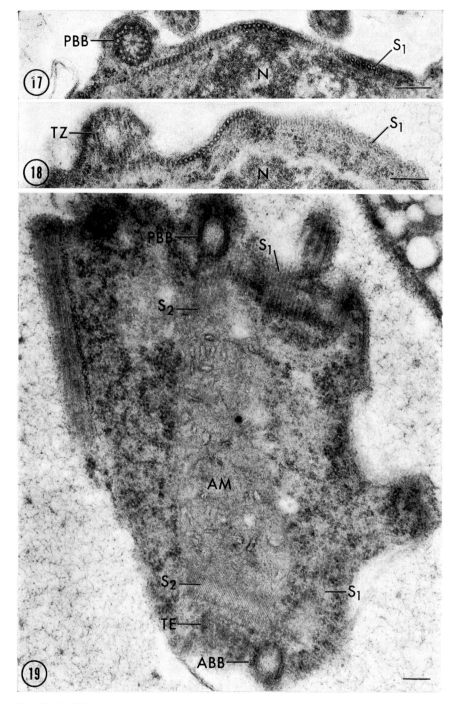

Figs 17, 18, 19

the spline is here about 25 microtubules wide. Figure 9 shows a transverse section through the proximal end of the anterior flagellum's transition zone. This portion of the zone is characterized by a circle of nine tubular doublets enclosing the stellate configuration of fine lines which contacts each doublet. On the left are three ventral triplets of the PBB; below, the spline is about 26 microtubules wide. At this level also the lamellar strip has attained its maximum width equivalent to about 19·5 microtubules.

The distal end of the anterior flagellum's transition zone may be recognized by its circle of nine doublets around an apparently unstructured interior (Fig. 10). It is at this level that the flagellum emerges above the surface of the cell to extend posteriorly as a long, whip-like organ of locomotion (cf. Fig. 5). The familiar nine-plus-two configuration of its axoneme is shown in Fig. 11. Beneath the anterior flagellum the spline is about 32 microtubules wide, and the lamellar strip is still at its maximum width.

The cylinder of 9 triplets belonging to the PBB (Fig. 16) begins about 2 μm behind the level at which the anterior flagellum emerges above the cell surface. Transverse sections along this portion of the blepharoplast (Figs 12–15) indicate that the three ventral triplets and central core of the PBB extend far forward of the cylinder. Additionally, there is a progressive widening of the spline and a concomitant narrowing of the lamellar strip (see also Fig. 19). These micrographs also reveal that the nucleus is situated beneath the portion of the spline that extends beyond the lamellar strip, and that the nucleus is becoming close-appressed to the spline (Figs 14, 15).

The distally tapering lamellar strip terminates beneath the ring of imbricated PBB triplets (Figs 16, 17), although the spline continues distally—at this stage in spermatid development, about halfway around the cell's periphery. Keels of osmiophilic material are present only where the S_1 microtubules are underlain by lamellar strip. Distal to the PBB is the transition zone of the posterior flagellum which, like the anterior one, has a stellate pattern in its interior (Fig. 18). Diagrammatic interpretations of these micrographic data are shown in Figs 20 and 21.

The greatest width observed for a *Haplomitrium* spline was 57 microtubules, or about 1·8 μm. The subtending lamellar strip is about 3 μm long, about 5 times longer than its greatest width and only slightly longer and narrower than the anterior mitochondrion. Centre-to-centre separation of the two basal body axes is about 0·3 μm. The length of the PBB including the long ventral triplets is calculated to be about 3·3 μm, nearly 5 times longer than that for the ABB. The distance between levels of flagellar emergence is about 2·8 μm.

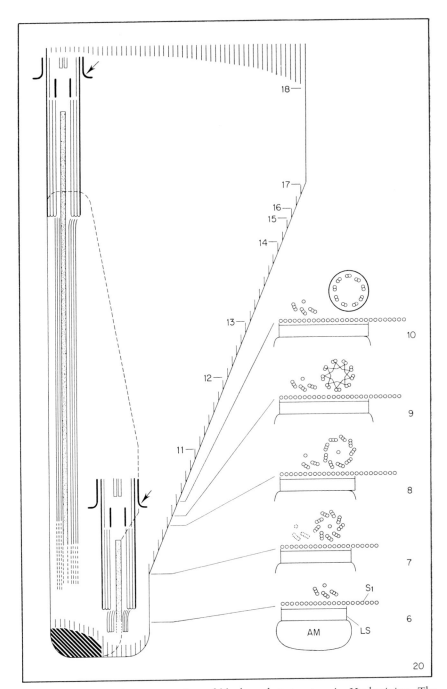

FIG. 20. Diagrammatic interpretation of blepharoplast structure in *Haplomitrium*. The five cross-sectional profiles on the right correspond to Figs 6–10. The numbers on the left correspond to the transverse sections shown in Figs 11–18 and 21. Longitudinal dimensions were derived from Figs 5, 19 and numerous other micrographs. A portion of the spline anterior has been omitted to show the orientation of the subtending S_2 lamellae. The anterior mitochondrion has been omitted from the drawing on the left; arrows indicate the levels at which the two flagella emerge from the cell surface.

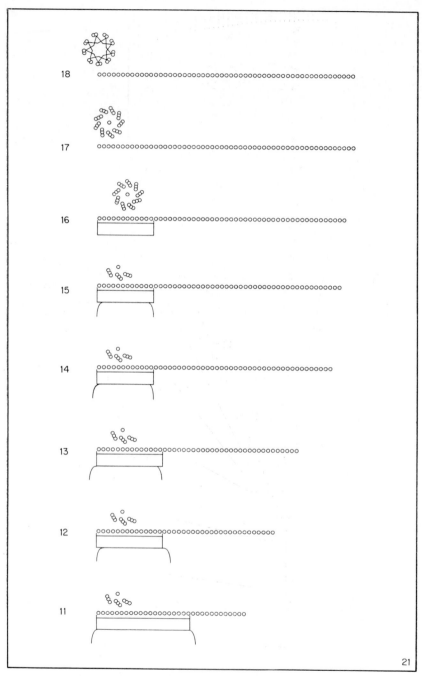

FIG. 21. Diagrammatic interpretations of the transverse sections shown in corresponding Figs 11–18. The levels at which these profiles reflect blepharoplast structure are indicated on the composite reconstruction in Fig. 20.

DISCUSSION

The presence of an MLS in the spermatid is a ubiquitous features of all bryophytes, just as it is of zooidogamous vascular plants. The same general anatomical organization of the MLS—a 4-layered complex comprising a dorsal spline of long, parallel microtubules and a subtending 3-layered strip of alternating dark and light lamellae oriented diagonally to the spline axis—occurs in all liverworts and hornworts examined to date. The many similarities in MLS ultrastructure include quantitative as well as qualitative details. For example, outside and inside diameters of S_1 microtubules are about 25 and 13 nm, respectively, with centre-to-centre spacing of about 32 nm. The lamellar strip's S_2 layer is approximately 40 nm thick, while the two lower layers (S_3 and S_4) are each about 20 nm in thickness. The diagonal orientation of the strip lamellae with respect to the spline axis is typically about 45°. Since variations in ultrastructure appear to be so slight we have excluded their citations from this paper. It is noteworthy, however, that *Haplomitrium* is virtually indistinguishable from other liverworts or hornworts in regard to details of MLS microanatomy.

The variations of systematic interest are embodied in the morphology of the blepharoplast. It is already clear from the few taxa studied that distinctive differences are expressed in at least 12 characters (Table II); these include spline shape and width; presence or absence of a spline aperture; shape, length and width of the lamellar strip; lengths of anterior and posterior basal bodies; position and orientation of basal bodies relative to the spline; and absence or presence and length of the stellate pattern in the flagellar transition zone. The following discussion of these characters concerns the three taxa for which blepharoplast morphologies have been diagrammatically set out in Fig. 22—*Haplomitrium hookeri*, *Marchantia polymorpha* and *Phaeoceros laevis* (=*Anthoceros laevis*). In order to preclude needless repetition the major sources of data on which these taxa are compared are identified here: *Haplomitrium* (the present study), *Marchantia* (Carothers and Kreitner, 1968), and *Phaeoceros* (Carothers et al., 1977).

1. The Multilayered Structure

(*a*) *Spline shape.* The band of parallel microtubules constituting the spline begins at the leading edge of the spermatozoid and extends posteriorly over the outer curvature of the nucleus to the distal end of the cell where it overlaps some or all of the terminal plastid. For comparative purposes the anterior portion of the spline exhibits the more distinctive outline (Fig. 22). The relatively massive

TABLE II. Summary of blepharoplast characters

	Haplomitrium	*Marchantia*	*Phaeoceros*
Spline shape (anterior end)	Unilaterally tapered	Spatulate	Rectangular
Spline width (microtubules)	55 (*c.* 1·7 μm)	17 (*c.* 0·5 μm)	12 (*c.* 0·4 μm)
Spline aperture	? Absent	Present, 3 tubules wide	Absent
Lamellar strip shape	Longitudinally elongated	Longitudinally elongated with posterior notch	Transversely elongated
Lamellar strip length (approx.)	3·0 μm	1·6 μm	0·5 μm
Lamellar strip max. width (approx.)	0·5 μm	0·5 μm	0·8 μm
Basal body position	Subapical, staggered	subapical, staggered	Apical, side by side (0·4 μm long each)
ABB approx. length (inc. ventral triplets)	0·7 μm	0·6 μm	
PBB approx. length (inc. ventral triplets)	3·3 μm	0·9 μm	
Basal body orientation	Parallel	Unequally divergent	Equally divergent
Stellate pattern	Present	Present	Absent
Stellate pattern length (approx.)	120 nm	230 nm	

spline of *Haplomitrium* has a rather blunt prolongation formed by some 21 microtubules; it is followed posteriorly by a gradual outward tapering on the right side to a width of 55 tubules. *Phaeoceros* exhibits the simplest, most symmetrical configuration with its 12 microtubules squarely aligned to form a straight leading edge. The spathulate anterior of *Marchantia*'s spline is more complex than that of the other two. From a rounded leading edge its parallel sides soon taper asymmetrically to a narrow band of 6 tubules, incorporating within it a narrow aperture which will be described later.

Examination of numerous species suggests that these ligulate splines become narrower more or less gradually toward their distal ends. For example, in *Southbya* the spline is about 16 microtubules wide at its anterior end, but it narrows progressively as it coils around the cell through more than two full gyres (see Figs 3, 5 in Carothers and Duckett, 1978). These illustrations of a midstage spermatid in cross section show the *Southbya* spline becoming reduced in width at half-gyre intervals in the following sequence: 16, 11, 7, 6, 4 micro-

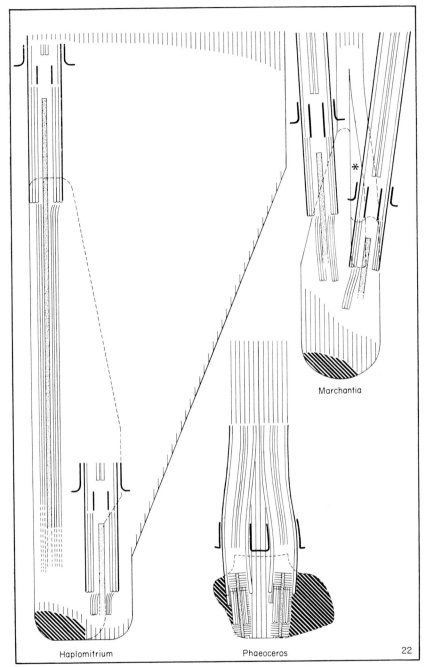

Haplomitrium

Phaeoceros

Marchantia

22

FIG. 22. A comparison of blepharoplast morphology in *Haplomitrium*, *Marchantia* and *Phaeoceros*. A portion of the spline's anterior end has been omitted from *Haplomitrium* and *Marchantia* in order to show the angular relationship between the spline axis and the diagonal plates of the lamellar strip. The spline aperture in *Marchantia* is indicated by an asterisk. Each illustration is drawn to the same scale.

tubules. Such gradual narrowing is a feature of the *Phaeoceros* spline, also. By contrast, *Marchantia*'s spline (Fig. 22) narrows abruptly at its anterior end in such a way as to form a slender (6 tubules) band with an asymmetric spathulate anterior end. Some evidence suggests that this narrowing is even more extreme in *Riella* in which the spline's maximum width of about 20 microtubules may be rather steeply narrowed to only two (Carothers, unpublished data).

(*b*) *Spline width.* The level at which a given spline must be sectioned in order to determine its maximum width accurately, depends on its overall shape. Figure 22 shows that the spline in *Haplomitrium* widens distally and unilaterally with its greatest width first reached under the posterior $\frac{2}{3}$ of the PBB cylinder of triplets. In *Marchantia* the maximum width is expressed only proximally to the cylinder of ABB triplets, while in *Phaeoceros*, owing to the absence of tapering at the anterior end, the spline displays its greatest width at its leading edge. Because of the variability in spline shape precise determinations of its width in microtubules can only be made after the morphology of the blepharoplast has been analysed from transverse and longitudinal sections. Nevertheless, it appears that for numerous taxa a good indication of the maximum width may be obtained from transverse sections at, or slightly posterior to, the level of the ABB.

Comparative data on maximum spline widths are summarized in Table I. By a considerable margin *Haplomitrium* with 57 microtubules has the widest spline yet encountered in any bryophyte. It is about 3 times wider than those representing the Sphaerocarpales and Marchantiales, 3–4 times wider than those of the Jungermanniales and Metzgeriales, and about 5 times wider than those of the Anthocerotae.

(*c*) *Spline aperture.* The spathulate anterior of the *Marchantia* spline depicted in Fig. 22 is 17 microtubules wide. Close inspection shows that its three adjacent tubules along the right-hand margin are long—in fact, they extend the full length of the cell. However, the set of three tubules adjacent to the left are short; these run only from the spline's leading edge to the level where the cylinder of ABB triplets begins. The next set of three to the left is also long, essentially matching the marginal set on the right. These two long sets gradually converge to form the narrow band of six tubules that extends the length of the cell. The proximal intercalation of three short tubules, plus the distal convergence of the six long ones, results in an off-centre, slot-like aperture. The remaining eight tubules are progressively shorter toward the left margin, thus imparting a gradually narrowing contour. A comparable spline aperture occurs in *Blasia* (Carothers, 1973), but is not found in *Pellia* (Suire, 1970) or *Phaeoceros*. There is some slight indication that a very narrow aperture, perhaps one microtubule wide (Figs 6, 7 and 9), may be present in *Haplomitrium* beneath the ABB; nevertheless, until verification can be obtained from later developmental stages we tentatively classify this taxon as lacking the feature.

(d) *Lamellar strip*. The transitory lamellar strip reaches its maximum size in the young spermatid and according to the taxon is varied in outline, but always located at what will become the cell's anterior end. In *Haplomitrium* it is about six times longer than wide and aligned evenly with the left margin of the spline which overlies it completely (Fig. 22). At a level about $\frac{1}{4}$ of its length behind the rather blunt anterior end it widens steeply to the right for a short distance, then tapers rearward from its region of greatest width to a blunt but narrower posterior end. The lamellar strip of *Marchantia* is about three times longer than wide and its anterior and lateral margins largely coincide with those of the spline. The right-hand margin in the posterior third of the strip is indented to form a notch-like recess. The only portion of the strip not superposed by spline tubules is that corresponding to the anterior third of the aperture.

An interesting departure from the form of lamellar strip seen in the liverworts occurs in *Phaeoceros*. At its maximum size the strip is actually about 1·6 times wider than long; furthermore, it is more than twice as wide as the spline. The latter is positioned to the left of the lamellar strip's midpoint, resulting in the unequal extension of the strip beyond the sides of the spline. But like the other taxa, the leading edge of the lamellar strip is evenly aligned with that of the spline, and the diagonal orientation of its constituent lamellae bears the same angular relationship to the spline axis.

2. The Basal Bodies

(a) *Length*. Each basal body is here construed to include the cylinder of triplets and those ventral triplets that extend proximally from it. Except for *Phaeoceros* in which the basal bodies are of equal length, the PBB is longer than the ABB—nearly five times longer in *Haplomitrium* and twice as long in *Marchantia*. Individual dimensions are given in Table II. The unequal lengths of anterior and posterior basal bodies are also reflected in the different levels at which the flagella emerge above the cell surface.

(b) *Position*. Figure 22 shows that among different taxa there is considerable variation in the positions of the basal bodies relative to each other and to the subtending spline. In *Phaeoceros* the basal bodies lie side by side with their proximal ends evenly aligned with the leading edge of the spline. However, in *Haplomitrium* and *Marchantia* the basal bodies are staggered and subapical, i.e. situated at some distance behind the spline's proximal edge.

(c) *Orientation*. Orientation here refers to the angular relationships of basal body axes between each member of the pair and between each basal body and the spline axis. As shown in Fig. 22, the axes of *Haplomitrium*'s anterior and posterior basal bodies lie parallel to each other and to the spline axis. Those in *Phaeoceros* and *Marchantia* are posteriorly divergent, that is, the posterior ends of the organelles are further from a midline separating the basal bodies than are the anterior ends. However, in early spermatids of *Phaeoceros* the individual divergence angle for each basal body is the same (about 4°) while in *Marchantia* the ABB angle is greater (about 8° to the right) than that of the PBB (about 3° to the left).

   Another type of orientation is exhibited by *Pellia* (Suire, 1970). In this taxon the PBB and spline axes are parallel, but the ABB axis is divergent to the right by about 18°.

(d) *Stellate pattern*. Strictly interpreted, the stellate pattern is not a feature of a basal body but of the transition region between the basal body and flagellar shaft; it is included here merely for convenience. This pattern (Fig. 9) is an angular configuration of dark staining lines that contact the tubular doublets. It is present in *Haplomitrium* and *Marchantia*, but absent from *Phaeoceros*. Further, its presence was not demonstrated in the studies of *Pellia* (Suire, 1970) or *Blasia* (Carothers, 1973).

3. Conspectus

Until very recently virtually all ultrastructural investigations on spermatogenesis in bryophytes emphasized one or more aspects of ontogenetic development. The initial attempt to seek data of potential phylogenetic value was a comparative study of the multilayered structure in 22 species of bryophytes (Carothers and Duckett, 1978). The present investigation, together with its companion study of mosses (Duckett and Carothers, this volume, Chapter 17), represents a second, larger step in the exploration of a new area of inquiry that, we believe, will ultimately provide much useful phylogenetic information.

The initial study referred to above showed that spline widths can be markedly dissimilar between distantly related taxa, as exemplified by *Haplomitrium* with 57 microtubules and *Marchantia* with 18, and that more closely related taxa may have similar widths, as in the Anthocerotales where *Anthoceros* and *Phaeoceros* each have 12. But for most liverworts examined the spline's width ranged only from 15 to 20 microtubules (Table I). Some understanding of this narrow range may be gained by considering the spline in relation to the morphology of the mature spermatozoid in which it serves as a cytoskeleton. The fully formed cell is characteristically much longer than wide; the cell body of *Marchantia*, for example, is 11·6–12·0 μm in length and 0·4–0·6 μm in width (Mehra and Sokhi, 1976). Its nucleus, which constitutes the greatest bulk of the cell, is a long, curved cone whose pointed anterior end is positioned below the spline aperture (Fig. 3D in Carothers, 1975). Over most of its length the spline is closely appressed to the greater (dorsal) curvature of the nucleus. Since the latter tapers anteriorly to a very slender apex, its diameter is less than the spline's width for a distance extending posteriorly about $\frac{1}{3}$–$\frac{1}{2}$ of the nucleus's length. The spline's gradual widening at its proximal end, where the unequally divergent basal bodies are situated in a staggered relationship (Fig. 22), tends to impart a more streamlined contour to the front of the gamete.

The foregoing brief description of the spline/nucleus relationship in a *Marchantia* spermatozoid applies broadly to liverworts; however, *Haplomitrium* is a notable exception: its spline is so wide that it extends well beyond the margin of the fully condensed nucleus (unpublished data). But, in general, it appears that the long, narrow spline is a reflection of the mature gamete's small size and characteristically elongated morphology. Furthermore, it seems likely that because of the narrow range of spline widths this particular feature, by itself, will have sharply limited utility as a diagnostic character.

As we have seen, the spline is but one constituent of the blepharoplast, that aggregation of cytoplasmic organelles which collectively expresses a high degree of structural complexity. Reaching its most elaborate state in the young spermatid it manifests numerous characteristics of potential systematic value. When the three ensembles of blepharoplast characteristics depicted in Fig. 22 for *Haplomitrium*, *Marchantia* and *Phaeoceros* are compared (Table II), it becomes immediately apparent that each taxon is distinctly different from the others.

Efforts to draw conclusions from the present study are hampered by too little information on too few species—a characteristic circumstance for any new approach in its opening stage. Nevertheless, on the basis of the data given here plus what we can glean from the literature, we would speculate that from the putatively primitive *Haplomitrium*-type blepharoplast there has been a trend toward reduction of structural complexity in the more advanced taxa. Many more examples of blepharoplast morphology are needed, as are data on the ranges of intraspecific variation in spline composition. It seems likely also that ultrastructural studies of mature spermatozoids would reveal hitherto unrecognized features of systematic value. But we would add a cautionary note regarding investigations such as these. The collection of antheridia at the proper developmental stage, the subsequent processing of tissue using "standard" procedures of transmission electron microscopy, and the micrographic analysis and reconstruction of blepharoplast morphology are laborious and time-consuming. Recognition of these technical complexities will, perforce, exert a strong influence in deciding which taxa warrant ultrastructural study. Accordingly, highest priority should be given to those species considered to have the greatest phylogenetic significance.

ACKNOWLEDGEMENTS

We wish to thank Mrs Alice Prickett, illustrator in the School of Life Sciences, University of Illinois, for her help in rendering Figs 20–22.

This investigation was supported by National Science Foundation grant PCM 76–04495.

REFERENCES

BAGCHEE, K. (1924). The spermatogenesis of *Anthoceros laevis* L. *Ann. Bot.* **38**, 105–111.

CAMPBELL, D. H. (1895). "The Structure and Development of the Mosses and Ferns." Macmillan, New York and London.

CAMPBELL, D. H. (1907). Studies on some Javanese Anthocerotaceae. I. *Ann. Bot.* **21**, 467–486.

CAMPBELL, D. H. (1908). Studies on some Javanese Anthocerotaceae. II. *Ann. Bot.* **22**, 91–102.

CAROTHERS, Z. B. (1973). Studies of spermatogenesis in the Hepaticae. IV. On the blepharoplast of *Blasia*. *Am. J. Bot.* **60**, 819–828.

CAROTHERS, Z. B. (1975). Comparative studies on spermatogenesis in bryophytes. *In* "The Biology of the Male Gamete" (J. G. Duckett and P. A. Racey, eds), pp. 71–84. Academic Press, London and New York.

CAROTHERS, Z. B. and DUCKETT, J. G. (1978). A comparative study of the multilayered structure in developing bryophyte spermatozoids. *Bryophyt. Biblthca* **13**, 95–112.

CAROTHERS, Z. B. and KREITNER, G. L. (1967). Studies of spermatogenesis in the Hepaticae. I. Ultrastructure of the Vierergruppe in *Marchantia*. *J. Cell Biol.* **33**, 43–51.

CAROTHERS, Z. B. and KREITNER, G. L. (1968). Studies of spermatogenesis in the Hepaticae. II. Blepharoplast structure in the spermatid of *Marchantia*. *J. Cell Biol.* **36**, 603–616.

CAROTHERS, Z. B., MOSER, J. W. and DUCKETT, J. G. (1977). Ultrastructural studies of spermatogenesis in the Anthocerotales. II. The blepharoplast and anterior mitochondrion in *Phaeoceros laevis*: later development. *Am. J. Bot.* **64**, 1107–1116.

DUCKETT, J. G. (1974). An ultrastructural study of spermatogenesis in *Anthoceros laevis* L. with particular reference to the multilayered structure. *Bull. Soc. bot. Fr.* **121** (Colloq. Bryologie), 81–91.

IKENO, S. (1903). Beiträge zur Kenntnis der pflanzlichen Spermatogenese: die Spermatogenese von *Marchantia polymorpha*. *Bot. Zbl.* **15**, 65–88.

KREITNER, G. L. and CAROTHERS, Z. B. (1976). Studies of spermatogenesis in the Hepaticae. V. Blepharoplast development in *Marchantia polymorpha*. *Am. J. Bot.* **63**, 545–557.

MEHRA, P. N. and SOKHI, J. (1976). Observations on the spermatozoids of some Hepaticae and Anthocerotae from India. *J. Hattori bot. Lab.* **41**, 359–376.

MOSER, J. W., DUCKETT, J. G. and CAROTHERS, Z. B. (1977). Ultrastructural studies of spermatogenesis in the Anthocerotales. I. The blepharoplast and anterior mitochondrion in *Phaeoceros laevis*: early development. *Am. J. Bot.* **64**, 1097–1106.

STRASBURGER, E. (1869). Die Geschlechtsorgane und die Befruchtung bei *Marchantia polymorpha* L. *Jahrb. wiss. Bot.* **7**, 409–422.

SUIRE, C. (1970). Recherches cytologiques sur deux Hépatiques: *Pellia epiphylla* (L.) Corda (Metzgériale) et *Radula complanata* (L.) Dum. (Jungermanniale). Ergastome, sporogénèse et spermatogénèse. *Botaniste* **53**, 125–392.

WOODBURN, W. L. (1911). Spermatogenesis in certain Hepaticae. *Ann. Bot.* **25**, 299–313.

19 | Chemotaxonomy of Bryophytes: a Survey

C. SUIRE

*Laboratoire de Botanique, Université de Bordeaux I, Avenue des Facultés,
33405 Talence cedex, France*

and

Y. ASAKAWA

*Institute of Pharmacognosy, Tokushima-Bunri University,
Yamashiro-cho, 770 Tokushima, Japan*

Abstract: The uses of biochemical data for all levels of bryophyte taxonomy are
reviewed. Biochemically, bryophytes appear to be more closely related to other land
plants than to green algae. Some taxa attain levels of biosynthetic complexity very close
to those in advanced families of angiosperms while others are much less advanced.
The chemical heterogeneity of the bryophytes, at all levels of classification, is at least
equal to their morphological diversity. Chemical data support the notion that the
Bryophyta is an unnatural and rather tenuous assemblage of the five groups Anthocerotae,
Hepaticae, Bryopsida, Andreaeopsida and Sphagnopsida. These five groups are clearly
separated by the distribution of lunularic acid and the varied metabolism of D-methionine.
   A wide range of flavonoids occurs in mosses and liverworts but they are most complex
in the sporophytes of both groups. In many taxa new compounds are synthesized during
sex organ formation. So far, we have more information on the biochemistry of the
hepatics than on the other groups of bryophytes, and flavonoids have so far provided
the most useful data for the chemotaxonomist. Flavonoid biochemistry supports the
inclusion of *Monoclea*, *Corsinia*, *Targionia*, *Cyathodium* and *Exormotheca* in the Marchantiales, indicates that the Ricciaceae is a very ancient group, and suggests that the Calobryales is more advanced than morphological data appears to indicate. Terpenoids
are exclusive to hepatics with oil-bodies. Isomeric differences between the sesquiterpenes

Systematics Association Special Volume No. 14, "Bryophyte Systematics", edited by
G. C. S. Clarke and J. G. Duckett, 1979, pp. 447–477, Academic Press, London and New
York.

of hepatics and spermatophytes indicate the independent evolution of biosynthetic pathways in the different groups. Races with different peroxidase enzymes have been identified in *Pellia*, and *Conocephalum* has races with different flavonoids.

INTRODUCTION

Whereas by 1960 a considerable body of biochemical information had been obtained on the spermatophytes and their principal biosynthetic sequences established, knowledge of the chemical composition of the cryptogams was at best fragmentary, and for bryophytes rudimentary. However, during the last two decades the analysis of natural compounds has been extended to cover all plant groups. Up to 1978, 350 publications, over half of them in the last five years, have been devoted to the chemical or biochemical results obtained on over 200 species of bryophytes.

The Bryophyta includes over 24 000 living species and rather less than 100 fossil species (Lacey, 1969). From studies of the morphology of extant taxa, since palaeobotanical records are very poor, most authors feel that the group is heterogeneous and probably had a polyphyletic origin (Howe, 1899; Jovet-Ast, 1967; Schuster, 1966; Steere, 1969). Thus the Andreaeopsida, Bryopsida, Sphagnopsida, Hepaticae and Anthocerotae may have no more in common than their similar life histories. It is therefore of considerable interest to discover how far chemical data support or contradict this notion. In addition, comparative biochemical studies on bryophytes, algae, and vascular plants may permit a more precise assessment of evolutionary relationships between the different phyla.

Rather than presenting a comprehensive review of the biochemistry of bryophytes, in this account we have attempted to pick out substances which are of particular interest to the taxonomist and to indicate profitable areas for future studies. The examples we have selected demonstrate that chemotaxonomy may be a powerful tool at all levels of bryophyte classification from the class down to the subspecies or even geographic races.

The majority of chemotaxonomic studies so far carried out on bryophytes have been devoted to hepatics. This account reflects this situation but at the same time indicates areas which might most profitably be investigated in the Musci. For a comprehensive review of the more biochemical aspects of many of the substances mentioned here, the reader is referred to Markham and Porter (1978b).

Constituents common to all the green plants (e.g. chlorophylls and plasto-quinones) or compounds whose fluctuations during the year are greater than interspecific variations (e.g. the tricarboxylic acids of the Krebs cycle, and alkanes) are not generally taxonomically useful. Other data such as the absence of lignin in the bryophytes (Erickson and Miksche, 1974; Miksche and Yasuda, 1978) have very limited systematic interest. We have therefore selected compounds, or groups of compounds, which have very varied distribution patterns.

BIOCHEMICAL EVIDENCE AT THE CLASS LEVEL

The occurrence and metabolism of lunularic acid and D-methionine highlight the profound biochemical differences which can occur between the various classes of bryophytes.

Lunularic acid (Fig. 2) occurs only in liverworts. It has not been found in mosses, *Sphagnum*, or *Anthoceros punctatus* and *Phaeoceros laevis* (Gorham, 1977).

Although bacteria and eucaryotic plants all deaminate the natural L-methionine, the unnatural D-isomer is metabolized in several ways. Tracheophytes and Anthocerotae conjugate it with malonic acid. Fungi, some lichens, *Andreaea* and *Sphagnum* convert it into N-acetyl-D-methionine. Algae and Hepaticae do not differentiate the two isomers and deaminate both. However, marine algae convert methionine into dimethyl-β-propiothetin, a metabolite which was not identified in the forty bryophytes so far tested. Mosses metabolize D-methionine by different pathways: *Plagiomnium cuspidatum*, *P. undulatum* and *Rhizomnium punctatum* deaminate it, one species of *Bryum* converts it into the N-acetyl conjugate, other species acetylate part of the D-methionine but deaminate the rest (Pokorny *et al.*, 1970; Pokorny, 1974).

Several important conclusions may be reached from these results:

(1) The synthesis of lunularic acid and the degradation pathway of D-methionine separate very distinctly the metabolism of the Hepaticae and Anthocerotae, thus supporting arguments based on morphological and cytological characters that the Anthocerotae should be elevated to the same rank as the Hepaticae rather than being placed as a subdivision of the latter class.

(2) The complete acetylation of D-methionine by *Andreaea* and *Sphagnum* supports their separation as classes distinct from the remainder of the mosses. However, it should be noted that this distinction is less clear-cut than that between the Hepaticae and Anthocerotae based on lunularic acid, since some Bryopsida also carry out partial acetylation of D-methionine.

(3) Musci are more varied in their metabolism of D-methionine than

any other group in the plant kingdom thus suggesting considerable heterogeneity.

1. Enzymes

According to Georgiev and Bakardjieva (1973), the number of isoenzymes increases with evolutionary advancement and the highest evolutionary levels are characterized by the appearance of isoenzymes of high molecular weight (a conclusion wrongly attributed to Shannon, 1968). Results obtained on the peroxidase isoenzymes of eight mosses are cited as corroborating this idea. However, the authors ignored the possibility of variations in enzymatic patterns depending on the ontogenic state of the plants, climatic conditions or the existence of geographic races.

2. Fatty Acids

Contrary to previous conclusions (see Suire, 1975, for review), arachidonic acid (C_{20} with four double bonds) is not a substitute of α-linolenic acid (C_{18} $3\varDelta$) in the bryophytes (Gellerman et al., 1975), the latter being one of the main constituents of lipids from plastid membranes in the spermatophytes. C_{20} and C_{22} polyunsaturated fatty acids occur in phycomcetes, algae, bryophytes and pteridophytes but not in basidiomycetes and spermatophytes, except for *Ginkgo biloba*. Phycomycetes and algae are largely aquatic and among the bryophytes it is chiefly those found in aquatic conditions or in moist places which have high concentrations of these acids. The main constituents of lipids from mosses growing in dry areas are C_{18} fatty acids, particularly α-linolenic acid. The production of C_{18} fatty acids may perhaps be an adaptation to dryness (Anderson et al., 1974) whereas highly unsaturated C_{20} or sometimes C_{22} fatty acids may perhaps be essential to the physiology of aquatic plants or plants growing in moist areas (Karunen, 1978).

Thus, the composition of fatty acids from lipids appears to reflect the ecological preferences of mosses. However, when more data become available, features of taxonomic interest could well come to light.

3. Flavonoids

These have now been found in just less than 70% of the acrocarpous mosses tested (McClure and Miller, 1967) but in more than 20% of the pleuro-

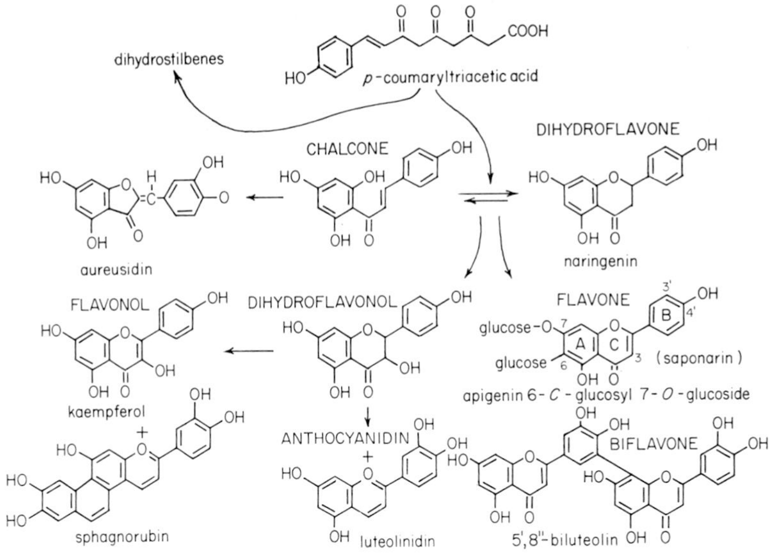

FIG. 1. Biosynthetic pathways of flavonoids

carpous ones (McClure and Miller, 1967; Vandekerkhove, 1977b). Thus their interest as taxonomic markers in the pleurocarps is somewhat limited.

The structure of the flavonoids has been determined in about a dozen species (Fig. 1). A flavone (luteolin) has been isolated from *Ceratodon purpureus* (Vandekerkhove, 1978c) and a biflavone from *Dicranum scoparium* (Lindberg *et al.*, 1974). The occurrence of a flavonol 3-O-diglycoside has been suggested in *Mnium arizonicum* Amann (Melchert and Alston, 1965) and two O-glucosides of luteolinidin (3-deoxyanthocyanidin) (Fig. 1) have been identified in *Bryum pallens*, *B. rutilans* Brid., *B. weigelii* and *B. tortilifolium* (=*B. cyclophyllum*) and two species of *Splachnum* (Nilsson and Bendz, 1974). Several flavones (apigenin or luteolin, Fig. 1) which are *C*- or *O*-glycosylated have been found in *Bryum*, *Dicranum*, *Mnium* (Nilsson and Bendz, 1974; Vandekerkhove, 1978a), *Hedwigia ciliata* (Österdahl and Lindberg, 1977), *Tetraphis pellucida* and *Hylocomium splendens* (Vandekerkhove, 1977a, b). In addition, bracteatin, an aurone, has been isolated from the sporophyte of *Funaria hygrometrica*, but not from the gametophyte (Weitz and Ikan, 1977). However, in the same species, Herzfelder (1921) reported the release of a yellow pigment from the jacket cells of mature antheridia. It would be of considerable interest to discover whether this pigment is also an aurone and, if so, whether or not it is bracteatin. This raises the question of whether biosynthetic capabilities change during sexual differentiation. To answer this dioecious and monoecious taxa should be compared; specific aurones might well occur in the male gametophytes of the former but only in the sporophyte of the latter (see also section on flavonoids in hepatics).

A great deal is known about the flavonoids of other green plants. Flavonoids have not been found in the algae, and biflavones occur only in groups considered to be primitive, such as the Psilotales, Selaginellales, Cycadales, Ginkgoales and some primitive angiosperms (Swain, 1974). The red colour of the juvenile fronds of seven Filicinae results from the presence of luteolinidin 5-*O*-glycosides (Harborne, 1966). Flavonol *O*-glycosides have been found in *Equisetum* (Swain, 1975) and *Ophioglossum* (Markham *et al.*, 1969). The charophytes are able to produce flavone *C*-glycosides (Markham and Porter, 1969; Swain, 1975) but the ability to *O*-glycosylate the flavones, as in the synthesis of *O*-heterosides and 3-hydroxylated aglycones, is restricted to the Isoetales and Lycopodiales, some advanced Filicales and the most advanced spermatophytes (Swain, 1974, 1975). Aurones have only been identified in eight advanced families of angiosperms where they most frequently occur in yellow flowers (Bohm, 1975). An extensive study of the available data on the higher plants enabled Harborne (1967) to distinguish the structural characteristics of the flavonoids of species considered either as primitive or as advanced on other criteria.

Bearing in mind the patchy nature of the data, it is still possible to draw some general conclusions about flavonoid biochemistry in mosses in relation to their evolution. Firstly, by comparison with the angiosperms, the structure of the flavonoids found so far in mosses indicates the existence of several levels of biosynthetic evolution which far surpass the range of morphological diversity in the group. Secondly, the flavonoids reach greater structural complexity in the sporophytes where they approach the levels of biochemical sophistication preferences of mosses. However, when more data become available, features tested (McClure and Miller, 1967) but in more than 20% of the pleuro- found in the angiosperms. Thirdly, the absence of flavonoids is possibly a characteristic of certain genera and does not necessarily indicate that the species are more primitive than those which synthesize flavonoids.

Despite their diversity and widespread occurrence, little attempt has yet been made to use flavonoids in moss taxonomy at the infrageneric level. However, Koponen and Nilsson (1978) found that the flavonoid patterns in various species of *Mnium*, as seen in thin layer chromatograms, corresponded to the relationships deduced from other criteria. Such findings suggest that this approach has considerable potential.

1. Flavonoids

Vowinckel (1975) demonstrated that the reddish colour of *Sphagnum magellanicum* is related to the production of a previously unknown anthocyanin, sphagnorubin (Fig. 1) which is closely associated with the cell wall. This red pigment was not found in the non-reddening species but no attempt was made to investigate the possible presence of flavonoids during sex organ formation. Many species which are not normally red assume a reddish or orange colouration in the autumn (e.g. *S. palustre*) during antheridial maturation (Schuster, 1966, p. 215; Duckett, personal communication). The chemotaxonomy of flavonoids appears to be a promising area which might shed light on interrelationships within *Sphagnum* and its position relative to other mosses. Many more features in common beyond leaf cell dimorphism must be found to establish a firm link between *Sphagnum* and *Leucobryum* as proposed by Gams (1959).

1. Stilbenoids

According to the preliminary results of Pryce (1972b), lunularic acid (Fig. 2), a dihydrostilbene derivative which induces dormancy in algae and Hepaticae, may be analogous in its mode of action to abscisic acid produced by mosses, ferns and higher plants. However, the more recent results of Gorham (1977) are at variance with this conclusion. Gorham found lunularic acid not only in Hepaticae but also in *Hydrangea macrophylla* (an angiosperm) and he found lunularin (Fig. 2e), the decarboxylation product of lunularic acid, in both Hepaticae and another angiosperm *Morus laevigata*. Neither of these two compounds was found in algae, where lunularic acid has probably been confused with a compound which has the same retention time in gas chromatography (Gorham, 1977). Moreover, the inhibitory effect of lunularic acid is closely comparable to that of many other similar compounds (Gorham, 1978) and thus can no longer be considered as the specific inhibitor of growth and inducer of dormancy in liverworts. Its presence in species both with and without oil bodies suggests that its elaboration is not related to secretory processes and that either it has an important physiological role which is still unknown, or it is an important intermediate in the synthesis of other compounds.

Swain (1974) compared the synthesis of lunularic acid to that of dihydrostilbenes and flavonoids, which are produced in spermatophytes by the ring enclosure of *p*-coumaryl triacetic acid (Fig. 1); the only initial difference is

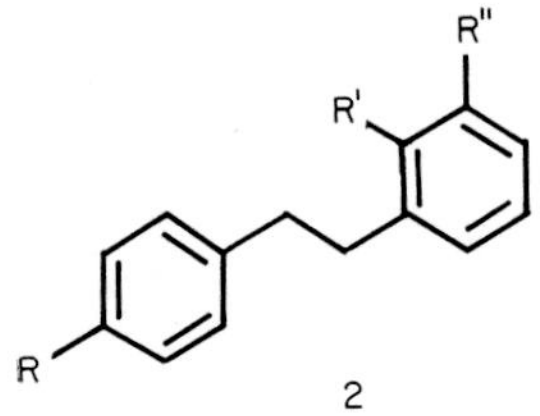

FIGS 2–5. Stilbenoids from liverworts

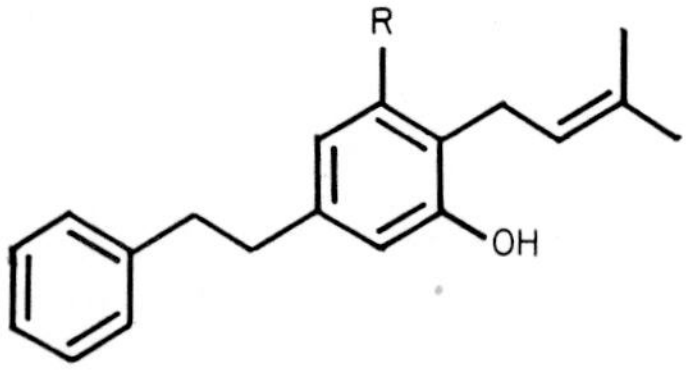

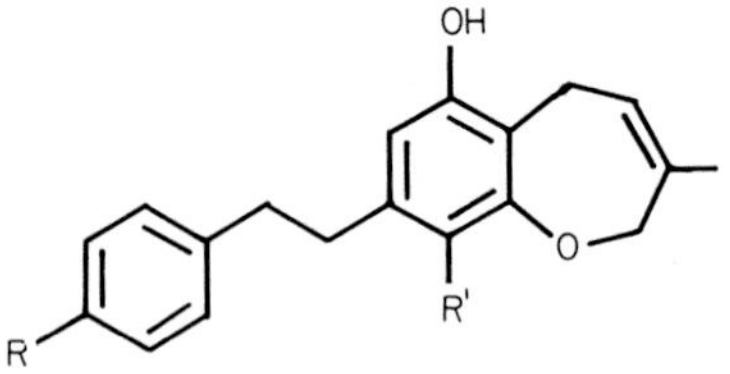

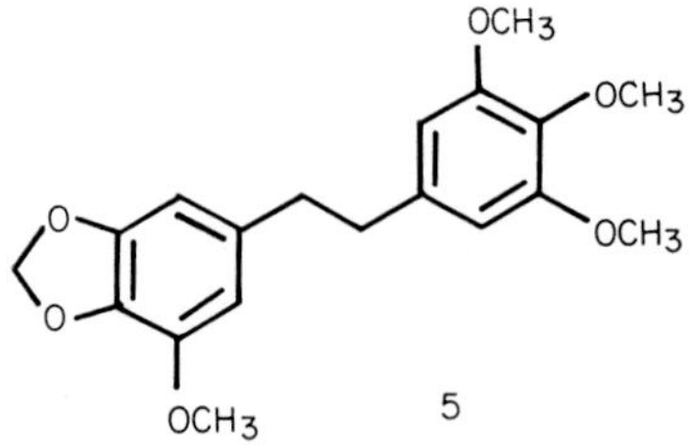

FIG. 2. 3-methoxybibenzyl (R=H; R′=OCH₃) (Suire, 1970); pellepiphyllin (R=
OCH₃; R′=OH; RI=OCH₃) (Benešová and Herout, 1970, 1972); 2,3,
4′-trihydroxybibenzyl (R=R′=R″=OH) (Huneck and Schreiber, 1975);
lunularic acid (R=OH; R′=COOH; R″=OH) (Valio et al., 1969) and lunularin (R=OH; R′=H; R″=OH) (Pryce, 1972a).

FIG. 3. Bibenzyl derivatives (R=OH and OCH₃) (Asakawa et al., 1978a).

FIG. 4. Bibenzyl derivatives (R=R′=H; R=OH, R′=H; and R=H, R′=COOH)
(Asakawa et al., 1978a).

FIG. 5. Brittonin-B(Asakawa et al., 1976b).

the mode of enclosure of this intermediate. However, no flavonoids were
found by Mues and Zinsmeister (1978) in 11 liverworts in which lunularic
acid was identified by Gorham (1977). Thus, the synthesis of flavonoids does
not appear to be related to that of lunularic acid.

On the other hand, liverworts are probably rich sources of bibenzyl derivatives such as the compounds illustrated in Figs 2–5. In *Radula complanata*, for
example, the compounds in Figs 2–4 are elaborated in larger amounts than
terpenes (Asakawa et al., 1978a). If they are, as we tentatively suggest, secretory
products, they should accumulate in oil bodies and lunularin might be a
possible precursor.

2. Enzymes

Critical studies by Krzakowa and Szweykowski (1977) and Krzakowa (1978) have demonstrated that in several liverworts the patterns of peroxidase isoenzymes, obtained by electrophoresis, were stable over long periods of time. They thus appear to be much more useful as taxonomic markers than dehydrogenases. Taylor and Elliott (1972) found that two months of mild conditions were required for *Eurhynchium* to re-establish normal activities of dehydrogenases after a period of dryness or intense cold. Indeed, peroxidase isoenzyme patterns have been used as evidence for geographic races in *Pellia endiviifolia* (Krzakowa, 1978). Gatherings of this species from various parts of Poland show three different peroxidase patterns while *P. epiphylla* and *P. neesiana* are uniform but clearly distinct from each other. Many authors consider *P. borealis* (*n* = 18) as merely a diploid form of *P. epiphylla* (*n* = 9) since both are monoecious and they differ only slightly in morphology and geographical distribution. The peroxidase patterns are closely in keeping with this notion; however, as far as we are aware, no attempt has yet been made to induce diploidy in *P. epiphylla* (for example by colchicine treatment) and then to compare the resulting plants with *P. borealis*. Until the origin of *P. borealis* has been firmly established, no final decision should be made on its taxonomic status in relation to *P. epiphylla*.

3. Flavonoids

In the Hepaticae, flavonoids are widespread and occur in all the orders.

(*a*) *Marchantiales.* According to Markham *et al.* (1978a), the preponderance of flavone O-glucuronides or flavone O-galacturonides is characteristic of the Marchantiales. However, at least one species is probably unable to synthesize flavones and elaborates only flavonol O-glycosides. This is an interesting finding which indicates the operation of a biosynthetic pathway different from that leading to the synthesis of flavones (Fig. 1).

Some genera such as *Corsinia*, *Cyathodium*, *Exormotheca* and *Targionia* have morphological characteristics which are substantially different from those of most of the other Marchantiales and preliminary results obtained on a few species support their exceptional position in the order. The flavonoids extracted from *Corsinia coriandrina* (Spreng.) Lindb. have been analysed by Reznik and Wiermann (1966). Schier (1974) and Markham and Porter (1978b). The aglycones consist only of flavonols and they are 3-O-glycosylated (Markham and Porter, 1978b). Schier (1974) has investigated in addition the flavonoids from *Exormotheca pustulosa* Mitt., *Targionia hypophylla* and *T. lorbeeriana* K. Müll.; he tentatively concluded that flavonol 3-O-diglycosides were present in all three species. According to Harborne (1967, p. 313), the synthesis of flavonols in place of flavones indicates a low evolutionary level. However, this is contradicted by the ability of these taxa to O-glycosylate the flavonols.

Most of the Ricciaceae belong to the genus *Riccia* which may be divided into three subgenera: *Thallocarpius*, *Ricciella* and *Euriccia*. Although few species have so far been investigated biochemically the results already indicate the presence of diverse flavonoid patterns. The main flavonoids of *R. crystallina* are naringenin, a dihydroflavone (Fig. 1), and its 7-O-glucuronide; the corresponding flavone (apigenin) together with its 7-O-glucoside and its 7-O-glucuronide, are far less abundant (Markham and Porter, 1975a). Two minor flavonoids are acylated derivatives of apigenin 7-O-glucuronide (Markham *et al.*, 1978b). Aquatic or terrestrial forms of *R. fluitans* have the same flavonoid pattern and both mainly produce several 7-O-glycosides of luteolin (Vandekerkhove, 1978b). Luteolin is a tetrahydroxylated flavone which is biosynthetically one step removed from the more basic trihydroxyflavone, apigenin (Fig. 1). Luteolin 7-O-glucuronide-3'-O-glucoside and its 4'-glucuronide derivative occur in unacylated and acylated forms (Markham *et al.*, 1978b). *R. duplex* synthesizes apigenin or luteolin 7-O-glucuronides, but only in unacylated form, and it also produces the 7-O-glucuronide of chrysoeriol, a 3'-methyl derivative of luteolin (Markham *et al.*, 1978b). The main flavonoids of nine *Euriccia* spp. have been tentatively identified by paper chromatography and colour reactions with different spray reagents (Schier, 1974). These preliminary results suggest the existence of various flavonoid patterns in the genus *Riccia* but further detailed investigations are necessary, particularly for the subgenera *Euriccia* and *Thallocarpus*.

Ricciocarpus natans produces an apigenin 7-O-glycoside but the major flavonoids are luteolin derivatives, particularly luteolin 7,3'-O-diglucuronide. This last compound is also elaborated by *Marchantia polymorpha*, a highly evolved species (see below), and indicates that a high level of biochemical sophistication has been reached by *Ricciocarpus*.

The notable differences in biosynthetic levels between *Riccia crystallina* which has the most primitive flavonoid pattern so far found in the Ricciaceae and *Ricciocarpus natans*, together with the various flavonoid patterns found in subgenus *Euriccia* are very much in accord with the view that the Ricciaceae is a highly diversified group with ancient origins (Schuster, 1966, p. 295).

The monotypic genus *Monoclea* was included in the Calobryales by Campbell (1936) and in the Jungermanniales by Evans (1939) while Schuster (1963) placed it in a new order, the *Monocleales*. More recently, Grolle (1972) included it in the Marchantiales. The occurrence in *Monoclea forsteri* Hook. of flavone O-glucuronides which are typical for the Marchantiales (Markham, 1972) is in accordance with Grolle's proposition. Moreover, the main compound is close to one of the constituents of *Marchantia berteroana* Lehm. and Lindenb. and perhaps indicates that *Monoclea forsteri* is extremely advanced biochemically.

The flavonoids isolated from *Riella affinis* Howe and Underwood, *R. americana* Howe and Underwood, and *Sphaerocarpos texanus* Aust. (Markham and Porter, 1976) are flavone O-glucuronides. This result is more in favour of including the Riellaceae and Sphaerocarpaceae in the Marchantiales (Grolle, 1972) than isolating them in the order Sphaerocarpales.

Turning attention to the lower taxonomic ranks, *Reboulia hemisphaerica* and *Asterella australis* (Tayl.) Verd., which are difficult to distinguish in the sterile state, have very distinct flavonoid patterns (Markham *et al.*, 1978a). At the subspecific level, the identity of the flavonoid patterns of *Marchantia polymorpha* var. *aquatica* and var. *alpestris* supports the notion that they are no more than two ecological forms of *M. polymorpha* (Markham and Porter, 1974). Substantial variations in the flavonoid pattern of samples of *Conocephalum conicum* collected in the USA and Germany show the existence of two geographic races with the same morphology. After two years of culture, the flavonoid pattern of the German sample was unchanged, so the differences noted cannot be due to variations in climate (Markham *et al.*, 1976a).

Finally, sex organ formation leads to large changes in the synthesis of flavonoids. Acacetin production in the thallus of *Marchantia berteroana* (acacetin is 4'-O-methylapigenin) stops during the development of sexual organs. The archegoniophores synthesize flavonoids not elaborated elsewhere by the vegetative thallus. The antheridiophores also produce these flavonoids and, in addition, aureusidin 6-O-glucuronide (Markham and Porter, 1978a), an aurone which has also been found in the antheridiophores of *M. polymorpha* and *Conocephalum supradecompositum* (Lindb.) Steph. (Markham and Porter, 1978a; Markham *et al.*, 1978c). Thus, the formation of sex organs appears to be particularly associated with the production of complex aurones, indicating a general increase in the level of biochemical capabilities above those found in the vegetative thallus. It would now be particularly interesting to discover whether a similar situation exists in *Riccia* (and especially *R. crystallina*) where the sporophyte is extremely reduced.

(b) Metzgeriales. Mues and Zinsmeister (1978) found flavonoids in more than half the species they tested. Although data on the Metzgeriales are scanty, they suggest the existence of different levels of biochemical sophistication. All the taxa so far tested synthesize exclusively C-glycosides (Markham *et al*,

1978a; Mues and Zinsmeister, 1978), whereas two *Hymenophyton* spp., *Blasia pusilla* and two Pelliaceae produce both *C*- and *O*-glycosides (Mues and Zinsmeister, 1978). Aglycones, such as dihydroflavonols, flavonols and flavones, have been identified but aurones have not. In terms of flavonoid chemistry, the Metzgeriales seem to be almost as diverse as the Marchantiales and it is likely that there is more variation still to be found. The presence of only *O*-glycosides in *Fossombronia* supports morphological interpretations which consider the Codoniaceae as highly evolved liverworts. The presence of *O*-glycosides is a valuable indicator that *Blasia pusilla*, often considered a primitive species, is similarly advanced.

The structure of flavonoids has been determined in *Hymenophyton flabellatum* (Lab.) Dum. and *H. leptopodon* (Tayl.) Steph., which has been considered either to be a variety of *H. flabellatum* or a species. The arguments advanced in favour of the second hypothesis are well supported by a substantial difference in the flavonoid patterns. According to Markham *et al.* (1976b), *H. flabellatum* elaborates exclusively *C*-glycosides while *H. leptopodum* also synthesizes kaempferol *O*-glycosides (Fig. 1).

(*c*) *Calobryales.* According to Schuster (1966, p. 633), the gametophyte of *Haplomitrium* recalls the fossil Naiaditinae and the Calobryales are considered to be the most primitive Hepaticae. However, two of the seven species of *Haplomitrium* produce flavone *O*-glycosides (Markham, 1977) indicating a surprisingly high biosynthetic level. *Haplomitrium gibbsiae* (Steph.) Schust. was found to contain previously unknown acylated 7-glucosides of apigenin and iso-scutellarein which is 8-hydroxyapigenin (Markham, 1977). 8-Hydroxylation is considered by Harborne (1967, p. 313) as an advanced character, and iso-scutellarein occurs as 8-*O*-glucuronide in *Marchantia berteroana*, an advanced species (Markham and Porter, 1975b). Thus, the claims of primitive status for the Calobryales, based on morphology, are at variance with interpretations of the biochemical data. Either one of these hypotheses is incorrect or biochemical evolution must have taken place without morphological evolution.

(*d*) *Jungermanniales.* Flavonoids have only been found in about half of the species tested. Some Jungermanniales seem to elaborate exclusively flavone *C*-glycosides, others only flavone *O*-glycosides (Mues and Zinsmeister, 1978), which are biosynthetically advanced in *Frullania* (Markham *et al.*, 1978a). The genus *Porella* is heterogeneous (see also terpenoids) and includes two types of species (Mues and Zinsmeister, 1978).

4. Terpenoids

Up to now, monoterpenes and sesquiterpenoids have been identified only in those bryophytes which also have oil bodies. These are present in the gameto-phyte cells of all the Calobryales, in about 90% of the Jungermanniales, 70% of the Metzgeriales and 65% of the Marchantiales (Schuster, 1966). However, in some Jungermanniales and most Marchantiales, they are restricted to specialized cells. In the Marchantiales the acquisition of such specialized secretory cells has been interpreted as advancement from ancestors whose cells all con-tained oil-bodies. This conclusion is very much in accord with that reached from the study of their flavonoids.

5. Monoterpenes and Related Compounds

The monoterpenes identified in liverworts (Hörster and Wiermann, 1976; Suire and Bourgeois, 1977; Asakawa and Takemoto, 1978b, for reviews) have the same general structure as those found in spermatophytes. Unfortunately, their optical activity has yet to be measured in the majority of taxa and only this can determine their stereochemistry. This is particularly important since most sesquiterpenes in liverworts belong to the *ent* series and are the optical antipodes of their counterparts in spermatophytes (*ent* is an abbreviation for Entgegen and indicates the configuration *trans* in which the distinctive sub-stituted group shows the position α (Fig. 5), as opposed to the most common *cis* or normal configuration in which the substituted group shows the β position. The two isomers are also called enantiomers and have a rotatory activity of the same value but opposed sign.) It is therefore most interesting that bornyl acetate, a monoterpene ester (Fig. 6), occurs in spermatophytes in its levo-rotatory form

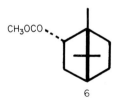

CH₃OCO

6

Fig. 6. (+)-bornyl acetate.

but in *Conocephalum conicum* in its dextro-rotatory form (Asakawa *et al.*, 1976a). Thus, we have evidence that similar biosynthetic pathways have evolved independently in the two groups.

6. Sesquiterpenoids

Apart from *Isotachis japonica* Steph. in which the main constituents of the essential oil are aromatic esters (Matsuo *et al.*, 1971), all the hepatics with secretory cells synthesize sesquiterpenoids. Among the sesquiterpenoids identified (Andersen *et al.*, 1978a), some have skeletons unknown in other plants (Figs 7–18) while others have skeletons already identified elsewhere (Figs 19–29). Among the latter, nearly all those of which the stereochemistry has been elucidated belong to the *ent* series and are the optical antipodes of their counterparts in spermatophytes. However, some compounds are exceptions to this rule:

(a) (—)-caryophyllene identified in *Scapania undulata* by Andersen *et al.* (1977a);

(b) some sesquiterpene lactones: costunolide and *Y*-cyclocostunolide (Fig. 32) from *Frullania tamarisci* (Connolly and Thornton, 1973); as well as the germacranolide (Fig. 37) and the two guaianolides (Fig. 38) isolated from *Conocephalum conicum* (Asakawa and Takemoto, 1979);

(c) the drimane derivatives (Figs 19–23). Their normal conformation confirms that they are biogenetically closer to triterpenes and sterols (normal structure) than to sesquiterpenes (usually *ent*).

As yet only two preliminary studies have been made on the biogenesis of sesquiterpenoids in the liverworts (Herout, 1971; Andersen *et al.*, 1977b) and our knowledge of it is largely speculative. However, the majority of compounds identified may still be arranged in a logical sequence by applying the general scheme of known biosynthetic patterns (Andersen *et al.*, 1977a; Asakawa and Takemoto, 1978b; Andersen *et al.*, 1978a). Unfortunately, in contrast to flavonoids, we do not yet know what the biochemical differences between sesquiterpenoids actually mean in evolutionary terms.

The distribution of those sesquiterpenoids which are known only from the Hepaticae and have been identified in more than one species, is given in Table I. As far as we know no species produces compounds both from the barbatene-anastreptene-bazzanene group (which are biosynthetically related) and from the pinguisane derivatives group.

At a subspecific or generic level, little of the available data can be used for taxonomic purposes since investigations dealing with more than three species belonging to the same genus are rare. However, the following examples suggest that sesquiterpenoids may well become extremely valuable in the future.

Aequilobene, a new sesquiterpene whose structure is now being elucidated,

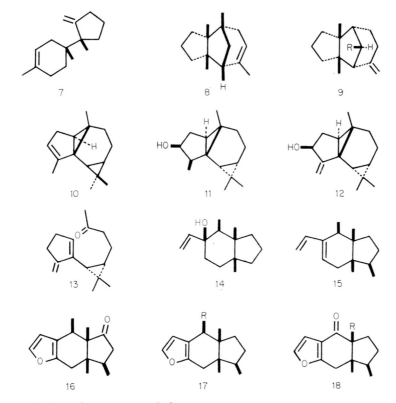

FIGS 7–18. Typical sesquiterpenoids from Hepaticae

FIG. 7. (+)-β-bazzanene (Andersen *et al.*, 1977a).

FIG. 8. (+)-α-barbatene (Andersen, 1977a).

FIG. 9. (—)-β-barbatene (R=H) (Anderson *et al.*, 1977a) and gymnomitrol (R=OH) (Connolly *et al.*, 1972).

FIG. 10. (+)-anastreptene (Andersen *et. al.*, 1978b).

FIG. 11. (—)-dihydromylione (Matsuo *et. al.*, 1977).

FIG. 12. myliol (Andersen *et al.*, 1977a).

FIG. 13. taylorione (Matsuo *et al.*, 1974).

FIG. 14. pinguisenol (Asakawa *et al.*, 1978).

FIG. 15. α-pinguisene (Asakawa and Takemoto, 1978b).

FIG. 16. pinguisone (Corbella *et al.*, 1974).

FIG. 17. deoxypinguisone (R=CH₃) (Krutov *et al.*, 1973) and deoxypinguisone methyl ester (R=OCOCH₃) (Asakawa *et al.*, 1978b).

FIG. 18. norpinguisone (R=CH₃) (Asakawa *et al.*, 1978) and norpinguisone methyl ester (R=OCOCH₃) (Asakawa *et al.*, 1978).

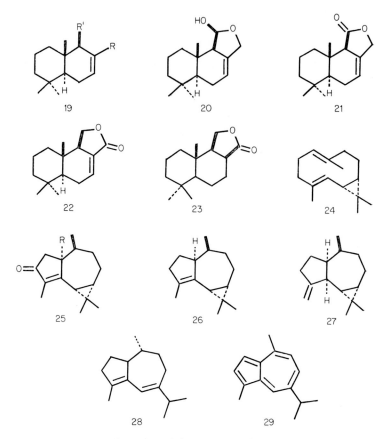

FIGS 19–29. Sesquiterpenoids with a skeleton previously known in plants other than Hepaticae

FIG. 19. (+)-tadeonal (R = R′ = CHO) (Asakawa and Takemoto, 1978a) and (—)-drimenol (R = CH₃, R′ = CH₂ OH) (Asakawa and Takemoto, 1978b).

FIG. 20. drimenol (Asakawa and Takemoto, 1978b).

FIG. 21. drimenin (Asakawa et al., 1976e).

FIG. 22. (—)-cinnamolide (Asakawa et al., 1976e).

FIG. 23. (+)-dihydrocinnamolide (Huneck and Schreiber, 1975).

FIG. 24. (—)-bicyclogermacrene (Asakawa and Takemoto, 1978b).

FIG. 25. (+)-cyclocolorenone (R = H) and 1-hydroxycyclocolorenone (R = OH) (Asakawa and Takemoto, 1978b).

FIG. 26. (+)-α-gurjunene (Asakawa and Takemoto, 1978b).

FIG. 27. (+)-β-gurjunene (Asakawa and Takemoto, 1978b).

FIG. 28. iso-α-gurjunene-B (Suire et al., 1978).

FIG. 29. Guaiazulene (Bourgeois and Suire, 1977).

TABLE I. Distribution of some typical sesquiterpenoids in liverworts

(+)-β-bazzanene (Fig. 7) bazzanenol	(+)-α-barbatene (Fig. 8) (—)-β-barbatene (Fig. 9) gymnomitrol (Fig. 9)	(+)-anastreptene (Fig. 10)	Pinguisane derivatives (Figs 14–18)
Bazzania—J—[a, b] *Dumortiera*—M—[c] *Scapania*—J—[d]	*Anastrepta*—J—[a] *Barbilophozia*—J—[e] *Bazzania*—J—[a] *Dumortiera*—M—[c] *Gymnomitrion*—J—[i] *Jungermannia*—J—[a] *Lepidozia*—J—[a] *Scapania*—J—[a]	*Anastrepta*—J—[a] *Barbilophozia*—J—[j] *Scapania*—J—[a]	*Aneura*—m—[f] *Porella*—J—[g] *Ptilidium*—J—[h]

Abbreviations: J = Jungermanniales; M = Marchantiales; m = Metzgeriales. [a] Andersen *et al.*, 1977a. [b] Hayashi and Matsuo, 1970. [c] Matsuo *et al.*, 1976. [d] Matsuo *et al.*, 1973. [e] Andersen *et al.*, 1973. [f] Benešova *et al.*, 1969a. [g] see Fig. 32. [h] Krutor *et al.*, 1973. [i] Connolly *et al.*, 1972. [j] Andersen *et al.*, 1978b.

is one of the major terpenes found in all four *Scapania* spp. so far investigated (Andersen *et al.*, 1977a). It could well prove useful in defining the limits of the genus.

(+)-Bazzanene (Fig. 7) is a major sesquiterpene common to all three *Bazzania* spp. investigated by Matsuo and Hayashi (1977) and Andersen *et al.* (1977a). Of the essential oil extracted from *B. tricrenata*, 72% is a mixture of sesquiterpenic alcohols, none of which corresponds to those of *B. trilobata* (Andersen *et al.*, 1977a). These observations corroborate the obvious morphological distinction between the two species.

In the genus *Pellia*, only the essential oil of *P. borealis* has yet to be analysed. Two of the other three species are distinguished by the presence of a characteristic compound. Guaiazulene (Fig. 29) occurs in *P. epiphylla* and pellepiphyllin (Fig. 2) in *P. neesiana*. *P. endiviifolia* lacks both these compounds (Table II). From a taxonomic point of view, these results reinforce morphological and cytological evidence for the specific status of the three taxa but they do not shed any new light on the affinities of *P. neesiana*.

Most *Porella* species so far investigated may be classified into two groups. The first group of species produces peppery tasting tadeonal (Fig. 19), other drimane derivatives and aromadendrane and pinguisane derivatives (Table III). The second group does not taste of pepper and synthesizes only derivatives of

TABLE II. Characteristic compounds of three *Pellia* spp.

Compound	*P. endiviifolia*	*P. epiphylla*	*P. neesiana*
pellepiphyllin (Fig. 2)	○[a]	○[b]	+[b]
guaiazulene (Fig. 29)	○[c]	+[c]	
iso-*a*-gurjunene-B (Fig. 28)	+[d]	+++[d]	

○ : absence verified.   + : present.   +++ : main constituant.

Abbreviations: [a] Pryce, 1972a. [b] Benešová and Herout, 1972. [c] Bourgeois and Suire, 1977. [d] Suire *et al.*, 1978.

pinguisane. *P. faurieri* has an intermediate pattern with only drimane derivatives. The seven species investigated probably do not reflect the whole range of biochemical variations present in the genus. *P. vernicosa* Lindb. and *P. macroloba* (Steph.) Hatt. and Inoue are closely related morphologically and their similar sesquiterpenes hardly support their specific status.

In *Frullania*, the species so far investigated can also be divided into two very different groups from the biosynthetic standpoint. *F. dilatata*, *F. tamarisci*, *F. usamiensis* Steph. and probably nine other species (Mitchell *et al.*, 1970) produce highly allergenic sesquiterpenic lactones belonging to the eudesmane (Figs 30–35) or the eremophilane series, and probably no bibenzyls. *F. brittoniae* Evans and *F. ericoides* (Nees) Nees are not allergenic; the main constituents of the extracts of the first species are bibenzyl derivatives (e.g. Fig. 5) (Asakawa *et al.*, 1976b). On the other hand, the recent revision of the genus *Frullania* by Hattori (1972) shows these biochemical divergencies are not accompanied by sufficient morphological differences to justify the erection of two genera.

Frullania obscura Steph. and *F. nisquallensis* Sull. are considered by Hattori (1972) as two subspecies of *F. tamarisci*. The sesquiterpene pattern indicates that *F. obscura* should be maintained as a species but that *F. nisquallensis* is best treated as a subspecies of *F. tamarisci*. (+)-Frullanolide (Fig. 30) is the main constituent of the allergenic fraction extracted from *F. nisquallensis* (Asakawa *et al.*, 1976c) and from *F. tamarisci* ssp. *tamarisci*, which contains in addition two lactones (Figs 32 and 33) (Connolly and Thornton, 1973). The lactone shown in Fig. 34 is the main constituent of the fraction extracted from *F. obscura* which also elaborates two additional compounds (Figs 35 and 36); it is not present in *F. tamarisci* (Asakawa *et al.*, 1976c).

The male gametophyte of *Conocephalum conicum* is not pungent. The characteristic pungency of the female is due to tulipinolide (Fig. 37), a sesquiterpene

TABLE III. Sesquiterpenoids of some *Porella* species

Fig. Numbers	Drimane derivatives						*ent*-bicyclogermacrane derivatives							Pinguisane derivatives			
	(19 R=R' =CHO)	(19, R=CH₃ R'= CH₂ OH)	(20)	(21)	(22)	(23)	(24)	(25, R=H)	(25,R =OH)	(26)	(27)	(14)	(15)	(17,R =CH3)	(17,R =OC OCH₃)	(18,R =CH₃ =OCH₃)	(18,R =OC OCH₃)
P. vernicosa	+	+[a]	+[a]	+	+		+[a]	+	+[c]	+[a]	+[a]	+[b]	+[c]	+	+	+	+
P macroloba	+		+[a]	+	+		+[c]		+[c]	+[a]	+[a]	+[c]	+[c]	+	+	+	+
P. gracillima	+		+[c]	+	+		+[c]	+	+[c]	+[c]	+[c]	+[c]	+[c]	+	+	+	+
P. arboris-vitae						+[d]											
P. faurieri	+				+												
P. densifolia												+[b]	+[a]	+	+	+	+
P. perrottetiana														+	+	+	+

Abbreviations: Not numbered—Asakawa *et al.*, 1976c. [a] Asakawa and Takemoto, 1978b. [b] Asakawa, 1976d. [c] Asakawa, 1978b. [d] Huneck and Schreiber, 1975.

lactone (Asakawa and Takemoto, 1978c). This is the only report of a sexual difference correlated with the synthesis of terpenes in the Hepaticae.

According to Stotler (1976), the pattern of sesquiterpenes cannot be considered to be characteristic of a species because it changes when the plant is placed in the dark. However, from our own experiments, particularly on *Pellia epiphylla* and *Conocephalum conicum*, the changes in the terpene pattern

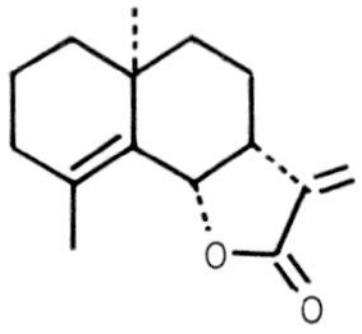

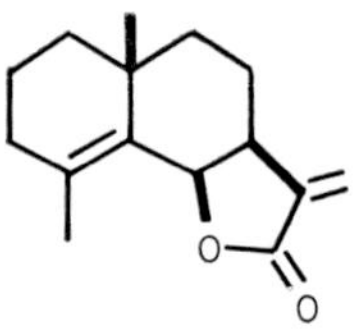

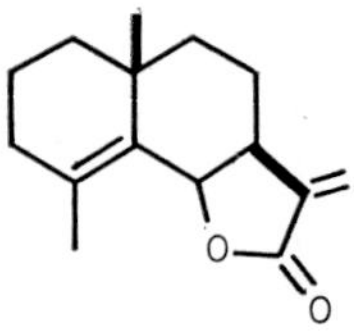

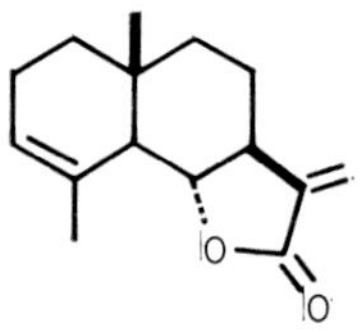

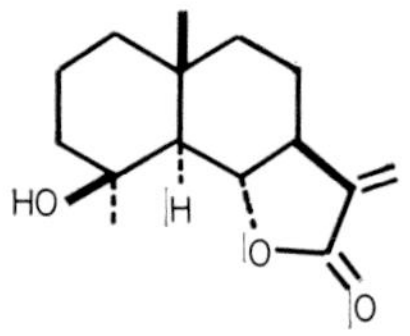

FIGS 30–38. Some sesquiterpenic lactones from liverworts
FIG. 30. (+)-frullanolide (Knoche *et al.*, 1969).
FIG. 31. (—)-frullanolide (Perold *et al.*, 1972).
FIG. 32. (+)-γ-cyclocostunolide (Connolly and Thornton, 1973).
FIG. 33. (+)-α-cyclocostunolide (Connolly and Thornton, 1973).
FIGS 34, 35. eudesmanolides (Asakawa *et al.*, 1975).
FIG. 36. eudesmanal (Asakawa *et al.*, 1975).
FIG. 37. tulipinolide (Asakawa and Takemoto, 1979).
FIG. 38. zaluzanin-C (R=OH) and zaluzanin-D (R=OCOCH₃) (Asakawa and Takemoto, 1979).

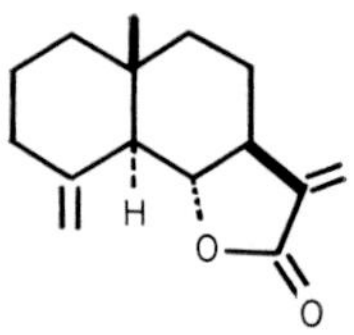

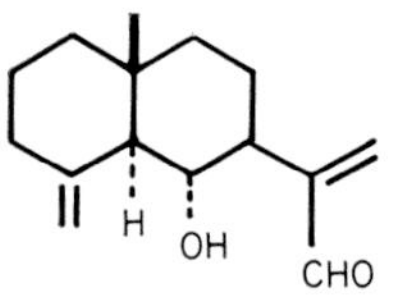

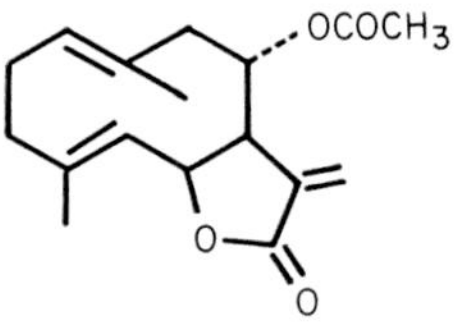

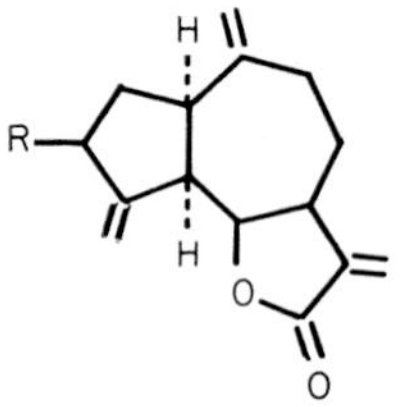

during a year and the differences between the patterns of plants collected in various parts of southwestern France, are limited; thus, the terpene pattern may be considered characteristic of the race collected in the same area. The terpene pattern of *C. conicum* collected in Japan is substantially different from that of French samples and provides evidence for the existence of at least two geographic races (Asakawa, unpublished data).

7. Diterpenoids

Diterpenoids are abundant in liverworts with oil bodies where they probably represent a resineous secretory product.

Two dials (Figs 39 and 40) responsible for the pungency of *Trichocoleopsis sacculata* (Mitt.) Okam. (Asakawa *et al.*, 1977) are biogenetically related to drimane and are of the normal series, as are all the drimane derivatives from hepatics. The other diterpenes are kaurane derivatives or are biosynthetically related to kaurane: they have a 10α structure and fall within the *ent* series, as opposed to the normal 10β series. In contrast to those from liverworts, the diterpenes from the resins of conifers and from the secretory products of higher plants belong to the two enantiomeric series (Ourisson, 1974).

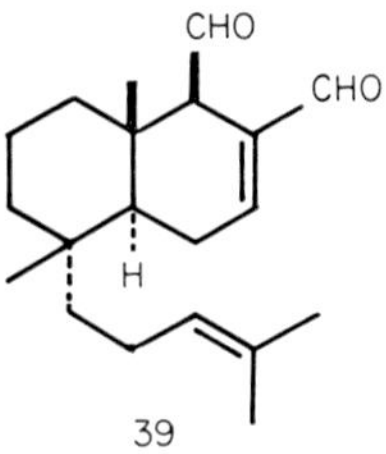

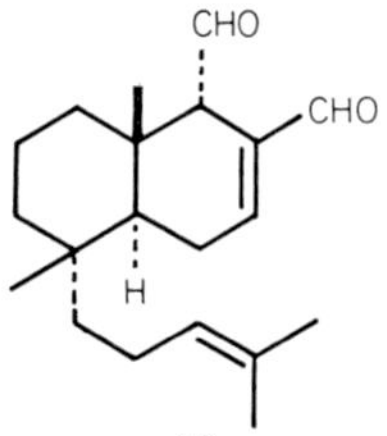

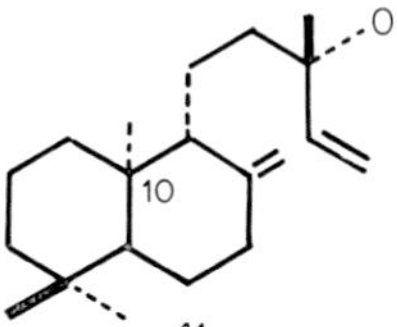

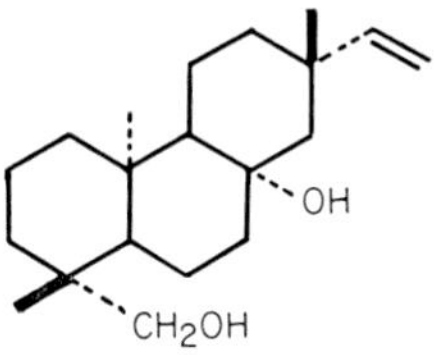

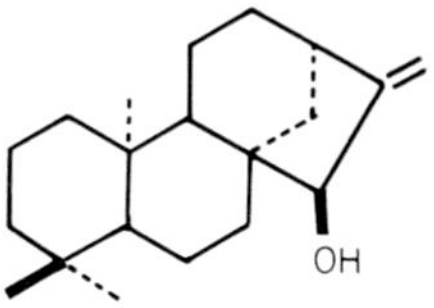

FIGS 39–43. Some diterpenoids from liverworts
FIG. 39. sacculatal (Asakawa *et al.*, 1977).
FIG. 40. isosacculatal (Asakawa *et al.*, 1977).
FIG. 41. (—)-manool (Matsuo *et al.*, 1978).
FIG. 42. (—)-thermarol (Matsuo *et al.*, 1978).
FIG. 43. *ent*-kauren-15α-ol (Matsuo *et al.*, 1978).

The identification of bicyclic, tricyclic and tetracyclic diterpenoids such as those illustrated in Figs 41–43 confirms that the biosynthetic pathway elucidated in higher plants is also present in liverworts. It is interesting to note that no species contains the three types of skeleton. Thus, only bicyclic diterpenes have been identified in *Jungermannia rosulans* Steph. and *J. torticalyx* Steph., only tricyclic diterpenes in *J. thermarum* Steph. and only tetracyclic forms in *J. atrovirens* Tayl. and *J. infusca* Mitt. (Matsuo *et al.*, 1978). The synthesis progresses by successive addition of new rings, so that the first two species are at a lower biogenetic level than the last two. At the level of gross morphology the first three species have erect stems and the others have creeping stems, perhaps suggesting that the latter may be a derived condition.

8. Alkaloids

Riccardia chamedryfolia (With.) Grolle (= *R. sinuata*), *R. incurvata* and *Aneura pinguis* were once united within the genus *Aneura* or within the genus *Riccardia*. At present however, the three species are divided into two genera (Grolle, 1976) and the chemical results tend to support this situation. The first two species produce alkaloids (Figs 44 and 45) which have not been found in

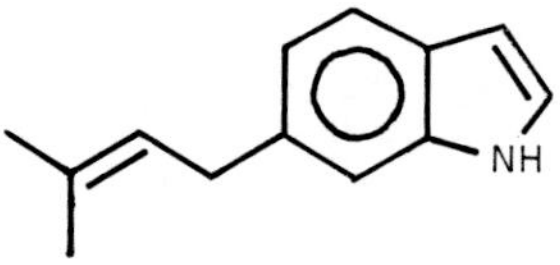

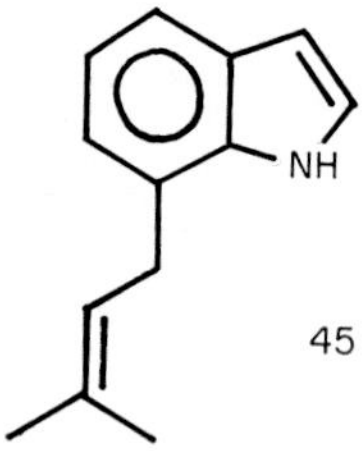

FIGS 44–45. Alkaloids from the Aneuraceae
FIG. 44. 6-(3-methyl-2-butenyl)-indole.
FIG. 45. 7-(3-methyl-2-butenyl)-indole

Aneura pinguis (Benešová *et al.*, 1969b). The latter synthesizes the sesquiterpenic ketone (Fig. 16) (Benešová *et al.*, 1969a) not found in *R. chamedryfolia* (Andersen *et al.*, 1977b).

GENERAL CONCLUSIONS

The selected examples in this account give an indication of the biochemical diversity found in Musci, Hepaticae and Anthocerotae. On the one hand they shed new light on the position of individual taxa, on the other hand comparisons with data from other plant groups enables us to assess the various levels of biosynthetic evolution in bryophytes. Biochemical information does not always fit with interpretations of bryophyte evolution based on more traditional criteria. Some taxa attain similar levels of biochemical sophistication to those found in the most advanced spermatophytes (e.g. the synthesis of aurones by some mosses and Hepaticae and the capacity of Anthocerotae to differentiate the two isomers of D-methionine), while others appear extremely primitive. The primitive flavonoid pattern in *Riccia crystallina*, otherwise generally considered to be a much reduced (and therefore advanced) taxon, is particularly noteworthy when taken against the fact that other members of the genus have far more advanced flavonoids. Possible interpretations include suggestions that the Ricciaceae are an ancient group, or are heterogeneous. Alternatively, flavonoid biosynthesis may be more capricious in bryophytes than appears to be the case in spermatophytes. A similar paradox is presented by the advanced flavonoid pattern in Calobryales, a group increasingly regarded as primitive (see Schuster, and Carothers and Duckett, this volume, Chapters 3 and 18).

Although bryophytes produce numerous compounds common to all green plants (e.g. amino acids, chlorophyll a, Krebs cycle enzymes) the list of substances elaborated exclusively by them is growing continuously. This suggests that biosynthetic evolution has taken place independently in several different groups of bryophytes along parallel pathways and with the same general trends as those seen in vascular plants. The logical conclusion from the biochemical data is that Hepaticae, Musci and Anthocerotae all appear to have evolved independently. There is also substantial evidence that the Andreaeopsida, the Sphagnopsida and the remainder of the mosses are polyphyletic. Thus, biochemistry adds to the argument that the phylum Bryophyta is an unnatural one and is perhaps better regarded as comprising five separate groups, interrelationships between which are extremely tenuous.

Regardless of whether or not the above splitting of the phylum is ultimately accepted, a further question of paramount importance is: from what did bryophytes evolve? Was it directly from algal ancestors or from other land plants, long since extinct but closer to extant pteridophytes than to the algae? Although this question will probably never be answered unequivocally, we should make some attempt to answer Steere's (1969, pp. 140–141) question "With which groups of plants do bryophytes share the greatest affinity in their biochemical behaviour?" Neither flavonoids nor monoterpenes have ever been found in the algae. The flavonoids of bryophytes are, except for those of *Sphagnum*, closely related to those of ferns and higher plants. Monoterpene skeletons produced by some liverworts are the same as those of higher plants. Some algae produce very small amounts of sesquiterpenes (Fenical *et al.*, 1972; Ohta and Takagi, 1977) but none of the sesquiterpene skeletons typical of the Hepaticae has so been found in algae. The other sesquiterpene skeletons produced by liverworts with oil bodies are identical to those of higher plants. Only linear diterpenoids have been found in some brown algae (Faulkner *et al.*, 1977; Robertson and Fenical, 1977; Francisco *et al.*, 1978) whereas liverworts elaborate polycyclic diterpenoids. The triterpenes and sterols of bryophytes (Suire, 1975) seem to be identical to those of ferns and higher plants and none of the sterols typical for algae, such as fucosterol from the Phaeophyceae, desmosterol from the Rhodophyceae, clionosterol from the Chlorophyceae or Xanthophyceae and sterols with β-orientation of the C-24 substituents from the Chlorophyceae (Mercer *et al.*, 1974) has so far been found. Thus, the possibility of a direct line from algae to bryophytes is not supported by any of the chemical data. On the contrary, bryophytes seem more closely related to higher plants than to algae.

However, data on the biochemistry of those green algae presently favoured as most closely related to land plants on the basis of male gamete and 200-spore ultrastructure, nuclear and cell division, and glycolytic enzymes (Frederick *et al.*, 1973; Moestrup, 1974; Pickett-Heaps, 1975) is almost non-existent. The enterprising chemotaxonomist could do worse than to turn his attention to these green algae. Another promising area for future evolutionary studies is geochemistry. If traces of compounds specific to bryophytes or vascular plants can be detected in fossil remains attributed to land archegoniates such as *Rhynia major* and *Nothia aphylla* (Hébant, 1978), we may be one step closer to under-standing the evolution of land plants.

REFERENCES

ANDERSEN, N. H., COSTIN, C. R., KRAMER, C. M., Jr, OHTA, Y. and HUNECK, S. (1973). Sesquiterpenes of *Barbilophozia* species. *Phytochemistry* **12**, 2709–2716.

ANDERSEN, N. H., BISSONETTE, P., LIU, C.-B., SHUNK, B., OHTA, Y., TSENG, C.-L., MOORE, A. and HUNECK, S. (1977a). Sesquiterpenes of nine European liverworts from the genera *Anastrepta, Bazzania, Jungermannia, Lepidozia* and *Scapania*. *Phytochemistry* **16**, 1731–1751.

ANDERSEN, N. H., OHTA, Y., LIU, C.-B., KRAMER, C. M., Jr, ALLISON, K. and HUNECK, S. (1977b). Sesquiterpenes of thalloid liverworts of the genera *Conocephalum, Lunularia, Metzgeria* and *Riccardia*. *Phytochemistry* **16**, 1727–1729.

ANDERSEN, N. H., OHTA, Y. and SYRDAL, D. D. (1978a). Studies in sesquiterpenes biogenesis: implications of absolute configuration, new structural types, and efficient chemical simulation of pathways. *In* "Bioorganic chemistry" (Second edition, E. E. Van Tamelen, ed.) Vol. 2, pp. 1–37. Academic Press, New York.

ANDERSEN, N. H., OHTA, Y., MOORE, A. and TSENG, C.-L.W. (1978b). Anastreptene, a commonly encountered sesquiterpene of liverworts (Hepaticae). *Tetrahedron* **34**, 41–46.

ANDERSON, W. H., HAWKINS, J. M., GELLERMAN, J. L. and SCHLENK, H. (1974). Fatty acid composition as criterion in taxonomy of mosses. *J. Hattori bot. Lab.* **38**, 99–103.

ASAKAWA, Y. and ARATANI, T. (1976). Sesquiterpenes of *Porella vernicosa* (Hepaticae). *Bull. Soc. chim. Fr., Paris* **1976** (Chim. Mol.) 1469–1470.

ASAKAWA, Y. and TAKEMOTO, T. (1978a). Le diterpenoid du goût piquant de *Pellia endiviaefolia. Phytochemistry* **17**, 153–154.

ASAKAWA, Y. and TAKEMOTO, T. (1978b). Chemical constituents of *Trichocolea, Plagiochila* and *Porella. Bryophyt. Biblthca* **13**, 335–353.

ASAKAWA, Y. and TAKEMOTO, T. (1979). Sesquiterpene lactones of *Conocephalum conicum. Phytochemistry* **18**, 285–288.

ASAKAWA, Y., OURISSON, G. and ARATANI, T. (1975). New sesquiterpene lactone and aldehyde of *Frullania tamarisci* subsp. *obscura* (Hepaticae). *Tetrahedron Lett.* **1975**, 3957–3960.

ASAKAWA, Y., TOYOTA, M. and ARATANI, T. (1976a). (+)–Bornyl acetate from *Conocephalum conicum. Phytochemistry* **15**, 2025.

ASAKAWA, Y., TANIKAWA, K. and ARATANI, T. (1976b). New substituted bibenzyls of *Frullania brittoniae* subsp. *truncatifolia. Phytochemistry* **15**, 1057–1059.

ASAKAWA, Y., OURISSON, G. and ARATANI, T. (1976c). Allergy-inducing substances of *Frullania. Miscnea bryol. lichen., Nichinan* **7**, 96–99.

ASAKAWA, Y., TOYOTA, M. and ARATANI (1976d). Un nouvel alcool sesquiterpénique de *Porella vernicosa* et *Porella densifolia* (Hépatiques). *Tetrahedron Lett.* **1976**, 3619–3622.

ASAKAWA, Y., TOYOTA, M., UEMOTO, M. and ARATANI, T. (1976e). Sesquiterpenes of six *Porella* species (Hepaticae). *Phytochemistry* **15**, 1929–1931.

ASAKAWA, Y., TAKEMOTO, T., TOYOTA, M. and ARATANI, T. (1977). Sacculatal and isosacculatal, two new exceptional diterpenedials from the liverwort *Trichocoleopsis sacculata. Tetrahedron Lett.* **1977**, 1407–1410.

ASAKAWA, Y., KUSUBE, E., TAKEMOTO, T. and SUIRE, C. (1978a). New bibenzyls from *Radula complanata. Phytochemistry* **17**, 2115–2117.

ASAKAWA, Y., TOYOTA, M. and TAKEMOTO, T. (1978b). Sesquiterpenes from *Porella* species. *Phytochemistry* **17**, 457–460.

BENEŠOVÁ, V. and HEROUT, V. (1970). Isolation and the structure of pellepiphyllin, 2-hydroxy-3-methoxy-4'-methoxydihydrostilbene, from liverwort *Pellia epiphylla* (L.) Dum. *Coll. Czech. chem. Comm.* **35**, 1926–1929.

BENEŠOVÁ, V. and HEROUT, V. (1972). The presence of pellepiphyllin in the liverwort *Pellia neesiana* (Gottsche) Limpr.; its absence in *Pellia epiphylla* (L.) Dum. *Coll. Czech chem. Comm.* **37**, 1764.

BENEŠOVÁ, V., SAMEK, Z., HEROUT, V. and ŠORM, F. (1969a). Isolation and structure of pinguisone from *Aneura pinguis* (L.) Dum. *Coll. Czech. chem. Comm.* **34**, 582–592.

BENEŠOVÁ, V., SAMEK, Z., HEROUT, V. and ŠORM, F. (1969b). Isolation and structure of two new indole alkaloids from *Riccardia sinuata* (Hook.) Trev. *Coll. Czech. chem. Comm.* **34**, 1807–1809.

BOHM, B. A. (1975). Chalcones, aurones and dihydrochalcones. *In* "The Flavonoids"

(J. B. Harborne, T. J. Mabry and H. Mabry, eds), pp. 422–504. Chapman and Hall, London.

BOURGEOIS, G. and SUIRE, C. (1977). Présence de guaiazulène chez *Pellia epiphylla* (L.) Corda (Metzgériales) et absence chez *P. fabbroniana* Raddi. *Revue bryol. lichén.* **43,** 343–346.

CAMPBELL, D. H. (1936). The relationships of the Hepaticae. *Bot. Rev.* **2,** 53–66.

CHADEFAUD, M. (1960). L'embranchement des Muscinées ou Bryophytes. *In* M. Chadefaud "Traité de Botanique systématique", Vol. I, pp. 903–972. Masson, Paris.

CONNOLLY, J. D. and THORNTON, M. S. (1973). Sesquiterpenoid lactones from the liverwort *Frullania tamarisci*. *Phytochemistry* **12,** 631–632.

CONNOLLY, J. D., HARDING, A. E. and THORNTON, I. M. S. (1972). Gymnomitrol, a novel tricyclic sesquiterpenoid from *Gymnomitrium obtusum* (Lindb.) Pears. (Hepaticae). *J. chem. Soc. chem. Comm.* **20,** 1320–1321.

CORBELLA, A., GARIBOLDI, P., JOMMI, G., ORSINI, F., DeMARCO, A., and IMMIRZI, A. (1974). Structure and absolute stereochemistry of pinguisone. *J. chem. Soc. Perkin Trans.*, Ser. 1, **15,** 1875–1878.

ERICKSON, M. and MIKSCHE, G. E. (1974). On the occurrence of lignin or polyphenols in some mosses and liverworts. *Phytochemistry* **13,** 2295–2299.

EVANS, A. W. (1939). The classification of the Hepaticae. *Bot. Rev.* **5,** 49–96.

FAULKNER, D. J., RAVI, B. N., FINER, J. and CLARDY, J., (1977). Diterpenes from *Dictyota dichotoma*. *Phytochemistry* **16,** 991–993.

FENICAL, W., SIMS, J. J., WING, R. M. and RADLICK, P. C. (1972). Zonarene, a sesquiterpene from the brown seaweed *Dictyopteris zonarioides*. *Phytochemistry* **11,** 1161–1163.

FRANCISCO, C., COMBAUT, G., TESTE, J. and PROST, M. (1978). Eleganolone nouveau cétol diterpénique linéaire de la Phéophycée *Cystoseira elegans*. *Phytochemistry* **17,** 1003–1005.

FREDERICK, S. E., GRUBER, P. J. and TOLBERT, N. E. (1973). The occurrence of glycolate dehydrogenase and glycolate oxidase in green plants. An evolutionary survey. *Pl. Physiol., Lancaster* **52,** 328–333.

GAMS, H. (1959). Remarques sur les affinités entre les Mousses primitives. *Revue bryol. lichén.* **28,** 326–329.

GELLERMAN, J. L., ANDERSON, W. H., RICHARDSON, D. G. and SCHLENK, H. (1975). Distribution of arachidonic and eicosapentaenoic acids in the lipids of mosses. *Biochem. Biophys. Acta* **388,** 277–290.

GEORGIEV, G. H. and BAKARDJIEVA, N. T. (1973). Activity and isoenzyme composition of mosses (class Musci) with different phylogenic position. *C. r. Acad. bulg. Sci.* **26,** 965–968.

GORHAM, J. (1977). Lunularic acid and related compounds in liverworts, algae and *Hydrangea*. *Phytochemistry* **16,** 249–253.

GORHAM, J. (1978). Effect of lunularic acid analogues on liverwort growth and IAA oxydation. *Phytochemistry* **17,** 99–105.

GROLLE, R. (1972). Die Namen der Familien und Unterfamilien der Lebermoose (Hepaticopsida). *J. Bryol.* **7,** 201–236.

GROLLE, R. (1976). Verzeichnis der Lebermoose Europas und benachbarter Gebiete. *Feddes Reprium* **87,** 171–279.

HARBORNE, J. B. (1966). 3-Desoxyanthocyanins and their systematic distribution in ferns and gesnerads. *Phytochemistry* **5**, 589–600.

HARBORNE, J. B. (1967). "Comparative Biochemistry of the Flavonoids." Academic Press, London and New York.

HATTORI, S. (1972). *Frullania tamarisci*-complex and the species concept. *J. Hattori bot. Lab.* **35**, 202–251.

HAYASHI, S. and MATSUO, A. (1970). Bazzanenol, a new sesquiterpene alcohol having a skeleton of bicyclo [5.3.1] undecane system from Hepaticae, *Bazzania pompeana* (Lac.) Mitt. *Experientia* **26**, 347–348.

HÉBANT, C. (1978). Les Bryophytes sont-elles des plantes vasculaires régressées. *Bryophyt. Biblthca* **13**, 21–28.

HEROUT, V. (1971). Biochemistry of sequiterpenoids. *In* "Aspects of Terpenoid Chemistry and Biochemistry" (T. W. Goodwin, ed.), pp. 53–94. Academic Press, London and New York.

HERZFELDER, H. (1921). Beiträge zur Frage der Moosfärbungen. *Beih. bot. Zbl.* **38**, 355–400.

HÖRSTER, H. and WIERMANN, R. (1976). Phytochemische Untersuchungen an thallösen Lebermoosen: Monoterpenen in *Conocephalum conicum* (L.) Wiggers (Marchantiales). *Nova Hedwigia* **27**, 183–186.

HOWE, M. A. (1899). The Hepaticae and Anthocerotes of California. *Mem. Torrey bot. Club* **7**, 1–208, pl. 88–122.

HUNECK, S. and SCHREIBER, K. (1975). Über die Inhaltsstoffe weiterer Lebermoose. *J. Hattori bot. Lab.* **39**, 215–234.

JOVET-AST, S. (1967). Bryophyta. *In* "Traité de Paléobotanique" (E. Boureau, ed.), Vol. 2, 17–186. Masson, Paris.

KARUNEN, P. (1978). Fatty acid composition of glycosyl diglycerides in *Ceratodon purpureus*, *Plagiothecium laetum* and *Barbilophozia barbata*. *Bryophyt. Biblthca* **13**, 365–377.

KNOCHE, H., OURISSON, G., PEROLD, G. W., FOUSSEREAU, J. and MALEVILLE, J. (1969). Allergenic component of a liverwort: a sesquiterpene lactone. *Science, N.Y.* **166**, 239–240.

KOPONEN, T. and NILSSON, E. (1978). Flavonoid patterns and species pairs in *Plagiomnium* and *Rhizomnium* (Mniaceae). *Bryophyt. Biblthca* **13**, 411–425.

KRUTOV, S. M., SAMEK, Z., BENEŠOVÁ, V. and HEROUT, V. (1973). Isolation and the structure of deoxypinguisone from the liverwort *Ptilidium ciliare*. *Phytochemistry* **12**, 1405–1407.

KRZAKOWA, M. (1978). Isozymes as markers of inter- and intraspecific differentiation in Hepatics. *Bryophyt. Biblthca* **13**, 427–434.

KRZAKOWA, M. and SZWEYKOWSKI, J. (1977). Peroxydases as taxonomic characters in two critical *Pellia* taxa (Hepaticae, Pelliaceae). *Bull. Acad. pol. Sci.* Ser. Sci. Biol., II, **15**, 203–204.

LINDBERG, G., ÖSTERDAHL, B. G. and NILSSON, E. (1974). 5′,8″-Biluteolin, a new bi-flavone from *Dicranum scoparium*. *Chemica Scripta* **5**, 140–144.

McCLURE, J. W. and MILLER, H. A. (1967). Moss chemotaxonomy. A survey for flavonoids and the taxonomic implications. *Nova Hedwigia* **14**, 111–125.

MARKHAM, K. R. (1972). A novel flavone-polysaccharide compound from *Monoclea forsteri*. *Phytochemistry* **11**, 2047–2053.

MARKHAM, K. R. (1977). Flavonoids and phylogeny of the "primitive" New Zealand hepatic, *Haplomitrium gibbsiae*. *Phytochemistry* **16**, 617–619.

MARKHAM, K. R. and PORTER, L. J. (1969). Flavonoids in the green algae (Chlorophyta). *Phytochemistry* **8**, 1777–1781.

MARKHAM, K. R. and PORTER, L. J. (1974). Flavonoids of the liverwort *Marchantia polymorpha*. *Phytochemistry* **13**, 1937–1942.

MARKHAM, K. R. and PORTER, L. J. (1975a). Evidence of biosynthetic simplicity in the flavonoid chemistry of the Ricciaceae. *Phytochemistry* **14**, 199–201.

MARKHAM, K. R. and PORTER, L. J. (1975b). Isoscutellarein and hypolaetin 8-glucuronides from the liverwort *Marchantia berteroana*. *Phytochemistry* **14**, 1093–1097.

MARKHAM, K. R. and PORTER, L. J. (1976). The taxonomic position of *Sphaerocarpos* and *Riella* as indicated by their flavonoid chemistry. *Phytochemistry* **15**, 151–152.

MARKHAM, K. R. and PORTER, L. J. (1978a). Production of an aurone by bryophytes in the reproductive phase. *Phytochemistry* **17**, 159–160.

MARKHAM, K. R. and PORTER, L. J. (1978b). Chemical constituents of the bryophytes. *Prog. Phytochem.* **5**, 181–272.

MARKHAM, K. R., MABRY, T. J. and VOIRIN, B. (1969). 3-O-methylquercetin 7-O-diglucoside 4′-O-glucoside from the fern, *Ophioglossum vulgatum*. *Phytochemistry* **8**, 469–472.

MARKHAM, K. R., PORTER, L. J., MUES, R., ZINSMEISTER, H. D. and BREHM, B. G. (1976a). Flavonoid variation in the liverwort *Conocephalum conicum*: evidence for geographic races. *Phytochemistry* **15**, 147–150.

MARKHAM, K. R., PORTER, L. J., CAMPBELL, E. O., CHOPIN, J. and BOUILLANT, M. L. (1976b). Phytochemical support for the existence of two species in the genus *Hymenophyton*. *Phytochemistry* **15**, 1517–1521.

MARKHAM, K. R., PORTER, L. J. and CAMPBELL, E. O. (1978a). The usefulness of flavonoid characters in studies of the taxonomy and phylogeny of liverworts. *Bryophyt. Biblthca* **13**, 387–398.

MARKHAM, K. R., ZINSMEISTER, H. D. and MUES, R. (1978b). Luteolin 7-glucuronide-3′-mono (*trans*)ferulylglucoside and other unusual flavonoids in the aquatic liverwort complex, *Riccia fluitans*. *Phytochemistry* **17**, 1601–1604.

MARKHAM, K. R., MOORE, N. A. and PORTER, L. J. (1978c). Changeover in flavonoid pattern accompanying reproductive structure formation in a bryophyte. *Phytochemistry* **17**, 911–913.

MATSUO, A. and HAYASHI, S. (1977). Revised structure and absolute configuration of the sesquiterpene (+)-bazzanene. *J. chem. Soc. chem. Comm.* **25**, 566–568.

MATSUO, A., NAKAYAMA, M. and HAYASHI, S. (1971). Aromatic esters from the liverwort, *Isotachis japonica*. *Z. Naturforsch.*, Teil B, **26**, 1023–1025.

MATSUO, A., NAKAYAMA, M. and HAYASHI, S. (1973). Sesquiterpene hydrocarbons of the liverwort, *Scapania parvitexta*. *Bull. chem. Soc. Japan* **46**, 1010–1011.

MATSUO, A., SATO, S., NAKAYAMA, M. and HAYASHI, S. (1974). Taylorione, a novel carbon skeletal ketone from the liverwort, *Mylia taylorii* (Hook.) Gray. *Tetrahedron Lett.* **1974**, 3681–3684.

MATSUO, A., UTO, S., NAKAYAMA, M. and HAYASHI, S. (1976). Sesquiterpene hydrocarbons of the liverwort, *Dumortiere hirsuta*. *Z. Naturf.* Teil C, **31**, 401–402.

MATSUO, A., NOSAKI, H., SHIGEMORI, M., NAKAYAMA, M. and HAYASHI, S. (1977).

(—)-Dihydromylione A, a novel tetracyclic sesquiterpene ketone containing two conjugated cyclopropane rings, from *Mylia taylorii* (liverwort). *Experientia* **33**, 991–992.

MATSUO, A., NAKAYAMA, M., HAYASHI, S., SEKI, T. and AMAKAWA, T. (1978). A comparative study of the diterpenoids from several species of the genus *Jungermannia*. *Bryophyt. Biblthca* **13**, 321–328.

MELCHERT, T. E. and ALSTON, R. E. (1965). Flavonoids from the moss *Mnium affine* Bland. *Science, N.Y.* **150**, 1170–1171.

MERCER, E. I., LONDON, R. A., KENT, I. S. A. and TAYLOR, A. J. (1974). Sterols, sterol esters and fatty acids of *Botrydium granulatum*, *Tribonema aequale* and *Monodus subterraneus*. *Phytochemistry* **13**, 845–852.

MIKSCHE, G. E. and YASUDA, S. (1978). Lignin of "giant" mosses and some related species. *Phytochemistry* **17**, 503–504.

MITCHELL, J. C., FRITIG, B., SINGH, B. and TOWERS, G. H. N. (1970). Allergic contact dermatitis from *Frullania* and Compositae. *J. invest. Dermatol.* **54**, 233–239.

MOESTRUP, Ø. (1974). Ultrastructure of the scale-covered zoospores of the green alga *Chaetosphaeridium*, a possible ancestor of higher plants and bryophytes. *Biol. J. Linn. Soc.* **6**, 111–125.

MUES, R. and ZINSMEISTER, H. D. (1978). Studies on phenolic constituents of Jungermanniales in relation to their taxonomy. *Bryophyt. Biblthca* **13**, 399–409.

NILSSON, E. and BENDZ, G. (1974). Flavonoids in bryophytes. *In* "Chemistry in Botanical Classification" (G. Bendz and J. Santesson, eds), pp. 117–120. Academic Press, New York and London.

OHTA, K. and TAKAGI, M. (1977). Halogenated sesquiterpenes from the marine red alga *Marginisporum aberrans*. *Phytochemistry* **16**, 1062–1063.

ÖSTERDHAL, B.-G. (1976). A new luteolin tetraglycoside from *Hedwigia ciliata*. *Acta Chem. scand.*, Ser. B, **30**, 867–870.

ÖSTERDAHL, B.-G. and LINDBERG, G. (1977). Luteolin 7-O-neohesperidoside-4'-O-sophoroside, another new tetraglycoside. *Acta Chem. scand.*, Ser. B, **31**, 293–296.

OURISSON, G. (1974). Some aspects of the distribution of diterpenes in plants. *In* "Chemistry in Botanical Classification" (G. Bendz and J. Santesson, eds), pp. 129–134. Academic Press, New York and London.

PEROLD, G. W., MULLER, J. C. and OURISSON, G. (1972). Structure d'une lactone allergisante: le frullanolide-I. *Tetrahedron* **28**, 5797–5803.

PICKETT-HEAPS, J. D. (1975). Structural and phylogenetic aspects of microtubular systems in gametes and zoospores of certain green algae. *In* "The Biology of the Male Gamete" (J. G. Duckett and P. A. Racey, eds), pp. 37–44. Academic Press, London and New York.

POKORNY, M. (1974). D-methionine metabolic pathways in Bryophyta: a chemotaxonomic evaluation. *Phytochemistry* **13**, 965–971.

POKORNY, M., MARČENKO, E. and KEGLEVIČ, D. (1970). Comparative studies of L- and D-methionine metabolism in lower and higher plants. *Phytochemistry* **9**, 2175–2188.

PRYCE, R. J. (1972a). Metabolism of lunularic acid to a new plant stilbene by *Lunularia cruciata*. *Phytochemistry* **11**, 1355–1364.

PRYCE, R. J. (1972b). The occurrence of lunularic and abscissic acids in plants. *Phytochemistry* **11**, 1759–1761.

REZNIK, H. and WIERMANN, R. (1966). Quercetin und Kämpferol im Thallusgewebe von *Corsinia coriandrina*. *Naturwissenschaften* **53**, 230–231.

ROBERTSON, K. J. and FENICAL, W. (1977). Pachydictyol-A epoxide, a new diterpene from the brown seaweed *Dictyota flabellata*. *Phytochemistry* **16**, 1071–1073.

ROTHMALER, W. (1951). Die Abteilungen und Klassen der Pflanzen. *Feddes Reprium* **54**, 256–266.

SCHIER, W. (1974). Untersuchungen zur Chemotaxonomie der Marchantiales. *Nova Hedwigia* **25**, 549–566.

SCHUSTER, R. M. (1963). Annotated keys to the genera of antipodial Hepaticae with special reference to New Zealand and Tasmania. *J. Hattori bot. Lab.* **26**, 185–309.

SCHUSTER, R. M. (1966). "The Hepaticae and Anthocerotae of North America East of the Hundredth Meridian." Vol. 1. Columbia Univ. Press, New York.

SHANNON, L. M. (1968). Plant isoenzymes. *A. Rev. Pl. Physiol.* **19**, 187–210.

STEERE, W. C. (1969). A new look at evolution and phylogeny in bryophytes. *In* "Current Topics in Plant Science" (J. E. Gunckel, ed.), pp. 134–143. Academic Press, New York and London.

STOTLER, R. E. (1976). The biosystematic approach in the study of the Hepaticae. *J. Hattori bot. Lab.* **41**, 37–46.

SUIRE, C. (1970). Recherches cytologiques sur deux Hépatiques: *Pellia epiphylla* (L.) Corda (Metzgériale) et *Radula complanata* (L.) Dum. (Jungermanniale). Ergastome, sporogénèse et spermatogénèse. *Botaniste* **53**, 125–392.

SUIRE, C. (1975). Les donnees actuelles sur la chimie des Bryophytes. *Revue bryol. lichén.* **41**, 105–256.

SUIRE, C. and BOURGEOIS, G. (1977). Les monoterpènes de *Conocephalum conicum*, *Frullania tamarisci* et *Porella platyphylla*. *Phytochemistry* **16**, 284–285.

SUIRE, C., BOURGEOIS, G., BARBE, B. and PETRAUD, M. (1978). L'iso-α-gurjunène B, principal constituant de la fraction sesquiterpénique de l'essence extraite de *Pellia epiphylla* (L.) Corda. *Bryophyt. Biblthca* **13**, 329–334.

SWAIN, T. (1974). Flavonoids as evolutionary markers in primitive Tracheophytes. *In* "Chemistry in Botanical Classification" (G. Bendz and J. Santesson, eds) (pp. 81–91. Academic Press, New York and London.

SWAIN, T. (1975). Evolution of flavonoid compounds. *In* "The Flavonoids" (J. B. Harborne, T. J. Mabry and H. Mabry, eds), pp. 1096–1129. Chapman and Hall, London.

TAYLOR, I. E. P. and ELLIOT, A. M. (1972). Dehydrogenases in a single population of the moss *Eurhynchium oreganum*. The effects of dehydration and low temperature on disc electrophoretic enzyme patterns. *Can. J. Bot.* **50**, 375–378.

VALIO, I. F. M., BURDON, R. S. and SCHWABE, W. W. (1969). New natural growth inhibitor in the liverwort *Lunularia cruciata* (L.) Dum. *Nature, Lond.* **223**, 1176–1178.

VANDEKERKHOVE, O. (1977a). Isolierung und Charakterisierung eines Dihydroflavonols bei dem Laubmoos *Georgia pellucida* (L.) Rabh. *Z. PflPhysiol.* **82**, 455–457.

VANDEKERKHOVE, O. (1977b). Apigenin-7-rhamnoglucosid bei *Hylocomium splendens* (Hedw.) Br. eur. *Z. PflPhysiol.* **85**, 135–138.

VANDEKERKHOVE, O. (1978a). Die Flavonoide von *Mnium undulatum* (L.) Hedw. *Z. PflPhysiol.* **86**, 135–139.

VANDEKERKHOVE, O. (1978b). Die Flavonoids von *Riccia fluitans* L. *Z. PflPhysiol.* **86,** 217–221.

VANDERKERKHOVE, O. (1978c). Luteolin aus dem Sporophyten von *Ceraton purpureus* (L.) Brid. *Z. PflPhysiol.* **86,** 279–281.

VOWINCKEL, E. (1975). Die Struktur des Sphagnorubins. *Chem. Ber.* **108,** 1166–1181.

WEITZ, S. and IKAN, R. (1977). Bracteatin from the Moss *Funaria hygrometrica*. *Phytochemistry* **16,** 1108–1109.

20 | Structure and Eco-physiological Adaptation in Bryophytes

M. C. F. PROCTOR

Department of Biological Sciences, University of Exeter,
Hatherly Laboratories, Prince of Wales Road,
Exeter EX4 4PS, Great Britain

Abstract: Compared with vascular plants, bryophytes represent an alternative strategy of adaptation to life on land. Endohydric bryophytes are analogous with vascular plants in possessing functional internal conducting systems and waxy cuticles. However, all bryophytes retain a need for peripheral water conduction, and much of the structure of bryophyte leaves is likely to be related to this.

Leaves of many Pottiaceae and Encalyptaceae are covered with hollow papillae; calculation suggests that water conduction in the interstices between these could balance high rates of evaporation. Leaf-surface papillae in *Hedwigia ciliata* and Orthotrichaceae are more widely spaced and are unlikely in themselves to provide a significant external conducting system. Porose hyaline cells, well known in *Sphagnum* and *Leucobryum*, occur also in the leaf bases of Calymperaceae and *Encalypta* and *Tortula* spp. Concave leaves may provide water conduction along the shoot under moist conditions, and significant water storage, while keeping the convex surface free for gas exchange. In species lacking external adaptations for capillary water movement conduction must be internal. A pathway through the cell walls appears only marginally adequate if the cell wall structure is similar to that reported in flowering plants and algae, but there is evidence that some incrassate bryophyte cell walls may contain larger spaces through which water can move. In species with large cells and thin walls water movement within cells is likely to be important, and may have provided a selection pressure towards the evolution of elongated cells. The special characteristics of liverworts, the relations of bryophyte growth-forms to their atmospheric environment and the possible roles of hair-points in reducing the rate of water-loss are briefly considered.

Systematics Association Special Volume No. 14, "Bryophyte Systematics", edited by G. C. S. Clarke and J. G. Duckett, 1979, pp. 479–509, Academic Press, London and New York.

INTRODUCTION

How far are the form and structure of bryophytes determined by function—by physiological imperatives—and how far is function dictated and constrained by inherent limitations of structure and development? A great deal has been written on the functional morphology of bryophytes in the course of the past century (see Goebel, 1905; Buch, 1947; Watson, 1971 for reviews), some of it well grounded in observation and experiment, some reasonable conjecture, and some certainly misconceived or over-simplistic and misleading. In this paper my intention is to reappraise some limited aspects of structure-function relationships. I shall be chiefly concerned with water relations, and with the peripheral conduction of water rather than with the internal conducting systems considered by Hébant (1977; this volume, Chapter 16). I believe that techniques and concepts that have developed over the last few decades now make possible a more soundly-based understanding of how bryophytes function and why they are organized as they are, and open up a potentially fruitful area of research. A firm foundation was laid thirty years or so ago by the admirable work of Buch (1945, 1947).

Broadly speaking, bryophytes and vascular plants represent two alternative strategies of adaptation to the uneven and intermittent distribution of water in the terrestrial environment. Vascular plants typically possess roots and xylem, bringing water from the soil where it is plentiful to the often dry above-ground environment in which the leaves must photosynthesize. Bryophytes typically show little evident adaptation to tap the reservoir in the soil or to control water loss. Some are sensitive to drying out and are confined to moist habitats, but many survive desiccation and resume normal metabolism when moist conditions return. They are poikilohydric—a pattern of adaptation which they share with lichens, many terrestrial algae, and with such animals as tardigrades, nematodes and rotifers (Crowe, 1971) but with few vascular plants. Bryophytes are limited by their characteristic mode of life, but also liberated; they can occupy hard substrates such as rock and bark which are impenetrable to roots and so untenable to vascular plants.

The adaptive differences between vascular plants and bryophytes are more subtle than the simple antithesis outlined above might suggest. The water relations of vascular plants involve adaptation to produce a balance of resistances to water movement through different parts of a plant and its surroundings—soil, root, vascular tissue, leaves and the air around them—appropriate to its normal environment. The stomata can be thought of as a variable trimming resistance in the system; stomatal control will not allow a vascular plant to grow in totally inappropriate conditions. A net absorption of energy by any organism will result in heating, evaporation or both (Gates, 1968; Monteith, 1973). The energy income necessary for photosynthesis therefore generally implies water loss, even in saturated air. Hence all bryophytes have some need for water conduction, and they too must evolve appropriate patterns of resistances to control the distribution and movement of water (Dilks and Proctor, 1979).

There are important differences in the physiology of vascular plants and bryophytes, but at a more basic level there are important similarities. It cannot be assumed that bryophyte parts are closely analogous in function to the apparently corresponding organs of flowering plants. Bryophyte physiology has suffered greatly in the past from preconceptions taken uncritically from studies of flowering plants. The need is for fundamental examination in the light of common physical first principles.

MAJOR ADAPTIVE PATTERNS IN BRYOPHYTES

Buch made the useful distinction between "endohydric" and "ectohydric" bryophytes. Endohydric species possess well-developed internal conducting systems, and their surfaces are not readily wetted or stained with aqueous solutions of basic dyes such as toluidine blue. In a number of species I have examined the leaves are covered with a layer of waxy material soluble in chloroform or carbon tetrachloride (Figs 1, 2). Green and Clayton-Greene (1977) found a similar waxy layer on the lamellae of *Dawsonia superba* Grev. and cuticular material of this kind may well be of general occurrence in endohydric mosses. Ectohydric species typically take up water freely over the whole surface of the plant and are rapidly stained by basic dyes. They usually have little obvious development of internal conducting tissues, but often show well-developed external capillary conducting structures. Buch recognized that some bryophytes (e.g. small *Barbula* spp., many liverworts) show the ectohydric feature of absorbing water readily over the whole surface, but clearly rely largely on internal water conduction. These he called "mixohydric". The ectohydric species he further divided into "pollacaulophytes" (from the Greek, implying "often dry", embracing the typical poikilohydric bryophytes), ectohydric hygrophytes, and submerged aquatics. His conclusions on the systematic distribution of these adaptive types may be summarized as follows (classifications of Dixon (1932) and Buch *et al.* (1938) for mosses and liverworts respectively).

Endohydric: All Nematodontae; Fissidentales (except a few aquatics);

Funariales; most Eubryales Acrocarpi; Anthocerotales; Marchantiales; Sphaerocarpales; Haplolaenaceae, Monocleaceae, Aneuraceae.

Ectohydric: Sphagnales (ectohydric hygrophytes); Andreaeales; Grimmiales; Syrrhopodontales; Encalyptales; Orthotrichales; most Pottiales; most Dicranales; most Eubryales Pleurocarpi; Ptilidiaceae, Schistochilaceae, Marsupellaceae, Physiotaceae, Radulaceae, Porellaceae, Lejeuneaceae, Frullaniaceae, Metzgeriaceae.

Mixohydric (mainly): Epigonianthaceae, Harpanthaceae, Plagiochilaceae, Scapaniaceae, Trigonanthaceae.

These groups are not sharply demarcated, particularly among the liverworts. Many of the families of leafy liverworts listed as primarily ectohydric show marked mixohydric tendencies; conversely there are clear examples of ectohydric species in families listed as mainly mixohydric.

SOME GENERAL COMMENTS ON ENDOHYDRIC BRYOPHYTES

The endohydric mosses and liverworts show the closest structural and functional analogies with vascular plants. These analogies are at their most obvious in the elaborate photosynthetic structures of the big Marchantialean liverworts (Figs 3, 4) and the leaves of Polytrichales (Fig. 5), and in the conducting tissues of the latter (Hébant, 1977). Dr T. G. A. Green (personal communication) has demonstrated that (as with the mesophyll of higher plants) the rate of photosynthesis is greatly depressed if these photosynthetic tissues are flooded. In most Marchantiales the air pores are surrounded by a hydrophobic cutinized ledge which prevents the entry of water, as in the stomata of higher plants (Schönherr and Ziegler, 1975); in the Polytrichales the waxy tops of the lamellae serve the same function. Quite high rates of internal water transport have been recorded from various endohydric bryophytes (Haberlandt, 1886; Bowen, 1931, 1933; Mägdefrau, 1935; Zacherl, 1956; Rifot and Barrière, 1974); if conduction appears inadequate to maintain the turgor of isolated shoots in a laboratory atmosphere even at a "high" humidity, this may mean

---

Fɪɢ. 1. *Plagiomnium undulatum*, Drewsteignton, Devon. Leaf cells, untreated. ×1670.

Fɪɢ. 2. *P. undulatum*, Drewsteignton. Leaf cells, CCl₄ washed. ×1670.

Fɪɢ. 3. *Preissia quadrata*, Cowside Beck, Arncliffe, Yorks. Vertical section and surface of thallus. ×170.

Fɪɢ. 4. *P. quadrata*, Cowside Beck. Pore in upper surface. ×420.

Fɪɢ. 5. *Pogonatum urnigerum*, Crib Goch, Snowdon, Gwynedd. Cross-section and upper surface of leaf. ×420.

Fɪɢ. 6. *Tortula laevipila*, Chudleigh, Devon. Leaf. ×190.

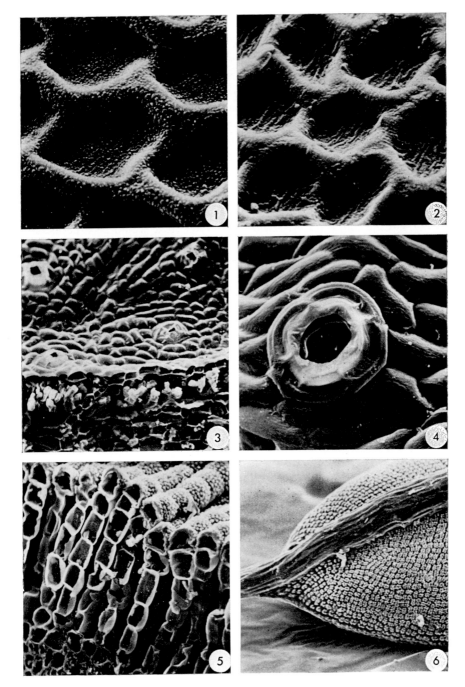

FIGS 1, 2, 3, 4, 5, 6

no more than that the experiments have been done in conditions quite irrelevant micrometeorologically to the plant's normal habitat. Field observation shows that rates of water conduction in, say, *Conocephalum conicum*, *Polytrichum commune* or *Plagiomnium elatum* are manifestly adequate to balance water loss under a wide range of normal conditions.

The resistance to water loss by evaporation from a liverwort thallus can be estimated in the same way as the "leaf resistance" (r_L) of a higher plant, by comparing the rate of water loss with that from a filter paper replica or from the same leaf or thallus wetted all over with water containing a trace of detergent. For thalli of two populations of *Conocephalum conicum* I have found values of *c.* 2·3 and 5·3 s cm^{-1} for R_L—within the range that would be expected for mesophytic flowering plants with their stomata open. By contrast, r_L for moist thalli of *Pellia epiphylla* was less than 0·5 s cm^{-1}. Similar experiments showed that the waxy covering of the leaves of *Mnium hornum* resulted in a leaf resistance around 1·5 s cm^{-1}.

EXTERNAL CAPILLARY CONDUCTING SYSTEMS

The water associated with a bryophyte shoot can be regarded as divided into three parts; (1) water within the cells ("symplast water"), (2) water in the cell walls ("apoplast water") and (3) water in external capillary spaces (Dilks and Proctor, 1979). The functional importance of this external capillary water was clearly appreciated by Goebel (1905), Bowen (1931, 1933) and Mägdefrau (1935). Buch (1945, 1947) divided bryophyte capillary systems into three categories:

  (1) inter-organ capillary systems, including spaces within sheathing leaf-bases or between leaves, and amongst paraphyllia or rhizoid tomentum;

  (2) epi-organ capillary systems, comprising spaces between papillae, ridges, plicae and other structures on leaf and stem surfaces; and

  (3) intra-organ capillary systems, best exemplified by the hyaline cells of *Sphagnum* but seen also in *Leucobryum* and in the inflated and often porose cells of the leaf bases of many Pottiaceae, Encalyptaceae and Calymperaceae.

This distinction is obviously of interest morphologically and taxonomically; functionally it may be of little moment, and the three types of system commonly coexist and interconnect.

1. Leaf-surface Papillae

If one end of a dry leaf of *Tortula ruralis* or *Tortella tortuosa* is dipped into water it quickly becomes moist throughout its length, as though the leaf surface were covered with a layer of blotting paper. The projecting surface of each cell forms a single, rather irregular, hollow, branched papilla; the individual papillae are separated by channels about 2 μm wide overlying the anticlinal cell walls. If the angle of contact of water with the cell walls is zero, the tension developed in a system of this kind can be calculated from the relation:

$$\Psi = -\sigma(1/r_1 + 1/r_2)$$

where Ψ is water potential (*Pa*), σ is the surface tension of water ($N\ m^{-1}$), and r_1 and r_2 are the two principal radii of the water surface (Nobel, 1974, p. 385). If we take r_1 as 1 μm and r_2 as ∞ (i.e. assume that the meniscus is cylindrical), we get a value for Ψ of -7.3×10^4 Pa (-0.73 bar). Assuming a water potential difference of 0.5 bar, and making what were probably rather conservative assumptions about the relevant dimensions of the capillary channels, Dilks and Proctor (1979) calculated a rate of water movement which would balance evaporation of about 200 μg cm^{-2} s^{-1} from the exposed apical parts of the leaves. Even if the surface of a moss cushion were to consist of a single layer of such leaves (and limited observations suggest that a "leaf area index" of 5 or more is probably common), this is more than enough to support the fastest rates of evaporation likely to be encountered in the field, and an order of magnitude higher than needed under normal circumstances (Appendix 1).

The water-carrying capacity of such a system is steeply dependent on size. If the whole system is scaled up or down, the amount of water that can be moved is proportional to $1/r^2$; if the spacing of the channels remains constant it is proportional to $1/r^3$. In either case, at large radii the system would become vulnerable to water potential differences in its surroundings; at small radii it would soon become very limited in capacity, or develop water potential differences large enough to influence cell function, or both.

The apparently prodigally generous provision for water conduction invites closer examination. Storage can hardly be significant; on a dry sunny day all the water in the surface capillaries could be evaporated in a minute or two, and much more would be held in larger capillary spaces within the moss cushion. A more likely explanation lies in the need to combine superficial capillary water conduction and gas exchange for photosynthesis in the same leaf surface. Even distribution of water over the leaves is important for optimal

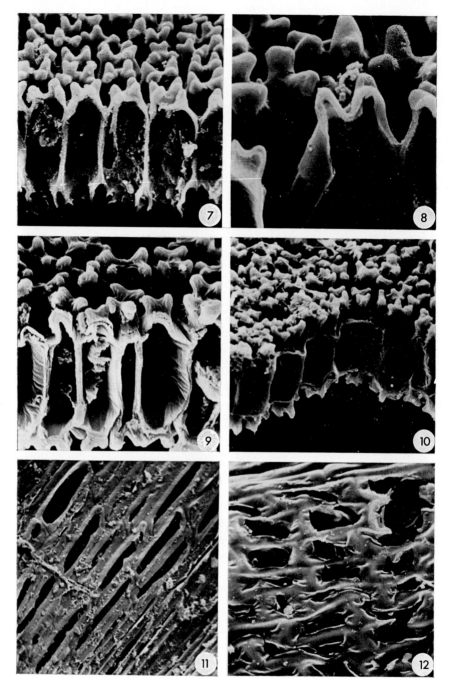

Figs 7, 8, 9, 10, 11, 12

assimilation (Dilks and Proctor, 1979). This would require response to small differences in water potential, so there may be less "surplus" capacity than appears at first sight.

Efficient capillary systems of this kind are characteristic of many Pottiaceae (e.g. *Tortula* (Figs 6, 7, 8; Casas de Puig and Molinas, 1974), *Desmatodon convolutus* (Magill *et al.*, 1974), *Pleurochaete squarrosa* (Fig. 8; Dilks and Proctor, 1979), *Cinclidotus mucronatus* (Fig. 13), Encalyptaceae (Fig. 9)), and occur in various genera of other families. It is presumably their efficiency which makes possible in exposed habitats such conspicuously large, lingulate leaves as those of the bigger *Tortula* and *Encalypta* species. The very similar structures of the leaves of *Tortula* and other Pottiaceae, *Encalypta* and *Anomodon viticulosus* (Fig. 10) are a remarkable example of evolutionary convergence. The less intricately ornamented abaxial surfaces of the leaves of *Thuidium tamariscinum* (Fig. 15) appear to function in essentially the same way. If they are moistened with a minute quantity of acid fuchsin solution, the papillae can be seen under the microscope as islands standing high and dry among the network of channels filled by the dye. Probably the effect is accentuated by a tendency for the papillae themselves to be water-repellent. This is strikingly the case in *Encalypta streptocarpa*, and Buch (1945, 1947) found a similar tendency in young leaves of *Tortula ruralis*. It is interesting that the papillae are confined to the abaxial surfaces of the leaves on the branches of *Thuidium tamariscinum*; the strongly concave inner surfaces are quite smooth. A similar state of affairs in *Thelia asprella* (B.S.G.) Sull. is illustrated by Magill *et al.* (1974). On the other hand, in *Distichium capillaceum* (with leaves of very different form) the limb of the leaf is covered with closely-set simple papillae which continue uninterrupted into the groove on its adaxial surface (Figs 14, 17).

Not all papillose leaves are organized in this way. Buch (1945, 1947) pointed out that leaves of *Hedwigia ciliata* do not rapidly become moist (though whole shoots do so) when one end is dipped in water, despite the apparently dense covering of papillae. Close examination of the leaves shows that the bases of the papillae, placed centrally over the cell lumina, are in fact rather widely

FIG. 7. *Tortula intermedia*, Chudleigh, Devon. Cross-section of leaf. ×1670.

FIG. 8. *T. intermedia*, Chudleigh. Cross-section leaf showing hollow papillae. ×4170.

FIG. 9. *Encalypta streptocarpa*, Bangor, Gwynedd. Cross-section of leaf. ×1670. Herbarium material, rehydrated and critical-point dried.

FIG. 10. *Anomodon viticulosus*, Chudleigh, Devon. Broken edge of leaf. ×1670.

FIG. 11. *Tortula intermedia*, Chudleigh. Leaf base. ×450. Freeze dried.

FIG. 12. *Encalypta streptocarpa*, Bangor. Leaf base. ×830. Herbarium material, rehydrated and critical-point dried.

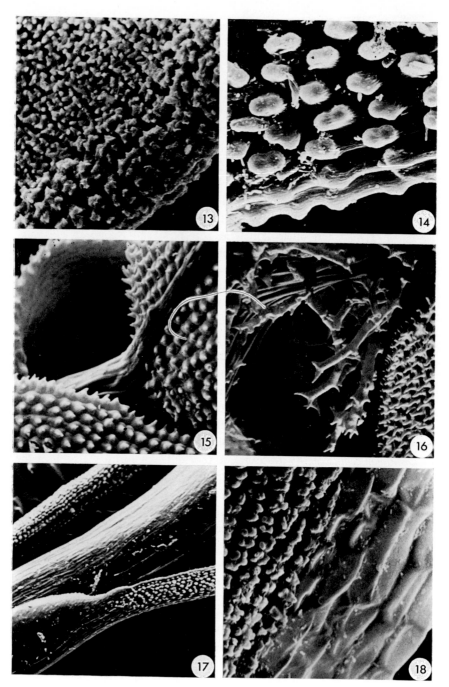

FIGS 13, 14, 15, 16, 17, 18

separated (Figs 19, 20, 21). The spaces between the papillae are much wider than the papillae themselves, and the system lacks the continuous narrow channels of *Tortula, Encalypta* or *Anomodon*. We may conjecture that the function of the papilla system is different from that of the *Tortula* type, and that in *Hedwigia ciliata* it is not the major pathway of water movement balancing evaporation. It could nevertheless bring about rapid redistribution of excess water once "primed" by raindrops. Consideration of the possible shape of water menisci amongst the papillae makes it clear that the system will develop much lower capillary tensions than that of *Tortula*. The papillae themselves are striking structures; those on the convex outer surface are much larger and more branched than those on the inner surface of the leaf. I have not been able to satisfy myself whether they are solid or have a narrow lumen. Certainly they do not show the clear relationship with gas exchange of the papillae of the *Tortula* type.

The leaf surfaces of the Orthotrichaceae I have examined—*Orthotrichum anomalum* (Figs 22, 23, 24), *Ulota crispa* and *Zygodon viridissimus*—have characteristic, rather widely separated, conical or peg-like papillae which probably function in essentially the same way as those of *Hedwigia ciliata*. Cells of *Orthotrichum anomalum* commonly bear two well-separated papillae, whose bases lie close to the cell margin, with a striking effect on the appearance of the leaf as the individual cells shrink on drying. A rather similar structure is shown in *Braunia secunda* (Hook.) B.S.G. (Hedwigiaceae) by Robinson (1971), and the varied patterns within *Orthotrichum* illustrated by Lewinsky (1974) appear to bridge much of the gap between *O. anomalum* and *Hedwigia ciliata*. Leaves of *Orthotrichum* are notably easily wetted, and the rather intricately patterned surface may allow a fair amount of superficial water movement.

It is worth emphasizing that there are some leaf papillae which may have little or nothing to do with water conduction, though they may be still of significance in maintaining gas exchange in moist habitats. Those of *Fissidens serrulatus* (Fig. 25) or *Heterocladium heteropterum* (Fig. 26), for example, could not stand as clear examples of the kind of mechanisms described here, nor

FIG. 13. *Cinclidotus mucronatus*, Chudleigh, Devon. Leaf margin near apex. ×920.

FIG. 14. *D. capillaceum*, Cowside Beck. Leaf surface. ×1700.

FIG. 15. *Thuidium tamariscinum*, Exeter, Devon. Branch leaves. ×450. Freeze dried.

FIG. 16. *T. tamariscinum*, Exeter. Paraphyllia. ×450. Freeze dried.

FIG. 17. *Distichium capillaceum*, Cowside Beck, Arncliffe, Yorks. Leaves and leaf bases. ×170.

FIG. 18. *Pleurochaete squarrosa*, Chudleigh. Leaf base. ×920.

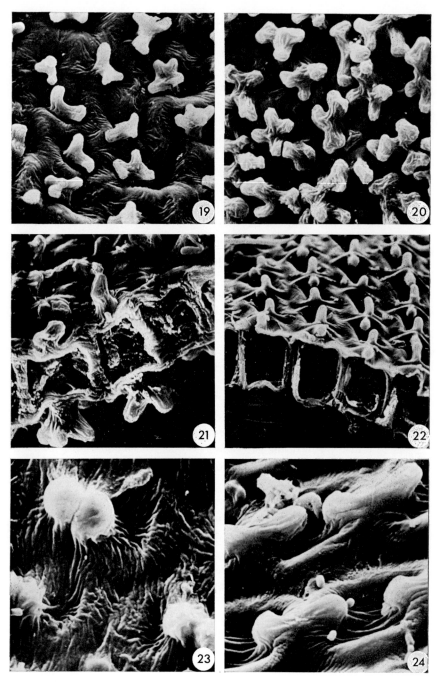

FIGS 19, 20, 21, 22, 23, 24

could the low papillae illustrated on the surfaces of various leafy liverworts by Duckett and Soni (1972a).

2. Other Capillary Structures

The hyaline cells of *Sphagnum* (Mozingo *et al.*, 1969; Castaldo and Di Martino, 1970) are too well known to need detailed description here. Their diameters are commonly in the range 20–40 μm, so they should drain under tensions around 0·1 bar. In practice, continuity for rapid water movement via the pores would depend on the presence of water in rather larger capillary spaces outside the hyaline cells, and water movement to replace evaporation would require some gradient of water potential. Thus the field observation that *Sphagna* are normally confined to habitats not more than a few decimetres above a free water surface is consistent with theoretical expectation. Some movement of water could certainly take place across the cell walls (as in the tracheids of higher plants), but would be relatively slow because of the small available water potential gradient.

The intra-organ capillary systems of the leaf bases of Calymperaceae, Encalyptaceae and some Pottiaceae provide an interesting comparison with *Sphagnum* (Richards and Edwards, 1972; Edwards, 1976). The basal cells of the mature leaves of these plants are large, thin-walled, and often empty and perforated by pores to the exterior and from cell to cell. Direct microscopical observation shows that the cells empty and fill rapidly as the leaf dries out or is remoistened. Leaf bases of this kind must provide a highly efficient external conducting system around the stem. Those with regular and numerous pores undoubtedly allow the most rapid movement of water, but water would pass fairly freely through the thin walls of unperforated cells provided these lack living contents and the barrier of a cell membrane. The hyaline bases of older leaves of *Tortella* or *Pleurochaete* (Fig. 8) may function in this way, but cells with living contents would allow only rather slow water movement. The maximum tensions that menisci in spaces the size of these cells (*c.* 10–15 μm

Fig. 19. *Hedwigia ciliata*, Craig Breidden, Powys. Upper surface of leaf. $\times$1670. Air dried.

Fig. 20. *H. ciliata*, Craig Breidden. Lower surface of leaf. $\times$1670. Air dried.

Fig. 21. *H. ciliata*, Drewsteignton, Devon. Cross-section of leaf. $\times$1670.

Fig. 22. *Orthotrichum anomalum*, Chudleigh, Devon. Cross-section of leaf near base. $\times$1670.

Fig. 23. *O. anomalum*, Chudleigh. Upper surface of leaf near apex. $\times$4170.

Fig. 24. *O. anomalum*, Chudleigh. Upper surface of leaf near base. $\times$4170.

diam.) could sustain is about 0·2–0·3 bar; ample for the movement of water over distances of a few centimetres under moist conditions, but half an order of magnitude less than that expected of a *Tortula*-type leaf papilla system.

Sheathing leaf bases without pores are common in many groups of mosses, and provide an obvious channel of capillary water conduction well attested by observation in *Polytrichum* and other genera. An elegant example in *Distichium capillaceum* is illustrated in Fig. 17, although Buch (1945, p. 25) regarded this species as primarily endohydric, and considered the leaf bases (which are often empty, giving the shoots their characteristic silvery sheen) to be of only limited significance for water conduction. All gradations exist between a well-circumscribed system like this (of rather sharply defined capacity) and the more indefinite systems of capillary spaces around the bases of leaves which overlap but do not conspicuously sheathe the stem. The conducting function of the latter in particular may be supplemented or taken over by rhizoid tomentum or paraphyllia (Fig. 16).

Leaves often show other adaptations of form related to the movement or localization of water. Plicae probably often provide channels for the longitudinal movement of water in the pleurocarpous mosses in which they occur (e.g. *Camptothecium sericeum*, Dilks and Proctor, 1979), supplementing the limited capacity of the parallel narrow channels along the cell boundaries. The channelled upper surface of the acumen of the leaf of many Dicranaceae may (as Buch (1945, 1947) suggested) serve a similar function, as may the ridges on the midribs of some *Campylopus* and *Dicranum* species, and some rolled leaf margins. However, the reservations have to be made that structures of this kind may also serve mechanical functions, and that a structure that will provide a capillary conducting channel if the surface is wettable will not do so if it is water-repellent.

Many ectohydric mosses have strongly concave leaves. If a shoot of *Pseudoscleropodium purum* is examined in moist weather, the concavities of the leaves will generally be found full of water. When we cultivated *Myurium hochstetteri*

Fig. 25. *Fissidens serrulatus*, near Ashburton, Devon. Edge of leaf close to apex. ×450.

Fig. 26. *Heterocladium heteropterum* var *flaccidum*, Black Head, Torquay, Devon. Leafy shoot. ×450.

Fig. 27. *Cryphaea heteromalla*, Whitlands Landslip, near Lyme Regis, Dorset. Leaf surface. ×1880. Air dried.

Fig. 28. *C. heteromalla*, Branscombe, Devon. Leaf surface. ×1670.

Fig. 29. *Leptodon smithii*, Chudleigh, Devon. Leaf surface. ×1800. Air dried.

Fig. 30. *Racomitrium lanuginosum*, Cronkley Scar, Teesdale, Yorkshire. Part of leaf apex. ×830. Air dried.

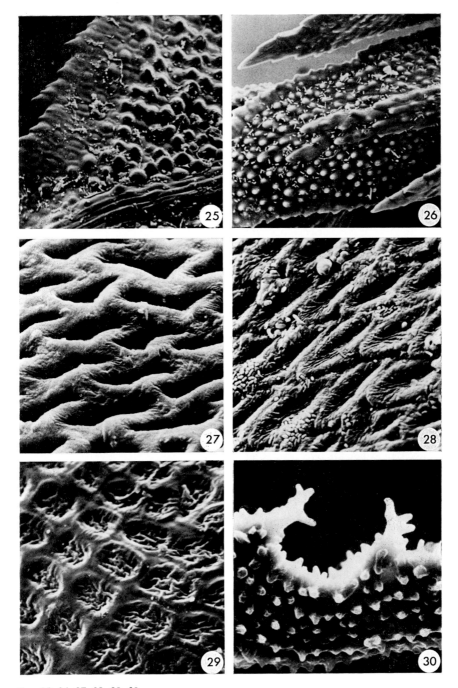

FIGS 25, 26, 27, 28, 29, 30

in a mist unit in the glasshouse at Exeter it was conspicuous that each leaf held a drop of water in its concave inner surface while the outer surface remained dry. This neatly combines the requirements of water supply and gas exchange, as surface tension will keep a convex surface free of surplus water even at high water potentials. This can be seen to be so in the many species that have leaves more or less of this type. At water potentials below *c.* -10^{-2} bar such leaves presumably come to rely upon internal water conduction, discussed in the next section. An interesting variant is *Thuidium*, mentioned already, which has strongly concave branch leaves with papillose outer surfaces, perhaps related to relatively inefficient internal conduction.

The overlapping leaves of *Pseudoscleropodium purum* should allow rapid water conduction (Bowen, 1933) as long as water is available from the environment at water potentials above about -3×10^{-2} bar, corresponding to a capillary rise of about 3 cm. Such conditions cannot persist for long after rain; have concave leaves a significant capacity for water storage under normal conditions of evaporation? Let us assume that each leaf measures $2 \cdot 5 \times 1 \cdot 5$ mm, giving an area of about 6 mm^2, and that it holds 500 μg of water. If the shoots effectively form a part of a continuous evaporating surface, and the leaf area index is 10, there will be about 8×10^4 μg of water to each square centimetre of evaporating surface. Under sheltered conditions in an open turf, with radiation income equivalent to about half noon summer sunlight, air temperature 20°C, relative humidity 70% and airspeed 20 cm s^{-1}, evaporation of about 10 μg cm^{-2} s^{-1} might be expected (Appendix 1). At that rate, the water would last for over 2 h. Real field conditions would certainly often be less severe than these assumptions, and water storage in the concave leaves could then be ecologically very important.

WATER CONDUCTION IN NON-PAPILLOSE LEAVES

Many bryophyte leaves lack any of the obvious external capillary conducting systems that we have been considering (Figs 27, 28, 29, 33, 35, 37, 39, etc.);

FIG. 31. *Racomitrium lanuginosum*, Cronkley Scar. Upper surface of leaf. ×1670. Air dried.

FIG. 32. *R. lanuginosum*, Crib Goch, Snowdon, Gwynedd. Cross-section of leaf. ×1670.

FIG. 33. *Neckera crispa*, Cowside Beck, Arncliffe, Yorks. Upper surface of leaf. ×1670.

FIG. 34. *N. crispa*, Cowside Beck. Slightly oblique section of leaf. ×4170.

FIG. 35. *Dicranum scoparium*, Drewsteignton, Devon. Upper surface of leaf. ×1670.

FIG. 36. *D. scoparium*, Drewsteignton. Cross-section of leaf. ×1670.

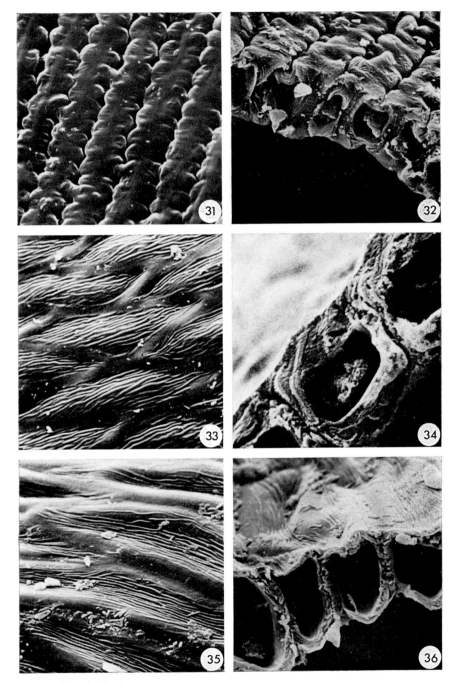

FIGS 31, 32, 33, 34, 35, 36

their needs for water movement must be met within the tissue of the leaf. There are two possible pathways to be considered. Water may move within the cell walls, or it may move within the cytoplasm, passing from cell to cell mainly across the cell membranes. In recent years it has become accepted that in higher plants much of the water movement in the root cortex and the mesophyll of leaves takes place within the "free space" of the cell walls and intercellular spaces. In general, the cell wall material is very much more permeable to water than the plasmalemma—the outer membrane of the cell itself (Nobel, 1974; Läuchli, 1976). Bryophytes generally lack intercellular spaces but they often have proportionately rather thick cell walls, especially in species of dry, exposed habitats where the need for water conduction is greatest.

   We can take as a starting point for a calculation on the adequacy of the cell wall as a route for water conduction within the leaf, the assumptions of Preston (1974, pp. 381–382), Briggs (1967, p. 91), or the hydraulic conductivity of the cell wall of *Nitella flexilis* (L.) Ag. measured by Tyree (1968). Let us accept Briggs's assumption that 50% of the cross-sectional area of a cellulose cell wall consists of channels of radius 5 nm, and assume a water potential gradient of 20 bar mm^{-1}, the same leaf dimensions as for the papillose leaf considered above (flow along a 1 mm $\times$ 1 mm length of leaf providing for evaporation from an area of 1 mm^2), and that the cell wall occupies 40% of the cross sectional area of a leaf 10 μm thick. The calculated flow would balance evaporation of about 0·6 μg cm^{-2} s^{-1}. Tyree's value for hydraulic conductivity gives a rate of 0·1 μg cm^{-2} s^{-1}, and Preston's basis of calculation gives a figure an order of magnitude less than that derived from Briggs's assumptions.

   At first sight these results are discouraging, especially in view of the high water potential gradient we have assumed. However, Preston indicates that his assumptions may be conservative even for vascular plant cell walls, and the hydraulic conductivity of bryophyte cell walls might be much higher than that of the walls of higher plants or of *Nitella* as the selection pressures operating in the various cases must be very different. Several of the scanning electron micrographs reproduced here show longitudinal surface striations *c.* 400 nm across which suggest similarly oriented structural elements within the cell walls (Figs 33, 35). In a cross-section of the leaf of *Neckera crispa* these appear to be the expression at the surface of a lamellate structure within the cell wall (Fig. 34). In *Dicranum scoparium* they appear to be coarse longitudinally-oriented fibrils forming the outer layer of a wall within which there are considerable spaces available for water conduction, and the walls between adjacent cells have a strikingly spongy texture (Fig. 36). To elucidate the structure of these walls adequately would require better techniques. However, the present micrographs do show clearly that these incrassate bryophyte cell walls are of coarser structure than normal vascular plant cell walls, and they suggest that, in some cases at least, spaces of the order of 100 nm diameter may make up a substantial proportion of the wall volume. The structure appears different from the "brooming" effect sometimes seen in scanning electron micrographs of cut surfaces of wood samples (Exley *et al.*, 1974). Many bryophytes undoubtedly have leaf cell walls of quite "normal" structure, as in *Tortula ruralis* (Tucker *et al.* (1975) and *Pleurozium schreberi* (Noailles, 1974). Coarse-textured walls could multiply by 50 or 100 the estimate of water movement calculated in the preceding paragraph. Further, the proportion of the total cross-section area of the leaf occupied by cell wall is often greater than the 40% assumed in the calculation. In the section of *Hedwigia ciliata* (Fig. 21) it is around 50%, in *Racomitrium lanuginosum* (Fig. 32) nearly 60% and in *Neckera crispa* about 70% of the cross-sectional area.

The remaining factor to be considered is the leaf area index (LAI) of the moss patch. Here we lack information, which could be quite readily obtained. I have found approximate values of 6 for *Tortula intermedia*, 18 for *Mnium hornum* and *c.* 20–25 for *Pseudoscleropodium purum*. If these latter figures seem high by flowering plant standards it has to be remembered that a unistratose bryophyte leaf is not equivalent to a higher plant leaf. Bryophyte LAIs could with some justification be considered alongside ratios of mesophyll area to leaf area for higher plants, which range from about 15 to 40 for mesophytes (Nobel, 1974, p. 327). For our purpose we can probably safely assume a representative LAI of 10.

Taking these considerations together, we could expect cell wall conduction at quite moderate water-potential gradients to balance rates of evaporation of at least $10-20 \mu g \, cm^{-2} \, s^{-1}$ from the surface of a moss carpet or cushion. This is adequate but not generous provision. The closeness of a physical limit is perhaps reflected in a tendency for such leaves to be narrowly triangular (reducing the evaporating area relative to the cross-section available for conduction; the Neckeraceae are a conspicuous exception to this) and for the development of supplementary capillary structures such as plicae. The furrows over the individual cells of *Racomitrium lanuginosum* (Figs 31, 32) could function to short-circuit neighbouring lengths of the higher-resistance cell walls, and, as Tallis (1959) has pointed out, in this species water is conducted from one part of the shoot system to another by the tangled papillose hair points (Fig. 30).

For the endohydric mosses, these expedients are not available. Conduction must be internal, but to the factors affecting the balance between conduction and evaporation we can add the resistance to water loss of the waxy cuticle. The almost isodiametric leaf cells of Mniaceae do not suggest strongly oriented water movement. Zacherl (1956), using fluorescent dyes as tracers, found rapid water conduction in the leaf midribs of *Plagiomnium undulatum, P. affine, P. cuspidatum* and *Rhizomnium punctatum*, with slow lateral movement from the midrib out into the lamina. By analogy with the observations of Mender (1938) on *Bryum capillare*, Buch (1945, p. 14) suggested that the elongated cells of the leaf margin in *Mnium (sensu lato)* might function as a water conducting structure. Zacherl found no evidence for this. My own observations on *Mnium hornum, Plagiomnium undulatum* and *Rhizomnium punctatum* support Zacherl's conclusions. However, there are indications that under some conditions water can move rapidly in the thickened leaf margin of *Mnium hornum*, which is clearly also an active site of evaporation; its function requires further study.

The possibility of water movement via the cells themselves remains to be considered. If we take the permeability of a cell wall 1 μm thick to be 100 times that of a cell membrane (Läuchli, 1976), then in a tissue composed of isodiametric cells 10 μm across with walls 1 μm thick, the resistance to water movement through the cells will be about 10 times as great as that to water movement between the cells via the cell walls. For cells 20 μm across with cell walls the same thickness, the resistance of the two pathways will not be greatly different, while for substantially larger cells the intracellular pathway will predominate. There are of course large uncertainties in these figures; the calculations of Raven (1977, p. 175) on vascular plant parenchyma give the apoplastic pathway a considerably greater advantage. Whatever the exact figures, it is clear that the resistance of the intracellular (symplastic) pathway could be reduced by reducing the number of membranes to be crossed and increasing the area of overlap between adjacent cells. The elongated areolation of many mosses appears to be an adaptation which does precisely this (e.g. Figs 26, 27, 28, 31–36, 39). Dicranaceae and Hypnaceae commonly combine elongated cells with thick cell walls, as if to maximize the potential of both pathways. Calculation suggests that for a hypnaceous moss with incrassate cells 50–100 μm long, the capacities of the two may well be similar. There are of course other factors involved in the shapes of cells, notably transport of nutrients and metabolites. Thus, passive accumulation of ions in the apical parts of the leaves by the transpiration stream through the free space must at least in part be balanced by downward translocation within the cells. However, these factors probably do not in themselves put such a premium on cell elongation as does the need for rapid water movement. It is an interesting reflection that optical microscopy emphasizes the cell-to-cell connections ("pores") probably chiefly concerned with the former, while scanning electron microscopy emphasizes the continuous longitudinally orientated structures most important to the latter.

SOME FEATURES OF LEAFY LIVERWORTS

The leafy liverworts fall into Buch's endohydric and mixohydric categories. Duckett and Soni (1972a, b) and Duckett (unpublished) examined representatives of a wide range of genera, and found no sign of superficial wax on the leaves of any of them (though observation suggests that the surfaces of at least the young leaves of many species are slightly water-repellent). Much of what has been said in the preceding sections about ectohydric mosses applies equally to liverworts, with some important exceptions and differences of emphasis.

Leafy liverworts are usually dorsiventral, and their leaf cells are (with few minor exceptions) always approximately isodiametric. Probably both features reflect a rather fundamental characteristic of the Hepaticae. The axis of a typical acrocarpous moss is oriented at right angles to the substratum, and parallel to the often steep humidity gradient above the surface. Correlated with this habit, the shoots and leaves are typically highly organized for acropetal water movement. Hepaticae typically grow more or less parallel with the substratum, and at right angles to the humidity gradient. Their provision for water movement is correspondingly diffuse. This might be seen as the consequence of a basic limitation of the adaptive equipment of the Hepaticae, or as the expression of a successful pattern of adaptation. The abundance of liverworts, the near-radial organization of many primitive hepatic genera, and the evident success of the rather comparably organized pleurocarpous mosses, argue for the latter view.

Many Jungermanniales, especially in those groups that Buch (1945, 1947) regarded as mainly mixohydric, show little external specialization for movement or localization of water (Fig. 40). They are commonly in more or less intimate contact with a substrate which remains equably moist for long periods at a time, and modest internal conduction either through the cell walls or from cell to cell must suffice to meet normal environmental demands. However, an apparent lack of specialization for directional water movement may hide a considerable capacity for more diffuse movement. Thus, in *Bazzania trilobata*, the cell walls appear to be of very open structure and well adapted to water movement, and this is especially true of the trigones (Figs 41, 42).

Many leafy liverworts have evolved capillary adaptations comparable to those already discussed in mosses. Thus *Nardia scalaris, Odontoschisma sphagni*

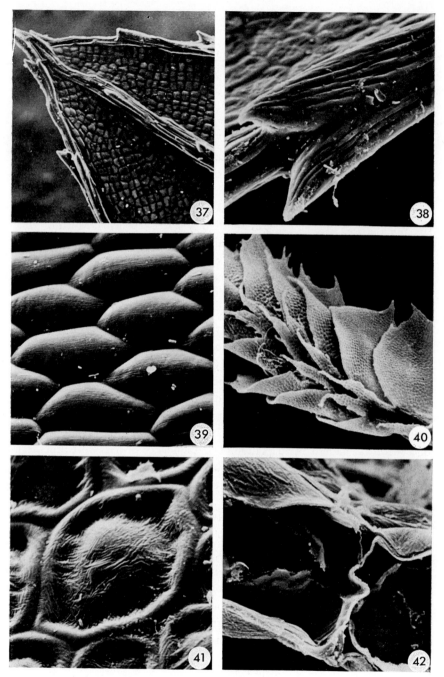

Figs 37, 38, 39, 40, 41, 42

or *Pleurozia purpurea* may be compared with the pattern of organization seen in *Pseudoscleropodium purum*; *Nowellia curvifolia* (Fig. 43) is an elegant smaller-scale example of essentially the same kind. The conduplicate leaves of Scapaniaceae and the lobules of Radulaceae, Porellaceae and Lejeuneaceae may be seen as parallels to the overlapping sheathing leaf bases of many mosses. Sometimes the overlap is rather slight. In *Radula complanata* (Fig. 44) it is easy to visualize capillary water movement under moist conditions bridging the short gaps between the lobules of successive leaves. With falling water potential, the rapid capillary channel of external conduction would soon break, leaving the individual leaves, each holding its own small volume of water, connected only by the slower conducting pathway of the stem tissue.

In the larger-leaved Lejeuneaceae (Figs 45, 46) the function of the lobule may often be mainly to distribute capillary water from around the underside of the stem to the leaf lamina, and in some it may in effect be a merely vestigial structure. In the small-leaved species the lobule often becomes a very prominent part of the leaf, and must play a proportionately important role in the water economy of the leaf as a whole. *Cololejeunea rossettiana* (Fig. 47) suggests a certain analogy with a branch of *Thuidium*—the lobules (like the concave leaves of *Thuidium*) conduct water at high water potentials, and store water, while the papillose outer surfaces of the leaves serve to regulate the distribution of water over different parts of the leaf surface. In many cases, the architecture of the shoots of these small Lejeuneaceae seems to rule out a primary conducting function for the lobules; the idea that they are essentially water-storing structures dates at least from the time of Goebel (1905). To explore how realistic this idea may be *Lejeunea ulicina* (Fig. 48) will serve as an example. The volume enclosed by the lobule is about $2.5 \times 10^5 \ \mu m^3$, or about $0.25 \ \mu g$ of water. The exposed area of the antical lobe is about $1.5 \times 10^{-4} \ cm^2$. The water available (c. $1700 \ \mu g \ cm^{-2}$) is therefore about 30 minutes' supply at an evaporation rate of $1 \ \mu g \ cm^{-2} \ s^{-1}$. This is hardly impressive, but could provide an ecologically significant buffer in fluctuating conditions in the sort of habitats in which these plants occur. However, it would be stretching credulity to

FIG. 37. *Mnium thomsonii*, Cowside Beck, Arncliffe, Yorks. Lower surface of leaf apex. ×170.

FIG. 38. *M. thomsonii*, Cowside Beck. Tooth on leaf margin. ×860.

FIG. 39. *Bryum pseudotriquetrum*, Cowside Beck. Upper surface of leaf. ×830.

FIG. 40. *Lophocolea fragrans*, Black Head, Torquay, Devon. Underside of leafy shoot. ×77.

FIG. 41. *Bazzania trilobata*, Holne, Dartmoor, Devon. Leaf cells. ×1670.

FIG. 42. *B. trilobata*, Holne. Vertical section of leaf. ×1700.

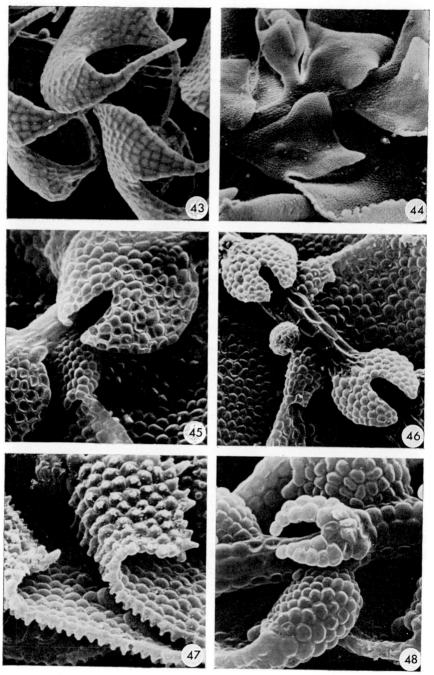

FIGS 43, 44, 45, 46, 47, 48

postulate a similar role for the small "water sacs" formed by the postical lobes in Frullaniaceae.

SHOOT AND LEAF GEOMETRY AND THE

ATMOSPHERIC ENVIRONMENT: CONCLUDING REMARKS

In interactions with its atmospheric environment, the habit and growth form of a bryophyte (and sometimes its substrate as well) are to all intents and purposes a part of its structure (Meusel, 1935; Gimingham and Robertson, 1950; Gimingham and Birse, 1957; Mägdefrau, 1969; Gimingham and Smith, 1971). Most bryophytes grow as cushions, turfs or mats; those that grow as individual shoots generally weave among the shoots of other species or grow closely appressed to a moist substratum. Growth form affects the capacity of a bryophyte shoot system for water storage, the distance which water must be conducted to the evaporating surface and the character of the capillary channels available for conduction, and the site and rate of water loss to the air.

The boundary layer of a leaf 5 cm across in a wind of 1 m s^{-1} is about 0·9 mm thick. In "still air" at a windspeed of 10 cm s^{-1} the thickness of the boundary layer is about 2·8 mm. (Monteith (1973, p. 80) gives the theoretically derived formula for the displacement boundary layer of a flat plate, $\delta = 1 \cdot 72 (lv/V)^{\frac{1}{2}}$, where l is distance (chord), V is velocity and v is the kinematic viscosity of air (c. 0·15 cm² s^{-1}). The empirically corrected relation for the thickness of the "unstirred layer" around a real leaf, $\delta = 0 \cdot 4 \sqrt{(l/V)}$, quoted by Nobel (1974, p. 307) gives the somewhat lower figures used here.) These dimensions are comparable with the size of moss leaves and the spaces between them. The consequence is that under many conditions a moss cushion behaves as a rough single "leaf", and its rate of water loss at low windspeeds can be estimated with fair accuracy from the tables of Gates and Papian (1971, see Appendix 1). The relation between windspeed and the thickness of the boundary layer provides a semi-quantitative basis for the common observation that moss cushions are more compact and have smaller leaves in exposed places, and a starting point for more precise experimental study.

Fig. 43. *Nowellia curvifolia*, Holne. Upperside of leafy shoot. ×150.

Fig. 44. *Radula complanata*, Chudleigh, Devon. Underside of leafy shoot. ×66.

Fig. 45. *Lejeunea cavifolia*, Chudleigh. Underside of leafy shoot. ×180.

Fig. 46. *L. lamacerina* var. *azorica*, Black Head, Torquay, Devon. Underside of leafy shoot. ×145.

Fig. 47. *Cololejeunea rossettiana*, Chudleigh, Devon. Leafy shoot. ×365.

Fig. 48. *Lejeunea ulicina*, Holne, Dartmoor, Devon. Underside of leafy shoot. ×365.

In earlier sections, rates of water movement along leaves were calculated on the assumption of uniform evaporation over the upper part of the leaf. However, rates of evaporation will rarely be uniform, especially if the leaf is oriented at right angles to the boundary layer of the moss cushion as a whole, when the apical part will be exposed to a steep water vapour concentration gradient. Under these conditions the leaf tips could be vulnerable to excessive drying out. Many species have evidently surmounted any adaptive problems that straight, erect leaves may pose. However, there does seem to be a potential selection pressure here to which the common occurrence of recurved and falcato-secund leaves among mosses might be related. Hair points also seem more likely to be explicable in terms of the whole moss cushion than of the individual leaf. Two possible functions for hair points suggest themselves. First, they will reduce the absorption of energy by the moss, by reflecting some of the incoming solar radiation, and this may help to keep down temperature (and evaporation). As Nobel (1974, p. 355) indicates in relation to flowering-plant leaves, even a small reduction in energy absorption may be significant. Second, it is found that water-soluble materials such as acid fuchsin or lead EDTA (Crowdy and Tanton, 1970) accumulate in the upper part of the lamina in *Tortula intermedia*, not the hair points showing that little water is evaporated by the hair points. They should therefore have the effect of separating the sites of momentum and water-vapour transfer, so increasing the diffusion path length for water vapour and reducing water loss.

The diversity of patterns of adaptation to water movement is a notable feature of the bryophytes, and it is closely linked to systematics. Some of it clearly reflects adaptation to different environmental regimes, while some apparently represents different means of arriving at the same adaptive ends, and in that sense is "non-adaptive". This need be no cause for surprise. Natural selection is analogous to the well known technological problem of computer optimization, which has been likened to seeking the highest summit in a cloud-covered range of mountains. It is easy to find *a* summit, and high summits, sometimes the same, sometimes different, may be reached from many different starting-points. Once on the path to one peak, it may be difficult or impossible to change to another. And in real life—whether of technology, evolution or mountain walking—Snowdon, Glyder Fawr and Carnedd Llewelyn may be near enough the same height!

A good deal of this paper has been conjectural, but most of the conjecture is open to (often simple) experimental test. In a short space I have been able to do no more than sketch a few outlines and suggest some theoretical possibilities and constraints in a large and fascinating field. It is to the advantage of the systematist to understand the function of the structures from which he draws his characters, as it is to the ecologist or physiologist to understand the structural and functional métiers and limitations of the systematic groups within which his material lies.

ACKNOWLEDGEMENTS

I am indebted to Dr S. R. Edwards for drawing my attention to the porose leaf bases of Calymperaceae and other mosses, and to Mr Peter Garlick for advice and technical assistance with the scanning electron microscopy.

REFERENCES

BOWEN, E. J. (1931). Water conduction in *Polytrichum commune*. *Ann. Bot.* **45**, 175–200.

BOWEN, E. J. (1933). The mechanism of water conduction in the Musci considered in relation to habitat. *Ann. Bot.* **47**, 401–422; 635–661; 889–912.

BRIGGS, G. E. (1967). "Movement of Water in Plants." Blackwell, Oxford.

BUCH, H. (1945, 1947). Über die Wasser- und Mineralstoffversorgung der Moose. *Commentat. biol.* **9**(16), 1–44; **9**(20), 1–61.

BUCH, H., EVANS, A. W. and VERDOORN, F. (1938). A preliminary check list of the Hepaticae of Europe and America (North of Mexico). *Annls bryol.* **10**, 3–8.

CASAS DE PUIG, C. and MOLINAS, M. L. (1974). Etude au microscope électronique à balayage de la surface des feuilles de *Tortula ruralis* (Hedw.) Gaertn. var. *hirsuta* (Vent.) Par. *Rev. bryol. lichén.* **40**, 267–270.

CASTALDO, R. and DI MARTINO, V. (1970). Le leucocisti di *Sphagnum recurvum* P. Beauv. studiate comparativamente al microscopio elettronico a scansione ed al microscopio elettronico a transmissione. *Delpinoa* N.S. **10–11**, 63–72.

CROWDY, S. H. and TANTON, T. W. (1970). Water pathways in higher plants. I. Free space in wheat leaves. *J. exp. Bot.* **21**, 102–111.

CROWE, J. H. (1971). Anhydrobiosis: an unsolved problem. *Am. Nat.* **105**, 563–573.

DILKS, T. J. K. and PROCTOR, M. C. F. (1979). Photosynthesis, respiration and water content in bryophytes. *New Phytol.* **82**, 97–114.

DIXON, H. N. (1932). Classification of mosses. *In* "Manual of Bryology" (F. Verdoorn, ed.), pp. 397–412. Martinus Nijhoff, The Hague.

DUCKETT, J. G. and SONI, S. L. (1972a). Scanning electron microscope studies on the leaves of Hepaticae. I. Ptilidiaceae, Lepidoziaceae, Calypogeiaceae, Jungermanniaceae, and Marsupellaceae. *Bryologist* **75**, 536–549.

DUCKETT, J. G. and SONI, S. L. (1972b). A scanning electron microscope study of leaf surfaces of some species of *Scapania* Dum. (Hepaticae). *J. Bryol.* **7**, 75–79.

EDWARDS, S. R. (1976). "A Taxonomic Revision of Two Families of Tropical African Mosses." Ph.D. Thesis, Univ. Wales.

EXLEY, R. R., BUTTERFIELD, B. G. and MEYLAN, B. A. (1974). Preparation of wood specimens for the scanning electron microscope. *J. Microsc. London* **101**, 21–30.

GATES, D. M. (1968). Transpiration and leaf temperature. *A. Rev. Pl. Physiol.* **19**, 211–238.

GATES, D. M. and PAPIAN, L. E. (1971). "Atlas of Energy Budgets of Plant Leaves." Academic Press, London and New York.

GIMINGHAM, C. H. and BIRSE, E. M. (1957). Ecological studies on growth-form in bryophytes. I. Correlations between growth-form and habitat. *J. Ecol.* **45**, 533–545.

GIMINGHAM, C. H. and ROBERTSON, E. T. (1950). Preliminary investigations on the structure of bryophytic communities. *Trans. Br. bryol. Soc.* **1**, 330–344.

GIMINGHAM, C. H. and SMITH, R. I. L. (1971). Growth form and water relations of mosses in the maritime Antarctic. *Bull. Br. Antarct. Surv.* **25**, 1–21.

GREEN, T. G. A. and CLAYTON-GREENE, K. A. (1977). The growth of *Dawsonia superba* Grev. *Bull. Br. bryol. Soc.* **29**, 5–6.

GOEBEL, K. (1905). "Organography of Plants. Part II. Special Organography" (trans. Isaac Bayley Balfour). Clarendon Press, Oxford.

HABERLANDT, G. (1886). Beiträge zur Anatomie und Physiologie der Laubmoose. *Jb. wiss. Bot.* **17**, 359–498.

HÉBANT, C. (1977). The conducting tissues of bryophytes. *Bryophyt. Biblthca* **10**, 1–157.

LÄUCHLI, A. (1976). Apoplasmic transport in tissues. In "Encyclopaedia of Plant Physiology. N.S.2. Transport in Plants. IIB Tissues and organs" (U. Lüttge and M. G. Pitman, eds), pp. 3–34. Springer, Berlin.

LEWINSKY, J. (1977). The genus *Orthotrichum*. Morphological studies and evolutionary remarks. *J. Hattori bot. Lab.* **43**, 31–61.

MAGILL, R. E., SEABURY, F. and MUELLER, D. M. J. (1974). Evaluation of the critical point drying technique and its application in systematic studies of mosses. *Bryologist* **77**, 628–632.

MÄGDEFRAU, K. (1935). Untersuchungen über die Wasserversorgung des Gametophyten und Sporophyten der Laubmoose. *Z. Bot.* **29**, 337–375.

MÄGDEFRAU, K. (1969). Die Lebensformen der Laubmoose. *Vegetatio* **16**, 285–297.

MENDER, G. (1938). Protoplasmatische Anatomie des Laubmooses *Bryum capillare*. I. *Protoplasma* **34**, 373–400.

MEUSEL, H. (1935). Wuchsformen und Wuchstypen der europäischen Laubmoose. *Nova Acta Leopoldina* N.F. **3**, 123–277.

MONTEITH, J. L. (1973). "Principles of Environmental Physics." Arnold, London.

MOZINGO, H. N., KLEIN, P., ZEEVI, Y. and LEWIS, E. R. (1969). Scanning electron microscope studies of *Sphagnum imbricatum*. *Bryologist* **72**, 484–488.

NOAILLES, M.-C. (1974). Comparaison de l'ultrastructure du parenchyme des tiges et feuilles d'une mousse normalement hydratée et en cours de dessiccation (*Pleurozium schreberi* (Willd.) Mitt.). *C. r. hebd. Séanc. Acad. Sci., Paris*, Sér. D, **278**, 2259–2262.

NOBEL, P. S. (1974). "Introduction to Biophysical Plant Physiology." Freeman, San Francisco.

PRESTON, R. D. (1974). "The Physical Biology of Plant Cell Walls." Chapman and Hall, London.

RAVEN, J. A. (1977). The evolution of vascular land plants in relation to supracellular transport processes. *Adv. bot. Res.* **5**, 153–219.

RICHARDS, P. W. and EDWARDS, S. R. (1972). Notes on African mosses, V. *J. Bryol.* **7**, 47–60.

RIFOT, M. and BARRIÈRE, G. (1974). La conduction dans le thalle de l'Hépatique *Cono-*

cephalum conicum (L.) Dum. I. Etude du transit de l'eau à l'aide d'une solution d'acétate de sodium ¹⁴C. *Revue bryol. lichén.* **40**, 45–52.

ROBINSON, H. (1971). Scanning electron microscope studies on moss leaves and peristomes. *Bryologist* **74**, 473–483.

SCHÖNHERR, J. and ZIEGLER, H. (1975). Hydrophobic cuticular ledges prevent water entering the air pores of liverwort thalli. *Planta* **124**, 51–60.

TALLIS, J. H. (1959). Studies in the biology and ecology of *Rhacomitrium lanuginosum* Brid. II. Growth, reproduction and physiology. *J. Ecol.* **47**, 325–350.

TUCKER, E. B., COSTERTON, J. W. and BEWLEY, J. D. (1975). The ultrastructure of the moss *Tortula ruralis* on recovery from desiccation. *Can. J. Bot.* **53**, 94–101.

TYREE, M. T. (1968). Determination of transport constants of isolated *Nitella* cell walls. *Can. J. Bot.* **46**, 317–327.

WATSON, E. V. (1971). "The Structure and Life of Bryophytes" Third Edition. Hutchinson, London.

ZACHERL, H. (1956). Physiologische und ökologische Untersuchungen über die innere Wasserleitung bei Laubmoosen. *Z. Bot.* **44**, 409–436.

APPENDIX 1

Evaporation Rates

It is easy to measure rates of evaporation from bryophytes under a range of "ordinary" conditions by simple weighing over a suitable time interval. However, this requires a reasonably sensitive balance (1 mg or better) close to the experimental site, so there are difficulties in obtaining measurements that span the whole range of conditions that might be of interest.

For the calculations in this paper I have taken some representative evaporation rates from the tables of Gates and Papian (1971). The justification for doing this is indicated on p. 503; some evidence of the extent to which it is justified in practice is given in Table I. The data required are wind speed, air temperature, relative humidity, absorbed radiation, leaf dimensions and stomatal resistance. The first three of these are easily measured by simple standard methods. Absorbed radiation is less easy to deal with, but typical values can be estimated adequately for our purposes from the information given by Gates and Papian (1971, p. 8 ff.) and Monteith (1973, p. 75), and limited field measurements. "Leaf" dimensions can be assigned easily for isolated moss cushions; a certain nicety of judgment is required for less clear-cut cases. "Stomatal resistance" has been taken as zero. In Table I, a–d are actual cases, while e–g represent three hypothetical situations. Example e gives some indication of the maximum rates that might be expected for e.g., small cushions of *Andreaea* spp. on exposed rocks. If the surrounding rocks were dry, the temperature would be higher than for an isolated leaf owing to conduction and advection of heat from the surroundings, so increasing evaporation. If they were wet, the effective "leaf" size would be much larger than 1 × 1 cm, with the opposite effect. The quoted figure may well be a fair compromise. Example f is representative of a range of common situations in overcast weather or under moderate shade. Higher humidity or lower radiation income will tend to reduce evaporation, and vice versa; the effect of "leaf size" will depend on the other parameters of the situation. Example g represents the sort of conditions that might be found with varying

TABLE I. Measured and estimated evaporation rates from bryophytes; for further explanation see Appendix 1

	(a)	(b)	(c)	(d)	(e)	(f)	(g)
Radiation absorbed (W m⁻²)	[600]	[1000]	[1000]	[1000]	1000	600	400
Short-wave ($\lambda < 3\,\mu$m) irradiance (W m⁻²)	230	900	900	900	[900]	[250]	[50]
Airspeed (cm s⁻¹)	20–50	50–100	50–100	50–100	100	100	10
"Leaf size" (cm) (* diameter)	5*	5*	5*	5*	1×1	10×10	10×10
Air temperature (°C)	16	20	20	20	20	20	10
Relative humidity (%)	77	40	40	40	50	50	95
Leaf resistance (s cm⁻¹)	~0	~0	~0·3	0	0	0	0
Evaporation (measured) (μg cm⁻² s⁻¹)	9·4	24·2	19·3	31·5	—	—	—
Evaporation (estimated) (μg cm⁻² s⁻¹)	8·5	23·4	23·4	23·4	45	12	1

(a) and (b), artificial cushion of *Tortula intermedia* in 5 cm petri dish; (c) and (d) turf of *Mnium hornum*, with approximately 1 cm projecting above top of plastic beaker 5 cm diameter, (c) fully turgid but untreated, (d) with leaves completely wetted. Columns (e), (f) and (g) represent hypothetical situations discussed in the text. Figures in brackets are estimated values given for comparison or taken as a basis for obtaining comparative estimates of evaporation rates.

degrees of constancy in shady, humid woodland, or in sheltered, shady ravines near the west coast of Britain and Ireland.

Scanning Electron Microscopy: Methods

Except where noted otherwise in the captions to the plates, fresh field material was fixed in formalin-acetic-alcohol, and transferred successively to 70% ethanol (at which point sections, if needed, were cut by hand, using a fresh single-edged safety-razor blade), 70% acetone and pure acetone. Dehydration was completed by placing the open vials containing the material, together with a dish of anhydrous calcium chloride, in a vacuum desiccator over pure acetone with anhydrous calcium sulphate for 12 h. The material was critical-point dried by transfer from acetone to liquid CO_2 in a Polaron E3000 critical-point drying apparatus, attached to specimen stubs with double-sided "Sellotape" and coated with about 50 nm of gold in a Polaron E5000 SEM coating unit.

Two batches of material were put through a standard glutaraldehyde-osmium tetroxide fixation sequence, and ethyl acetate used as a transfer fluid for critical-point drying. For the cell-wall detail examined here, this more complicated procedure appeared to give no advantage over that outlined above. In either case, it was striking that delicate plants and tissues retained their *in vivo* appearance better than those whose form is determined largely by the hygroscopic changes of a thick cellulose cell wall.

The micrographs were taken on a Cambridge "Stereoscan 600" scanning electron microscope, at an accelerating voltage of 15 kV.

21 | Climatic Adaptation of Bryophytes in Relation to Systematics

R. E. LONGTON

Department of Botany, University of Reading, Whiteknights, Reading RG6 2AS

Abstract: This is a review of climatic adaptation in bryophytes with particular emphasis on studies of relevance to systematics. Phenotypic plasticity and acclimatization have been shown to play a major role in the adaptation of certain moss and hepatic species to a range of climatic conditions. Despite a paucity of concerted investigation, however, genecological differentiation with respect to morphological and physiological characteristics has also been reported in several species, and phenotypic and geneotypic responses have been shown to interact in some cases. The available results are discussed in relation to certain features of the reproductive biology of bryophytes which theoretically could influence evolutionary processes within the group.

INTRODUCTION

It is self-evident that climate varies between different parts of the world, that microclimatic conditions vary between habitats at a given site, and that living organisms show adaptation to differences in climate and other features of their environment. Such adaptation may be recognized as habitat correlated variation in morphological or physiological characteristics between populations of a species, or between species or higher groups, which confers an advantage on the organisms concerned in terms of survival in their particular habitats. Studies on flowering plants have shown that adaptive variation may be either genetically based or the result of phenotypic plasticity, and that genetic variation and phenotypic plasticity may interact in fitting a population to local conditions.

Systematics Association Special Volume No. 14, "Bryophyte Systematics", edited by G. C. S. Clarke and J. G. Duckett, 1979, pp. 511–531, Academic Press, London and New York.

Adaptation to climate and other environmental factors is of interest to the systematist as it occurs in response to natural selection, and may thus provide a major impetus for the gradual evolutionary differentiation of populations leading to speciation and possibly to the origin of higher groups. Thus in flowering plants the occurrence of ecotypes within species is viewed as representing, potentially at least, a stage in the divergence of one species into two or more.

A combination of studies on such topics as climatic adaptation, reproductive biology and genetics has provided a reasonable insight into the evolutionary processes operating in flowering plants, at least at the species level, but comparable work on bryophytes is considerably less advanced. The present paper reviews those aspects of climatic relationships in bryophytes which have greatest relevance to systematics and, in particular, on mechanisms which permit certain species to survive in a range of climatic environments. The results will then be discussed in relation to features of the bryophyte life cycle which may conceivably influence the development of adaptive variation patterns, and ultimately the processes of speciation, within the group.

SIGNIFICANCE OF STUDIES ON MICROCLIMATE

Bryophytes generally occur at the interface between air and substrate, where climatic conditions may differ markedly from those recorded during routine meteorological observations. Variation in microclimatic conditions between different parts of a bryophyte colony at any given time is commonly less marked than that experienced by different parts of taller-growing plants, but substantial differences in, for example, temperature and saturation deficit may occur between the surface and the interior of the colonies. The collection and interpretation of microclimatic data thus forms a significant component of many contemporary studies on such topics as bryophyte phenology (Busby et al. (1978); Johnsen, 1969), the control of local distribution (Billings and Anderson, 1966; Sholl and Ives, 1973; Zehr, 1977), and the environmental relationships of metabolic processes (Hicklenton and Oechel, 1976; Collins, 1977).

Of particular interest in the present context is the comparison of microclimatic regimes affecting closely related species, or different populations of individual, widely distributed taxa. Unfortunately, no comprehensive microclimatic comparison appears to be available for geographically separated populations of any bryophyte species. However, details of the annual temperature regimes in turfs of *Polytrichum alpestre* at Antarctic, sub-Arctic and boreal sites may be found in Longton (in press a), and a marked contrast between the temperature regimes experienced by different populations of the cosmopolitan moss *Bryum argenteum* was reported by Longton and MacIver (1977). Based on readings at 30 min intervals over 24 days in mid-summer, it was shown that the greatest frequency of readings in *B. argenteum* turfs at a continental Antarctic site on Ross Island fell between 2·5°C and —2·5°C both during the day and at night. The mean of the day-time readings was only 4°C, and the mean daily maxima were consistently below 17·5°C. However, a mean day temperature of 18°C, with daily maxima up to 55°C, was recorded for *B. argenteum* near the southern boundary of the boreal forest region Pinawa in west-central Canada.

Microclimatic conditions may also differ substantially between habitats within a given climatic region. Zehr (1977) emphasized the variation in environment that may occur within what is normally regarded as a single microhabitat, citing Clausen's (1952) observation that relative humidity was 85% above *Nardia scalaris* on a rotting log but only 55% above *Frullania tamarisci* growing on a small protuberance 5 cm away. Billings and Anderson (1966) investigated the habitats characteristic of endemic mosses and disjunct species of tropical affinity in the southern Appalachian mountains. They reported that such taxa occurred in gorges with high and seasonally reliable rainfall, and a moderate temperature range with lower maxima and higher minima than recorded at other sites in the area. It was suggested that the taxa concerned might have occupied these equable microenvironments throughout the Pleistocene.

In the Bryophyta, as in other groups of plants, the related species comprising many given genera show similar preferences with respect to climate and other environmental factors. In *Andreaea* for example, most species are characteristic of acidic rocks in polar-alpine regions. It is not uncommon, however, for closely related species, or infra-specific groups within a single species, to be found in contrasting habitats. Familiar examples from the British moss flora include species of *Polytrichum*, and the varieties of *Hypnum cupressiforme*, some of which occur at different positions along the vertical light intensity and humidity gradients occurring within deciduous woodland. Further examples were cited by Richards (1932), who also drew attention to the fact that the habitats occupied by a given species may vary between different parts of its geographical range. The origin of such differences in habitat preferences between related plant populations is of clear relevance to the systematist.

MORPHOLOGICAL ADAPTATION

The adaptive significance of morphological variation among bryophytes is imperfectly understood. The structure of *Sphagnum* shows many features whose effectiveness in water retention can hardly be doubted, and the compound lobing characteristic of leaves in many hepatic genera may have a similar function, though occurring in plants characteristic of a range of both wet and relatively dry habitats. In mosses it was traditionally believed that the presence of such morphological features as hyaline leaf tips, thickening of the cell walls and papillosity was correlated with xeric habitats, and that these characters were of value in controlling rates of water loss. Their significance in this role has been questioned by more recent authors (Patterson, 1964), although they may be important in water movement (Proctor, this volume, Chapter 20). Desiccation resistance of individual moss shoots is now believed to be based more on physiology than on morphology (Lee and Stewart, 1971), except perhaps in *Polytrichum* spp. and other partially endohydric species (Hébant, 1977).

There is ample evidence, however, that aggregation of moss shoots into colonies may influence the rate of water loss. Correlations between bryophyte growth form and habitat have been reported in many areas (Hamilton, 1953; Gimingham and Birse, 1957; Birse, 1958) with short, compact turfs, cushions and mats generally predominating in dry situations. Working on Antarctic mosses, Gimingham and Smith (1971) demonstrated that variation in rates of water uptake and loss by intact colonies of different growth forms under experimental conditions corresponded with field distribution in relation to water availability. These authors noted that although growth form is largely species-specific it may vary with habitat in certain cases, *Drepanocladus uncinatus*, for example, occurring as carpets in wet habitats and as mats on drier ground. Experimental investigation of the genetic basis of such variation would be of interest.

Bazzaz *et al.* (1970) have demonstrated rather subtle morphological differences between material of *Polytrichum juniperium* from a warm, relatively dry forest environment in Indiana and from a cooler, moist, alpine site on Mt Washington. Plants from the cooler environment were small and they also had a lower chlorophyll : dry weight ratio than those from Indiana. The Indiana plants showed such features as relatively rapid hygroscopic leaf movements, and leaves closely adpressed to the stem when dry, which could be regarded as adaptations to a dry environment. In the Mt Washington plants the leaf margins were sufficiently wide to form a vented "microgreenhouse" over the photosynthetic lamellae of hydrated leaves which, it was suggested, could influence lamella temperature under conditions of low ambient temperature and strong insolation. Unfortunately, the genetic basis of these differences, or of the physiological variation between the populations (page 521), is not clear.

Bryophytes are well known to exhibit pronounced phenotypic plasticity. Schuster (1966; this volume, Chapter 3) has stressed the importance of such "malleability" as an adaptive mechanism in hepatics. Plastic and inherent variation within species can frequently be distinguished only by experiment, and Smith (this volume, Chapter 9) has reviewed reports of collateral cultivation studies involving bryophytes. As might be expected the results have confirmed the occurrence of substantial, inherent variation in morphology within certain species while demonstrating that in others inter-population variation, or in some cases the differences between what had been regarded as pairs of related species, disappeared in the new growth formed when contrasting plants were cultivated in the same environment.

Experiments with clonal material derived from polar, temperate and tropical populations of *Bryum argenteum* have demonstrated the occurrence of well-marked inherent, morphological variation within the species, but it is not yet clear whether the differences are of adaptive value. Despite the contrast between field temperature regimes experienced by *B. argenteum* (page 513), all clones so far tested have shown a similar relationship between temperature and growth under experimental conditions, a regime of 22°C day/15°C night proving optimal. There is some evidence, however, that Antarctic clones are the most vigorous under a range of conditions (Longton and MacIver, 1977; Longton, unpublished data).

In contrast to flowering plants, relatively few examples of genetically determined, ecologically correlated variation, or genecological differentiation as understood by Heslop-Harrison (1964), have yet been confirmed in mosses and liverworts. On the basis of preliminary experimental study Richards (1959) suggested that *Tortula ruraliformis* (= *T. ruralis* subsp. *ruraliformis*) should perhaps be regarded as a sand-dune ecotype of *T. ruralis*. Richards also referred to unpublished work by Chamberlain which indicated that many of the characters distinguishing the varieties of *Hypnum cupressiforme* are genetically fixed so that these taxa may also be regarded as ecotypes.

More recently, detailed observations have been made on the growth and morphology of *Polytrichum alpestre* at a range of climatically contrasting sites within its bipolar range (Longton, 1970, 1974; in press a). *P. alpestre* is characteristic of *Sphagnum* bogs in temperate and boreal regions, but occurs in drier habitats in some Arctic and Antarctic regions. Field studies have demonstrated the occurrence of topoclinal variation in morphology as there is a progressive reduction in the length and dry weight of annual growth segments, in leaf length, and in the number of leaves produced per year, with increasing latitude in both northern and southern hemispheres. It was shown experimentally that these features are all subject to phenotypic plasticity, with larger plants developing under warmer conditions. However, marked differences in plant height and in leaf length between material from contrasting natural populations persisted in cultures where shoots were raised by vegetative propagation under uniform conditions. Thus plastic and inherent responses may reinforce each other to give the clinal variation recorded in the field. The decrease in plant size in polar colonies is accompanied by an increase in shoot density sufficient to outweigh the reduction in annual dry matter increase per shoot, as the highest rates of colony production have been recorded in the Antarctic. The polar colonies thus have a shorter, more compact growth form than those at lower latitudes, and this feature could well be of adaptive value in cold, relatively dry habitats.

PHENOLOGICAL ADAPTATION

Seasonal periodicity in bryophyte growth and reproductive development has been recognized since the early studies of Grimme (1903), Arnell (1905), Hagarup (1935), Lackner (1939) and Jendralski (1955), and more recent workers have sought to explain the phenological cycles in terms of experimentally determined relationships between plant development and environmental factors. Growth and reproductive processes in *Anthoceros* and in many hepatics appear to be influenced by interaction between temperature and day length (Wann, 1925; Voth and Hamner, 1940; Benson-Evans, 1964; Fredericq and De Greef, 1966; Ridgway, 1967). Photoperiodic effects on reproductive development have also been reported in some mosses (Hughes, 1962; Benson-Evans, 1964; Newton, 1972a), but in others gametangial initiation appears to be influenced by temperature independently of day length (Monroe, 1965; Nakosteen and Hughes, 1978). Vegetative growth of moss gametophytes appears generally to be less dependent on photoperiod than in the case of hepatics, with water availability and temperature important in its regulation (Johnsen, 1969; Busby *et al.*, 1978). Bryophyte growth is often considered to be largely opportunistic, but growth (Longton, in press b), and reproductive development (Newton, 1972a), may be controlled by interaction between environment and endogenous factors in some species.

Growth in field populations occurs primarily in summer in certain mosses (Longton and Greene, 1969; Busby *et al.*, 1978; Longton, in press b), and in winter in others, at least in temperate latitudes (Johnsen, 1969; Pitkin, 1975). Hagarup (1935) considered that many European mosses have two major growth periods, in spring and in autumn, and initiation of growth under snow cover has been reported in *Calliergon* cf. *austro-stramineum* in the Antarctic (Collins *et al.*, 1975). There is evidence that the seasonal pattern of growth in several species may vary between climatically diverse localities. For example, *Pleurozium schreberi* appears to grow throughout the summer in British populations (Longton and Greene, 1969) but principally in autumn in more continental sites in Finland (Kallio and Heinonen, 1975). Similarly, several *Sphagna* show maximum rates of stem elongation during summer in Britain, but in spring and autumn at drier, European sites (Clymo, 1970), and Pitkin (1975) noted that *Hypnum cupressiforme* grows principally in winter at several stations in southern Britain, but extends its growing season into summer in the more oceanic southwest. In these cases the variation in seasonal growth patterns appears to be related to water supply, and there is no evidence as yet that it represents genetically determined climatic adaptation. In *Lunularia cruciata*, however, Benson-Evans and Hughes (1955), working with British material, and Schwabe and associates, using plants from Israel (Schwabe and Nachmony-Bascombe, 1963; Schwabe and Valio, 1970), have reported differences in the relationships between photoperiod and growth which could be interpreted as adaptations to the contrasting climatic regimes of the natural habitats (Longton, 1974b). Further comparative experimental study of these populations could prove rewarding.

It is clear from the work of Greene (1960), van der Wijk (1960) and others that gametangial and sporophyte development show distinct seasonal cycles in field populations of many mosses. The cycles vary considerably between species, suggesting that different patterns of climatic control may be operating, and in certain cases, for example in *Bryum argenteum* (van der Wijk, 1960), there may be no clearly defined seasonal periodicity. In some species essentially similar reproductive cycles have been demonstrated in widely separated, climatically diverse localities, although variation in detail may occur in response to climatic differences between sites (Longton and Greene, 1969), and between successive years at a given station (Benson-Evans and Brough, 1966).

Of particular interest are bipolar mosses with ranges extending into both temperate and polar regions. These may show a period of winter dormancy, or extremely retarded growth and reproductive development, at north temperate sites at temperatures comparable with those during the summer period of greatest activity in Antarctic regions. Such a situation is strongly suggestive of inter-population climatic adaptation, of some measure of endogenous control over the seasonal cycle, or of a greater degree of photoperiodic control than has yet been demonstrated experimentally for most mosses.

Some insight is provided by Clarke and Greene's (1970, 1971) comprehensive study of *Pohlia nutans* and *P. cruda*. The former was shown to have essentially similar patterns of reproductive development at north temperate, Arctic and sub-Antarctic sites. At the polar sites, however, development was delayed by winter snow cover, but following the spring thaw gametangial maturation proceeded more rapidly than at the temperate sites. Sporophyte development was shown to be more rapid under Arctic than either sub-Arctic or temperate conditions. Comparable adaptation, involving particularly rapid reproductive development during summer at sites with prolonged winter snow cover, has recently been reported in *Polytrichum alpestre* (Longton, in press a).

Cultivation of *Pohlia nutans* in a range of constant environments for a period of 80 weeks provided evidence of inherent adaptation to local climate (Clarke and Greene, 1971). Thus, compared with temperate plants, material from the sub-Antarctic population had a lower temperature optimum for gametangial maturation and a higher degree of frost resistance. The rates of reproductive development in plants from a given population showed variation indicative of acclimatization effects during the course of the experiments, suggesting that the success of *P. nutans* in a range of field environments is due to both inherent and plastic responses. *Pohlia cruda* appeared to be less adaptable than *P. nutans* and, in contrast to the latter, it retained seasonal periodicity in gametangial development when grown under stable environmental conditions.

METABOLIC ADAPTATION

Understanding of plant adaptation to climate may be facilitated by information on the influence of environmental factors on the basic metabolic processes of photosynthesis and respiration. Several workers have investigated mosses from this viewpoint, particularly in respect of adaptation to polar conditions, and many of the results have been reviewed by Kallio and Kärenlampi (1975). As Dilks and Proctor (1975) have pointed out, however, extreme caution is necessary in the interpretation of these data, due to the variety of experimental procedures adopted by different workers, and the difficulty of extrapolating from short-term laboratory experiments to long-term performance in the field. The influence of different experimental techniques, and possibly of seasonal variation in plant activity on, for example, the temperature/net photosynthesis response curve is demonstrated dramatically in a comparison of data for *Racomitrium lanuginosum* from southern Britain: Kallio and Heinonen (1973) found a broad curve with an optimum temperature of 5°C and an upper compensation point of 25°C, Tallis (1959) reported a narrow curve with an optimum temperature at 15°C and Dilks and Proctor (1975) a broad curve with an optimum temperature of 35°C and an upper compensation point of 45°C. Moreover, due to acclimatization effects the occurrence of inherent variation between populations can only be confirmed by comparing material of different provenances grown in a common environment for a substantial period prior to experimentation.

Considering first the effects of temperature, the results discussed by Kallio and Kärenlampi (1975) suggest adaptive, inter-specific differences in photosynthetic responses, optimum temperatures of 15–20°C commonly being quoted for temperate mosses such as *Bryum sandbergii* (Rastorfer and Higinbotham, 1968), and of 5–10°C for some polar species. Most species considered were able to maintain positive net assimilation down to −5 to −10°C. Generally, higher minimum and optimum temperatures were reported in a range of mosses tested at unnaturally high CO_2 concentrations by Dilks and Proctor (1975), and in these data there was not complete correlation between temperature/photosynthesis response and the geographical distribution of the species concerned.

Inter-specific differences in the mechanism of temperature adaptation are demonstrated in Collins's (1977) studies on *Polytrichum alpestre* and *Drepanocladus uncinatus* on Signy Island in the maritime Antarctic. *P. alpestre* showed a relatively narrow temperature/photosynthesis response curve and evidence of a substantial capacity for acclimatization. The optimum temperature, the upper compensation point, and probably also the lower compensation point, were raised by warm pretreatment compared with pretreatment in a simulated Antarctic temperature régime. Collins suggested that these plastic responses would permit *P. alpestre* to survive in the temperature regimes experienced throughout its wide geographical range (Longton, in press a) without genetic adaptation to local conditions, but experimental comparison of plants from different populations remains desirable. In contrast, Antarctic material of *Drepanocladus uncinatus*, which also occurs widely in temperate and polar regions, showed a higher optimum temperature for net photosynthesis (*c.* 20°C), but the response curve was broader, with effects of acclimatization less marked, than in *P. alpestre* (Collins, 1977).

Temperature responses of photosynthesis in mosses may vary seasonally in relation to changing environmental conditions, and the significance of such

seasonal acclimatization has been studied in detail in sub-Arctic and Arctic populations of *Dicranum fuscescens* (Hicklenton and Oechel, 1976; Oechel, 1976). The
optimum temperature for net photosynthesis in both field populations increased
during the early part of the growing season, and in the sub-Arctic population
at Schefferville, Quebec, the optima showed close agreement with the
rising mean daily moss tissue temperatures recorded in the field. At Barrow,
Alaska, optimum temperatures for net assimilation were slightly above the
prevailing mean tissue temperatures throughout much of the summer, although
the data suggested that substantial rates of assimilation could be maintained
under field conditions. Rather surprisingly, the temperature optima in the
Barrow plants were slightly higher than those recorded at Schefferville at
comparable periods of the growing season.

   Hicklenton and Oechel (1976) have also compared photosynthetic responses
in plants from both low and high altitude populations of *D. fuscescens* at
Schefferville after 1·5 months in collateral cultivation under two temperature
regimes. Further acclimatization effects were demonstrated in these experiments, but certain inter-population variation was also recorded. For example,
as compared with the low altitude material, plants from the high altitude
population showed a slightly lower optimum temperature for photosynthesis
after cultivation under cool conditions, and a substantially higher maximum
assimilation rate after growth under warmer conditions. Hicklenton and
Oechel considered that these observations pointed towards genecological
differentiation within *D. fuscescens*, but that the results were not conclusive.
Differences in the responses of net photosynthesis to water and light have also
been demonstrated between four species of *Dicranum*, and between two
varieties of *D. fuscescens*, in Finland (Kellomäki *et al.*, 1978). In this case the
material had been allowed to acclimatize under comparable conditions before
the experiments, but only one population of each taxon was investigated.

   The physiological ecology of different populations of *Racomitrium lanuginosum* has been investigated by Kallio and Heinonen (1973), who reported
strikingly similar optimum temperatures for net photosynthesis in freshly
collected plants from a range of polar and temperate localities. Maximum
assimilation rates were generally higher in plants from the warmer natural
environments, although a Welsh population proved an exception, and dark
respiration rates at all temperatures were lower in plants from polar than from
temperate localities. Kallio and Heinonen (1973) originally suggested that
little genecological differentiation had occurred in *R. lanuginosum*, but these
authors later (1975) showed that the difference in maximum rates of photosynthesis between plants from northern and southern Finland persisted, though

to a reduced degree, two years after the latter had been transplanted to the northern site. In contrast, respiration rates were then similar in the transplants and in the northern material beside which they had been grown.

Kallio and Heinonen (1973) also showed that plants from a boreal forest population of *Pleurozium schreberi* had a higher rate of net photosyntheses at all temperatures tested, with a slightly higher optimum, than plants from an Arctic population. Similar results were obtained in a comparison of temperate forest and alpine populations of *Polytrichum juniperinum* by Bazzaz *et al.* (1970), who pointed out that the higher rates of photosynthesis in the forest plants would enable this population to sustain positive rates of net assimilation at higher temperatures than the alpine population. Further evidence of adaptation to local conditions was indicated, as the forest plants showed a greater response to elevated CO_2 concentration. In neither *Pleurozium schreberi* nor *Polytrichum juniperinum*, however, was any attempt made to distinguish between genetic and plastic responses.

Results reviewed by Longton (in press b) indicate that moss species may differ substantially in the degree of hydration required for maximum net assimilation. Light compensation and saturation levels are also known to differ between species, and the differences are often suggestive of adaptation to regional or local conditions as in Hosokawa and Odani's (1957) study of corticolous epiphytes along the vertical light intensity gradient in a beech forest. Nevertheless, few cases of intra-specific variation in these characteristics have been reported. Indeed, the photosynthesis/light intensity curves proved similar, at a range of temperatures, in different populations of *Pleurozium schreberi* and *Racomitrium lanuginosum* (Kallio and Heinonen, 1973). In *Polytrichum juniperinum*, however, plants from the forest population showed saturation at a lower light intensity than their alpine counterparts (Bazzaz *et al.*, 1970).

Kallio and Valanne (1975) have investigated the effects of continuous light on moss photosynthesis from the viewpoint of adaptation to polar climates. Plants from temperate and boreal forest populations of several mosses, including *Pleurozium schreberi* and *Racomitrium lanuginosum*, showed reduced rates of photosynthesis after cultivation in continuous light, and this was related to reduction in chlorophyll content, to slight changes in the chlorophyll a : b ratio, and to ultrastructural modification. In some species, e.g. *Pleurozium schreberi*, Arctic populations showed a similar response. Both ecotypic and inter-specific adaptation was indicated in other cases, however, as continuous light had little effect on chlorophyll content, or on the rate of photosynthesis, in plants from Arctic populations of *Racomitrium lanuginosum* or of the Arctic moss *Dicranum elongatum*, after the material had been in cultivation for up to one year.

ADAPTATIONS TO STRESS

Of the climatic factors capable of causing stress in bryophytes high temperature, frost and drought are among the most significant. Bryophytes differ from higher plants in that individual shoots exert little control over water loss to dry air, while the cytoplasm itself may show a high degree of desiccation tolerance (Hinshiri and Proctor, 1971). Rates of water uptake and loss may, however, be influenced by variation in chemical composition of the cell walls (Roberts and Haring, 1937). Drought tolerance can be assessed from the effects of desiccation on either shoot survival or on rates of photosynthesis and respiration. These features may vary between species, and laboratory studies on both tropical and temperate bryophytes have demonstrated that the degree of drought tolerance may be ecologically correlated, generally being highest in plants from dry habitats (Clausen, 1952, 1964; Johnson and Kokila, 1970; Dilks and Proctor, 1974). For example, studies on epiphytic mosses in Japanese deciduous forests have shown inter-specific variation in osmotic pressure and desiccation resistance in relation to the vertical gradient in relative humidity and evaporation rates (Hosokawa and Kubota, 1957). However, two aquatic mosses survived isolation from stream water for up to one year in field experiments (Glime, 1971).

Intra-specific adaptation to desiccation has been investigated by Lee and Stewart (1971), who demonstrated that resistance was less in plants from wet than from dry habitats in *Calliergon cuspidatum*, *Climacium dendroides* and *Hypnum cupressiforme*. The latter comparison involved plants of var. *cupressiforme*, collected from a woodland boulder, and of var. *filiforme* (=*H. mamillatum*), growing epiphytically on tree trunks where they were frequently not wettened significantly during rainfall. The former showed a greater reduction in photosynthesis in relation to decreasing water content during the course of desiccation, and a slower recovery on remoistening, than var. *filiforme*. As noted earlier (page 515) there is some experimental evidence that the morphological variation between varieties of *H. cupressiforme* is in part genetic, and some of the characters may have adaptive significance. Thus the compact mat growth form of var. *filiforme* may be expected to reduce the rate of water loss from the colony as compared with the looser growth form of var. *cupressiforme*, and Lee and Stewart's data (1971, Table 5) provide supportive evidence. However, these experiments were apparently conducted on freshly collected material and thus, as the authors point out, they do not provide evidence of genetically based physiological adaptation. Comparative experiments on plants previously maintained for substantial periods under similar conditions are particularly desirable here, as desiccation resistance may vary seasonally within a population (Hosokawa and Kubota, 1957; Dilks and Proctor, 1976a) and it may be enhanced by intermittent desiccation (Proctor, 1972; Dilks and Proctor, 1976b), such as the var. *filiforme* may experience in the field.

It is well known that the tolerance of plant tissue to frost and to desiccation may be related, and that rapid freezing is more damaging than the slower freezing that generally occurs in nature. The reasons are discussed by Levitt (1972), and by Dilks and Proctor (1975) who also briefly review the literature on frost resistance in bryophytes. At least in short-term experiments, partially dehydrated plants of many species have proved tolerant of temperatures down to —20°C, i.e. to below the minimum likely to be encountered in the field during normal conditions given the insulating effects of snow cover. Dilks and Proctor (1975) and Clausen (1964) have noted that both mosses and liverworts are often more susceptible to damage when frozen in a moist condition, with extensive shoot mortality occurring in many species at —5 to —10°C.

There have been few investigations of intra-specific adaptation to frost in bryophytes, with the exception of Kallio and Heinonen's (1973) observations on *Racomitrium lanuginosum*. These authors studied net assimilation in material from a range of temperate and polar populations during the course of slow freezing from 0 to —30°C in the light, maintenance at —30°C for 12–16 h, and subsequent return to 0°C. The results to some extent paralleled those of desiccation in other mosses, and could conceivably have been related to withdrawal of water from the cells during freezing. Net assimilation rate fell to zero at —30°C, showed negative values indicative of enhanced respiration on return to 0°C, and subsequently recovered to positive values. There was no evidence of frost damage during these experiments, nor of significant differences between the populations examined.

There is good evidence that mosses are more resistant to heat when dry than when wet. For example, Nörr (1974) reported lethal limits of 85–110°C for dried material of eight European mosses, compared with only 42–51°C for turgescent plants, with some evidence of seasonal variation in tolerance. Similarly, Lange (1955) found that dried mosses survived temperatures ranging from 70°C to over 100°C. In these experiments species from open, lowland habitats were tolerant of exposure to somewhat higher temperatures than others from shaded or alpine sites, but apparently there have been no investigations of intraspecific variation in heat resistance in bryophytes.

Many of the studies considered in this paper can only be regarded as pre-liminary. Quite apart from the problem of distinguishing between genotypic and phenotypic variation, comparisons based only on two or three populations are of limited value in the present context as it is not clear whether any differences detected are random or the result of selection. In addition, it is not always appreciated by systematic bryologists that the failure of differences observed in the field to persist under a given set of inevitably unnatural experimental conditions, although suggestive, does not constitute evidence that the variation has no genetic basis. Nevertheless, bearing these points in mind, certain conclusions can be drawn.

It is clear from the results reviewed here, and by Smith (this volume, Chapter 9), that bryophyte species differ substantially in their degree of genetic variation. In *Pohlia proligera* (Lewis and Smith, 1977) and *Lophocolea heterophylla* (Hatcher, 1967), as well as in *Bryum argenteum* (page 515), geographically isolated populations proved morphologically distinct in collateral cultivation. Several other bulbiliferous *Pohlia* species studied by Lewis and Smith proved morphologically uniform, however, and Szweykowski and Vogel (1966) showed that European and North American plants of *Geocalyx graveolens* were also morphologically similar following cultivation under uniform conditions. Szweykowski and Vogel concluded that little genetic divergence had occurred since geographical divergence of the populations, probably during the Tertiary. In other, comparable experiments Szweykowski and his associates were able to demonstrate genetically fixed differences between pairs of closely related hepatic species (Szweykowski and Mendelak, 1964; Szweykowski and Koźlicka, 1969), including *Plagiochila asplenioides* and *P. major* (=*P. asplenioides* var. *major*) (Szweykowski and Krzakowa, 1969). The latter result has been supported by the demonstration of differences in peroxidase isozymes between the two taxa (Krzakowa and Szweykowski, 1977). In addition, Krzakowa (1978) has drawn attention to intrapopulation variation in peroxidase isozymes in *P. major* which suggested in some cases that a given carpet might have developed from genetically different diaspores. Geographically correlated variation in flavonoids has also been reported within hepatic species (Markham *et al.*, 1976).

The relative paucity of reports indicating inherent, habitat-correlated variation within bryophyte species may well be due in part to a lack of concerted investigation. The probable occurrence of genecological differentiation between populations from contrasting climatic environments has been established in some species with respect to morphological characters as in *Polytrichum alpestre*, and to physiology as in *Pohlia nutans* and *Racomitrium lanuginosum*. Genetic adaptation to naturally occurring gamma radiation has been demonstrated in *Marchantia polymorpha* (Sarosiek and Wozakowska-Natkaniec, 1968). Plastic responses to environmental variation are also of undoubted importance in bryophytes, but it must be remembered that the capacity of an organism to respond to environmental pressure through phenotypic plasticity is itself genetically determined and thus subject to selection.

While the occurrence of genecological differentiation within bryophytes may have been established, its pattern within species has generally not been adequately documented, nor is it yet possible to compare its frequency in the bryophytes with that in the more extensively studied flowering plants. Similarly, we do not yet know how many genes are responsible for the inherent variation within bryophyte species, nor whether adaptive variation has generally arisen through selection acting on isolated mutations or on the recombination products of meiosis resulting from effective sexual reproduction. Apart from the work of V. Proctor (1972) on *Riella*, we are also largely ignorant of the degree of interfertility between populations within moss and hepatic species.

There is speculation in the literature that bryophytes may be limited in their speed of evolution as a result of genetic restrictions imposed by aspects of their reproductive biology and life history (Anderson, 1963; Crum, 1972). It has been pointed out that their evolutionary potential may be limited by the occurrence of a free-living haploid gametophyte, by the rarity of sporophytes in some species and the possibility that spores may seldom give rise to new gametophytes in others, and by a high incidence of inbreeding in monoecious forms. Other authors have been reluctant to accept these arguments fully (Longton, 1976; Smith, this volume, Chapter 9; Szweykowski, 1978), noting for example, that current cytological data suggest that many bryophytes may be at least diploid in the gametophyte, and that insufficient evidence is yet available to assume obligate inbreeding in monoecious species. It has also been suggested that evolutionary flexibility may be lower in certain groups of hepatics than in mosses (Khanna, 1964; Schuster, 1966).

It may thus be significant that inherent, adaptive variation apparently occurs in *Polytrichum alpestre*, in which reports of a chromosome number of $n=7$ from several geographically isolated populations (Newton, 1972b) suggest that the gametophytes are probably haploid, as well as in other species with higher chromosome numbers possibly derived through polyploidy. The studies discussed here indicate the occurrence of inherent variation both in dioecious species, including *P. alpestre*, *Bryum argenteum* and *Racomitrium lanuginosum*, and in monoecious forms such as *Lophocolea heterophylla* and

Pohlia nutans. In contrast to the apparently more uniform bulbiliferous species of *Pohlia*, these taxa commonly produce sporophytes in at least part of their range. There remain, however, few reports of the establishment of natural populations of bryophytes from spores, and an attempt to grow *Bryum argenteum* from spores at a field site in Scotland failed to give conclusive results (Longton, unpublished data).

There is a significant number of moss and liverwort species in which sporophytes are unknown. Nevertheless, as argued more fully in Longton (1976), it would appear premature to suppose that the bryophytes as a whole, and particularly the mosses, have lost the capacity to adapt to climate and other environmental variables through selection processes akin to those believed to operate in flowering plants. Clearly, however, much more fascinating work remains to be done before these problems are resolved.

ACKNOWLEDGEMENTS

This paper is an outgrowth from work supported by the National Research Council of Canada and the University of Manitoba Northern Studies Committee, to whom grateful acknowledgement is made.

REFERENCES

ANDERSON, L. E. (1963). Modern species concepts: mosses. *Bryologist* **66,** 107–119.

ARNELL, H. W. (1905). Phaenological observations on mosses. *Bryologist* **8,** 41–44.

BAZZAZ, F. A., PAOLILLO, D. J., Jr and JAGELS, R. H. (1970). Photosynthesis and respiration of forest and alpine populations of *Polytrichum juniperinum*. *Bryologist* **75,** 579–585.

BENSON-EVANS, K. (1964). Physiology of the reproduction of bryophytes. *Bryologist* **67,** 431–445.

BENSON-EVANS, K. and BROUGH, M. C. (1966). The maturation cycles of some mosses from Fforest Ganol, Glamorgan. *Trans. Cardiff Nat. Soc.* **92,** 4–23.

BENSON-EVANS, K. and HUGHES, J. G. (1955). The physiology of sexual reproduction in *Lunularia cruciata* (L.) Dum. *Trans. Br. bryol. Soc.* **2,** 513–522.

BILLINGS, W. D. and ANDERSON, L. E. (1966). Some microclimatic characteristics of habitats of endemic and disjunct bryophytes in the southern Blue Ridge. *Bryologist* **69,** 76–95.

BIRSE, E. M. (1958). Ecological studies on growth-form in bryophytes, III. The relationship between growth-form of mosses and ground water supply. *J. Ecol.* **46,** 9–27.

BUSBY, J. R., BLISS, L. C. and HAMILTON, C. D. (1978). Microclimate control of growth rates and habitats of the boreal forest mosses, *Tomenthypnum nitens* and *Hylocomium splendens*. *Ecol. Monogr.* **48,** 95–110.

CLARKE, G. C. S. and GREENE, S. W. (1970). Reproductive performance of two species of *Pohlia* at widely separated stations. *Trans. Br. bryol. Soc.* **6,** 114–128.

CLARKE, G. C. S. and GREENE, S. W. (1971). Reproductive performance of two species of *Pohlia* from temperate and sub-Antarctic stations under controlled environmental conditions. *Trans. Br. bryol. Soc.* **6,** 278–295.

CLAUSEN, E. (1952). Hepatics and humidity, a study of the occurrence of hepatics in a Danish tract and the influence of relative humidity on their distribution. *Dansk bot. Ark.* **15**, 5–80.

CLAUSEN, E. (1964). The tolerance of hepatics to desiccation and temperature. *Bryologist* **67**, 411–417.

CLYMO, R. S. (1970). The growth of *Sphagnum*: methods of measurement. *J. Ecol.* **58**, 13–49.

COLLINS, N. J. (1977). The growth of mosses from two contrasting communities in the maritime Antarctic: measurement and prediction of net annual production. *In* "Adaptations within Antarctic Ecosystems" (G. A. Llano, ed.), pp. 921–933. Smithsonian Institution, Washington.

COLLINS, N. J., BAKER, J. H. and TILBROOK, P. J. (1975). Signy Island, maritime Antarctic. *Ecol. Bull., Stockholm*, **20**, 345–374.

CRUM, H. A. (1972). The geographic origins of the mosses of North America's eastern deciduous forest. *J. Hattori bot. Lab.* **35**, 269–298.

DELAY, J. (1974). Prospection carylogique dans une population de *Bryum argenteum* L. *Bull. Soc. bot. Fr.* **121**, 125–127.

DILKS, T. J. K. and PROCTOR, M. C. F. (1974). The pattern of recovery of bryophytes after desiccation. *J. Bryol.* **8**, 97–115.

DILKS, T. J. K. and PROCTOR, M. C. F. (1975). Comparative experiments on temperature responses of bryophytes: assimilation, respiration and freezing damage. *J. Bryol.* **8**, 317–336.

DILKS, T. J. K. and PROCTOR, M. C. F. (1976a). Seasonal variation in desiccation tolerance of some British bryophytes. *J. Bryol.* **9**, 239–247.

DILKS, T. J. K. and PROCTOR, M. C. F. (1976b). Effects of intermittent dessication on bryophytes. *J. Bryol.* **9**, 249–264.

FREDERICQ, H. and DE GREEF, J. (1966). Red (R), far-red (FR) photoreversible control of growth and chlorophyll content in light-grown thalli of *Marchantia polymorpha* L. *Naturwissenschaften* **53**, 337.

GIMINGHAM, C. H. and BIRSE, E. M. (1957). Ecological studies on growth-form in bryophytes. I. Correlations between growth-form and habitat. *J. Ecol.* **45**, 433–445.

GIMINGHAM, C. H. and SMITH, R. I. L. (1971). Growth form and water relations of mosses in the maritime Antarctic. *Bull. Br. Antarct. Surv.* **25**, 1–21.

GLIME, J. M. (1971). Response of two species of *Fontinalis* to field isolation from stream water. *Bryologist* **74**, 383–386.

GREENE, S. W. (1960). The maturation cycle, or stages of development of gametangia and capsules in mosses. *Trans. Br. bryol. Soc.* **3**, 736–745.

GRIMME, A. (1903). Über die Blüthezeit deutscher Laubmoose und die Entwickelungs-dauer ihrer Sporogone. *Hedwigia* **42**, 1–75.

HAGARUP, O. (1935). Zur Periodizität im Laubwechsel der Moose. *Biol. Meddr* **11**, 1–88.

HAMILTON, E. S. (1953). Bryophyte life forms on slopes of contrasting exposures in central New Jersey. *Bull. Torrey bot. Club* **80**, 264–272.

HATCHER, R. E. (1967). Experimental studies of variation in Hepaticae. I. Induced variation in *Lophocolea heterophylla*. *Brittonia* **19**, 178–201.

HÉBANT, C. (1977). The conducting tissues of bryophytes. *Bryophyt. Biblthca* **10**, 1–157.

HESLOP-HARRISON, J. (1964). Forty years of genecology. *Adv. ecol. Res.* **2**, 159–247.

HICKLENTON, P. R. and OECHEL, W. C. (1976). Physiological aspects of the ecology of *Dicranum fuscescens.* in the subarctic. I. Acclimation and acclimation potential of CO_2 exchange in relation to habitat, light and temperature. *Can. J. Bot.* **54**, 1104–1119.

HINSHIRI, H. M. and PROCTOR, M. C. F. (1971). The effects of desiccation on subsequent assimilation and respiration of the bryophytes *Anomodon viticulosus* and *Porella platyphylla. New Phytol.* **70**, 527–538.

HOSOKAWA, I. and KUBOTA, H. (1957). On the osmotic pressure and resistance to desiccation of epiphytic mosses from a beech forest, south-west Japan. *J. Ecol.* **45**, 579–591.

HOSOKAWA, T. and ODANI, N. (1957). The daily compensation period and vertical ranges of epiphytes in a beech forest. *J. Ecol.* **45**, 901–915.

HUGHES, J. G. (1962). The effects of day-length on the development of the sporophytes of *Polytrichum aloides* Hedw. and *P. piliferum* Hedw. *New Phytol.* **61**, 266–273.

JENDRALSKI, U. (1955). Die Jahresperiodizität in der Entwicklung der Laubmoose im Rheinlande. *Decheniana* **108**, 105–163.

JOHNSEN, A. B. (1969). Phenological and environmental observations on stands of *Orthotrichum. Bryologist* **72**, 397–403.

JOHNSON, A. and KOKILA, P. (1970). The resistance to desiccation of ten species of tropical mosses. *Bryologist* **73**, 682–686.

KALLIO, P. and HEINONEN, S. (1973). Ecology of *Rhacomitrium lanuginosum* (Hedw.) Brid. *Rep. Kevo Subarctic Res. Stat.* **10**, 43–54.

KALLIO, P. and HEINONEN, S. (1975). CO_2 exchange and growth of *Rhacomitrium lanuginosum* and *Dicranum elongatum. Ecol. Stud.* **16**, 138–148.

KALLIO, P. and KÄRENLAMPI, L. (1975). Photosynthesis in mosses and lichens. *In* "Photosynthesis and Productivity in Different Environments" (J. P. Cooper, ed.), pp. 393–423. Cambridge University Press. Cambridge.

KALLIO, P. and VALANNE, N. (1975). On the effect of continuous light on photosynthesis in mosses. *Ecol. Stud.* **16**, 149–162.

KELLOMÄKI, I., HARI, P. and KOPONEN, T. (1978). Ecology of photosynthesis in *Dicranum* and its taxonomic significance. *Bryophyt. Biblthca* **13**, 485–507.

KHANNA, K. R. (1964). Differential evolutionary activity in bryophytes. *Evolution* **18**, 652–670.

KRZAKOWA, M. (1978). Isozymes as markers of inter- and intraspecific differentiation in hepatics. *Bryophyt. Biblthca* **13**, 427–434.

KRZAKOWA, M. and SZWEYKOWSKI, J. (1977). Peroxidases as taxonomic characters. II. *Plagiochila asplenioides* (L.) Dum. sensu Grolle (=*P. maior* S. Arnell) and *Plagiochila porelloides* (=*P. asplenioides* Aucti, non Grolle; Hepaticae, Plagiochilaceae). *Bull. Soc. Amis Sci. Lett. Poznán*, Sér D., Sci. Biol., **17**, 33–36.

LACKNER, J. (1939). Über die Jahresperiodizität in der Entwicklung der Laubmoose. *Planta* **29**, 534–616.

LANGE, A. (1955). Untersuchungen über die Hitzeresistenz der Moose in Beziehung zur ihrer Verbreitung. I. Die Resistenz stark ausgetrockneter Moose. *Flora, Jena* **142**, 381–399.

LEE, J. A. and STEWART, G. R. (1971). Desiccation injury in mosses I. Intra-specific differences in the effect of moisture stress on photosynthesis. *New Phytol.* **70**, 1061–1068.

LEVITT, J. (1972). "Responses of Plants to Environmental Stress." Academic Press, London and New York.

LEWIS, K. and SMITH, A. J. E. (1977). Studies on some bulbiliferous species of *Pohlia* section *Pohliella*. I. Experimental investigations. *J. Bryol.* **9**, 539–556.

LONGTON, R. E. (1970). Growth and productivity of the moss *Polytrichum alpestre* Hoppe in Antarctic regions. *In* "Antarctic Ecology" (M. W. Holdgate, ed.), Vol. 2, pp. 818–837. Academic Press, London and New York.

LONGTON, R. E. (1972). Reproduction of Antarctic mosses in the genera *Polytrichum* and *Psilopilum* with particular reference to temperature *Bull. Br. Antarct. Surv.* **27**, 591–596.

LONGTON, R. E. (1974). Genecological differentiation in bryophytes. *J. Hattori bot. Lab.* **38**, 49–65.

LONGTON, R. E. (1976). Reproductive biology and evolutionary potential in bryophytes. *J. Hattori bot. Lab.* **41**, 205–223.

LONGTON, R. E. (in press a). Growth, reproduction and population ecology in relation to microclimate in the bipolar moss *Polytrichum alpestre*. *Bryologist*.

LONGTON, R. E. (in press b). Physiological ecology of mosses. *In* "Bryophytes of North America" (R. J. Taylor, ed.). American Association for the Advancement of Science, Washington.

LONGTON, R. E. and GREENE, S. W. (1969). The growth and reproductive cycle of *Pleurozium schreberi* (Brid.) Mitt. *Ann. Bot.* N.S. **33**, 83–105.

LONGTON, R. E. and MacIVER, M. A. (1977). Climatic relationships of Antarctic and northern hemisphere populations of a cosmopolitan moss, *Bryum argenteum*. *In* "Adaptations within Antarctic Ecosystems" (G. A. Llano, ed.), pp. 899–919. Smithsonian Institution, Washington.

MARKHAM, K. R., PORTER, L. J., MUES, R., ZINSMIESTER, H. D. and BREHM, B. G. (1976). Flavonoid variation in the liverwort *Conocephalum conicum*: evidence for geographic races. *Phytochemistry* **15**, 147–150.

MONROE, J. H. (1965). Some factors evoking formation of sex organs in *Funaria*. *Bryologist* **68**, 337–339.

NAKOSTEEN, P. C. and HUGHES, K. W. (1978). Sexual life cycle of three species of Funariaceae in culture. *Bryologist* **81**, 307–314.

NEWTON, M. E. (1972a). An investigation of photoperiod and temperature in relation to the life cycles of *Mnium hornum* Hedw. and *M. undulatum* sw (Musci) with reference to their histology. *Bot. J. Linn. Soc.* **65**, 189–209.

NEWTON, M. E. (1972b). Chromosome studies in some South Georgian bryophytes. *Bull. Br. Antarct. Surv.* **30**, 41–49.

NÖRR, M. (1974). Hitzeresistenz bei Moosen. *Flora, Jena* **163**, 388–397.

OECHEL, W. C. (1976). Seasonal patterns of temperature response of CO_2 flux and acclimation in arctic mosses growing *in situ*. *Photosynthetica* **10**, 447–456.

PATTERSON, P. M. (1964). Problems presented by bryophyte xerophytism. *Bryologist* **67**, 390–396.

PITKIN, P. H. (1975). Variability and seasonality of the growth of some corticolous pleurocarpous mosses. *J. Bryol.* **8**, 337–356.

PROCTOR, M. C. F. (1972). An experiment on intermittent desiccation with *Anomodon viticulosus* (Hedw.) Hook. and Tayl. *J. Bryol.* **7**, 181–186.

PROCTOR, V. W. (1972). The genus *Riella* in North and South America: distribution, culture and reproductive isolation. *Bryologist* **75**, 281–289.

RASTORFER, J. R. and HIGINBOTHAM, N. (1968). Rates of photosynthesis and respiration of the moss *Bryum sandbergii* as influenced by light intensity and temperature. *Am. J. Bot.* **55**, 1225–1229.

RICHARDS, P. W. (1932). Ecology. *In* "Manual of Bryology" (F. Verdoorn, ed.), pp. 367–395. Nijhoff, The Hague.

RICHARDS, P. W. (1959). Bryophyta. In "Vistas in Botany" (W. B. Turrill, ed.), pp. 387–420. Pergamon, Oxford.

RIDGWAY, J. E. (1967). Factors initiating antheridial formation in six Anthocerotales. *Bryologist* **70**, 203–205.

ROBERTS, E. A. and HARING, I. M. (1937). The water relations of the cell walls of certain mosses as a determining factor in their distribution. *Annls bryol.* **10**, 97–114.

SAROSIEK, J. and WOZAKOWSKA-NATKANIEC, H. (1968). The effects of low-level acute gamma radiation on the growth of *Marchantia polymorpha* L. originating from various ecological populations. *Acta Soc. Bot. Pol.* **37**, 413–426.

SCHUSTER, R. M. (1966). "The Hepaticae and Anthocerotae of North America." Vol. 1. Columbia University Press, New York.

SCHWABE, W. W. and NACHMONY-BASCOMB, S. (1963). Growth and dormancy of *Lunularia cruciata* (L.) Dum. II. The response to daylength and temperature. *J. Exp. Bot.* **14**, 357–378.

SCHWABE, W. W. and VALIO, I. F. M. (1970). Growth and dormancy in *Lunularia cruciata* (L.) Dum. VI. Growth regulation by daylength, by red, far-red, and blue light, and by applied growth substances and chelating agents. *J. Exp. Bot.* **21**, 122–137.

SHOLL, R. D. and IVES, J. D. (1973). The microenvironment of *Climacium americanum* *Trans. Illinois St. Acad. Sci.* **66**(3–4), 97–104.

SZWEYKOWSKI, J. and KOZLICKA, M. (1969). The variability of Polish *Ptilidium* L. species as grown in parallel cultures under identical conditions. *Bull. Soc. Amis Sci. Lett. Poznán*, Sér. D, Sci. Biol., **9**, 71–84.

SZWEYKOWSKI, J. and KRZAKOWA, M. (1969). The variability of *Plagiochila asplenioides* (L.) Dum. *s.l.* as grown in parallel cultures under identical conditions. I. The status of *Plagiochila major* (Nees) S. Arnell. *Bull. Soc. Amis Sci. Lett. Poznán*, Sér. D, Sci. Biol., **9**, 85–103.

SZWEYKOWSKI, J. and MENDELAK, M. (1964). Experimental investigations on the variability of *Riccia gougetiana* and *Riccia ciliifera* from Czechoslovakia. *Acta Soc. Bot. Pol.* **33**, 359–369.

SZWEYKOWSKI, J. and VOGEL, S. (1966). Experimental comparisons of variability in American and European populations of *Geocalyx graveolens* (Hepaticae, Harpanthaceae). *Fragm. florist. geobot.* **7**, 51–57.

TALLIS, J. H. (1959). Studies on the biology and ecology of *Rhacomitrium lanuginosum* Brid. II. Growth, reproduction and physiology. *J. Ecol.* **47**, 325–350.

VOTH, P. D. and HAMNER, K. C. (1940). Responses of *Marchantia polymorpha* to nutrient supply and photoperiod. *Bot. Gaz.* **102**, 169–203.

WANN, F. B. (1925). Some of the factors involved in the sexual reproduction of *Marchantia polymorpha*. *Am. J. Bot.* **12**, 307–318.

Wijk, R. van der (1960). De periodiciteit in de ontwikkeling der bladmossen. *Bux-baumia* **14**, 25–39.

Zehr, D. R. (1977). An autecological investigation of selected bryophytes in three sandstone canyons in southern Illinois. *Bryologist* **80**, 571–583.

Author Index

The numbers in *italics* indicate the pages on which names are mentioned in the reference lists

Taxonomic Index

Detailed classifications of the Musci appear on pp. 31–39 and the Hepaticae on pp. 68, 69, 71–78. The taxa therein are not listed here and similarly the genera of the Lejeuneaceae on pages 85–87 and the species in the *Sphagnum capense* group on pp. 112

Subject Index

A

Aber, 4
abscisic acid, 453
acacetin, 457
accessory chromosomes, 207, 212, 221–223
acclimatization, 512, 518–520
acetic anhydride, 238
acetic-orcein, 212, 214, 218
aceto-carmine, 212
acetolysis, 238, 239, 258, 261
N-acetyl-D-methionine, 449
acetylation of methionine, 449
acid fuchsin, 487, 504
acid phosphatase, 368
acrocarpous mosses, 318, 499
   chemotaxonomy, 450, 451
   conducting strands, 374
   rhizoids, 349
acrocentric chromosomes, 216
acroscopic leaves, 99–102
acumen, 492
acylated derivatives, 456, 458
adaptation, 204
   climatic, Ch. 21 (511–531)
   eco-physiological, Ch. 20 (479–509)
   metabolic, 518–522
   morphological, 514–516
   to stress, 522, 523
adaptive patterns, 418, 482
aequilobene, 460, 463
after ripening, spores, 270
Africa, 83, 92
   Lejeuneaceae, 93–95
   Sphagna, 109, 111, 113, 115, 117–121
   tropical mosses, Ch. 8 (185–193)
Afro-alpine regions and *Sphagnum*, 116
aggressive species, 135

aglycones, 452, 455, 458
air chambers, 52
alar cells, 20
Alaska, 55, 124, 126, 127, 128, 130, 133–138, 142–152, 520
Alberta, 137, 147
Aleutian Islands, 161
alcian blue, 256
alkaloids, in hepatics, 468
allergenic sesquiterpene lactones, 464
alpine flora, North Wales, 2
alpine populations, 513, 514, 521, 523
altitude, high and low populations, 520
Amazon, 89
amateurs, role in tropics, 190
Americans, in Japan, 163
America, tropical, 83, 89, 92–94
amino acids, 469
amphithecial layers, 320, 337
amylochloroplasts, 257
amyloplasts, 257, 271, 273, 275 (*see also* starch)
anaphase, 209, 210, 214–216, 218, 220–222
anatomy, Ch. 16 (365–383)
   Lejeuneaceae, 92, 95
(+)-anastreptene, 460, 461, 463
ancestry, of hepatics, 45–53
Andes, 89
Andrewsian *Sphagnum* species, 111
androecia, 54, 65, 67
aneuploidy, 208
Angara flora, 25
Anglesey, 2–7
aniline blue, 255
anisomorphous sexes, 188, 189, Ch. 13 (281–316) (*see also* sexual dimorphism)
anisophylly, in hepatics, 66–70